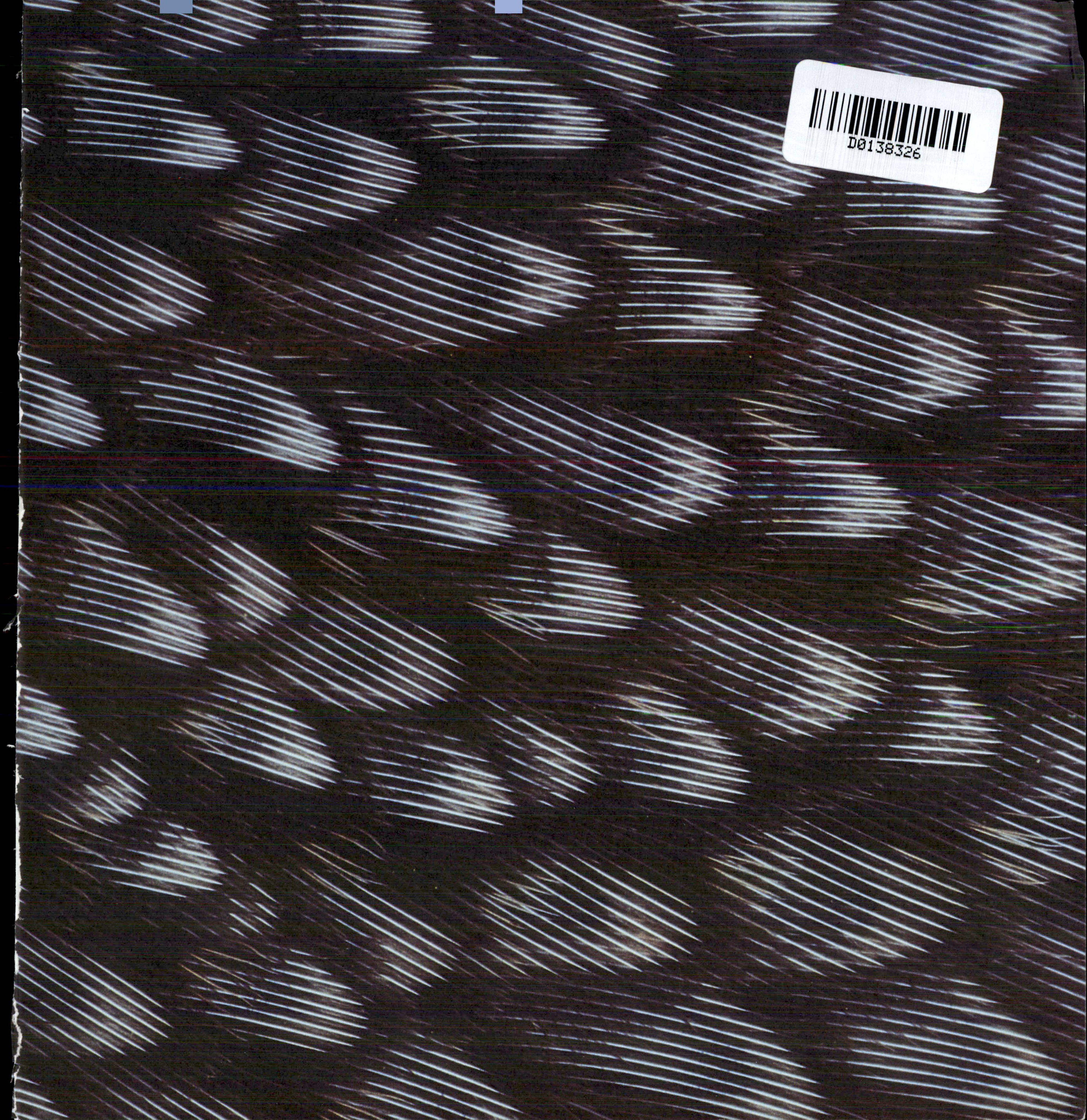
D0138326

# The Cambridge Encyclopedia of Ornithology

# The Cambridge Encyclopedia of ORNITHOLOGY

Edited by Michael Brooke and Tim Birkhead

CAMBRIDGE UNIVERSITY PRESS
CAMBRIDGE · NEW YORK · PORT CHESTER · MELBOURNE · SYDNEY

Published by the Press Syndicate of the University of Cambridge
The Pitt Building, Trumpington Street, Cambridge CB2 1RP
40 West 20th Street, New York NY 10011-4211, USA
10 Stamford Road, Oakleigh, Melbourne 3166, Australia



First published 1991

Printed in Spain by Mateu Cromo Artes Gráficas, S.A., Madrid

*A catalogue record for this book is available from the British Library*

*Library of Congress cataloging in publication data applied for*

ISBN 0 521 36205 9

A CAMBRIDGE REFERENCE BOOK

*Editor:* Peter Richards

*Designer:* Julian Smith

*Original artwork:* Julian Smith; Paul Richardson; Jones Sewell Associates

*Picture research:* Callie Kendall

*Index:* Kay Ollerenshaw

*Title spread illustration:* Graham Robertson

*Half-title (Brent geese):* Robert Glover/Aquila

VN

# Contents

## SPECIAL FEATURES

*Information panels on topics of special interest are provided throughout the Encyclopedia*

## *Contributors*

RRB Dr R. R. Baker
*University of Manchester*

CJB Dr Chris Barnard
*University of Nottingham*

PB Professor Peter Berthold
*Max-Planck-Institut für Verhaltensphysiologie, Vogelwarte Radolfzell*

PMB Peter Bircham
*University of Cambridge*

TRB Dr Tim Birkhead
*University of Sheffield*

JHB Dr J. H. Brackenbury
*University of Cambridge*

MdeLB Dr Michael Brooke
*University of Cambridge*

DB Dr Donald Bruning
*Ornithology Department, New York Zoological Society*

RWB Robert Burton
*Freelance writer, Huntingdon*

PJB Professor P. J. Butler
*University of Birmingham*

CKC Dr Clive Catchpole
*Royal Holloway and Bedford New College, University of London*

NEC Professor N. E. Collias
*University of California, Los Angeles*

BC Brian Cresswell
*Biotrack, Wareham, Dorset*

AD Dr A. Dawson
*Institute of Terrestrial Ecology, Huntingdon*

AWD Dr Tony Diamond
*Canadian Wildlife Service, Saskatoon*

PRE Professor P. R. Evans
*University of Durham*

CJF Dr Chris Feare
*Ministry of Agriculture, Fisheries and Food, Worplesdon*

CJOH Dr Colin Harrison
*formerly British Museum (Natural History)*

IH Ian Hepburn
*Royal Society for the Protection of Birds*

AMH A. M. Hutson
*formerly British Museum (Natural History)*

TAMJ T. A. M. Jack
*British Falconers' Club*

DRJ Professor David R. Jones
*University of British Columbia*

PJJ Dr P. J. Jones
*University of Edinburgh*

JK Dr Janet Kear
*Wildfowl and Wetlands Trust, Slimbridge*

MCM Dr Mary C. McKitrick
*University of Michigan*

JM Dr J. McLelland
*University of Edinburgh*

RMcNA Professor R. McNeill Alexander
*University of Leeds*

GRM Dr Graham Martin
*University of Birmingham*

ALAM Professor A. L. A. Middleton
*University of Guelph*

IN Dr Ian Newton
*Institute of Terrestrial Ecology, Huntingdon*

UMN Dr Ulla Norberg
*University of Gothenburg*

RJO Dr Raymond J. O'Connor
*University of Maine*

JMVR Dr Jeremy Rayner
*University of Bristol*

RS Dr Roland Sossinka
*University of Bielefeld*

EJLS Professor E. J. L. Soulsby
*University of Cambridge*

JGS Professor J. G. Strauch, Jr.
*University of Colorado, Boulder*

AKT Dr Angela Turner
*University of Sussex*

JAW Professor John Wiens
*Colorado State University*

RCY Dr R. C. Ydenberg
*Simon Fraser University*

# 1 INTRODUCTION

## What is ornithology?

The aim of this encyclopedia is to stimulate interest in ornithology, the study of birds.

Within the broad compass of ornithologists is found a remarkable array of people. They range from the person who notices which species visit the garden bird table, to the shivering scientist recording the courtship behaviour of a throng of Antarctic Adélie Penguins, and to the computer buff trying to describe mathematically the fate of some host population subject to the parasitic attentions of cowbirds or cuckoos. All, in their separate ways, are asking questions about birds – the very essence of ornithology.

The encyclopedia aims to survey the sort of questions ornithologists ask and to provide at least some of the answers. Not all the answers can be provided, for two basic reasons. Firstly, there remain areas of substantial ignorance: for example, how does a bird decide on a moment-by-moment basis whether it should sing or preen or feed? Secondly, there is the sheer volume of our knowledge about wild birds. This information could fill a score of encyclopedias.

Birds are so well studied at least in part because they are much more visible to us every day in streets and gardens, and are aesthetically more attractive than many other animals. As a result, ornithologists have often led the search for answers to general questions about the biology, behaviour and evolutionary history of wild animals, themes that are examined in detail in the pages that follow.

## The questions ornithologists ask

The simplest questions in ornithology can often be prefaced by a 'What?'. What species is that flying overhead? What species occur in such-and-such a mountain range? What is the earliest bird in the fossil record? What muscles does a bird use to flap its wings? Many of these basic questions were answered, in North America and Europe at least, by the collectors and anatomists of the nineteenth century. Yet even today, when new bird species are discovered almost annually in South America, there remain 'What?' questions to be answered. In our encyclopedia the first four chapters cover this ground.

From 'What?' questions the ornithologist proceeds to ask 'How?' and 'Why?'. Let us take 'How?' first. How does a hummingbird use its flight muscles to hover? (The answer can be found on p. 62.) At the start of the breeding season, how does a male penguin find his mate from last year amid a million-strong colony? (Usually by recognizing her voice: see p. 90.) Or how does a migrating bird find its way from the breeding grounds to its winter quarters? (Apparently by a mental map and magnetoreception: p. 193.)

This class of questions is quite distinct from that prefaced by a 'Why?'. Using the same examples as in the last paragraph we might ask, for instance: Why does the hummingbird feed from a fuchsia rather than a nearby hibiscus? Why does the penguin seek out his former mate rather than pairing up with the nearest available female? Or why does the migrating bird undertake a hazardous 5000 km long journey anyway?

The simple answer is that such activities are likely to enhance the bird's lifetime reproductive success. In other words a bird that migrates will leave on average more young than one that stays put, and over time it will be favoured by natural selection. This important topic, which underpins the way many basic ornithological questions are asked, is discussed in detail below. But first we need to consider what evidence we can use to answer 'Why?' questions.

## Adaptation

Why, we may ask, does a woodpecker have a stiffened tail? A biologist would answer that the stiffened tail is an adaptation that serves as a prop helping the woodpecker to scurry up trees

*Woodpeckers, like the Gila Woodpecker (Melanerpes uropygialis) of the south-western United States and northern Mexico, characteristically ascend tree trunks using their stiff tail as a prop*

more efficiently. But that very plausible answer is as yet unsupported by firm evidence. How therefore might we determine that stiffened tail feathers are not some fortuitous metabolic consequence of the woodpeckers' diet or, alternatively, that the value of stiff tail feathers is not as a tree-climbing prop but rather as a weapon in territorial squabbles between males?

To start with, we could survey other tree-climbing birds. The largest such group, the woodcreepers of South and Central America, likewise possess stiffened tail feathers. Conversely, some near-relatives of woodpeckers, the wrynecks, which perch crosswise and so do not use their tails as props, have more flexible tail feathers. This comparative evidence strongly suggests that stiff tail feathers are associated with tree-climbing and assist woodpeckers with this specific activity. Next comes observational evidence. The tail *is* used as a prop; it is not deployed in territorial combat.

The final and often the best evidence for an adaptive explanation comes from experiments. It would in theory be possible to cut off a woodpecker's tail feathers and replace them with the more flexible feathers of another bird. Happily this drastic experiment has never been undertaken; but if it were, the probable outcome would be a soft-tailed woodpecker unable to feed efficiently because it was unable to hold itself against a tree trunk at the required angle.

One classic instance where experimental evidence has demonstrated the adaptive value of this sort of behaviour sequence concerns the Black-headed Gull. Shortly after their young hatch, the gulls remove the eggshells from the vicinity of the nest. The ornithologist Niko Tinbergen wondered whether this behaviour might help protect the young from predators attracted to the nest by the conspicuous white inner surface of the hatched shells.

To test the idea Tinbergen's experimenters set eggs out on the ground. Those eggs with an empty hatched shell placed nearby were more likely to be taken by predators, mostly crows, than eggs without a nearby shell. Presumably hatched shells would also attract predators to newly-hatched chicks. Thus the parent gulls which removed shells would lose fewer young to predators. The shell removal behaviour is said to confer a selective advantage because the parents undertaking this behaviour will generally rear more young. The parents therefore will be favoured by natural selection.

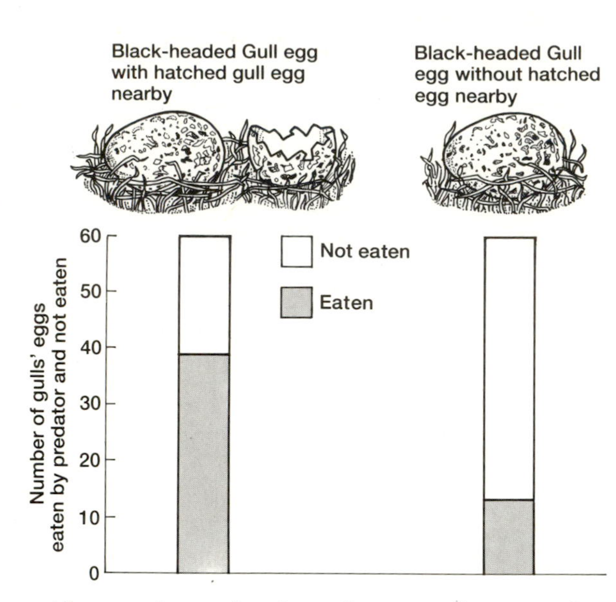


*Eggs of the Black-headed Gull (Larus ridibundus) are much more vulnerable to predation when a hatched egg shell is placed nearby*

## Natural selection

Natural selection is inevitable. It is a process so obvious and so inexorable that, from an historical perspective, it now seems remarkable that nobody realized its importance before Darwin so powerfully spelled out its implications in his *Origin of Species*, first published in 1859.

Populations of birds and other animals could in theory multiply very rapidly. If each pair of Blue Tits or Black-capped Chickadees produced eight young in a season and all survived, there would be five pairs breeding the following season. Continue this process over ten years and the population has multiplied almost ten million times! This does not happen in practice, of course, as bird populations are broadly held in check because so many individuals die. On average, those that die tend to be less well-equipped to survive in the prevailing environment. When the better-equipped survivors breed, they will therefore, thanks to inheritance, pass on to their young the characteristics which enhanced their own chances of survival. With the (partial) elimination of the less fit and the survival of the fitter, natural selection has effected evolutionary change.

## The unit of selection

A population of individuals of superior genetic constitution will be likely to produce more offspring in the next generation than a population of individuals of less fit genetic constitution. Is natural selection acting on the population, or the individual, or the genes?

Darwin himself always treated the individual organism as the fundamental unit of selection. The single exception was the social insects, such as ants and bees, in the case of which he argued that selection would favour the most productive colonies. For all other animals Darwin believed that natural selection favours individual organisms that produce as many young as possible. Birds should therefore possess adaptations that enable them to produce the maximum number of young.

This turns out to be a powerful way of answering 'Why?' questions. When trying to understand why birds sing, or forage in the way they do, we can ask as a first step how it enables them to leave more offspring.

Since Darwin's time, as the complex mechanism of inheritance has been elucidated, biologists have come to realize that it is more useful for some purposes to treat the gene, rather than the individual organism, as the unit of selection. Richard Dawkins has advocated this view particularly persuasively in *The Selfish Gene*. It is therefore often more useful, when posing a 'Why?' question, to ask why genes for doing X (say, singing) are favoured over genes for not doing X. For many purposes it does not matter in practice whether the question is posed in terms of individuals or genes, because genes are only passed on by the individual's reproduction. But in cases where birds help or hinder each other's reproduction the gene approach often gives us clearer insights.

Another, ultimately less fruitful, idea about the unit of selection did gain currency after Darwin. This was the idea of 'group selection'. Some biologists, such as Vero Wynne-Edwards of Aberdeen, suggested that organisms should maximize not their own reproduction but that of their group, even if this meant restraining their own reproduction. Nowadays, however, such ideas have largely fallen from favour.

## Absence of reproductive restraint

With the knowledge that natural selection acts on individuals, it becomes easier to resist sentimental arguments that an animal is acting 'for the good of the species'. Yet such claims still bedevil television scripts and popular writing.

Consider, for instance, the Red Grouse which establish autumn territories on heather moorland. Birds which fail to secure a territory leave the moor and probably die over winter. Are such birds quitting the heather moorland in order to ensure that the territory holders will have ample space to breed successfully next spring and so propagate the species? No. Having lost the territorial battle they simply retreat on the off-chance that a territory owner will die they can return to the moorland in his place. By retreat, a losing grouse gives himself a chance of breeding, which certainly would not be the case if he fought with, and was killed by, an existing territory owner.

*A male Red Grouse (Lagopus lagopus scoticus) surveys the heathery terrain of his territory, ownership of which is crucial to his chances of successful breeding*

Consider also reproduction and an odd-ball 'prudent' bird programmed by its genes to produce less than the maximum number of young of which it is capable in order to conserve resources for the population at large. Let this prudent bird live alongside other 'reckless' birds busy rearing as many young as they can. Certainly conserving resources could be regarded as an advantageous strategy in the long term, judged over scores of generations. Nevertheless, more birds from the reckless lineage will always go on to enter the next generation. Such birds will be favoured by natural selection, which acts over each and every generation and can never favour any sort of reproductive restraint that reduces an individual's lifetime reproductive output.

Do not birds which refrain from breeding themselves, and instead altruistically help others, show reproductive restraint and so break the above rule? The usual answer is no. Since the topic is discussed further in Chapter 10, one example will here suffice to show how apparent altruism need not imply reproductive restraint. Pairs of Long-tailed Tits sometimes receive help feeding their young. It turns out that the helpers are male tits, the brothers of the male of the breeding pair. However, the helpers are birds whose own nests have failed. Having lost the opportunity of rearing their own young, the helpers are not behaving altruistically. They are adopting the best alternative evolutionary strategy: helping rear nephews and nieces, with which they share genes.

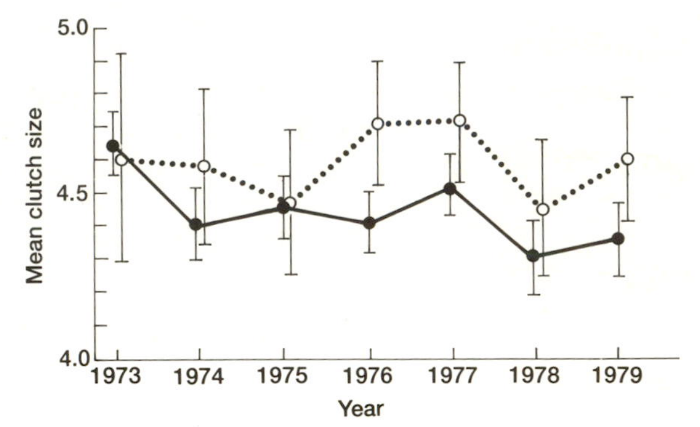


*An experiment on European Starlings (Sturnus vulgaris) in New Zealand. In one group (○) small clutches were deliberately destroyed. Over the course of the experiment this group then came to lay larger clutches than those in an untouched control group (●). Vertical bars give an index of variation*

## Heritability

For natural selection to operate it is essential that the characteristics of the parents that enhanced their survival be passed to, and inherited by, the young. In the broadest terms, it is quite obvious that birds' features are inherited. Goslings are goose-shaped. Owlets have the features, the large eyes and binocular vision, that enabled their parents to catch voles by night. But evidence is fast accumulating that many more subtle features important in birds' lives are also inherited.

When individual birds differ in some measurement like beak length, there is said to be phenotypic variance. Such differences arise because of variations in the genetic constitution of the individuals (genotypic variance) and because of differences in the environment experienced while the beak was growing. The ratio of the genotypic to the phenotypic variance is the heritability. Where heritability approaches 100%, the resemblance between parents and offspring is high, regardless of whether they were reared in a similar environment.

In a variety of bird species, Great Tits, Song Sparrows and Pied Flycatchers, the heritabilities of such structural traits as tarsus length and beak shape have been measured in detail, and often found to be around 60–70%, occasionally up to 90%. With these high values we conclude that most of the differences between individuals are due to genetic differences.

Other ecologically important traits, such as clutch size, are under genetic control, and can be inherited. In a notable seven-year experiment on European Starlings in New Zealand, the study population was divided into two groups laying clutches of equal size. In the unmanipulated control group, no action was taken. In the experimental group, all clutches smaller than average were destroyed for a number of years. The result was that in the later years of the experiment the mean clutch size of the experimental group became larger than that of the control group. This showed that, over a few generations, it was possible to alter the average clutch size of the starling. Such alteration would not have been possible if clutch size had not been, at least partly, under genetic control. In fact heritability was estimated at 33%.

Some behavioural characteristics are also known to be heritable. For example, Blackcaps from southern Germany show migratory restlessness for a longer period in the autumn than those from the Canaries since the former have further to fly to their winter quarters. Crossbred Blackcaps, with parents from both countries, are restless for an intermediate length of time. This strongly suggests that the period of migratory restlessness, which probably determines how far a migrant flies, is under genetic control.

## Natural selection in action

Darwin's visit to the Galápagos Islands aboard HMS Beagle in 1835 helped crystallize the thinking that resulted in the theory of natural selection. It is therefore wonderfully appropriate that the best verified case of natural selection acting on a wild bird population relates to a species of Darwin's Finch living in the Galápagos Islands.

The Medium Ground Finch is a seed-eater. During a severe two-year drought small seeds dwindled (because they were eaten by birds) until only large seeds remained. Since large seeds could be eaten only by the birds with the

largest bills, the smaller-billed birds starved to death. When the rains returned, the population consisted of birds having, on average, 4% larger beaks than before the drought. As beak shape and size in this ground finch are highly heritable, the demise of the smaller-beaked birds resulted in a change in the genetic composition of the population.

As the climate returned to normal, vegetation sprouted and seeds of all size classes again became abundant. Large-billed birds were no longer particularly favoured. Indeed, in some circumstances, those birds with smaller-than-average bills were favoured. Consequently the range of bill size within the population reverted to its pre-drought pattern. Natural selection had occurred. Indeed, it had operated very rapidly and led to evolutionary change – although in this case the process was reversible.

*The 13 species of Galápagos finch played a significant part in stimulating Charles Darwin's thoughts on natural selection. Particularly detailed studies have been carried out on the Medium Ground Finch (Geospiza fortis), a group in which individuals with bigger or smaller bills are favoured by natural selection in drier or wetter years*

## The unequal battle of the sexes

In sexually reproducing species like birds, reproduction requires that two individuals of opposite sexes come together, however briefly, in order to achieve fusion of their gametes. At first sight it may seem axiomatic that such reproduction should be an amicable affair characterized by unreserved cooperation towards a common goal. Nothing could be further from the truth.

Individuals maximize their reproductive success, and evolutionary success means reproducing more prolifically than others. In any one breeding episode an individual bird is seeking both to maximize its reproductive output and to minimize the costs and risks involved in breeding. These costs and risks should, if possible, be off-loaded onto other birds, provided this can be done without undue prejudice to reproductive success. In other words, male and female will act in concert as long as it is in their interest to do so – but they will be selected to place as many of the costs as possible onto their partner.

Cooperation ceases the moment either bird – and it is usually the male, as will be explained below – gains from defecting and pursuing another strategy, for example initiating another breeding episode with a new partner. Not even between pair members can we anticipate generosity. In this way arises the aptly-named battle of the sexes.

In nearly all sexually reproducing organisms the sexes produce gametes of markedly different size. There are good theoretical reasons why this should be so. In birds the individuals which produce relatively few big gametes (big enough in some cases to enjoy for breakfast) are females. Conversely, males produce enormous numbers of tiny gametes: sperm. Since sperm so overwhelmingly outnumber eggs, it is inevitable that most sperm will fail to fertilize eggs,

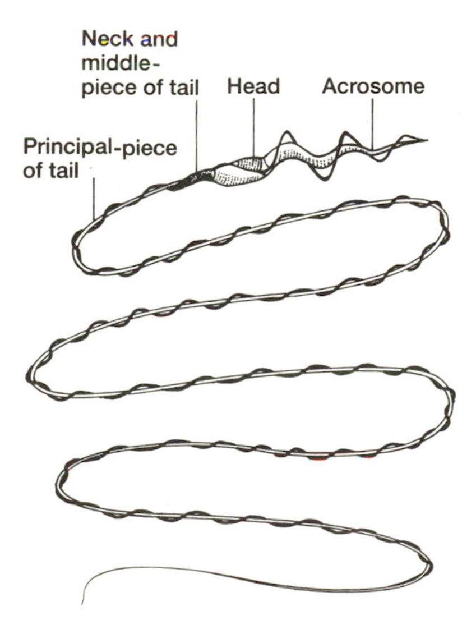


*The egg of a Song Thrush (Turdus philomelos) shown life-size, alongside the sperm of the Greenfinch (Carduelis chloris), magnified 10,000 times*

although most eggs will be fertilized. This asymmetry means that a male can often enhance his reproductive success by fertilizing as many females as possible. On the other hand a female, assured that every one of her few eggs will be fertilized, can best increase her reproductive success by giving every egg the best possible start in life: by providing it with plenty of yolk, by protecting the clutch against predators and so forth. In ornithological jargon, females are resource-limited while males are female-limited.

## Sexual selection

Because males are often female-limited they commonly compete for females. This competition is sexual selection. Because both sexual selection and natural selection are sieving processes, weeding out individuals less successful in the race to procreate, it may be best to think of sexual selection as a special case of natural selection. Sexual selection, a subject which has generated much intense debate, is certainly responsible for such extravagant ornamentation as peacocks' tails (properly speaking, tail coverts) or the dazzling plumage of male birds of paradise. The matter is further treated in Chapter 10. Here we will just briefly consider the two main paths of sexual selection.

*Intrasexual selection.* One avenue open to males striving to secure mates is to subdue other males; hence *intra*sexual selection. Large size often evolves in this process. At communal mating grounds or leks – for instance those of the Sage Grouse – certain males in the central territories typically secure the lion's share of matings. The peripheral males mate only rarely.

Alternatively, males can compete among themselves for the resources needed by breeding females. For example, male Indian Honeyguides do not help defend the nest or feed the young but they do defend bees' nests, and the female honeyguides need beeswax to feed on at the start of their breeding season. In a study of this species, one male honeyguide that defended a bees' nest was seen to mate 46 times with at least 18 different females. Other males, without a bees' nest, the resource critical to females, did not mate at all.

*Intersexual selection.* Males may not compete directly among themselves. Instead they may display characteristics which females find attractive. As females choose the more attractive males, so *inter*sexual selection occurs.

What features might a female find attractive? William Hamilton of Oxford University champions the idea that brighter males might be more attractive to females because the very brilliance of their plumage indicates resistance to parasites and diseases. The shimmering peacock's tail is in these terms a 'Health Certificate' to be flaunted in front of the female. By choosing such a mate, a female is possibly securing for her offspring genes which will enable them to resist the infections stalking the environment. This idea also offers a possible explanation for an otherwise puzzling observation, that birds endemic to islands are usually less brightly coloured than their continental counterparts. Is this because islands are relatively disease-free, making it more advantageous for males to adopt drab camouflage plumage rather than flaunt their resistance to non-existent diseases?

Another idea proposes a genetic coupling between a certain male trait and variable female preference for that trait. It then pays the male to develop the trait, even if it is detrimental to his survival, because he will be preferred by females. And it also pays a female to mate with such males because her sons will in turn develop the trait which will render them attractive to females. This runaway selection process may be responsible for the highly impractical tail of the African Long-tailed Widow Bird, and other equally extravagant displays of plumage throughout the world of birds.

MdeLB

## THE MECHANISM OF INHERITANCE

A bird's body is composed of countless thousands of cells. Within each cell lies a nucleus, within which are the chromosomes. The cells of different bird species differ somewhat in the number of chromosomes: for example, the Herring Gull has 66 and the Mallard 80. The chromosomes normally occur in homologous pairs with one member of each pair originating from the mother and the other from the father.

Every chromosome is a helical molecule of deoxyribonucleic acid (DNA). It is divided into a series of units, or genes. Each gene carries different information, coded by the precise sequence of building blocks of the DNA molecule. When that information is translated by the cell's metabolic machinery into, say, a digestive enzyme or a feather protein, the DNA has served as the blueprint for the living bird.

When eggs and sperm (gametes) are formed during the process known as meiosis, homologous chromosomes pair up and exchange pieces. After this process of recombination, the cell splits into two, and each new cell contains only half the number of chromosomes of the original cell. Thus either a gene copy from the mother or a gene copy from the father gets into the gamete. Both cannot make it.

Thanks to recombination and the enormous number of ways a complete set of chromosomes can split into two prior to entering a gamete, the genetic make-up of each gamete is unique. Consequently, when egg and sperm meet at fertilization (and so restore the normal number of chromosomes) each individual is unique. It is this variation, inherited from mother and father, on which natural selection works.

**Sex determination**

As mentioned above, chromosomes normally occur in pairs that look identical. But this is not true for one pair, the sex chromosomes, which determine whether a fertilized egg will develop into a male or a female. The two sex chromosomes, called *w* and *z*, look decidedly different.

In birds, unlike mammals, the female is the heterogametic sex, meaning that a female bird's cells carry one *w* and one *z* chromosome. When eggs are formed, they either carry a *w* or a *z* chromosome. Meanwhile the cocks, having two *z* chromosomes, produce sperm that all carry *z*. If a sperm fertilizes an egg carrying *w*, the resulting *wz* embryo will develop into a female; but if the egg has a *z* chromosome, the embryo will be *zz* and become a male.

MdeLB

A bird can be thought of as a flying machine in which all the aspects of its physiology and anatomy are integrated for maintenance, growth, and reproduction. Even the simplest movement of a bird is the result of a complex interaction of many systems. For example, when a bird spots a cat and flies away, sensory organs detect and transmit the cat's presence to the brain, where the danger is recognized and a decision for action is made. Orders are then relayed to the muscular system to jump up and fly away. These steps depend in turn on a host of cellular actions in all of the organs involved, as well as on the nutritional and developmental history of the bird. The workings and interactions of some systems are fairly well understood, but even now much remains unknown.

## Skeleto-muscular system

The skeleto-muscular system is the mechanical part of a bird. The bones and muscles work together. If they were not moved and linked together by muscles and ligaments, bones could serve no useful purpose; likewise, muscles not attached to bones could only contract and relax. By means of the skeleto-muscular system a bird moves from place to place, acts on the environment, and interacts with other organisms. To understand how the system works, the nature of bones, muscles, and articulations must be understood.

### Bone

Bone is a living tissue that can adapt when subjected to different conditions and repair itself when damaged. It consists of needle-like crystals of an inorganic mineral (hydroxyapatite) in an organic matrix of fibrous protein (collagen). This combination of components is more versatile (more rigid than collagenous fibres, more flexible than mineral) than either component alone in the kinds of stresses it can withstand. A mature bone usually has an exterior of compact, dense, ivory-like bone and an interior of spongy bone. As the diagram to the right shows, spongy bone is a meshwork of thin sheets (trabeculae) the orientation of which corresponds to the patterns of stresses on the bone. Spongy bone usually contains a cavity filled with marrow, fat, or outgrowths of the air sacs.

Most of the skeleton is laid down in the embryo as cartilage which then ossifies into bone. Ossification of most bones is not complete until maturity. For example, in a typical limb

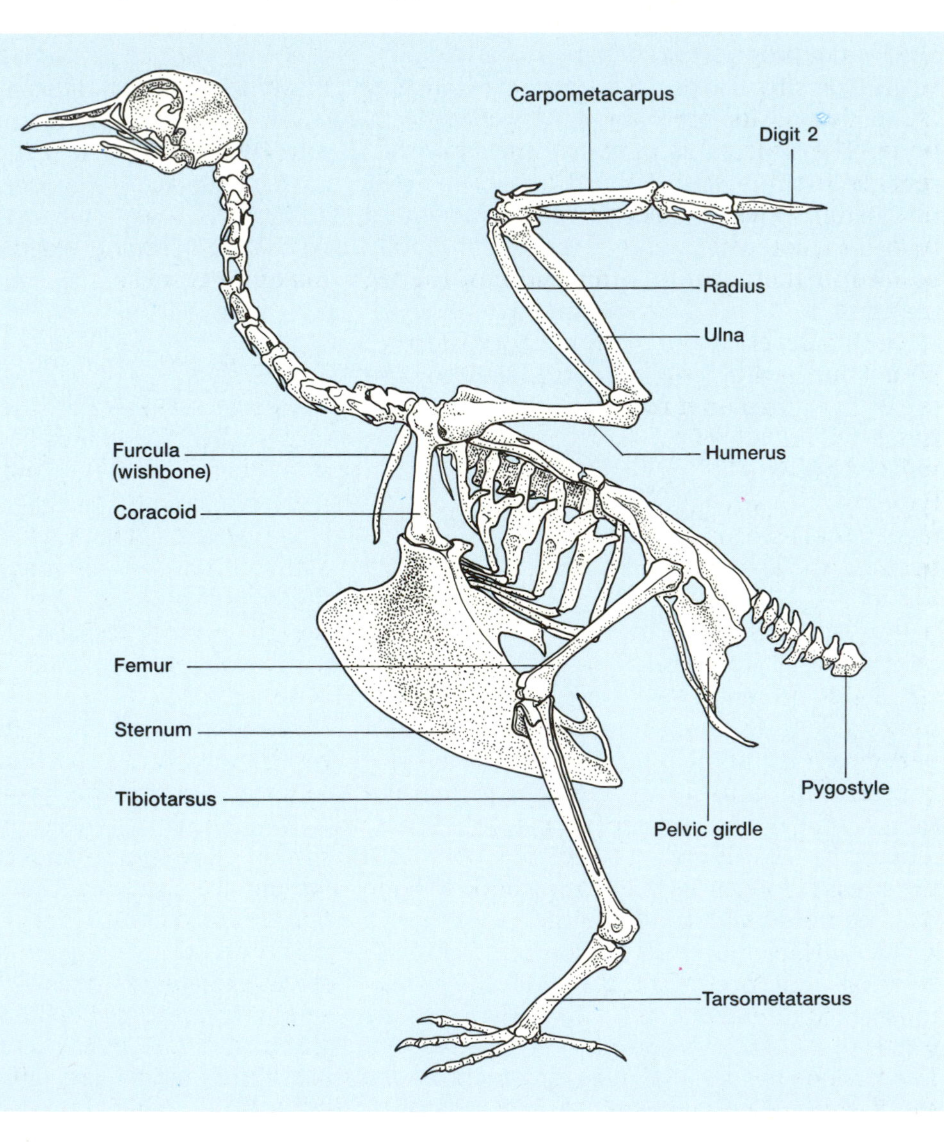


*A pigeon skeleton showing the principal post-cranial bones*

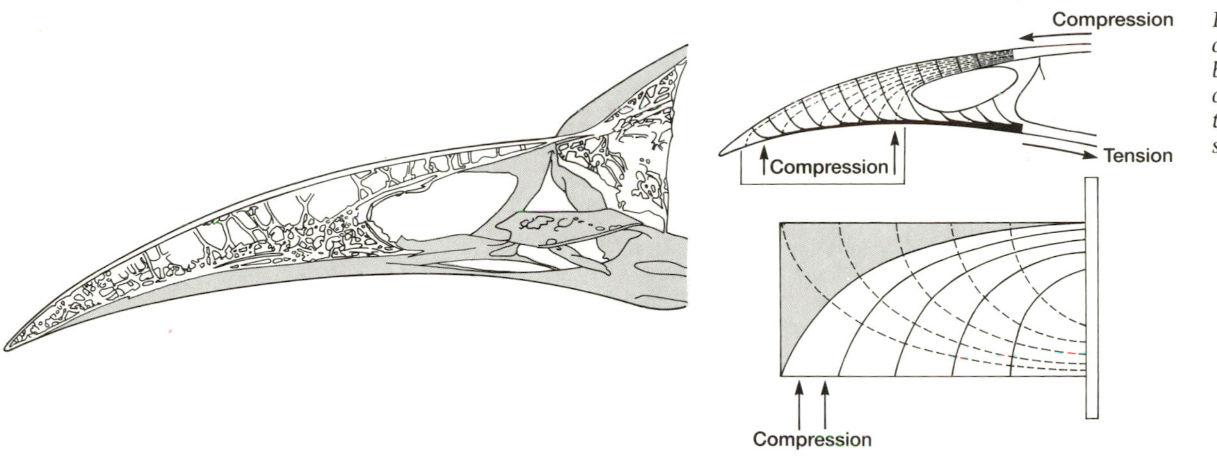


*Far left. Cross-section of the bill of a crow showing the arrangement of its bony internal struts or trabeculae. The orientation of the trabeculae corresponds to the patterns of stress on the bill, as the schematised diagrams near left illustrate*

bone ossification begins in the middle of the shaft and progresses towards the ends. The bone grows by the proliferation of cartilage at the ends and its subsequent conversion into bone. The process stops when the bone has reached its adult size. Ossification is completed at different rates in different bones. In young migrant waders the bones of the wing are fully ossified in the autumn while those of the leg may not be.

During development of the embryo, the form of a bone is influenced by the tension and pressure exerted on it by the muscles. If a bone is isolated but allowed to continue to develop, it will not attain its normal form. The ability of a bone to change continues into adult life. The form of a bone is maintained by a balance between the actions of cells that lay down bone (osteoblasts) and those of cells that resorb it (osteoclasts). Stresses on bones change this balance and lead to changes in shape. A stressed bone generates an electric current that may guide the actions of these cells. They also participate in the healing process of a broken bone and will cause distortion if the bone is not set and subjected to normal stresses. (In space, where the lack of gravitational pull decreases the stress on bones, the rapid resorption of bone has been observed in astronauts.)

The surface markings on bones indicate the location and nature of the attachments of muscles and tendons and give evidence of the forces to which the bones have been subjected. Depressions (fossae) indicate the attachment of the fleshy muscles whereas elevations (crests, processes, tuberosities, etc.) indicate the attachment of tendons. It is from the details of these markings that paleontologists can sometimes 'reconstruct' an entire animal from one bone and even theorize how it moved.

The females of domestic and at least some wild birds store calcium inside their bones before the breeding season and later use it for making eggshells.

## Muscle

Vertebrate muscle consists of three types: cardiac muscle, found only in the heart, consists of a network of fibres and contracts spontaneously; smooth muscle, found in the gut, blood vessels, glands, and at the bases of the feathers, has few identifiable fibres and is not under voluntary control; striated, or skeletal, muscle, usually associated with the skeleton, is under voluntary control. Skeletal muscle alone is discussed here.

Skeletal muscles are composed of fibres held together by connective tissue. The fibres consist mainly of contractile proteins which shorten when stimulated. There appear to be two major sets of fibre types, the relationships among which are unclear. One set includes *twitch* fibres, which contract rapidly when stimulated and stop contracting as soon as stimulation stops, and *tonus* fibres, which contract more slowly when stimulated and relax more slowly when stimulation stops. The other set includes red fibres, which are difficult to fatigue, and white fibres, which fatigue rapidly. Attempts to equate red with tonus fibres and white with twitch fibres have been unsatisfactory in explaining the properties of bird muscles.

The red fibres contain a higher concentration of the oxygen-carrying pigment myoglobin and a richer blood supply than the white fibres. Most muscles contain red and white fibres; their relative proportions are correlated with the way in which a given muscle is normally used. Thus, the breast muscles of a sandpiper or songbird that migrates several thousand kilometres without stopping consist mostly of red fibres whereas those of a grouse, which flies only in short bursts, consist mostly of white fibres. The claim that a grouse can be caught by hand if flushed four times in rapid succession is supported by the rapid fatigue of white fibres.

Skeletal muscles vary greatly in size, shape, fibre arrangement, and form of attachment to bones. When stimulated, skeletal muscles contract and develop a pulling force. Though they are said to contract when active, the actual degree of shortening of the muscle depends on its structure and the load against which it works. In an intact animal about 30% to 40% of the fibres are contracted at one time when a muscle has developed its maximum force. The maximum force that a muscle develops, the amount that it shortens, and the speed with which it shortens depend on the number of fibres it contains, their length, and their arrangement within the muscle. In general, a short, thick muscle can move heavier loads than a long, thin muscle, but the latter can move a lighter load over a longer distance.

Long-fibred muscles shorten faster. A muscle cannot lengthen unless an external force is applied to it. For this reason pairs of opposing, or antagonistic, muscles are often found controlling the movement of a given bone. The balance between the forces developed by the opposing muscles can be adjusted to move a bone anywhere within its normal range. Slow, exact motion is usually the result of a delicate balance between antagonistic muscles.

Muscle also responds to the physiological needs of a bird. As a bird adds fat before migration, the flight muscles grow larger to carry the extra fuel (Chapter 7). Some arctic breeding birds are believed to store protein for making eggs by building up their muscle mass on the wintering grounds (Chapter 9). During periods of food shortage muscles may be converted into energy to help a bird survive.

## Ligaments and tendons

Ligaments join bones to other bones whereas tendons join muscles and bones. Both are usually made of collagen fibres, which are highly elastic (returning to their original shape when distorted) and resistant to changes in shape (non-compliant). They act like ropes or chains, which have strength against a pulling force but which collapse if they are pushed on. Rarely tendons are made of yellow elastic fibres, which act much like rubber bands.

Ligaments consist of two types: articular, which occur at joints and prevent disruption of an articulation and limit its movement, and linkage, which coordinate or limit the movement of some of the bones in the head but are not part of specific joints.

Tendons allow several muscles to attach on a restricted surface area of a bone, allow muscles to span large distances without excessive length, permit muscles to be located away from a joint, reduce the moment of inertia around a joint, and permit changes in the direction of pull. Tendons are about 30 times as strong as muscles.

The juncture of two bones is called a suture if the bones do not move relative to one another and a joint or articulation if they do move. Joints may be filled with fibrous tissue, cartilage, or bone, in which cases they have restricted motion. Most freely moving joints, on the other hand, contain a fluid-filled cavity and are known as synovial joints. In synovial joints the ends of the bones are covered with a layer of cartilage, and movement between them is lubricated with synovial fluid, the entire joint being enclosed in a fibrous capsule. The two major types of articular surface in synovial joints are ovoid (concave or convex in all directions), as in the familiar ball-and-socket joint, and saddle-shaped, in which movement is largely restricted to two planes.

## Special avian features

If the skeleto-muscular system of a typical bird is compared to that of a mammal of similar size, many striking differences can be seen. A bird is more compact, and its body considerably shortened. The neck, however, may be quite long. A bird has a light, horny beak instead of a heavy, toothed muzzle. The muscles of a bird are concentrated around the breast and the bases of the legs and wings, a bird's skeleton is mostly hollow and less massive, and in a bird many individual bones have become fused. These characteristics combine to produce the light weight and high power required for flight. In fact a bird's skeleton constitutes only some 5% of its weight.

The concentration of the muscle mass near the centre of gravity leads to aerodynamic streamlining and, more important, allows the limbs to move faster with the expenditure of less energy. Movements at the ends of the limbs are controlled by long tendons. Shortening of the body, replacement of a heavy tail by feathers, and reduction in the mass of the head result in overall lightness while retaining balance. Fusion and hollowing of the bones produce a strong, light frame.

A distinctive feature of a bird's skeleton is pneumatization, the presence in many of the bones of air-filled cavities. In pneumatized bones the hollow parts communicate with an air-sac system. The skull is pneumatized via the air sacs of the nasal and tympanic cavities, the post-cranial skeleton via the air sacs of the respiratory system. In general, the skeletons of large flying birds are extensively pneumatized whereas those of diving birds, in which increased buoyancy would hinder staying under water, have little or no pneumatization. Many primitive cultures have fashioned flutes from the long, hollow limb bones of large birds such as swans, cranes, and vultures.

Whether these features arose specifically for flight is still a subject of debate since some of them, such as the horny bill and hollow bones, are also found in some of the birds' dinosaur ancestors. On the other hand, our hypothetical mammal used for comparison could not fly even if its fur were replaced with feathers.

## Skull and beak

Modern birds have no teeth, although some such as shrikes and falcons have bony notches extending from the jawbone. The beak is covered with horny plates comprising the rhamphotheca. Usually the rhamphotheca's shape follows that of the underlying bone, but it may be modified by local thickening into ridges, knobs, or plates. Thus some waterfowl have serrated bills. In most birds the rhamphotheca is horny and hard, but it is leathery in waterfowl and flamingos and soft in scolopacids. In many birds of prey, pigeons, and parrots the base of the upper rhamphotheca is soft and swollen, forming the cere. Generally the rhamphotheca grows continuously, wearing off at the tip. However in some species the entire rhamphotheca (in ptarmigan and some other grouse) or a part of it (in puffins and auklets) is seasonally shed. The shape of the rhamphotheca, but not the underlying bone, may change with season and diet. The bills of wild and captive individual oystercatchers have been observed to change within a few weeks from stout and blunt when feeding on hard-shelled prey to long and narrow when feeding on soft prey.

A bird's skull is light, often extensively pneumatized, with large cavities (orbits) for the eyes and a large, rounded brain case. Most of

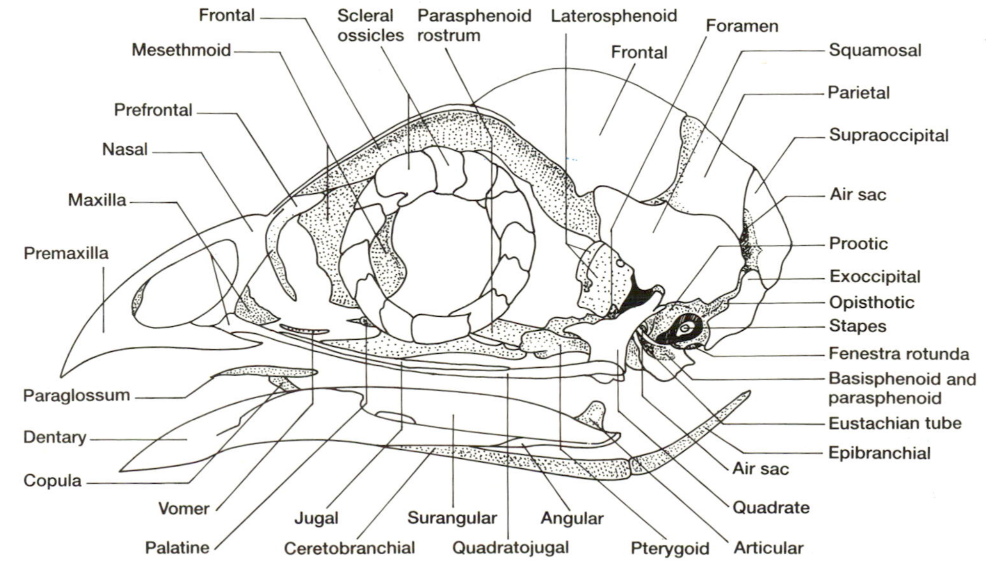


*The skull of a domestic chicken at 2–3 days old showing the individual bones before fusion takes place*

the individual bones of the cranium are fused and cannot be discerned in the adult bird. The nostrils are set near the base of the bill, except in kiwis in which they are at the tip. In some birds, such as gannets, the nostrils are completely covered by bone. Gannets breathe through the mouth; when the bill is closed, the proximal (i.e. nearer the skull) plate of the upper rhamphotheca is flared to provide an air passage between the jaws. This is automatically closed by external pressure when the birds dive and swim under water.

A dried bird's skull in a museum case seems to be a rigid structure. In life, however, a bird's skull is quite flexible and contains many joints for moving both jaws. The upper jaw and skull of all birds are jointed, usually with a thin, flexible bony hinge; some large parrots have synovial joints between the cranium and upper jaw. The ability to move the upper jaw involves a linkage system consisting of several jaw bones and bending zones in the bill. This ability, known to ornithologists as cranial kinesis, allows birds to widen their gape to handle large food items and to close their jaws faster; it also creates a shock absorber between bill and skull. Additional bending zones at the bill tips permit many sandpipers, snipe, and woodcock to open just the tip of the bill to grasp prey while probing in the ground.

The lower jaw is composed of six bones, many of which are fused. The rear process of the lower jaw is enlarged for increased muscle attachment in birds that forcibly open their jaws while gaping. This enlargement is found in the European Starling and North American meadowlark; the forcible opening of the jaws can easily be seen in birds at bird feeders as a jerky opening of the jaws even though the birds are not probing in the ground. In many fish-eating birds, gulls, pigeons, owls and nightjars, which eat large prey or have specialized feeding mechanisms, there are bending zones in the sides of the lower jaw that allow the mandibles to spread apart, further increasing the size of the gape.

*Right. The hyoid horns of a woodpecker are attached at one end to the tongue. At the other they attach to the base of the bill (left, as in piculets), enter the cavity of the right upper mandible (centre, as in flickers), or encircle the eye (right). When the genio-hyoid muscle that surrounds the horns for most of their length contracts, the tongue can be greatly extended, enabling the woodpecker to extract food from deep crevices*

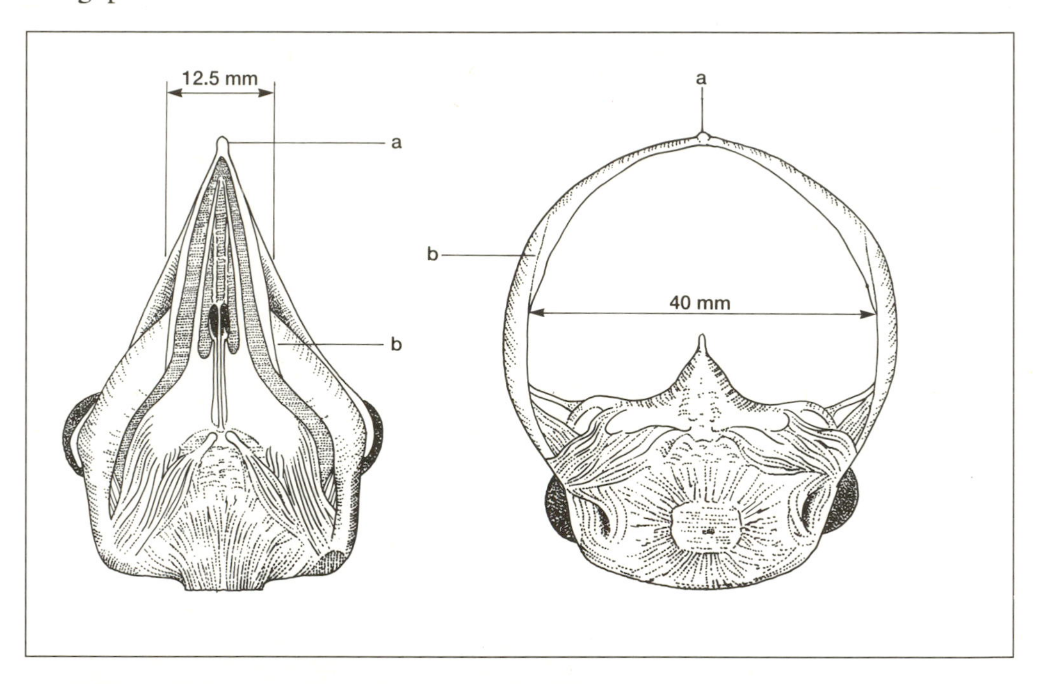


*Above. Two views of the lower jaw of a nightjar to show, left, the jaws closed in the resting position and, right, the jaws fully opened with the lower mandible spread apart*

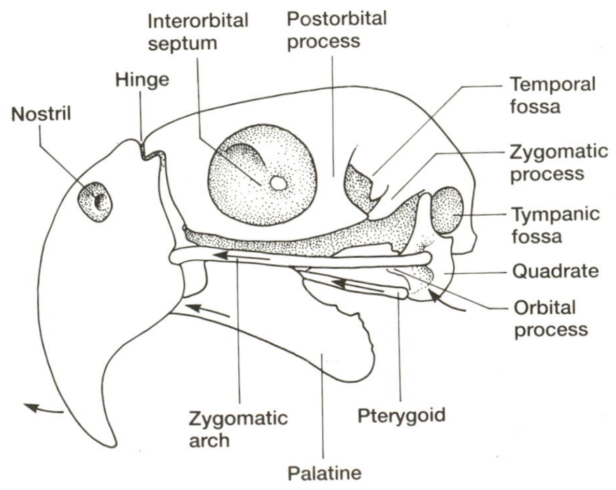


*Far left. A macaw skull showing the mechanism of cranial kinesis. Arrows show direction of movement. As the pterygoid and palatine move forward, so the upper jaw rotates about the hinge*

*Left. The horny outer sheath or ramphotheca of the bill of the Sulphur-breasted Toucan (Rhamphastos sulphuratos) is amongst the most vivid belonging to any bird. Since large toucans have rather similar plumage, bill colour may aid species recognition*

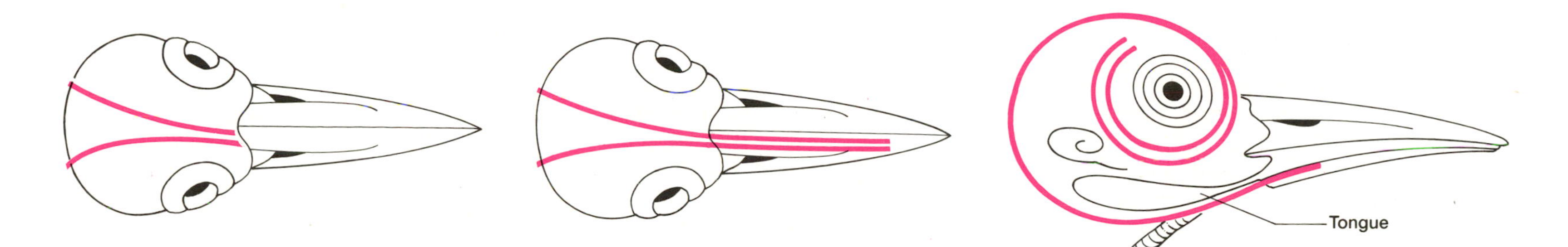


The tongue is supported and extended by the hyoid apparatus, which can be bony or cartilaginous. In woodpeckers and hummingbirds the hyoid horns are greatly elongated and may curve around the back and top of the skull and enter the nostrils.

In the ear, a bird has only a single bone, the columella (the equivalent of the stapes or 'stirrup' in humans). This conducts sound from the eardrum to the inner ear. Details of hearing are discussed in the section on birds' senses.

The eyes of birds, extremely important for guidance in flight and for the detection of prey and predators, are relatively larger than those of other vertebrates and may meet in the centre of the skull. To accommodate these large eyes the orbits are large, and the bony partition between them is very thin or absent. Kiwis, however, are an exception. These strange birds, which do not fly and which find their food at night by touch and smell, have quite small eyes and large nasal chambers that extend between the orbits.

Birds generally have a ring of 16 to 18 small, bony plates, the scleral ossicles, embedded in the eyeball which help to stabilize the shape of the eye. Because the movement of the eyeballs is restricted by their large size and flattened shape, a bird frequently moves its entire head when visually scanning the landscape. Owls cannot move their eyes at all but are able to swivel their head through a 180° arc.

## Post-cranial axial skeleton

The spinal column of birds can be divided into the neck, chest, loin, hip, and tail regions. The number of vertebrae varies from about 40 to 60, most of the variation occurring in the neck. Small birds have 15 or fewer neck vertebrae

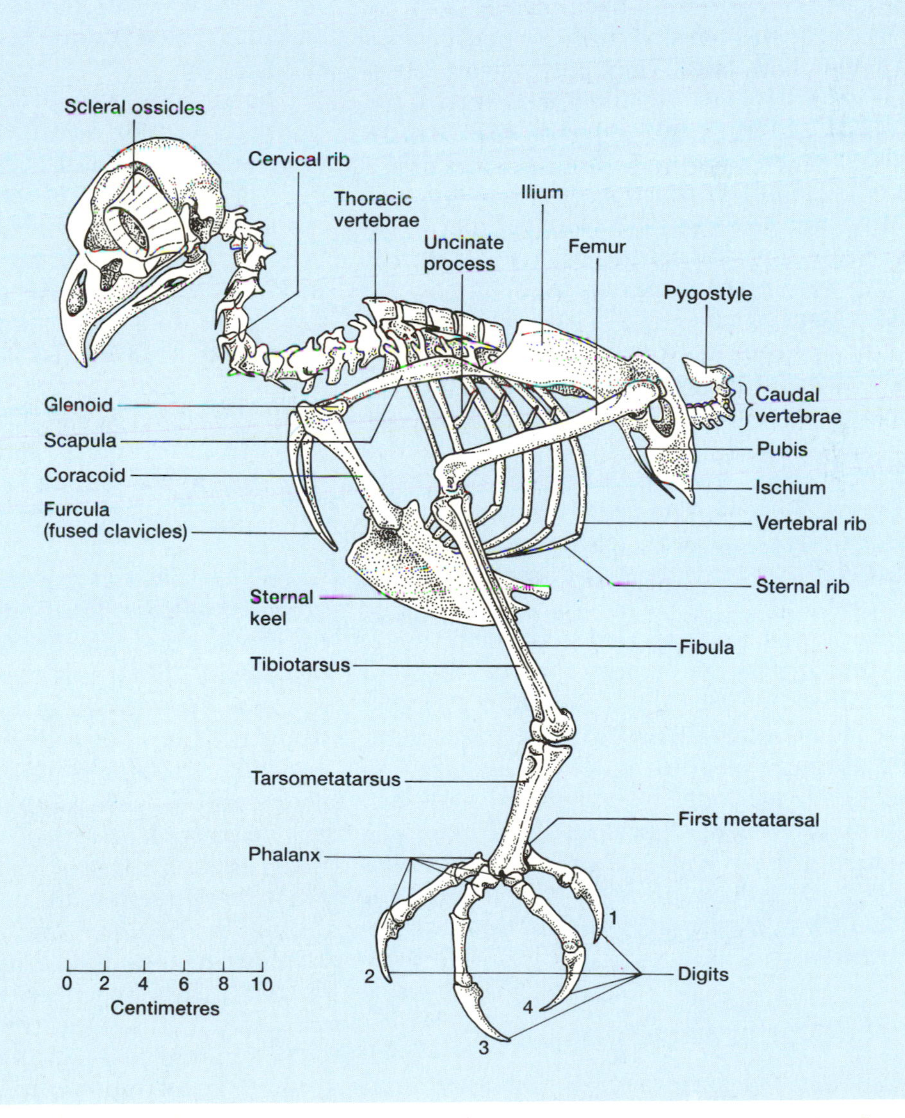


*The skeleton of an Eagle Owl (Bubo bubo) with the left wing removed*

whereas large birds may have more than 20; the Mute Swan has the most – 23. The neck vertebrae move freely and usually have saddle-shaped joints. The other vertebrae in the spinal column are more or less immobile and in many cases are fused.

The necks of darters and herons are permanently kinked. This kink is part of a 'trigger' or 'spear mechanism' that allows the head suddenly to dart forward to capture fish. The neck acts much as a sling used to hurl a javelin.

The chest vertebrae have little movement, and some may be fused; they bear the ribs. Most of the ribs articulate with the sternum and are jointed near their midpoint. Most of them bear a flat flange (uncinate process). These processes overlap one or more adjacent ribs and are attached to them with ligaments and muscles, reinforcing the rib cage so that it is like a woven basket. They are most highly developed in diving birds such as divers (loons) and alcids (auks and their relatives) and are believed to protect the internal organs from pressure in deep water.

The sternum is large compared to that of other vertebrates. A distinctive feature of the sternum of modern flying birds is a well-developed keel, to which are attached the strong pectoral muscles. The front end of the sternum of swans and some cranes is hollow and contains loops of the trachea. In ratites and other flightless birds the keel is greatly reduced or absent.

The synsacrum is formed from the fusion of loin and hip vertebrae and some of the chest and tail vertebrae. The synsacrum is firmly attached to or fused with the pelvis into a light, rigid structure that distributes the weight widely along the backbone when a bird is on its feet.

The tail consists of several small vertebrae and in most birds ends in a structure known as the pygostyle (plough-share bone), which is formed from the fusion of several vertebrae in the embryo.

## Appendicular skeleton

The shoulder or pectoral girdle is formed from three paired bones, the scapula, coracoid, and clavicle. This girdle acts like a tripod on each side of the bird to support the wings and distribute the stresses developed in flight. The scapula is a long, thin bone that extends rearward and is attached to the ribs. In penguins, which use their wings to swim under water, it is wide and robust. The coracoid extends from the front end of the scapula to the front of the sternum; it forms a stout brace for the wings. The juncture of the coracoid and scapula forms a socket, the glenoid cavity, with which the head of the humerus articulates. The ends of the clavicles are attached to the front of the coracoid and scapula, where the three bones enclose an opening, the foramen triosseum, which is important in the functioning of the wings.

The two clavicles are usually fused at the midline, forming the furcula ('wishbone' or 'merrythought'), which acts as a curved strut or stiff spring to brace the wings apart. Although the furcula is thought to be important for flight, it is absent in some parrots and owls; in these birds the ends of the clavicles are connected only by cartilage. The clavicles are reduced in flightless birds and are completely absent in kiwis and the extinct moas. In contrast, in the Pelecaniformes the bones of the pectoral girdle and/or sternum show considerable fusion; the extreme condition is found in the frigate birds, in which all these bones are fused.

The bird's wings are greatly modified vertebrate front limbs. Their most striking feature is the loss and fusion of the bones in the hand. The head of the humerus is expanded and has crests for attachment of the pectoral muscles. At the elbow the humerus articulates with the radius and ulna. The ulna is the stouter of the forearm bones, and the secondary flight feathers attach to it, usually on a series of quill knobs. Only two wrist bones (carpals), called the radiale and ulnae, are present in the adult. The rest are lost during development or fuse with the palm bones (metacarpals) to form the carpometacarpus. Only three fingers are found in birds. These are generally thought to represent digits 1, 2, and 3 of the five-digit vertebrate hand; the human thumb represents digit 1. Digit 1 in the bird has some freedom of movement and carries the bastard wing (alula). Digits 2 and 3 and

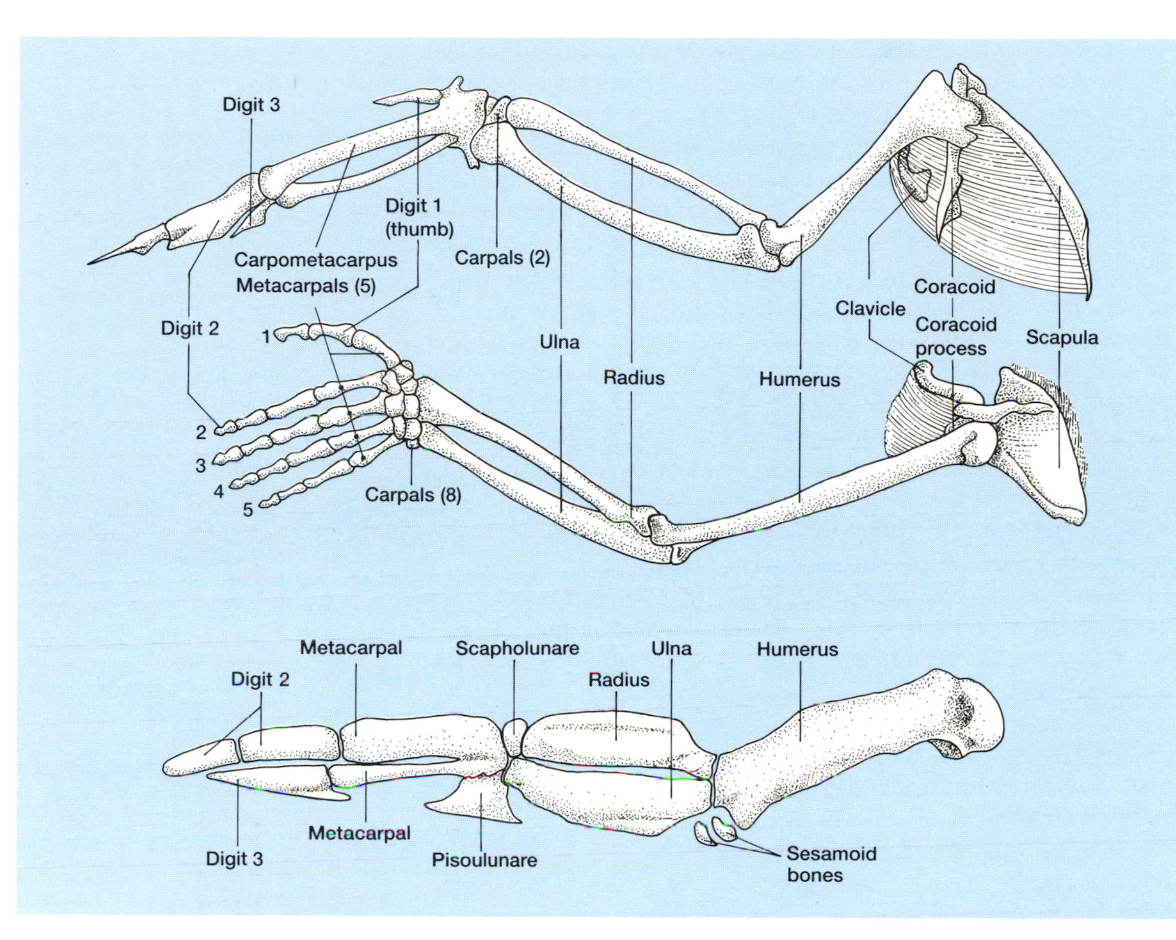


*An exercise in homology. The forelimb of a bird and a human are compared here to show the correspondence between the individual bones. The highly flattened wing of a penguin is also illustrated*

the carpometacarpus support the primary flight feathers. In some birds digit 1 is clawed, but the claw is usually small and easily overlooked. The claws of digits 1 and 2 are well developed in the nestling Hoatzin, but are lost by adulthood.

The pelvis is formed, as in other vertebrates, from three paired bones, the ilium, ischium, and pubis. The pubis lies parallel to and beneath the ischium, an orientation found elsewhere only in some dinosaurs. Except in the Ostrich the pubic bones do not meet to enclose the pelvic outlet. This condition is thought to be related to the size and hardness of birds' eggs. The three pelvic bones meet to form a socket, the acetabulum, into which the head of the femur fits.

The pelvis is wide in running and perching birds, and quite narrow in diving birds such as divers and grebes. It is also narrow in rails, as is the rest of their body, leading to the expression 'thin as a rail'.

The leg consists of three segments and appears to have one more segment than the human leg. The uppermost segment, the femur, is characterized by a large, spherical head and a prominent crest, the trochanter. At the knee the femur articulates with the tibia and fibula. The bird's knee is the joint between the thigh and the 'drumstick' and is usually concealed by feathers in living birds. Many species have a kneecap, or patella. The tibia is much larger than the fibula, which usually ends in a thin splint about two-thirds of the way down. In penguins, however, it extends to the ankle.

Like the hand, the foot of birds is highly modified from the condition found in reptiles and mammals. In most vertebrates a series of small bones, the tarsals, is found between the leg bones and the toes. In humans these are incorporated into the foot, where they form the heel and back part of the foot. In birds some of the tarsals have disappeared; the others have fused with the end of the tibia, which is properly called the tibiotarsus. Other tarsals have fused with some of the metatarsals of the foot to form the tarsus (properly called the tarsometatarsus). Thus, the ankle joint in birds occurs between sets of tarsals rather than between leg bones and tarsals as in humans. Because a bird's ankle joint is exposed (it is covered with feathers in grouse and some raptors) and occurs between what are mistaken for the long bones of the leg, it is sometimes mistaken for the knee. This has led to the mistaken popular notion that the knee of birds bends backwards. What is popularly called a bird's foot is really the tip of its toes.

Birds have two to four toes. In a typical perching bird the second, third, and fourth toes point forward, while the first toe (hallux) points backward, opposing the other toes. Several different arrangements of the toes are found: in swifts all four toes face forward; some birds can move one or another toe forward, to the side, or backward; others hold two toes forward and two backward.

The first toe is often very small or completely absent. Usually it has been lost in birds that run on hard surfaces. The Sanderling, which runs on hard, sandy beaches more than any other sandpiper, is the only sandpiper that has only three toes. Some shorebirds regularly feed on rocky shores, where the first toe is probably used to help grip slippery rocks. Only the Ostrich has but two toes.

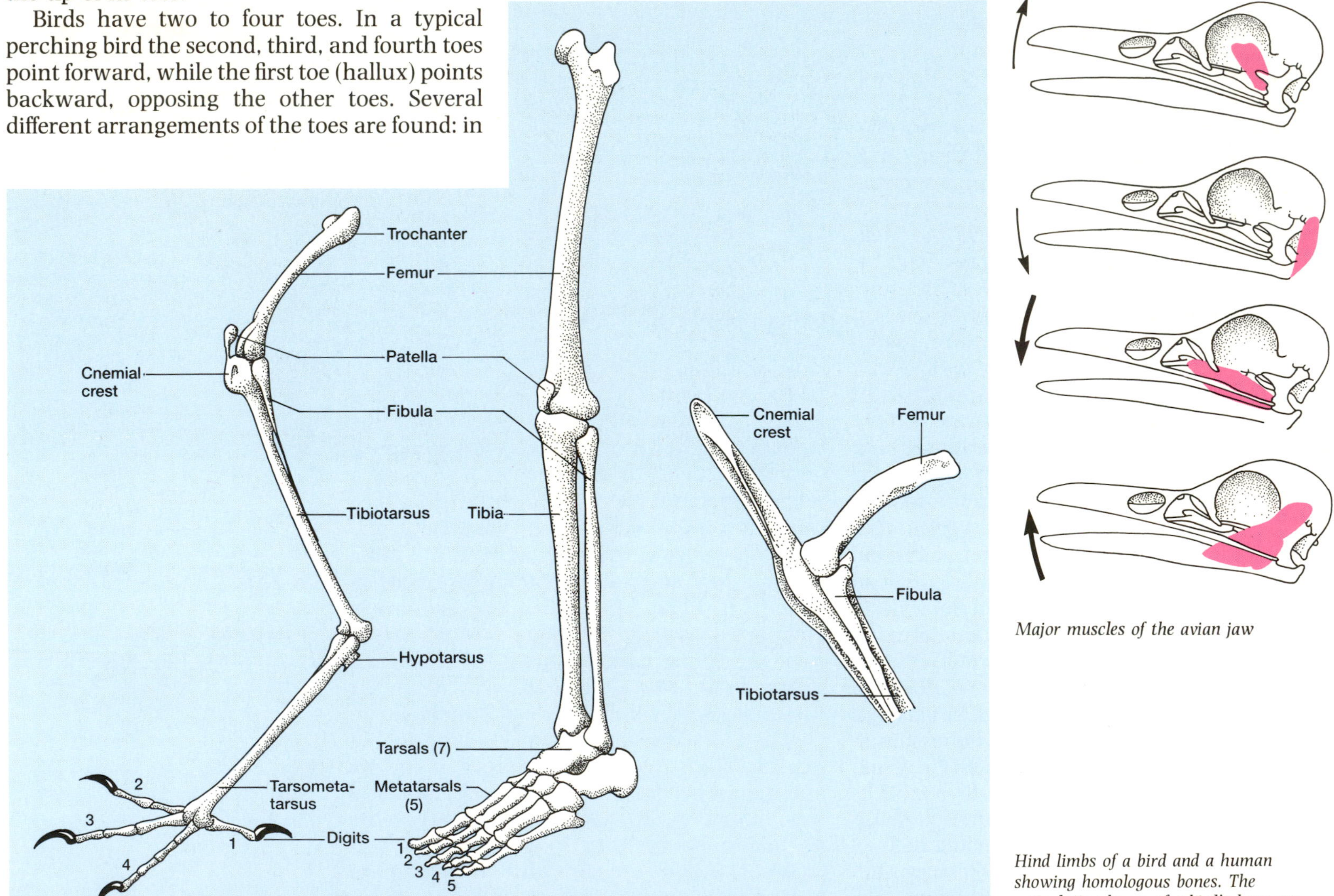


*Major muscles of the avian jaw*

*Hind limbs of a bird and a human showing homologous bones. The articulating bones of a bird's knee are detailed right*

## Muscular system

Muscles account for 30% to 60% of the mass of wild birds; they are concentrated in the flight and leg muscles. The proportion of flight to leg muscles is correlated with the mode of life of the species. Strong fliers such as hummingbirds, swallows, and doves have relatively heavy flight muscles and light leg muscles; species that use wings and feet about equally, like waterfowl and raptors, have about equal muscle mass in each area; whereas those that mainly run on the ground, like Ostriches and rails, have most of their muscle mass in the legs. In game and domesticated species the muscles account for a large portion of the body weight.

The jaw, tongue, and hyoid muscles show considerable variation according to feeding method; they have received considerable study. The muscles that close the jaws are notably well developed in species that break open hard seeds or fruit, grasp strong, active prey such as fish, or tear flesh.

The muscles that open the jaws are well developed in species that eat large prey, gape in the ground, or pry bark from trees. The jaw muscles are often small and weak in species that feed on tiny insects or nectar. The tongue and hyoid muscles are particularly complex in parrots, which manipulate fruit and flowers with their tongue, and woodpeckers, which probe extensively with theirs. The muscles of the voice box (syrinx) are highly developed in birds with loud and complex songs.

The neck muscles of birds are well developed and complex, especially in species that pursue active, fast-moving prey. In many species one of the muscles connecting the head and neck becomes quite large just before hatching; it helps the chick to break the eggshell with its bill. Birds have few and small trunk muscles since this area is stiffened by ligaments and fusion of the vertebrae. Much variation is found in the muscles of the tail, which is used extensively in flight and courtship. Flying birds generally lower and spread the tail when taking off and landing to help maintain lift at low speeds. The tail muscles are particularly well developed in lyrebirds, which raise heavy tails during courtship displays.

The flight muscles, which include those of the breast, pectoral girdle, and wing, are well developed in birds that use their wings for flight and swimming. The breast muscles provide most of the power for locomotion whereas the

*A Feral Rock Dove (Columba livia) dissected to show the main muscles. Notice the size of the pectoralis muscle which constitutes the main means of flight*

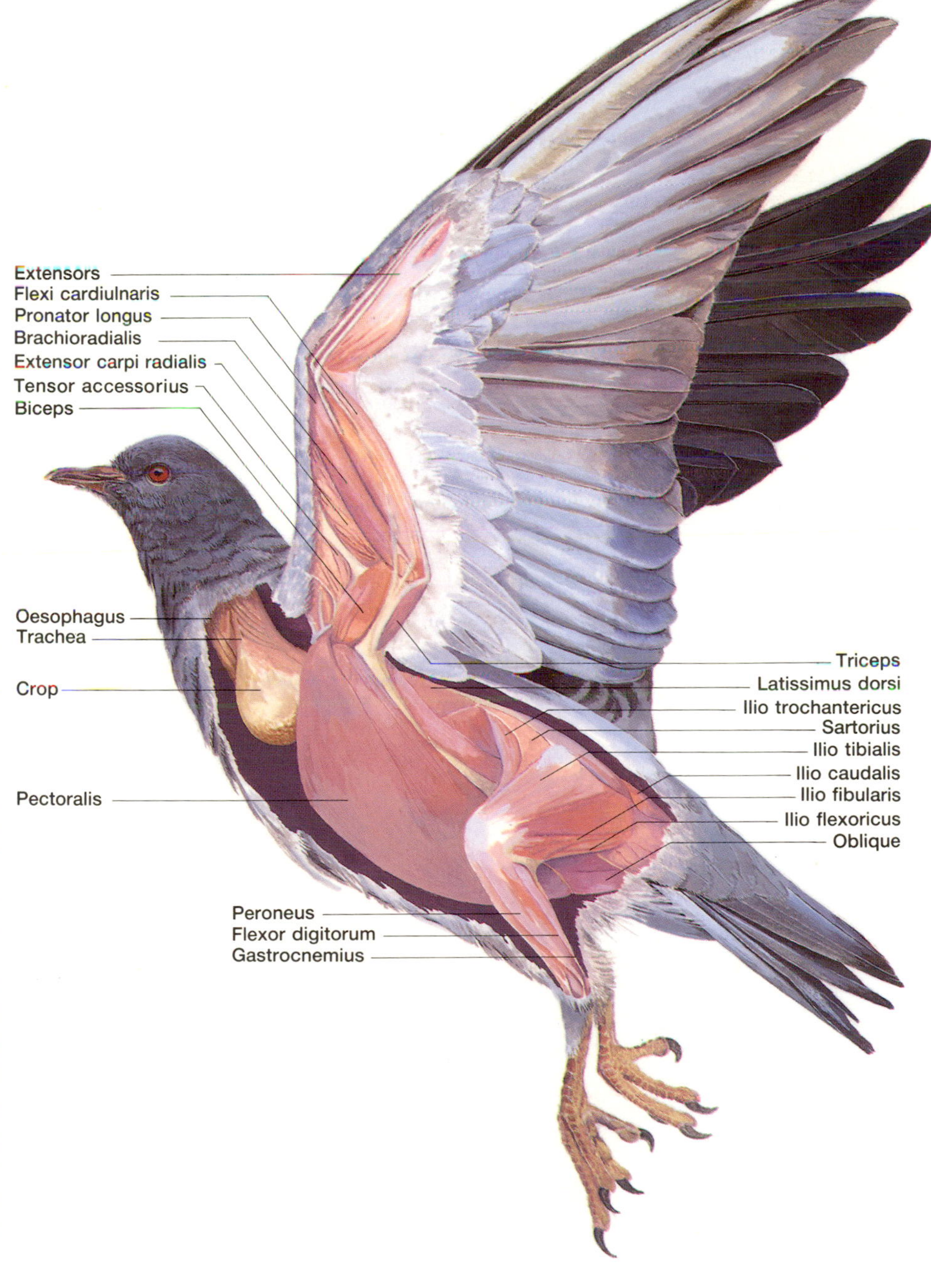

other muscles control the intricate movements such as the extension and folding of the wing, depression and elevation of its leading edge, and positioning of the feathers.

The largest flight muscle, and usually the largest in the body (left and right muscles together average 15% or more of the body weight) is the pectoralis or pectoral muscle, which powers the downward thrust of the wing. It is the large, outer muscle of the breast, attaching by a stout tendon to a crest on the lower side of the humerus. In soaring birds the inner part of the pectoral muscle functions to keep the wing motionless in a horizontal plane against the varying force of the wind. This part is well developed in albatrosses, which soar over the tops of waves, where the wind is continually changing.

The other large muscle of the breast, which lies under the pectoralis muscle, is the supra coracoideus. This muscle raises the wing and is particularly large in birds like penguins that use their wings to swim under water. A tendon from the supra coracoideus muscle passes through the foramen triosseum, which acts as a pulley to change the direction of the pull, and attaches to the upper surface of the humerus. Both of the breast muscles, their tendons, and attachments can easily be seen when carving a chicken.

Motion of the individual elements of the wing is restricted by the arrangement of the muscles and ligaments and the interlocking of the bones. The humerus can move in any direction at the shoulder. Motion at the elbow and wrist is almost entirely restricted to the horizontal plane, thus giving the wing considerable vertical stiffness. In addition, the wing always opens and closes as a unit whether used in flight, displays, or comfort movements. This is the result of the elbow and wrist articulations acting like a parallel rule; any opening of the elbow forces the simultaneous opening of the wrist.

The muscles of the legs are much more variable than those of the wings, reflecting the much wider variety in use of the legs. Most of the approximately 35 muscles of the leg are found in its upper two-thirds. Only in a few birds such as hawks and owls, which strongly grasp prey, are the muscles along the tarsus well developed. Not only does having the muscles close to the body provide mechanical advantages, it also places them where they are insulated by feathers instead of lying exposed on the usually naked tarsus.

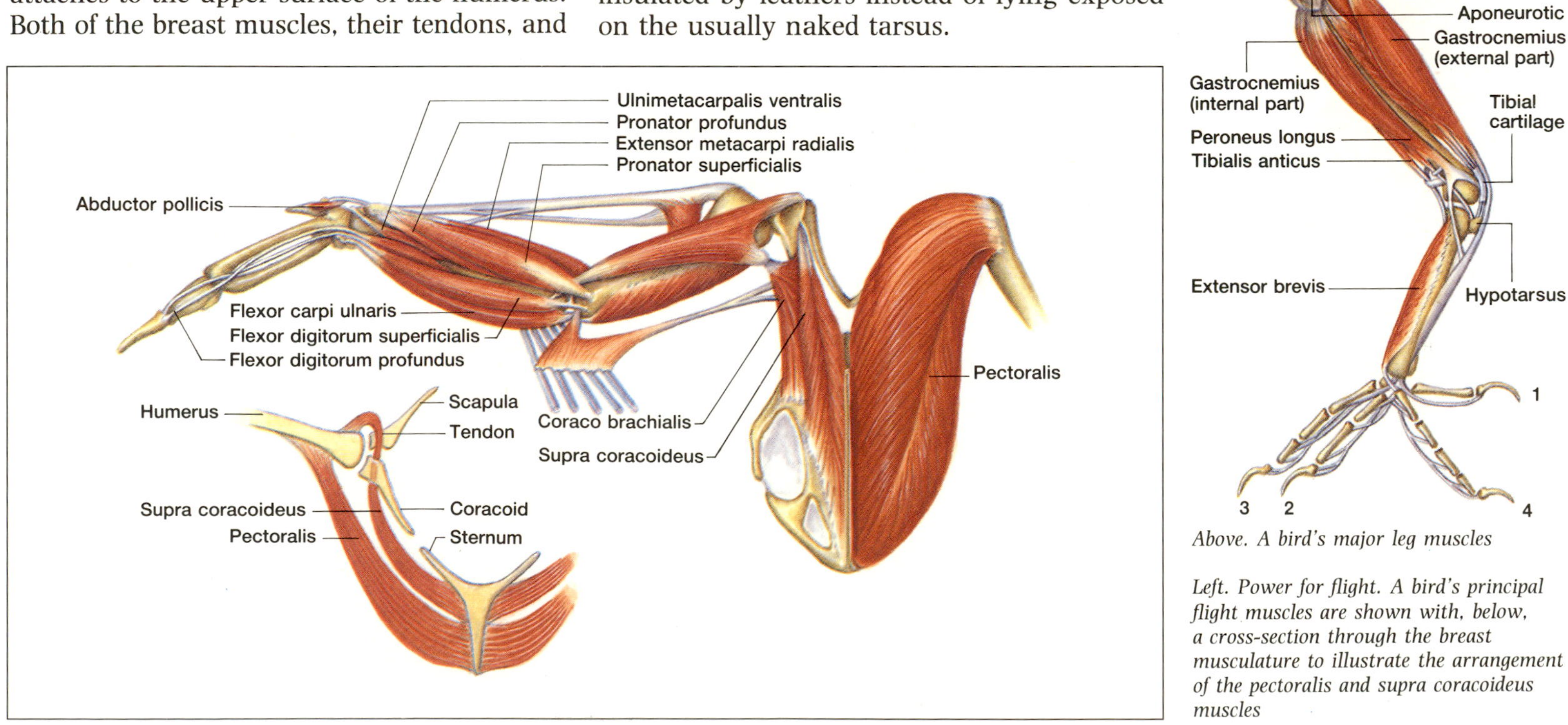


*Above. A bird's major leg muscles*

*Left. Power for flight. A bird's principal flight muscles are shown with, below, a cross-section through the breast musculature to illustrate the arrangement of the pectoralis and supra coracoideus muscles*

The position of the legs and proportions of their three segments are constrained by the need to maintain proper balance when the bird is walking, perching, or settling on the nest. In birds that use their legs extensively for terrestrial locomotion, the femur is short relative to the tibiotarsus and tarsometatarsus, which are of almost equal length. The femur is usually held close to the body and nearly horizontal; most of the motion for walking and running occurs in the lower two parts of the leg. This keeps the centre of gravity over the feet. In diving birds the legs are shifted more toward the rear, and these birds are quite awkward on land. The extreme condition is found in divers, in which the short femur is closely attached to the body and quite restricted in motion. In the foot-propelled diving birds (divers, grebes, cormorants, and, to a lesser extent, some ducks and shearwaters) large thigh muscles are attached to a well-developed cnemial crest and develop a strong thrust while swimming.

Almost all the actions of the foot and toes are controlled, through a complex series of tendons, by muscles in the upper part of the leg. The number, arrangement, and interconnections of these tendons vary widely and are related to the use of the foot as well as to the genealogy of different groups of birds. The details of these tendons and their associated muscles have been extensively studied in an effort to understand the functioning and evolution of the avian foot.

In many perching birds the tendons of the flexor muscles pass around behind the ankle so that when the joint is bent the toes are automatically flexed around a branch. This allows a strong grip to be maintained without the exertion of continuous muscular force. In addition, some birds have ridges on the inner side of their toe pads that lock against the toe tendons and ensure a strong grip even when the bird is sleeping.

JGS

*Left below. Comparison of the leg muscles of a pheasant (left) and diver (right)*

*Below. How birds perch. The simplified diagrams at the top show how bending of the ankle joint automatically causes flexion of the toes in a perching bird. Ridges on the toe pads and tendons, below and detail, ensure a firm grip when the bird is sleeping*

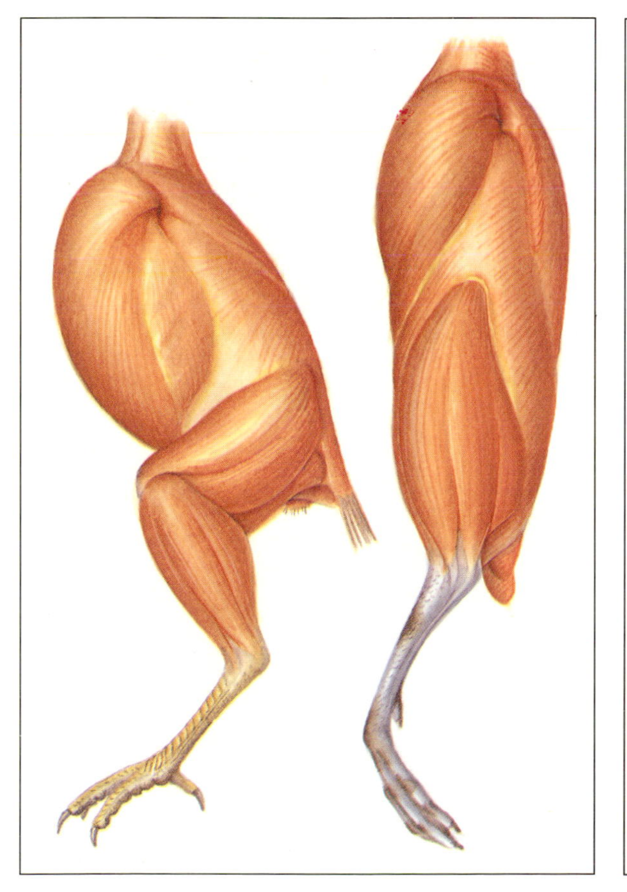

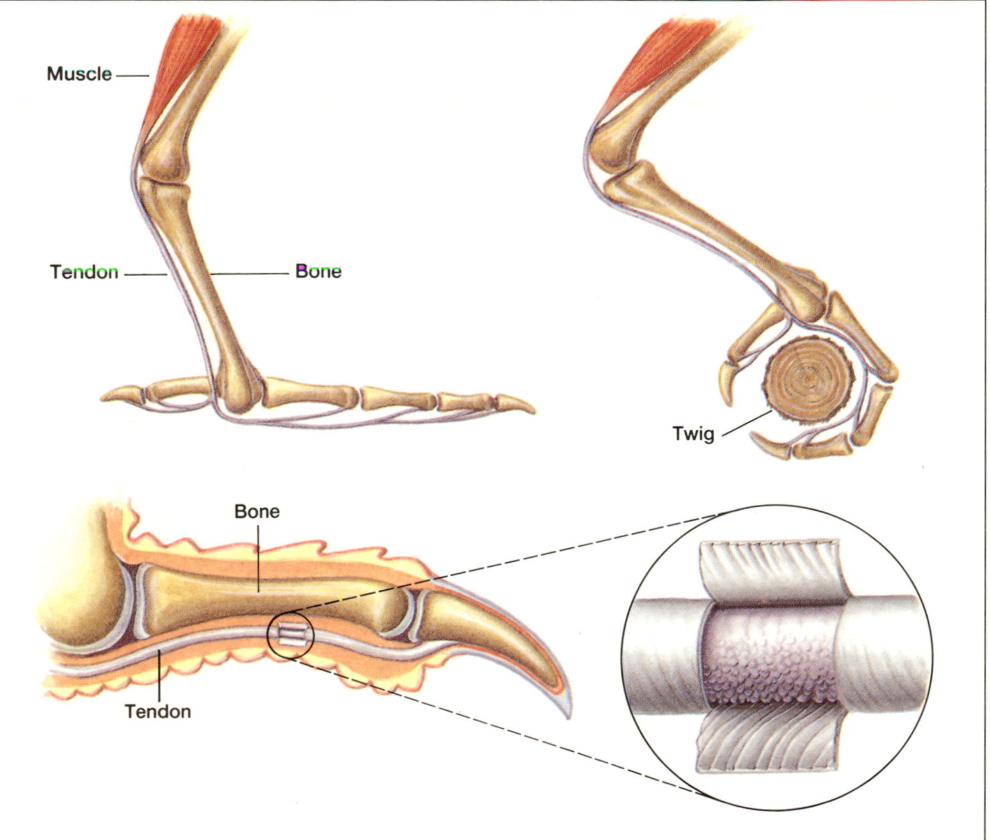

## Feathers

It has been frequently stated that having feathers defines an animal as a bird, as all modern birds have feathers. Recent speculation on whether some dinosaurs were feathered and uncertainty whether *Archaeopteryx*, the earliest known feathered animal, is a direct ancestor of birds have called this 'definition' into question. Evidence has not been found in the fossil record to resolve these issues. On the other hand, birds are well defined by numerous skeletal characters, and there is no doubt that any living animal with feathers is a bird.

Feathers provide a bird with insulation, a flight surface, protection of the skin, and their external appearance, and they lower the bird's specific gravity. Some birds, such as snipe, woodcock, and neotropical manakins, use the feathers to make sound; others, like sandgrouse, carry water to their young with them.

Feathers are epidermal structures that first appear as little bumps on the skin of the embryo; each soon develops into a backward-projecting cone anchored in a small pit, the feather follicle. At the base of the follicle is a small hump of cells, the dermal papilla, which is covered by an epidermal cap. The epidermal cap produces a ring of cells, the epidermal collar, from which a feather arises. The cone of tissue arising from the epidermal collar differentiates into three layers which become the feather sheath, the feather proper, and temporary nutrient pulp.

At first the middle layer is a cylinder marked with parallel longitudinal ridges, but because of the differential growth patterns around the cylinder, it develops a large dorsal ridge, a smaller ventral one, and a series of branches coming off them.

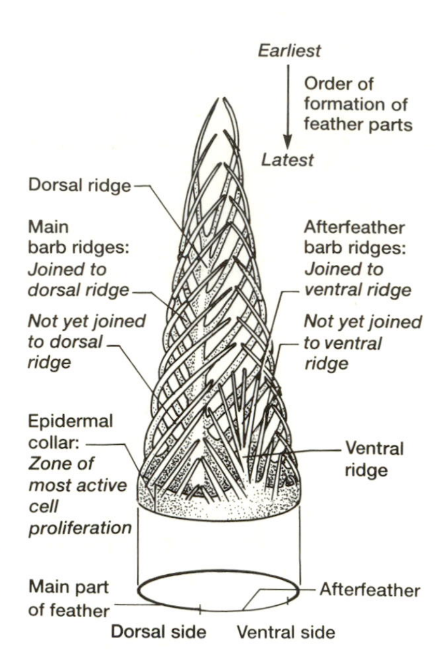


*A growing contour feather showing the development of the major parts. The sheath and pulp have been omitted for clarity*

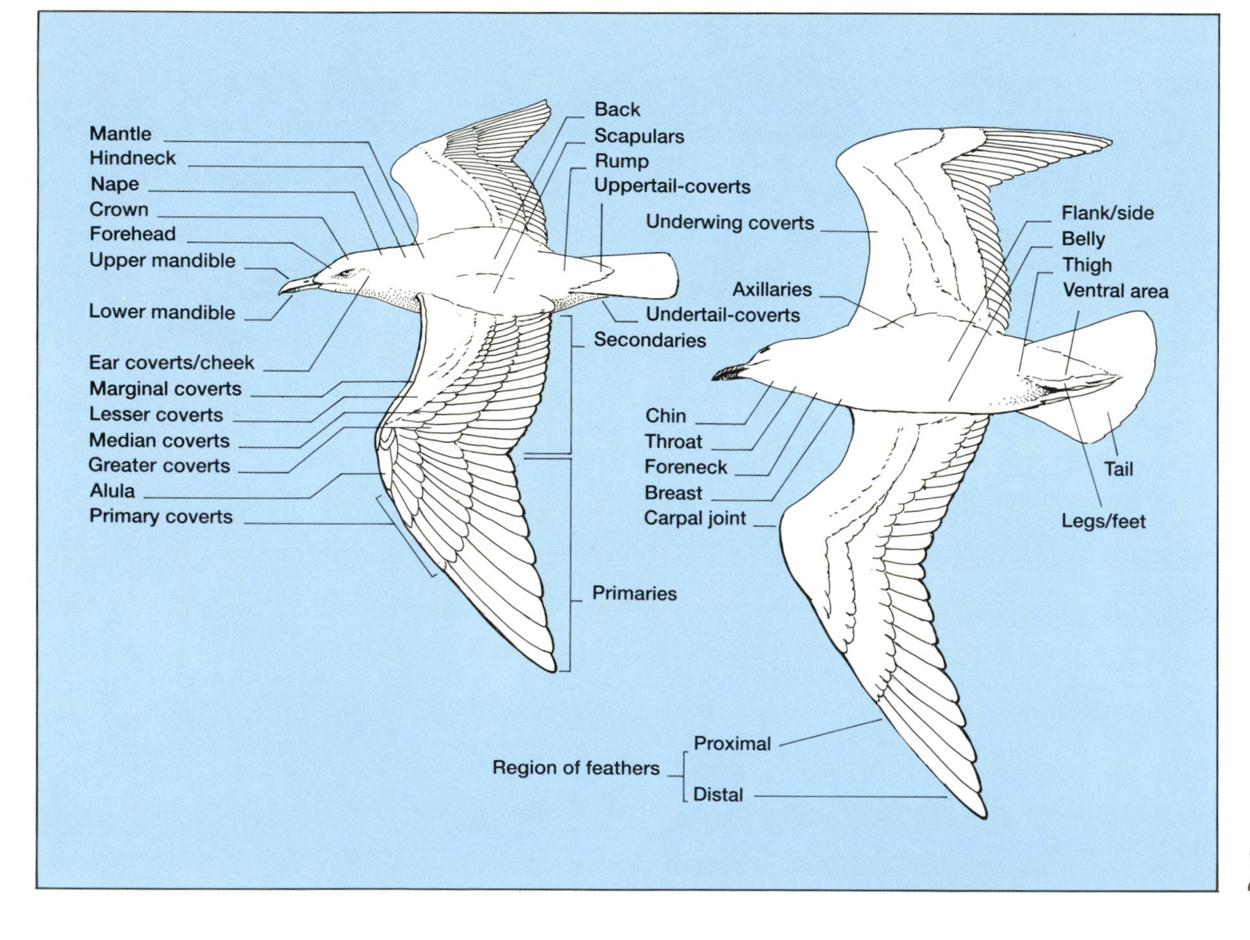


*The surface shape of a seabird seen in dorsal view (left) and ventral view (right)*

The dorsal ridge becomes the main shaft (rachis and calamus) of the mature feather; the ventral ridge, the aftershaft; and the branches, the barbs. An additional set of branches develops on each of the barbs; these become the barbules.

A feather grows only at the base; thus, the tip is produced first followed by lower parts. As the tip is pushed out of the skin final development is completed, the cells die, and the protein in them polymerizes to form β-keratin. The finished part of a feather may extend several centimetres from the skin while the base is still developing. Growth rates vary from 1 mm/day up to 13 mm/day in the case of the Wattled Crane. A completed feather consists only of dead tissue. A feather remains furled in its sheath until the latter breaks away or is preened off. Then the feather unrolls, finishes drying, and assumes its final shape.

As illustrated right, feathers are of six types: contour, flight, down, semiplume, bristle and bristle/eyelash, and filoplume. The structure of a typical contour feather consists of a main shaft, the further end of which (the rachis) is filled with spongy tissue and has a flat vane extending from it. The vane is composed of numerous barbs which, in turn, have small barbules radiating out from them. On the stiff, or pennaceous, part of the vane the distal barbules are covered with tiny hooks (barbicels) and smooth cilia whereas the proximal barbules are variously notched or ridged.

The hooked barbules of one barb overlap and hook onto the ridged barbules of the barb anterior to it, thus forming a solid interlocking structure. If the barbules become disengaged or tangled a bird can restore the interlocking structure by preening. The lower part of the vane in many cases has a downy or plumulaceous part in which the barbules are smooth and do not interlock. The lower end of the main shaft (the calamus) is hollow and free of barbs. Extending from the junction of the rachis and calamus is a small downy branch called the after feather.

Small pores, the superior and inferior umbilici, are found at either end of the calamus and mark the course of the living pulp that once filled the centre of the growing feather.

*Feather types: bristle, semiplume, flight, and (bottom row) contour, down, bristle/eyelash, and filoplume*

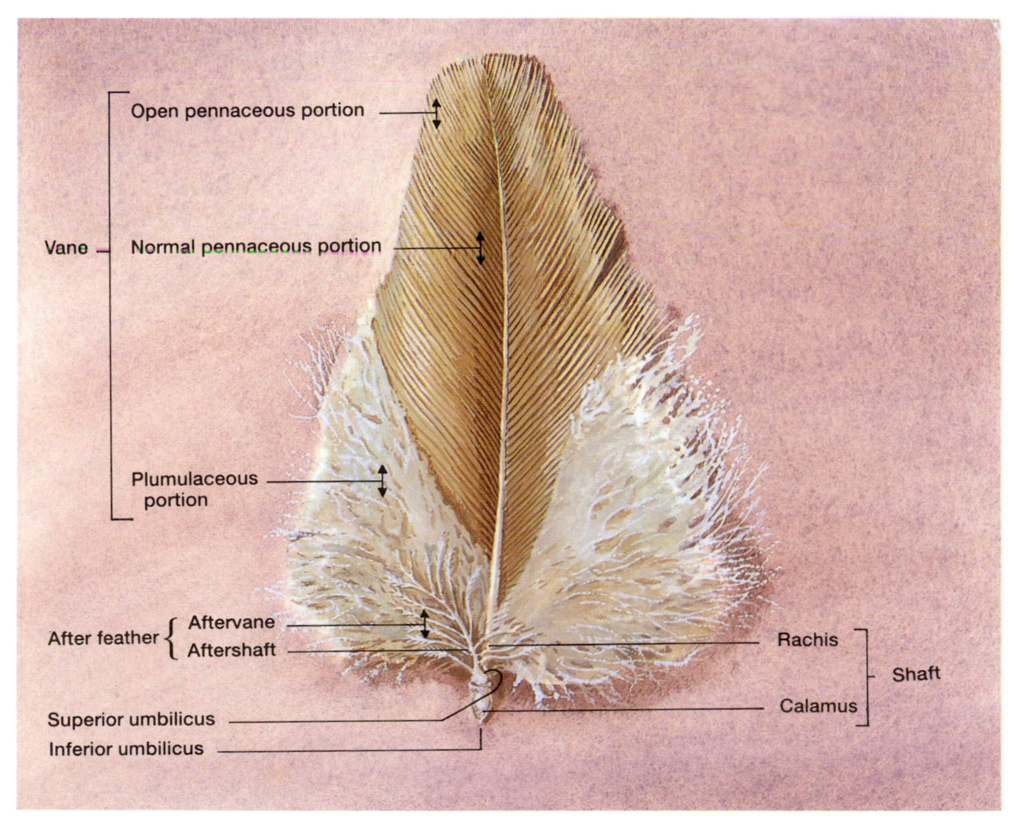


*The main parts of a typical contour feather*

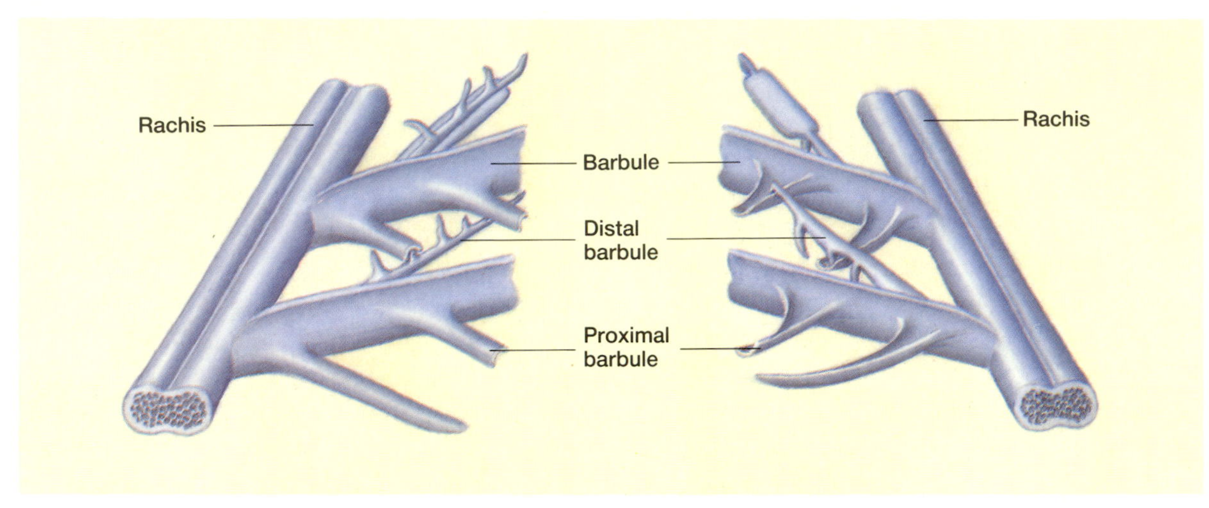


*Detail of barbules interlocking at the point where flight feathers overlap*

Contour feathers cover all of the body except the beak and the scaled parts of the legs and feet. They protect the skin from physical damage from vegetation and blowing grit and shield it from the harmful rays of the sun. Contour feathers also streamline the body and increase the efficiency of flight. The lower, downy portions of the contour feathers contribute to thermal insulation and fill out the body contours. In aquatic birds the tips of the contour feathers are frizzled and repel water.

Flight feathers differ from contour feathers in being stiffer and longer, with little or no down at their base and with a greatly reduced or absent afterfeather. The flight feathers of the wing (remiges) and those of the tail (rectrices) are anchored by connective tissue to bone. The interlocking mechanism of the barbules is more highly developed, giving greater cohesiveness and strength. In some cases the ventral edges of the barbs are expanded and act as flap valves, preventing air from passing upward between them. The vanes are of unequal width, which causes the primaries (the remiges of the hand) and outer rectrices to twist when subjected to air pressure, thus forming surfaces with higher lift. In many birds the outer primaries are narrowed or notched at the tip so that slots are formed between them when they are outstretched. The slots allow birds to fly at slow speeds without stalling. Many birds have friction barbules with modified barbicels in the areas where the flight feathers overlap in flight. These rub against the barbs of overlapping vanes and prevent the feathers from slipping apart. Most owls (fishing owls are the exception) and nightjars have modifications of the wing feathers which allow quiet flight. These include feather combs along the leading edge, downy filaments on the dorsal surface of the wing, and fringed tips on the trailing edge of the wing.

Down feathers have a greatly reduced or absent rachis and reduced barbicels. They are soft and fluffy and function as insulation underlying the contour feathers. Many birds such as herons have powder downs, a specialized down that breaks off into a fine powder. This powder is preened onto the contour feathers and acts as a water repellent. The powder downs of herons grow continuously.

Semiplumes are feathers intermediate in structure between contour feathers and down. They have a well-developed rachis but entirely plumaceous (soft) vanes. They are mostly hidden under the contour feathers and act as insulation and help fill out the body contours.

Bristles are spiny feathers with a stiff rachis and with usually only a few barbs clustered around their base. They grade into contour feathers; intermediate forms are sometimes called semibristles. In most birds they are found mainly on the head and neck, but they also occur on the feet of Barn Owls. They function as guard hairs especially around the eyes and nostrils, as sensory organs very much like the whiskers of a cat, and as a net around the open gape of insectivorous birds to help catch prey. Crowned Cranes have a large, ornamental tuft of bristles on the back of their head.

Filoplumes are fine, hair-like feathers with a tuft of barbs at the tip. They do not grade into any other type of feather and are usually not exposed. Unlike other types of feathers they never occur alone, but are always found next to contour feathers. Also unlike other feathers they have no muscles at their base, but their follicles are endowed with abundant nerve endings (lamellar corpuscles) which are sensitive to changes in pressure and vibration. Filoplumes function as sensory organs for keeping the feathers in place and adjusting them in flight, insulation, and bathing.

Feathers appear to have evolved from scales. They are composed of a protein, β-keratin, found elsewhere in the animal kingdom only in lizard skin. Developmentally, feathers and scales are quite similar. Experimental transplants have shown that bird epidermis can produce feathers or scales depending on its location on the embryo. During development bird scales often develop small downy tips which last for a few weeks after hatching in some breeds of chickens.

During the course of the annual cycle feathers become bleached, worn, broken, and infected with parasites, and need to be renewed. The renewal process, moult, occurs in most birds at least once a year. This topic is further considered in Chapter 5.

*The mouths of birds that catch flying insects are often surrounded by bristles which act as a scoop. This feature is pronounced in nightjars and relatives such as the Large Owlet-Nightjar (Aegotheles insignis) of New Guinea*

## The distribution of feathers over the body

Although the exterior of a bird's body is uniformly covered with contour feathers, these feathers grow only from distinct tracts, or pterylae, fanning out to cover any bare areas. Pterylae may also contain semiplumes, down, bristles, and filoplumes among the contour feathers. The spaces between the pterylae (apteria) may be bare or covered with semiplumes or down. The relative size and configuration of the tracts vary: aquatic birds tend to have wider pterylae and small apteria; penguins have an almost uniform covering of contour feathers with only a few small apteria. Apteria around the joints may increase the mobility of the limbs, but their major function seems to be to cool the body when they are exposed.

The bases of the feathers in a tract are connected by a complex network of muscle fibres. These fibres allow complex movement of the plumage: fluffing up to increase insulation, sleeking down for cooling or lowering the specific gravity, parting the contour feathers to expose bare areas of the skin, arranging the feathers for flight, and displaying.

## Feather numbers

Most birds have 10 primaries on each wing, with extremes of 9 and 12. Hummingbirds have the fewest secondaries (6) whereas albatrosses have as many as 32 and the Bateleur Eagle has 35. Rectrices (tail feathers) number from 6 (several small songbirds) to 32 (Bulwer's Pheasant) but are typically 10 or 12.

The total number of feathers varies with the species, size of the bird, sex, age, health, season, and geographic distribution. Most songbirds have between 1500 and 3000 feathers, which account for 6% to 7% of their body mass. The smallest recorded number is 940 for a hummingbird and the greatest is 25 216 (40% of which were on the head and neck) for a swan. Birds which winter in cold regions have more feathers in the winter than in the summer. The plumage of redpolls has been found to be about 30% heavier in the winter than in the summer. Feathers may make up a considerable part of the body mass; for instance, one 4082 g Bald Eagle studied in detail had feathers that weighed 677 g (17%) whereas its dried skeleton weighed only 272 g (7%).

*A bird's major feather tracts (pterylae)*

## Colours and pigments

A striking feature of birds is their diversity of plumage colours and patterns. Often these are used to increase their conspicuousness in courtship and threat displays or serve as camouflage for concealment from prey and predators.

Feather pigments may also protect the bird from the ultraviolet rays of the sun, absorb radiant energy, and increase the strength and wearing qualities of the feathers. As a rule it is only the exposed part of the feather that is coloured and patterned, the concealed part being dull. However, some birds such as Goldcrests and New World flycatchers have brightly coloured concealed feathers which they expose during displays.

The colours of feathers are produced in two different ways, by coloured pigments embedded in the feather or by special structural features. Two kinds of pigments are found in feathers, melanins and lipochromes. Melanins produce blacks (eumelanin) and browns, red-browns, and yellows (phaeomelanin), of which there may be several chemical forms. Some melanins are related to dietary amino acids; black-plumaged birds, for instance, require high amounts of riboflavin in their diet. Lipochromes produce yellows, oranges, reds, blues, and greens, and are chemically related to vitamin A and blood pigments. The red pigments of several finches are derived from carotenoids in their food: for example, the plumage of canaries can be changed from yellow to red if they are fed paprika during their moult. Lipochromes also tend to fade easily.

Structural colours are usually classified as iridescent or non-iridescent. Iridescent colours, like those found in the wing specula of ducks or the throat gorgets of hummingbirds, can be seen only at certain angles to the feather surface. These colours result from the interference of light by different layers of melanin pigment in the barbules. The reflective surface can be either that between melanin and keratin or that between melanin and air. The size and geometry of the reflecting particles influence the shade of the reflected colour. Non-iridescent colours can be seen from any angle. They have long been attributed to the scattering of light by minute air pockets in the feather structure, a process known as Tyndall scattering, which also causes the blue colour of the sky. Recently it has been suggested that these colours are the result of interference from air-keratin interfaces.

A given feather may have different pigments and structures on different areas. The overall colour of an area may also be a combination of pigment and structure: in the Green Magpie the green colour is the result of a yellow pigment overlying a blue structural colour. Because the yellow fades in the sun, magpies which live in open areas are noticeably bluer than those living in shady areas.

The difference in wear resistance of pigmented and unpigmented areas of a feather can cause apparent colour changes in some birds. For example, the feathers of House Sparrows and Snow Buntings that appear to be solid black in the late spring emerge in the autumn moult with broad, light-coloured tips. The tips wear away during the winter resulting, by spring, in the solid black of the breeding plumage.

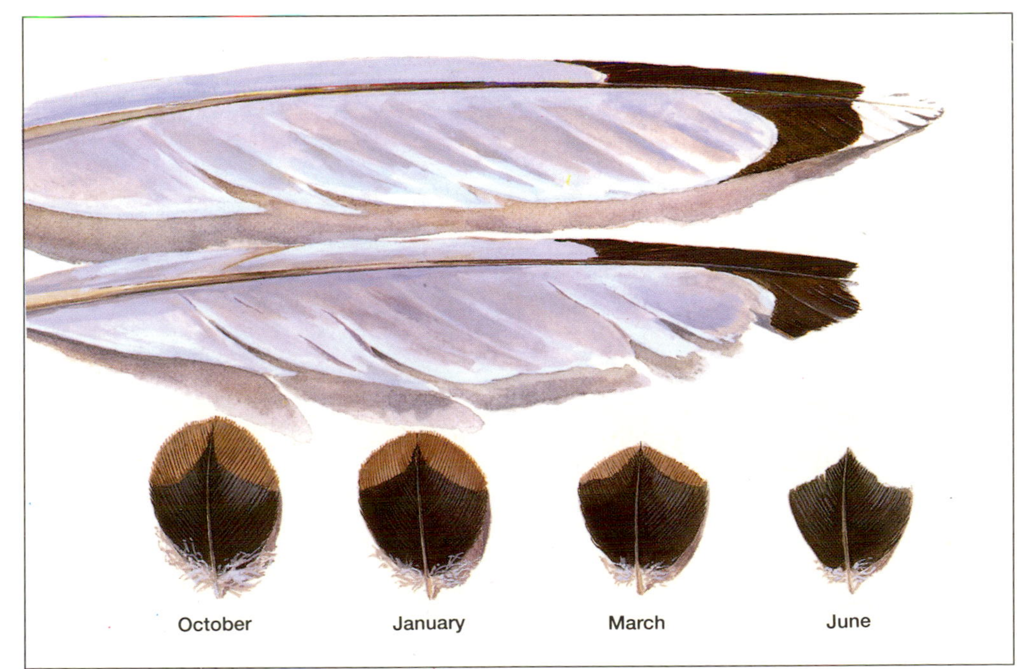


*Fresh and worn primary feathers from a Herring Gull (Larus argentatus) showing the resistance to wear of the black tip relative to the unpigmented remainder. Below are feathers of a Snow Bunting (Plectrophenax nivalis) to illustrate how the bird changes its colour from brown in October to black in June because of the erosion of its feather tips over the course of the winter*

## Aberrant plumages

Abnormalities in bird plumage can be caused by disease or poor diet, or they can have a genetic basis. The abnormalities may be the result of changes in the amount and distribution of pigments or in feather structure. In wild birds pigmentation abnormalities are occasionally observed, but structural ones are rarely seen because they are usually fatal.

Many abnormalities of pigmentation appear to be genetic as they usually appear with a consistent form and they can in some cases be produced by controlled breeding. An increase or decrease in the normal amount of a pigment is called heterochroism. Melanism is an excess of black or brown pigment, in erythrism chestnut-red replaces other melanins, while in flavism (xanthochroism) there is an excess of yellow. A reduction in the intensity of all pigments is called dilution or leucism, while albinism is the complete absence of pigment.

Schizochroism is the absence of one or more normally occurring pigments. The 'buff', 'fawn', and 'cinnamon' varieties of captive birds often result from the absence of black melanin. In rare cases of schizochroism one colour is replaced by another.

One of the most striking types of aberrant plumage is that found in so-called 'gynandromorphs' in which all the feathers of the bird are male on one side and female on the other. Such individuals have been reported among pheasants, falcons, parrots, woodpeckers, and several songbirds.

Because the plumage pigments of many birds depend on proper diet, abnormally dull and colourless plumages often appear in captive birds. The loss of the pink colour in captive flamingos, easily restored by a diet of small crustaceans, is a well-known example. Wild finches such as crossbills often exhibit considerable variation in their plumage colour, even sometimes on the same individual, presumably because of variations in diet.

The colour pattern of the breeding plumage often depends on hormone levels. The male plumage has been experimentally produced in the females of several gamebirds by reducing the level of female hormone.

Feather structure abnormalities are more frequently observed in domestic birds; frizzled and silky breeds have even been produced in fowl and pigeons. In wild birds a structural weakness of the feather vanes that leads to rapid wear resulting in 'needle-tailed' rectrices and flightlessness has been occasionally found in guillemots (*Uria* and *Cepphus*). Its cause is unknown.

In the proper light, growth bars can be seen on the feathers of many birds. These are believed to result from changes in the daily metabolic cycle during moult. If a bird suffers from an inadequate diet while moulting, however, narrow zones of weakness (fault bars) may develop and cause a feather to break prematurely.

Birds with abnormal plumage appear poorly suited to survive in the wild. They may be more conspicuous to predators or suffer increased physiological stress. In addition they may be unable to breed since they are not recognized by conspecifics or are attacked by them.

JGS

*Schizochroism is the absence of one or more normally occurring pigments. This beige-backed Royal Penguin (Eudyptes schlegeli) lacks the melanin which contributes to the blue-black backs of its fellows. However, such plumage abnormalities do not always exclude birds from the joys of parenthood. The upper picture shows a partially albino female Blackbird (Turdus merula) raising a brood fathered by a normal black male*

## The digestive system

The high metabolic rate of birds demands an alimentary tract which is geared to the digestion of food as rapidly and as efficiently as possible. Towards this goal a number of important adaptations have evolved, the most striking of these being the conversion of part of the stomach in many species into a chamber where food is ground down to a fine consistency to permit swift penetration by the gastric juice, and the development of storage facilities in the upper alimentary tract which allow rapid ingestion and holding of food in bulk despite the absence of teeth and the limitations of flight. Further down in the lower alimentary tract most of the enzyme digestion of food takes place and here there are striking modifications in gut length and morphology which can also be related to the type of food ingested.

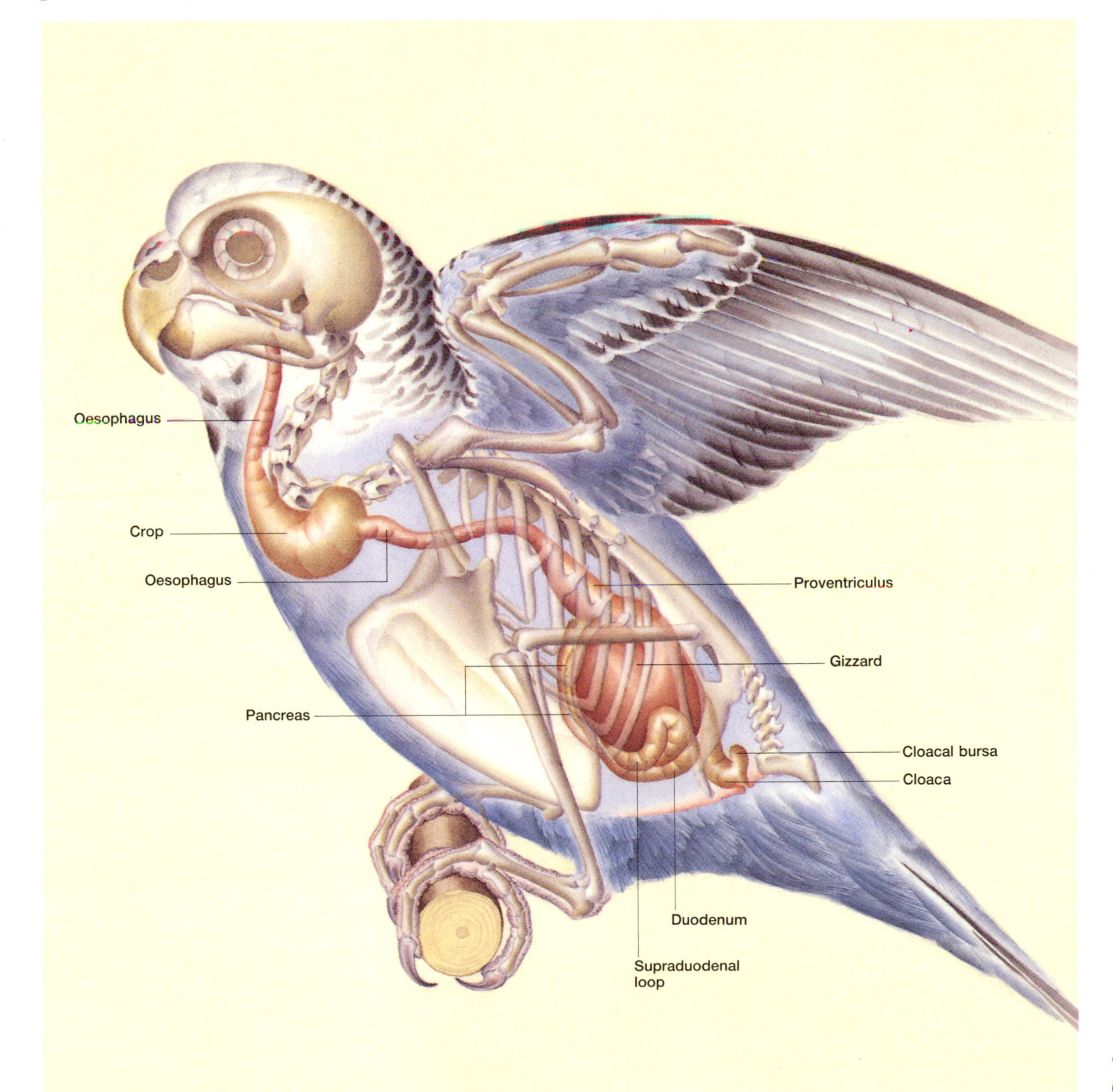


*The digestive tract of the Budgerigar (Melopsittacus undulatus)*

## Upper alimentary tract: mouth, oesophagus and crop

Flight has enabled birds to penetrate a wide range of habitats and there are numerous variations in the form and structure of the bill, mouth and tongue which are closely related to feeding habits. The absence of teeth, and the concomitant reduction in chewing musculature, is believed to be a weight-saving adaptation for flight, as well as permitting faster ingestion and digestion of food.

The bulk of food storage in birds occurs in the oesophagus. The avian oesophagus is a much more capacious tube than in mammals and may store large quantities of food without any discomfort. There is a close correlation between the calibre of the oesophagus and the size of the items of food which are swallowed. Insect feeders like swifts, for example, have an extremely narrow oesophagus, whilst in seabirds it may be large enough to be crammed with fish head to tail. The inner surface of the oesophagus is lined by a thick protective membrane so that no damage is caused by the stored food, even though it may still be alive.

In a relatively small number of birds, including gallinaceous species, ducks, pigeons, falcons, parrots and some songbirds, food is also stored in a diverticulum or swelling of the oesophagus called the crop. Whilst striking variations in both the shape and size of the crop occur, their adaptive significance is not known. The crop of pigeons is particularly interesting, not only because it consists of two sacs, but because its inner lining is sloughed off as 'crop milk' which is regurgitated by both parents and fed to the chick. Nutritive juice for the chick is also provided by glands of the oesophagus in the Greater Flamingo and the male Emperor Penguin.

In a very few birds food is stored in sacs suspended from the floor of the mouth below the tongue. In the Rosy Finch and Pine Grosbeak the sacs occur only in the adult bird and only during the breeding season, but in nutcrackers they appear to occur at all ages and to be permanent. In pelicans the entire floor of the mouth is massively enlarged to form a gular pouch that is used to catch fish.

## The stomach

The avian stomach consists essentially of two chambers. The first is the glandular stomach or proventriculus, the walls of which are packed with glands which manufacture the gastric digestive juice. The second chamber is the muscular stomach or gizzard which in many species is the site of physical digestion.

Gastric juice is a mixture of hydrochloric acid and the protein-digesting enzyme, pepsin. Although originating in the proventriculus, it appears to act on the food mainly in the gizzard

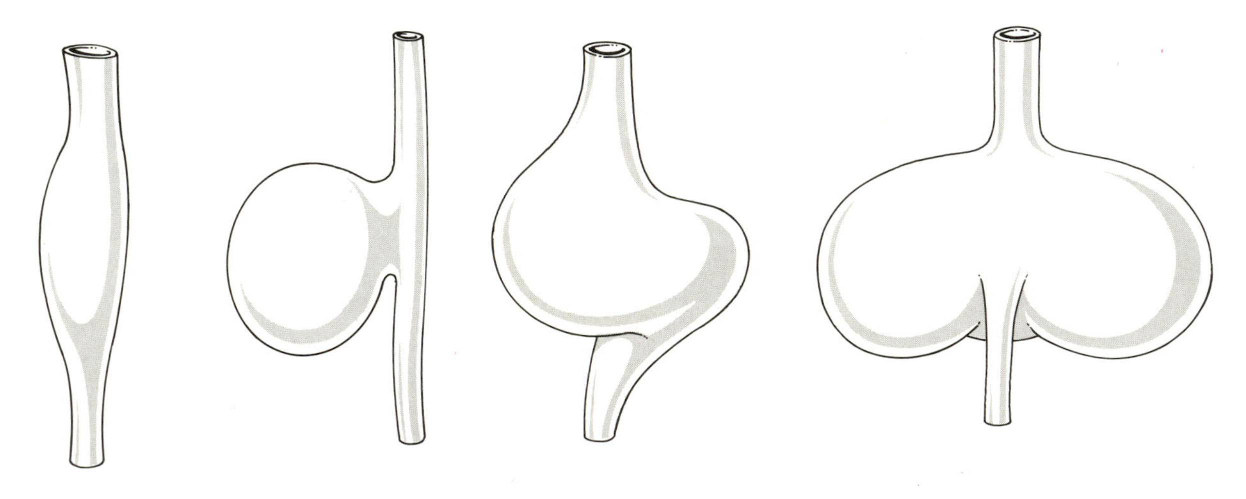

*The crops of birds differ markedly in shape, but the reasons for this are not known. Variants shown are, from left, Great Cormorant (Phalacrocorax carbo), Peafowl (Pavo cristatus), Budgerigar (Melopsittacus undulatus), Rock Dove (Columba livia)*

where there is low acidity. Contrary to popular opinion, the 'stomach oil' spat out by petrels at intruders or as food for the chick is not a product of the proventricular glands but is derived mainly from the marine invertebrates on which the birds feed.

Gizzard development is closely related to the nature of the bird's food. Thus in species such as fish- and meat-eaters which feed on relatively soft food, the gizzard is thin-walled and sac-like, and functionally appears to be little more than a container where food is acted on by the digestive juice. In contrast, species feeding on such relatively hard items of food as insects, vegetable matter and grain, have a gizzard which is specially adapted to reduce the food to a fine consistency. In this type of gizzard the musculature is massively developed, and unlike in other parts of the digestive tube, is asymmetrically arranged. Consequently, when the gizzard contracts there are both rotatory and crushing movements and a tremendous pressure (100–200 mm Hg in the chicken) may be generated within the lumen. Assisting the musculature is a tough membrane or cuticle secreted by glands in the wall of the gizzard, and lining the right and left internal surfaces. This forms two grinding plates between which the food is milled. Indeed, in the older literature the term 'gastric mill' was sometimes applied to this chamber of the stomach. An interesting adaptation of the cuticle to diet occurs in some exotic pigeons such as Peale's Pigeon and the Island Imperial Pigeon which feed on nutmegs and other extremely hard fruit. In these birds the inner surface of the cuticle is raised into a number of extremely hard, pointed, conical processes. Opposing sets of processes interdigitate with one another like cogs and thus provide a highly effective device for crushing the fruit. In some species including the European Starling and the Common or Black-billed Magpie the cuticle is periodically loosened from the wall of the gizzard and excreted via the vent. Why this should occur is not certain. During the breeding season the cuticle in some male hornbills is also lost, but in these birds it is regurgitated orally and is delivered to the female sitting on the nest in the form of an envelope enclosing digested seeds. Another feature which is essential to the physical breakdown of food is the presence of substantial quantities of grit, pebbles or sand in the gizzard lumen which form an abrasive background for the pulping action of the muscle and cuticle. The coarseness of the stomach stones is generally related to the coarseness of the food items to be acted upon. In the Ostrich the gizzard may be packed with up to two pounds of pebbles.

The process of gastric digestion involves a highly complex cycle of contraction in which food is shunted backwards and forwards between the proventriculus and gizzard. Raptors such as owls and falcons appear to have a very special form of contraction cycle which is related to the fact that the gizzard in these birds is also involved in separating out the indigestible portions of the diet, for example teeth, bones, claws and fur, into compact pellets which are then regurgitated. Dissection of the pellets reveals a great deal about feeding habits.

Connecting the gizzard to the small intestine is the pyloric part of the stomach which is usually very small. However, in some species such as the Grey Heron, the Great Cormorant, the Great Crested Grebe and darters, which take in large volumes of water with their food, this region is a distinct compartment. It is believed that an enlarged pyloric region may be a device to prolong the stay of food in the stomach until the hard items like fish bones, which might damage the vulnerable inner surface of the intestines, are rendered harmless. In grebes, the cavity of the pyloric compartment is crammed full of feathers plucked from the bird's own

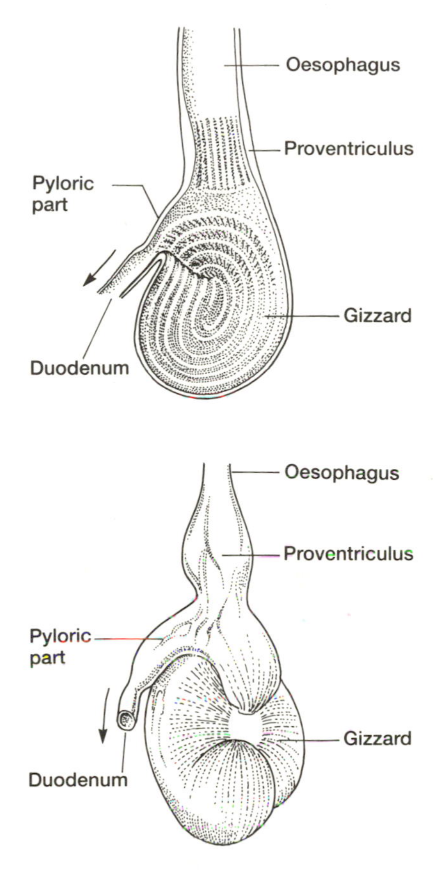


*Stomachs of, above, the Little Owl (Athene noctua) and the Peafowl (Pavo cristatus)*

*Raptors' pellets. Dissection of this Barn Owl (Tyto alba) pellet reveals that the bird had been hunting small rodents (left) and shrews (right)*

breast and these may also assist in delaying sharp objects. Interestingly, feathers are also found in the stomach of grebe chicks, the parents feeding them to the young birds as soon as they are able to take adult food.

### Lower alimentary tract: small intestine, caeca and rectum

A bird's small intestine is arranged into a number of U-shaped loops, the pattern of the looping having some taxonomic importance. Here enzyme digestion of the food is completed and the products of digestion are absorbed through the gut wall and into the bloodstream. For efficient absorption, the available surface area is increased by lengthening of the tube and the development of folds and finger-like processes on its inner wall. In some species, including game birds and several ratites, further digestion, especially of cellulose by bacterial fermentation similar to that in the cow's rumen, may take place in the two caeca. In these birds the caeca are enormously enlarged and sometimes even divided into sacs. Whilst the rectum appears to be unimportant in food absorption, it does reabsorb water from urine which has been excreted by the kidneys and retrogradely enters the rectum *via* the cloaca. This may be especially important in desert birds which have to conserve fluid.

In general the intestinal tract tends to be relatively long in birds feeding on grain, vegetable matter and fish, and short in meat and insect-eaters. Why this is so has not been established. Within some species intestinal length has been shown to vary with sex and age, juvenile and female birds tending to have relatively longer intestines than male birds. Seasonal changes in intestinal length have also been observed in a number of species including California Quail and Spruce Grouse, and this has been correlated with changes in the rate of food consumption and with the fibre content of the diet. Thus the gut of European Starlings is longer in winter when they eat more vegetable matter than in summer when they consume mostly animal material. There are also some indications that changes in diet over several generations can significantly influence the length of the intestine. JM

## Respiration and circulation

### The lung and air sacs

As active warm-blooded animals, birds need a supply of oxygen. It is the role of the respiratory system to assure that supply, and the circulatory system to carry oxygen (and other nutrients) to the tissues. The upper respiratory tract (external and internal nares, glottis and trachea) is similar in birds and mammals. A few birds, especially those that plunge into water from some altitude, do not have external nares but breathe through their mouths. In mammals the trachea divides into two, 20 or more times, yielding over a million tubes which end in thin-walled sacs, the alveoli, where the exchange of oxygen and carbon dioxide takes place. In birds, however, the trachea divides evenly once (into two mesobronchi) and unevenly twice more, giving rise to secondary (dorso- and ventrobronchi) and then tertiary bronchi (parabronchi). Air capillaries arising from the parabronchi form gas exchange sites.

This basic bronchial arrangement constitutes the *paleopulmo* and is in fact seen only in penguins and emus. In other birds, particularly songbirds and chickens, a further set of parabronchi arises from the main bronchus and the dorsobronchi and connects to the caudal air sacs, forming the *neopulmo*. The parabronchi are not neatly arranged in the neopulmo but form an open meshwork (shown in Diagram (*b*) opposite).

In mammals the whole lung expands and contracts with each breathing cycle. However, birds lack a diaphragm. Instead, air is drawn through the lung by the expansion and contraction of the air sacs which surround the lung like bellows and connect to the bronchi. A cranial group of air sacs (cervical, unpaired clavicular and cranial thoracic) connects to the ventrobronchi while a caudal group (caudal thoracic and abdominal) connects to the main bronchus (Diagram (*a*)). The air sacs are expanded by contraction of the inspiratory muscles which move the ribs and sternum up and out, causing a sub-atmospheric pressure in the sacs so that air rushes in to equalize the pressure with the outside. Air is expelled by contraction of the expiratory muscles. The

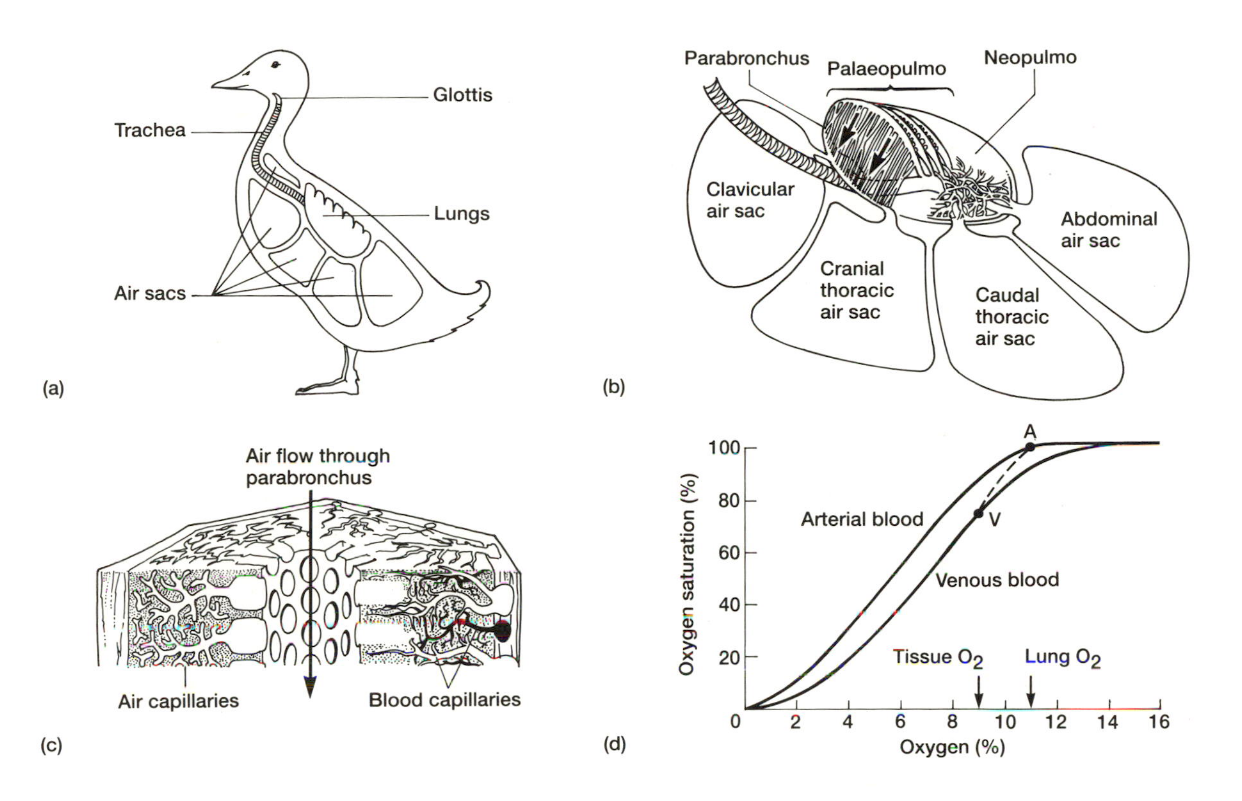


*The avian respiratory system. The paired lungs are small and located along the vertebral column, surrounded by thin-walled air sacs, as shown in (a). The lung/air sac system is shown in (b). In the palaeopulmo the parabronchi are straight parallel tubes, while in the neopulmo they form a meshwork. Air flows through the parabronchi, in the direction of the arrows, throughout the breathing cycle. The network of air (left) and blood capillaries (right) which constitute the site for exchange of oxygen and carbon dioxide are illustrated in cut-away form in (c). Although drawn separately, air and blood capillaries are in close association all around each parabronchus. Graph (d) gives oxygen saturation curves for arterial and venous blood in the duck. Blood is 100% saturated in the lungs (A) and, at rest, loses 25% of its oxygen to the tissues. $CO_2$ is added to the blood in the tissues, shifting the curve to the right (V). In the body, the curve actually follows the dotted line between points A and V*

breathing rhythm is generated by nerve cells in the hind brain which are connected by nerves to the breathing muscles.

Although air flows in and out of the trachea tidally, air flow through the paleoparabronchi is nearly continuous and in the same direction during inspiration and expiration. The air pathway during inspiration and expiration is shown by arrows in Diagram (*b*) above. The precise mechanism responsible for the unidirectional flow through the parabronchi is not known. Because no valves are present, it is probably a consequence of the airway's shape.

The paleopulmonic parabronchi are of similar length and diameter (varying with the size of the bird) and may be several hundred in number. Along their length the parabronchi give off fine air capillaries (1 µm diameter: much smaller than the mammalian alveolus) which run at right angles to the parabronchial lumen (the central space of the parabronchus); in some orders of birds they end blindly while in others they connect with those of neighbouring parabronchi (Diagram (*c*)). The gas in the air capillaries is changed passively by diffusion with gas in the parabronchial lumen.

### *Blood supply to the lungs*

Blood low in oxygen and high in carbon dioxide is pumped from the right ventricle of the heart through the pulmonary arteries to the parabronchial region. Pulmonary arterial pressures are low (2 kPa), reflecting the low resistance of the lung circulation to blood flow. Although blood leaves the heart in pulses, flow in the lung circuit is relatively smooth because of two mechanisms: (1) a slower ejection of blood from the right than from the left ventricle (2) a highly elastic pulmonary artery which stores blood in the contraction phase and recoils in the period between heart beats. This ensures a smooth flow through the gas exchange capillaries. Unlike mammals, the major branches of the pulmonary arteries and veins are not associated with the bronchi in the lung. Large arterial branches divide repeatedly to give vessels running in parallel to the parabronchi; small arteries lead from these vessels and run towards the parabronchial lumen, splitting up into a network of blood capillaries (about 5 µm diameter), each of which contacts air capillaries from only a small region of a parabronchus (Diagram (*c*)). The capillaries join veins which run close to the parabronchial lumen, and these veins unite to form the pulmonary vein opening into the left atrium.

Because the pumping mechanism of the bird lung (the air sacs) is separate from the gas exchange structure (the capillaries), the latter need not be robust. In fact the tissue barrier between the air and blood in the lung is thinner (0.2 to 0.3 μm) in birds than in any other vertebrates. The volume of the major blood vessels of the lung is small and most of the pulmonary blood resides in the gas exchange capillaries. Also, air is continually refreshed in the parabronchi (parabronchial volume is about 10% of the total volume of the respiratory system). In consequence, the bird lung is more efficient than that of the mammal and removes 25% more oxygen from each breath, expired air having 11% oxygen in a bird compared with 13% in a mammal (previous page, Diagram (*d*); atmospheric air has 21% oxygen).

*Gas transport by blood*

Carbon dioxide is 30 times more soluble in plasma than oxygen and needs no special mechanisms for its transport. The oval, nucleated, red blood cells are packed with the iron-containing pigment haemoglobin which increases the oxygen capacity of a unit volume of blood by 30–40 times compared with plasma.

When blood leaves the lung all the binding sites for oxygen on the haemoglobin are filled and the blood is fully saturated. The pulmonary artery carries blood which has had about one-quarter of the oxygen removed by the body tissues. In the lung, oxygen moves from air capillaries into the blood and carbon dioxide moves from blood to air. Since carbon dioxide decreases the solubility of oxygen in blood, the elimination of carbon dioxide in the lungs aids oxygen uptake. Conversely, carbon dioxide is added to the blood in the tissues, unloading of oxygen is favoured there.

## The heart

The heart sits in the thoraco-abdominal cavity surrounded by the membranous pericardial sac which contains lubricating fluid. The veins empty into the thin-walled atria, the right being much larger than the left. The ventricles, which provide the propulsive power to circulate the blood, are filled by contraction of the atria. The left atrio-ventricular valve is attached to the ventricular muscle to prevent it being turned inside out when the powerful left ventricle contracts. The muscular right atrio-ventricular valve is a flap-like structure unique to birds; it is possible that the valve closes actively before right ventricular contraction, as shown in the illustration below. The right ventricle wraps around the left and its outer wall is considerably thinner than that of the left ventricle, as cross-section A-B shows. The more muscular left ventricle in fact generates pressures that are four to five times higher (10 kPa) than those generated by the right.

The heart's rhythm is governed by a group of specialized muscle cells located in the dorsal wall of the right atrium, the sino-atrial node or pacemaker. The pacemaker is regulated both by acceleratory (cardiac sympathetic) and inhibitory (parasympathetic vagus) nerves (see Diagram (*b*) opposite).

The signal to start cardiac contraction passes first across the atria and, after a delay period which allows both the atria to empty, passes

*A duck's heart. This section through the long axis of a duck's heart shows the position of the right and left atrio-ventricular valves. The right valve is muscular and the left is membranous, tied down by tendons (chordae tendineae) to prevent it turning inside out as the ventricle contracts. The cross-section at the level shown by the dotted line A–B, illustrates how the right ventricle wraps around the left*

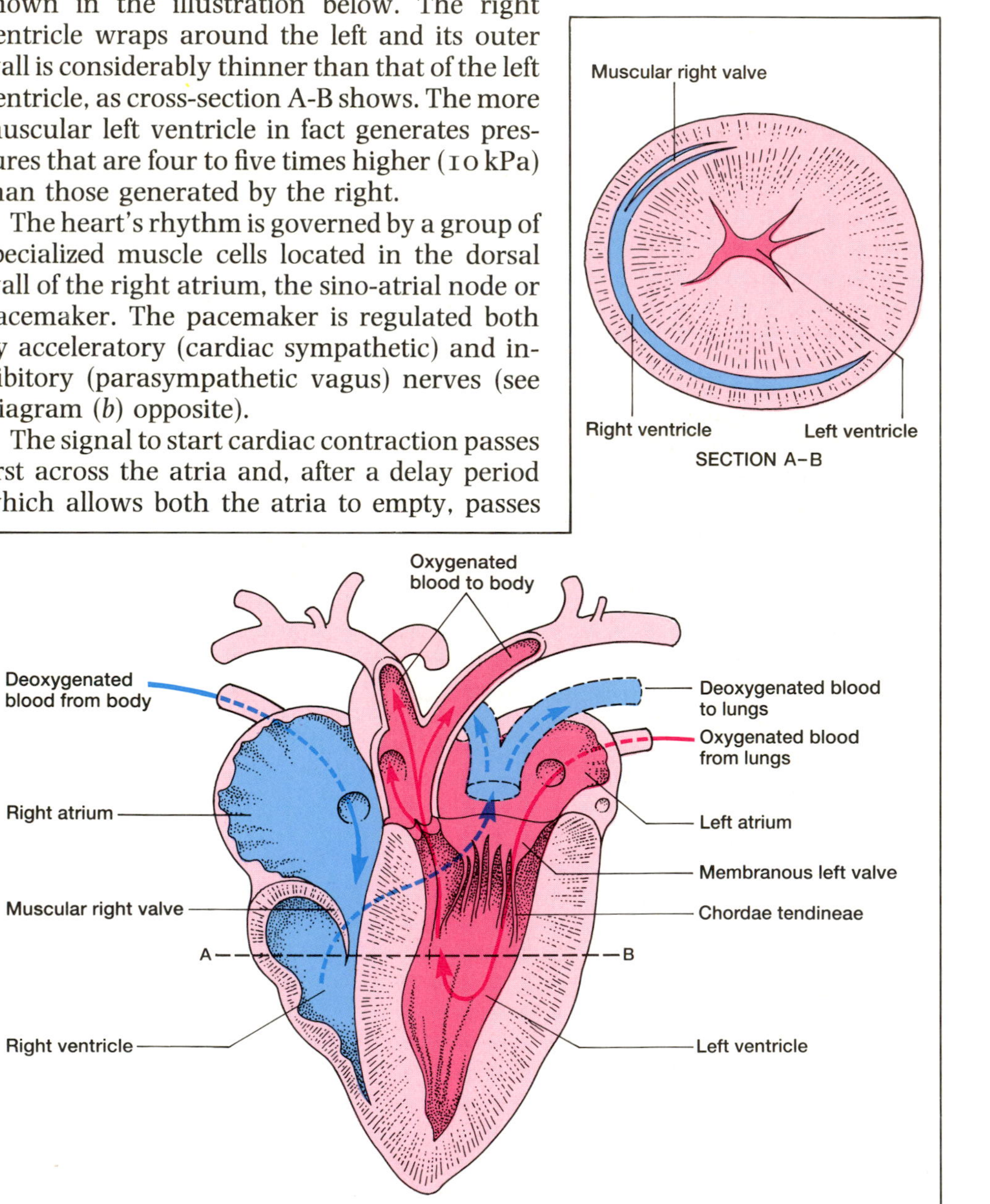

through the ventricles on muscle cells specialized for impulse conduction (Purkinje fibres). The ventricles contract more or less simultaneously. Avian cardiac muscle cells are about half the diameter (2–7 μm) of those in mammals. However, muscle cells of the specialized conducting pathways have diameters five times those of ordinary heart muscle cells in birds.

## Arteries and veins

Of the six embryonic arterial arches only three, numbers 3, 4 and 6, remain as the carotid artery, aorta and pulmonary artery, respectively, in adult birds (Diagram (*a*)). In birds only the right branch of the 4th arch remains as the aorta. The major arteries are elastic, smoothing blood flow into the peripheral vessels. Smaller arterial vessels are muscular and by their contraction regulate flow to the vascular beds.

The first major vascular bed, supplied by the aorta, is that of the heart. Right and left coronary arteries arise from the base of the aorta and branch out to form capillary beds, which are drained by five groups of veins opening in the right atrium or ventricle.

Surprising variation exists in the pattern of the carotid arteries close to the heart. As Diagram (*a*) indicates, the most common arrangement is two vessels of equal size. Other patterns are: (1) a single artery formed by fusion of both carotids (herons, bitterns and kingfishers); (2) a single vessel due to loss of the right (passerines) or left (plovers) carotid; or (3) two arteries of unequal size (flamingos, Sulphur-crested Cockatoo).

Blood flow to the brain must not be interrupted. The carotids lie in a groove in the base of the neck vertebrae, close to the axis of rotation, and are protected from obstruction caused by neck movements. Other safety measures are provided by interconnections between the carotid and vertebral arteries and, at the base of the brain, by either an X-, I- or H-shaped junction between both carotids, found respectively in geese, owls and sparrows. Blood returns from the brain via an unpaired vertebral venous sinus and the thin-walled vertebral and jugular veins, as shown in Diagram (*b*).

The liver, kidney and pituitary organs in birds have a dual blood supply. Blood comes at high pressure from the arteries and at low pressure from the veins. This arrangement, when another vascular bed is interposed between the venous return and the heart, is a *portal system*. In the hepatic portal system the liver capillaries, called sinusoids, are supplied both arterially and venously. A renal portal system carrying venous blood from the hind limbs to the kidney is present in birds but not mammals. Blood-flow is controlled by the renal portal valve, and the kidney can be by-passed depending on whether the valve is open or shut. An open valve allows all blood to by-pass the kidney by the shunts shown as 1, 2 and 3 in Diagram (*b*). Blood in the caudal mesenteric vein (or coccygeomesenteric) can flow either towards the liver or kidney depending on whether the renal portal valve is open or shut.

The pituitary is the master endocrine gland, although its anterior portion has no nervous connections with the rest of the brain. Hormone releasing factors, formed in the median eminence at the base of the brain, are transported by veins to the anterior lobe. In some birds these portal veins are divided into anterior and posterior groups so that the activities of each region of the anterior lobe can be precisely controlled.

## Heat exchange

Heat exchangers in the head and legs of many birds are called *retia mirabilia* (singular *rete mirabile*) or 'wonderful nets'. In a *rete* the artery divides into a large number of vessels which reunite further on; the vein also splits up into vessels which run along and between the arterial vessels. Blood flow in the arteries and veins is in opposite directions (called counter current). The advantage of this arrangement is that heat can be exchanged all along the length of the artery and vein. The opthalmic rete lies between the eye and inner ear and serves to cool blood going to the brain, for brain temperature is generally 1 or 2°C below deep body temperature.

If a bird, for example a gull, is standing in ice-cold water, venous blood may return from the

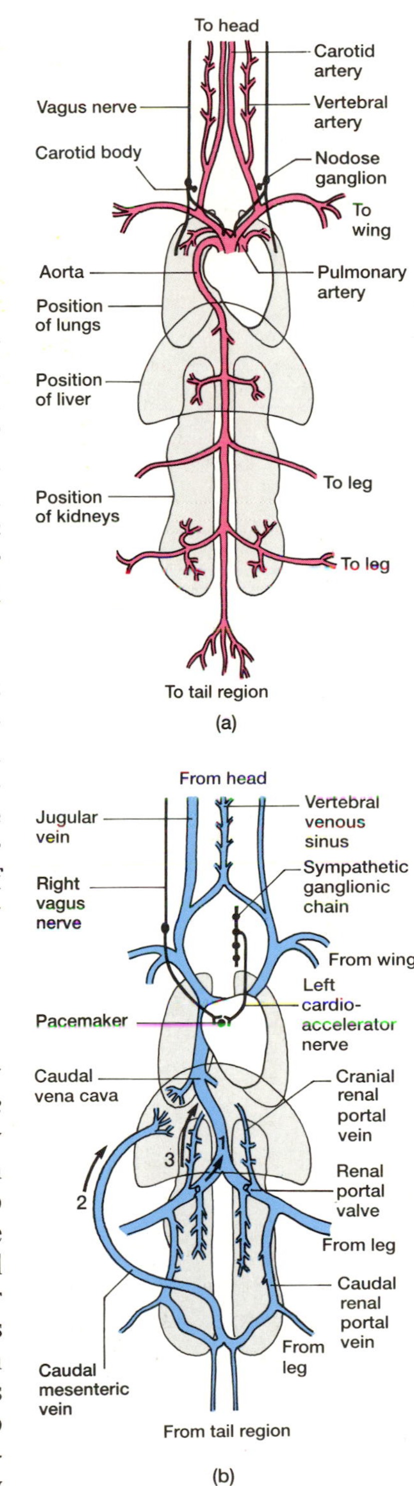

foot at 4 or 5°C. Arterial blood entering the rete will be at 40–41°C and can be cooled to 5 or 6°C before it leaves for the foot. Hence the temperature gradient in the foot (inside to outside) is reduced and the bird can prevent massive heat loss. Not all wading or swimming birds have these heat exchangers so, in these, other mechanisms must be available to conserve body heat.

## Control of breathing and the circulation

A series of reflexes matches circulatory and respiratory activities to the body's demands. The most obvious example of this matching is that, in birds, there are about nine heart beats per breath. Breathing is controlled by cells (chemoreceptors) located in the carotid bodies and brain stem which measure levels of oxygen or carbon dioxide in the blood. Decreases in oxygen or increases in carbon dioxide stimulate these cells which cause breathing to increase.

Since the lung is inexpansible, controlling the volume of each breath presents difficulties. Volume receptors could be placed in the air sacs but these would send all sorts of strange messages when air sacs were compressed or expanded by body movements during flying, or external pressures when diving. Actually, receptors that monitor carbon dioxide in the lung airway regulate each breath. Carbon dioxide is washed out of the lung by overbreathing while it is retained by underbreathing, so that information from these receptors can be used to adjust the volume of each breath, reflexly, volume being proportional to the level of carbon dioxide in the airway.

Blood pressure is sensed by stretch receptors in the wall at the base of the aorta (there may also be receptors in the pulmonary arteries controlling pulmonary blood pressures). These receptors are stretched by each cardiac ejection and send a series of nerve impulses to the brain. The frequency of impulses increases when blood pressure rises because the walls are stretched more, while frequency decreases when pressure falls. The regulatory centres in the brain adjust either heart output or the diameter of the peripheral blood vessels to restore blood pressure to the normal level.

DRJ

# Regulation of body temperature

## How birds maintain a controlled body temperature

Birds, like mammals, generate their own body heat by the oxidation of absorbed nutrients. By controlling both the rate at which heat is produced and the rate at which it is lost, birds normally maintain a body temperature of 40±2°C, some 3–4°C above that of most mammals. Passerines maintain their body temperature at the higher end of this range, other birds slightly lower. In all birds there is a small daily variation; the daytime body temperature in diurnal species is one or two degrees higher than at night, whereas in nocturnal species the pattern is reversed.

*Previous page. The circulation of the blood. The avian arterial system is dissected diagrammatically in (a). The vagus nerve innervates chemoreceptors in the carotid body, blood pressure receptors in the base of the aorta (just outside the heart) and lung receptors. The venous system is shown in (b). The pacemaker (sino-atrial node) in the right atrium is innervated by the regulatory nerves (cardioaccelerator/cardiac sympathetic; cardioinhibitor/parasympathetic vagus). For clarity the acceleratory nerve is shown only on the right side, and the inhibitory vagus is shown only on the left. When the renal portal valve is open (as shown), venous blood bypasses the kidney via three shunts: 1, renal portal valve and caudal vena cava; 2, caudal renal portal vein and caudal mesenteric vein; 3, cranial renal portal vein and internal vertebral venous sinus. Generally, these shunts are partial with part of the renal portal flow bypassing the kidney, the remainder entering it*

*Far left. Thanks to the retia mirabilia in its legs, a Slaty-backed Gull (Larus schistisagus) minimizes heat loss through its bare feet by a heat-exchange system which ensures that the arterial blood flowing to the feet is already cool; it has given up much of its warmth to the venous blood returning to the core of its body*

*Left. Such is the efficiency of the avian respiratory system that migrant birds, such as these Canada Geese (Branta canadensis) traversing the Rockies, are able to cross the highest mountain ranges. The record is held by geese seen at Dehra Dun, India at almost 9000 m*

Unless birds go torpid, the metabolic heat production at rest is normally never less than the level known as the basal metabolic rate (BMR); if a bird is active it may be 10–20 times greater. To maintain a stable body temperature it is necessary that heat production and heat loss remain in balance. The rate of heat loss depends on body size and on ambient temperature. Small birds lose heat more rapidly than large ones because their surface area (through which heat is lost) is large relative to their body mass (where the heat is produced). They must therefore have high metabolic rates in compensation.

Heat loss also increases as ambient temperature decreases. However, over a range of ambient temperatures known as the thermoneutral zone, the heat production of a resting bird is able to remain at its basal level, because any changes in ambient temperature can be countered by altering the insulative properties of the plumage and other parts of the body. By a complex neuromuscular arrangement connecting adjacent feather follicles, a bird can fluff out its feathers to their greatest extent, so trapping a greater amount of air between them and increasing their insulation value. Birds can almost double their effective plumage volume by this means and reduce their heat loss by more than one-third. Many species acquire denser plumage in winter, making a 10–15% saving in energy expenditure. And postural changes, such as hiding the face under the scapular feathers or standing on one leg, also reduce heat loss. On sunny days in winter, sunbathing allows birds to absorb solar radiation and reduce metabolic heat production. Nevertheless, small birds are still disadvantaged in winter and many increase their basal metabolic rate, on average by about 40%. This can be reduced when roosting by huddling in groups; two Long-tailed Tits huddled next to each other can save 27% of their energy expenditure each, while three birds can save 39%. A tight ball of roosting birds, often seen in treecreepers, is even more effective. Also effective is roosting in the cover of a tree cavity or snow-hole (e.g. by redpolls), where rapid radiative losses to the cold night sky are avoided and there is shelter from the wind; convective heat losses increase by 10% for every one metre per second increase in wind speed.

Any further drop in ambient temperature below the thermoneutral zone must be matched by an increase in heat production. Active birds, which generate greater amounts of heat, can tolerate higher rates of heat loss but all species are eventually limited in their tolerance of cold conditions by the maximum amount of additional metabolic heat they can produce; if this does not match heat loss, hypothermia and death result.

At higher ambient temperatures insulation must be reduced. The feathers are sleeked back, blood circulation to the skin is increased and any bare areas may be exposed. Zebra Finches and coursers, for example, partly spread their wings to allow radiation of heat from their sparsely-feathered under-surfaces. Incubating plovers may sit with their backs to the wind and raise their feathers, allowing the breeze to break the insulating layer of air in the plumage. However, as ambient temperature increases towards body temperature, conduction and radiation become less effective at dissipating heat and evaporative cooling becomes more important. Unlike mammals, birds do not possess sweat glands, so that water is usually evaporated through the respiratory tract. Here the rate of evaporation is greatly increased by panting or, in non-passerine birds, by gular flutter, a rapid oscillation of the thin floor of the mouth and upper throat. Nevertheless, unlike mammals, evaporative water loss by birds at high ambient temperatures rarely dissipates more than half of the total heat production. To augment respiratory evaporation some birds, such as the American Wood Stork, excrete down their legs to provide an additional cooling site. To reduce heat stress further birds seek shade or, in the case of large soaring birds, may take to the wing to reach cooler air at high altitude. Some species reduce heat production by lowering the metabolic rate; tropical birds have metabolic rates some 30% lower than similar-sized temperate species, while tropical birds that forage in the sun reduce their metabolic rate still further compared with similar birds foraging in the shade. Others, especially desert birds such as the American

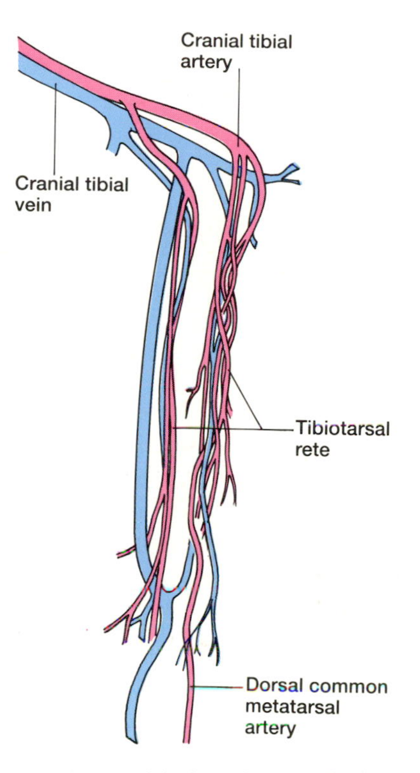


*In a rete mirabile, here shown in the leg, networks of small arteries (red) and small veins (blue) are elaborately interlaced*

*A huddle of eight Treecreepers (Certhia familiaris) means that each bird saves almost half the energy it would expend overnight if it roosted alone*

Mourning Dove, may allow their body temperature to rise above normal to 45°C, re-establishing a temperature gradient to their surroundings so that heat loss by conduction and radiation can continue. However, the capacity to do this is limited; death from hyperthermia intervenes at body temperatures of 46–48°C.

## Birds which go torpid or hibernate

Although, by these behavioural and physiological adjustments, most birds maintain a high, stable body temperature, there are circumstances when it may be advantageous for a bird not to do so. The metabolic rate may have to be reduced when there is simply not enough energy available to the bird to sustain it. The body temperature drops and torpor ensues. Hummingbirds have very high metabolic rates and so, despite their small size, normally maintain body temperatures of about 40°C, similar to other birds. However, during periods when food is in short supply and they are unable to accumulate sufficient night-time energy reserves during the day's foraging, the body temperature drops at night. Some high-altitude Andean species, such as the Andean Hillstar, may allow their body temperature to drop by as much as 30°C to near ambient, though not all species need to show such extreme responses in order to achieve worthwhile energetic savings; a reduction of 10°C can decrease the metabolic demands by 25%. A similar strategy is adopted by some sunbirds living at over 3000 m in East Africa, and by several species of American and Eurasian tits, but the phenomenon is not confined to birds in cold climates. Some South American forest manakins, such as the Red-capped and Golden-collared Manakins, can allow their body temperatures to fall by 8–9°C during the night, and the rather larger Speckled Mousebird of Africa may enter torpor at 15°C, although it is perfectly capable of regulating its body temperature normally at freezing air temperatures.

Other species, such as martins and swifts, may enter torpor for much longer periods than just overnight if they are prevented from feeding for days at a time by prolonged bad weather. Sand Martins, unable to feed during a four-day storm, have been observed to huddle together and allow their body temperatures to drop to 8°C, resulting in a saving of 62% of their metabolic costs. Adult White-throated Swifts also undergo intermittent hibernation during bad weather and nestling Common Swifts have the ability to go torpid and suspend growth during food shortages lasting a few days. The desert-dwelling American nightjar, the Poorwill, truly hibernates in a torpid state, while regulating its body temperature at 18–20°C.

The physiological process by which birds arouse from torpor is not fully understood, since birds seem to lack the brown fat deposits that in mammals become metabolically very active and produce large amounts of heat during arousal.

## What sets a lower limit to bird body size?

Because an increased metabolic heat output is necessary to maintain a high body temperature at small body sizes, and because the smallest birds must often go torpid to avoid such high energy expenditure, it is sometimes suggested that the difficulty of thermoregulation sets a lower limit to bird body size. However, this is probably not the case. Insects such as bumblebees and moths, which may be as large as hummingbirds, also need to maintain high body temperatures well above ambient in order to remain active, and are quite capable of doing so; the thermal properties of their bodies are no different to the smallest birds. The limitation to small size in birds (and very small mammals such as shrews) seems to be the high rate at which the heart needs to deliver oxygenated blood around the circulatory system to sustain the high metabolic rate. Even though the hearts of hummingbirds (and shrews) are relatively large in partial compensation for this, they must still pump at a rate of 1200–1400 beats per minute. Thus one heart beat, during which the heart must fill with blood, contract and relax again, lasts only 40–50 milliseconds. It is unlikely that the design of the vertebrate heart could allow this to occur any more quickly in order to sustain even higher metabolic rates.

PJJ

# Excretory, reproductive and endocrine systems

## Excretion

In all vertebrates the composition of internal body fluids has to be closely regulated. This means that water, salts and waste products must be excreted in exactly the right amounts to compensate for those taken up in the diet. Some water and salts are excreted in an uncontrolled way through breathing and through the skin. Actively controlling the remaining excretion is the role of the kidneys, and, in birds in particular, the salt glands.

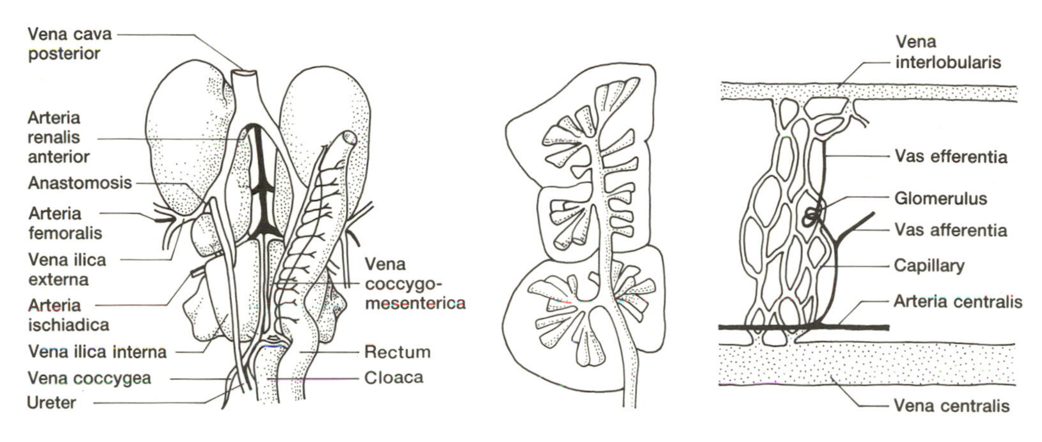


*Avian kidneys. Typical whole kidneys are shown above left, with a detail of the ureter and medullary cones centre. The right hand diagram schematizes the vascular system in the cortex of a medullary cone or lobule*

### *The kidneys*

The kidneys of birds have some features resembling those found in reptiles, and others which are similar to those found in mammalian kidneys. They are paired, and each normally consists of three lobes lying in depressions in the dorsal side of the pelvis. In a mammalian kidney there is a central region, called the medulla, and a region almost completely surrounding this, the cortex. In birds the structure is more complicated. The medulla is divided into many conical shaped regions called lobules. One end of each lobule is attached to several regions of cortex, and the other end is attached to a branch of one of the ureters. There are two ureters, and these collect the urine. They pass under the three lobes of each kidney but, unlike mammals, they do not drain into a bladder; the urine is collected instead in a region at the end of the large intestine, the cloaca.

The kidneys receive blood from the legs, tail and the mesentries of the body cavity. This blood is filtered in the kidney cortex; water and small molecules pass from the blood into kidney tubules. The filtrate then passes to the medulla where some of the water is reabsorbed. The final concentration of urine depends on how much water is reabsorbed, and this is determined by the permeability of parts of the tubules in the medulla. Permeability is controlled by hormones called anti-diuretic hormones, which are secreted from the pituitary gland. The major anti-diuretic hormone in birds is arginine vasotocin (AVT). The more AVT is released from the pituitary, and the more permeable the tubules become, the more water is reabsorbed, and the more concentrated the urine becomes. Because this concentration of urine occurs in the medulla, desert birds, which have to conserve water, have a large volume of medulla. By varying the amount of blood filtered and the amount of water reabsorbed, urine flow in such birds can be varied by as much as 100 times.

Birds are uricotelic, which means that the nitrogen produced from protein metabolism is excreted as uric acid, as opposed to mammals which are ureotelic and excrete nitrogen as urea. The uric acid is mainly produced in the liver, and passes in the blood to the kidneys, where it is actively excreted into the tubules. It is so insoluble that it forms a precipitate in the collecting ducts and ureters. Consequently, the ureters are peristaltic; they contract rhythmically to force the semi-solid urine along their length and into the cloaca. The ureters also produce mucus to aid the passage of urine. In the cloaca, the urine is often mixed with faeces before being voided through the anus. Because of this mixing, bird droppings typically include both white urine and darker faecal material.

### *The salt glands*

All birds possess paired nasal glands in the orbit of the eye. In terrestrial and freshwater species these may be non-functional, but in birds from marine environments the glands are well developed, being most pronounced in petrels and

*Seabirds use salt glands situated above the eye to excrete excess salt. The glands are especially well-developed in tube-nosed birds such as the Southern Giant Petrel (Macronectes giganteus) which shed salty liquid through their nostrils*

auks. The glands produce a secretion containing large amounts of sodium chloride. By this means, a bird may excrete more than 90% of the sodium chloride in its diet.

Salt secretion is probably controlled by hormones produced in the adrenal glands; the adrenal glands of marine birds are larger than those of terrestrial or freshwater birds, and in species able to live in both environments, the adrenal glands increase in size as an adaptation to seawater.

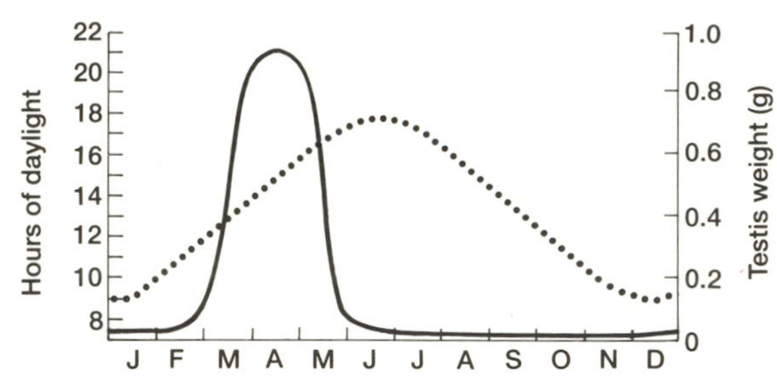


*Left.* Change in testis size in European Starlings during the year (solid line) and change in day length (dotted line) at latitude 52°N

## The reproductive system

### *The male reproductive tract*

Unlike most mammals, the testes of birds are internal, suspended from the dorsal wall of the body cavity adjacent to the kidneys. Outside the breeding season they are small, weighing as little as 0.005% of total body weight, but at the beginning of the breeding season they increase substantially in size, up to 1000 fold. They consist of a mass of tubules containing two types of cells: germ cells, which, by a series of cell divisions, give rise to sperm, and Sertoli cells which assist with the maturation of the sperm. Mature sperm are carried from the testes, in seminal fluid, along a duct called the ductus deferens to the cloaca. Close to the cloaca the duct expands into a seminal sac which everts into the cloaca during copulation. In most birds there is a small erectile phallus, but this is well developed and protrusible in only a few species, for example in the ratites (e.g. the Ostrich) and Anseriformes (ducks and geese).

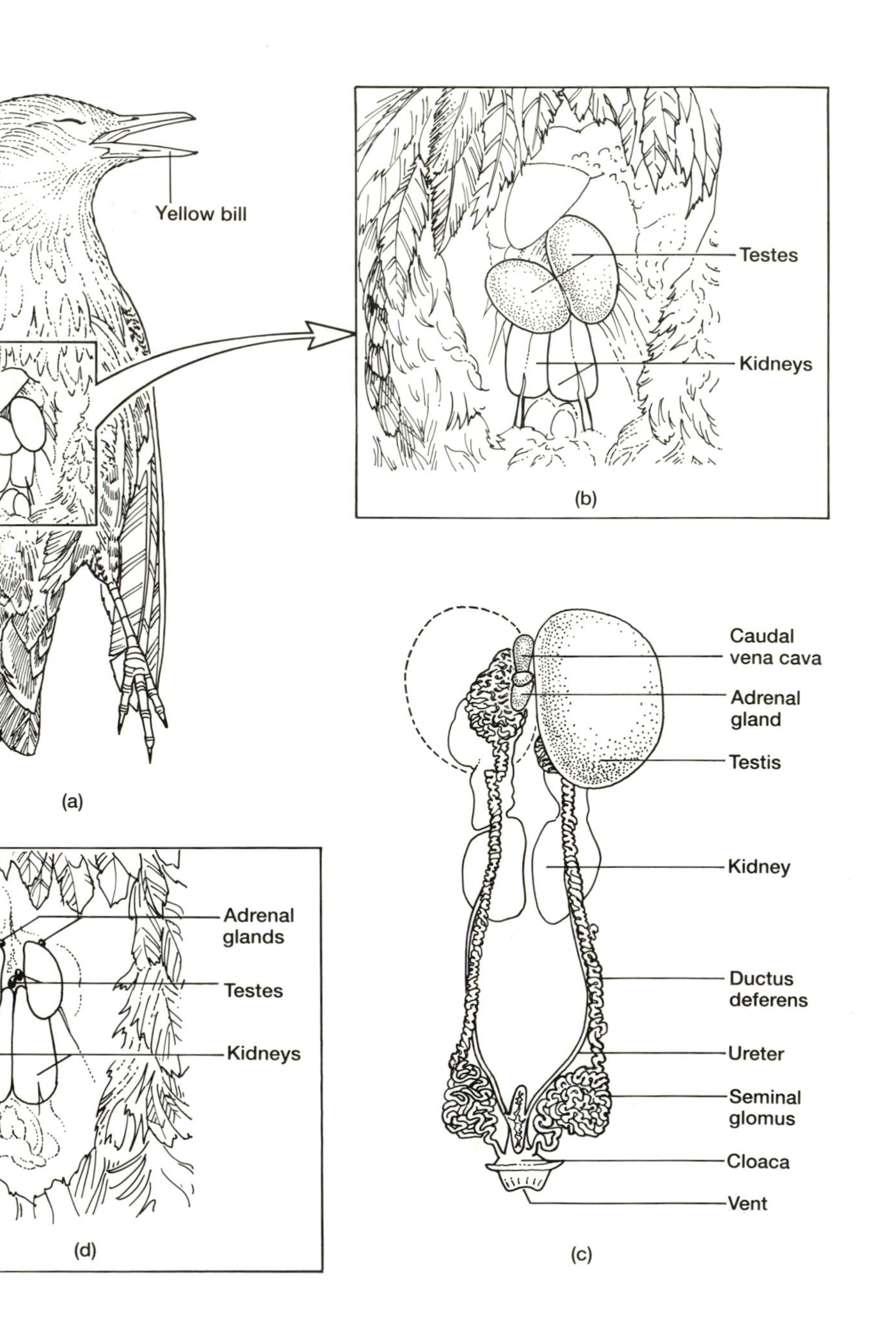


### *The female reproductive tract*

In nearly all species of birds only the left ovary develops, and, like the testes, it is suspended from the dorsal wall of the body cavity near the kidneys. It consists of a very large number of oocytes of which only very few will ever be ovulated. Unlike mammals, the sex of the offspring is determined by the oocyte rather than the sperm; it is the oocyte which contains either the male or female sex chromosome (see Chapter 1). During the breeding season, an oocyte destined to be ovulated is first surrounded by a protective follicle which extracts yolk materials, produced in the liver, from the

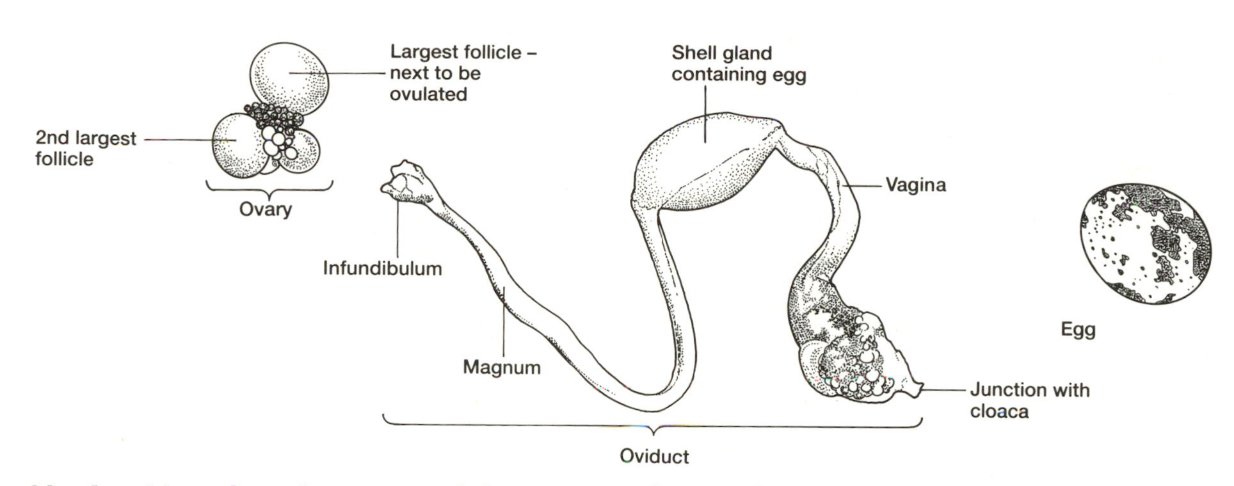


*The reproductive system of a breeding female Common Quail (Coturnix coturnix)*

blood and lays them down around the oocyte. The ovary will contain several such developing oocytes but each will be at a slightly different stage of development. The one with most yolk will normally be ovulated first and the others will follow, in most species at approximately 24-hour intervals. The number ovulated in a sequence determines the clutch size.

At ovulation, the follicle ruptures and the oocyte passes into the infundibulum, the first part of the oviduct. Fertilization occurs in the infundibulum usually within an hour of ovulation. Most, if not all, female birds can store active sperm in the sperm storage tubules, located near the junction of the shell gland with the vagina. Fertilization can then occur some time after mating: the interval may reach two months in petrels and turkeys. As the egg passes down the oviduct, a layer of albumen (egg-white) is laid around it. At the lower end of the oviduct, in the shell gland, water is taken up by the albumen, and then the outer layer of the egg becomes calcified to produce the shell. Now pigments are deposited on the shell. These pigments themselves are derived from two sources, the reddish-brown ones from blood haemoglobin and the blue-green ones from cyanin compounds in the bile. In general, most of the ground colour of the egg is secreted in the upper oviduct whilst the spots, marks and scrawls overlaying this are deposited only shortly before laying. Finally the egg passes from the shell gland to the cloaca and is eventually laid. The whole process, from the follicle rupturing, to the egg being laid, takes about 24 hours in most birds, but in large birds it may take 48 hours or more.

## The endocrine system

A large number of body processes are controlled by hormones. The mechanisms are complex, but largely similar to those of mammals. The following description is therefore brief and serves mainly to highlight differences between birds and mammals.

The whole endocrine system is controlled by the hypothalamus, a region of the brain behind the point where the optic nerves cross (the optic chiasma). Cells in this region synthesize very small amounts of hormones called releasing hormones which pass along a short system of blood vessels to the pituitary gland lying just beneath the brain. Releasing hormones cause the pituitary gland to secrete much larger amounts of other hormones and these pass in the blood to their target organs; in other words, weak signals from the brain are amplified by the pituitary gland. There are several releasing hormones, each controlling a different hormone system.

### *The endocrine system and reproduction*

The releasing hormone which controls the reproductive system is gonadotrophin-releasing hormone (GnRH). In birds there are two GnRH's which differ slightly, in their amino acid sequence, from each other and from mammalian GnRH. GnRH stimulates the pituitary gland to secrete luteinizing hormone (LH)

*Opposite page. The reproductive organs of a breeding male European Starling are seen in (a), in closer view in (b), and dissected clear of the body with the overlying left testis removed in (c). Outside the breeding season, the testes are markedly smaller, as indicated in (d)*

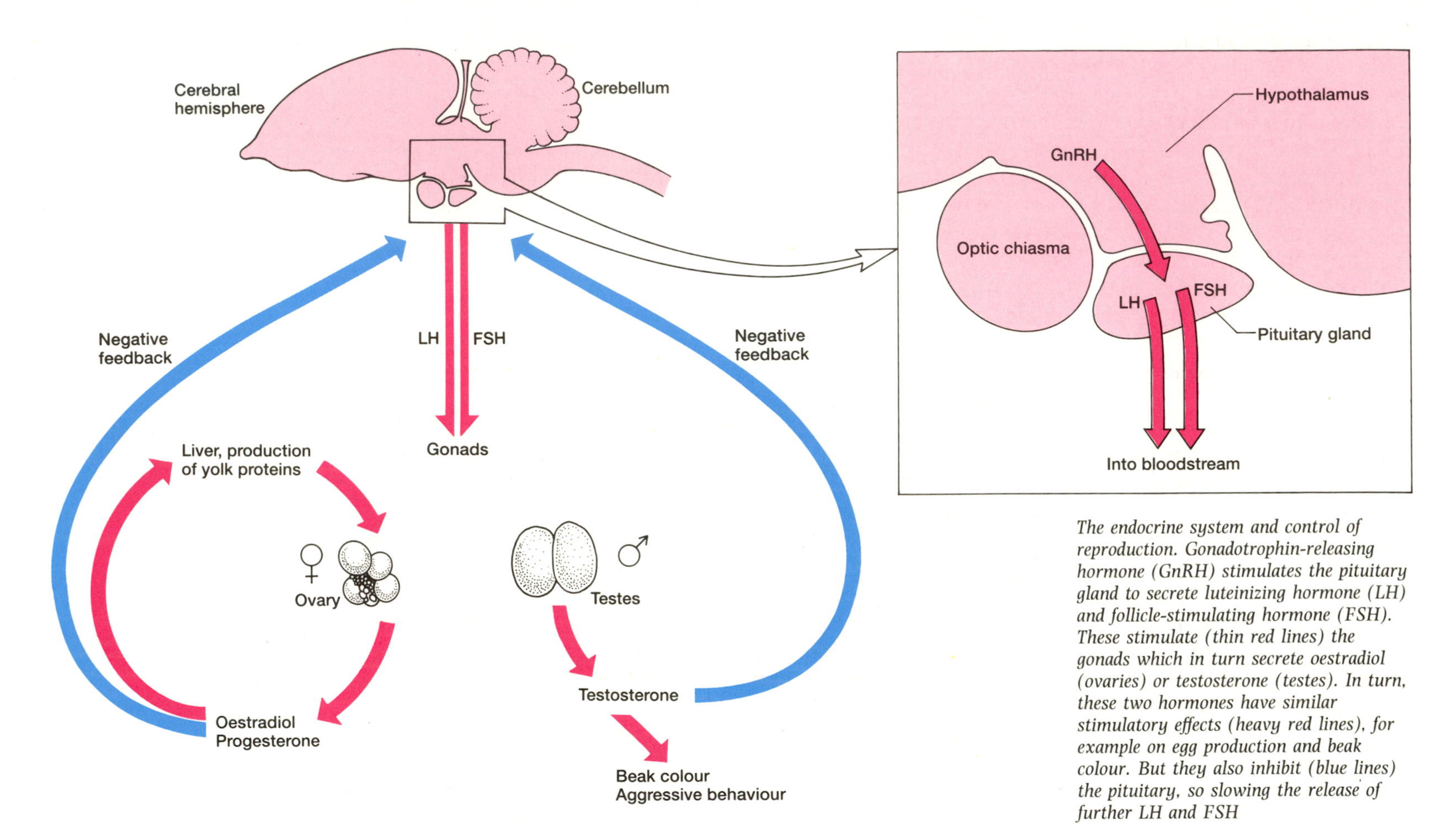


*The endocrine system and control of reproduction. Gonadotrophin-releasing hormone (GnRH) stimulates the pituitary gland to secrete luteinizing hormone (LH) and follicle-stimulating hormone (FSH). These stimulate (thin red lines) the gonads which in turn secrete oestradiol (ovaries) or testosterone (testes). In turn, these two hormones have similar stimulatory effects (heavy red lines), for example on egg production and beak colour. But they also inhibit (blue lines) the pituitary, so slowing the release of further LH and FSH*

and follicle-stimulating hormone (FSH). These hormones are gonadotrophic, i.e. they cause growth and development of the gonads, the testes in the male and the ovary in the female.

Within the gonads of both sexes, between the cells directly concerned with the production of sperm or oocytes, there are interstitial cells. These cells are stimulated by LH to produce steroid hormones, which are comparatively small hormones, based on the structure of cholesterol.

In females, the most important of these gonadal steroids are oestradiol and progesterone and in males the most important is testosterone. They have a variety of functions, affecting both physiology and behaviour. They can affect secondary sexual characters such as colour of the plumage or the bill. Testosterone induces aggressive behaviour, and oestradiol can induce nest building. Oestradiol also stimulates the liver to produce yolk material which is secreted into the blood and then taken up by the follicles surrounding the growing oocytes in the ovary. Another function of gonadal steroids is to inhibit release of LH and FSH from the pituitary; the higher the concentration of steroid hormones in the blood, the less LH and FSH is released. This system, called a negative feedback loop, damps out short term fluctuations in hormone concentrations. An exception occurs when the eggs are being laid. At specific times, instead of inhibiting LH, progesterone stimulates the release of LH so that there is a sudden increase in LH concentration in the blood. This causes the largest follicle in the ovary to rupture and release its oocyte, which is then laid as an egg the following day.

### *Thyroid glands*

The thyroid glands are located in the neck adjacent to the carotid artery. They secrete two hormones, thyroxine and triiodothyronine, which are notable in that they contain iodine. The thyroid glands are stimulated to secrete these hormones by thyroid-stimulating hormone (TSH) which is produced by the pituitary gland, and production of which is in turn stimulated by thyrotrophin-releasing hormone from the hypothalamus. As in mammals, one effect of thyroid hormones is to control metabolic rate. In birds, thyroxine also has important effects on the reproductive system and moult. The process which ends the breeding season for most birds, the inhibition of GnRH synthesis (see Chapter 8), requires the presence of thyroxine. If thyroxine is absent, for example if the thyroid glands have been removed, the gonads do not regress as they would normally at the end of the breeding season, and the post-nuptial moult, which should then occur, is prevented.

*Adrenal glands*

As well as being controlled by thyroid hormones, metabolic rate is also influenced by hormones produced in the adrenal glands, which, as the name implies, are situated next to the kidneys. The adrenal glands of birds consist of two types of tissue, medullary and cortical, which are mixed together. This situation is intermediate between that found in primitive vertebrates, where the two types of tissues are completely separate, and in mammals where all the medullary tissue is concentrated at the centre of the glands, and the cortical tissue forms a layer around the medulla. As in mammals, the medullary tissue is connected to the nervous system, and secretes the hormone adrenalin. Particularly large amounts are secreted immediately after stress. The cortical tissue secretes hormones in response to adrenocorticotrophic hormone (ACTH) which is produced by the pituitary gland. Large amounts of ACTH are also produced immediately following stress. The cortical tissue produces a range of hormones called corticosteroids, and these have a structure based on that of cholesterol, as do the gonadal steroids. Unlike most mammals, where the major corticosteroid is cortisol, the major hormone in birds is corticosterone.

The main function of corticosterone is the control of metabolism; it causes blood sugar levels to increase and promotes breakdown of muscle protein. Another important corticosteroid is aldosterone, which is important in the control of kidney function.

*Prolactin*

Prolactin is produced by the pituitary gland, and like LH and FSH, it consists of a long chain of amino acids. Control of prolactin secretion is very different to that in mammals. In mammals, hormones from the hypothalamus inhibit secretion of prolactin; in their absence the pituitary secretes large amounts of the hormone. In birds, prolactin secretion is stimulated by some factor from the hypothalamus, but what this may be is not yet known. Prolactin is found in all vertebrates but was first discovered in birds; it was found to cause the production of 'crop milk', a substance produced in the crops of pigeons which they use to feed their young. This is why the hormone was called prolactin; it was later that the same hormone was found to stimulate milk production in mammals. In birds, secretion of prolactin is often stimulated by long days, but also much more so by the presence of eggs. Prolactin causes the development of the brood patch, an area on the underside of birds where the feathers are lost and blood supply increases so that heat can be transferred to the eggs more efficiently. Prolactin also makes birds become broody. However, in birds where the male does not incubate the eggs, but does help to feed the young, prolactin levels are also high; prolactin is therefore considered to be a 'parental' hormone. Prolactin levels are high while eggs or young are present, and during this time it causes the gonads to regress, so preventing further reproductive effort until the young have fledged.

AD

## The egg

All birds reproduce by laying eggs. An egg is a life-support system for the embryo, through its development from ovum to hatchling. It protects the new animal until it is developed enough to cope with its environment and contains everything needed for the differentiation and growth of the embryo's tissues. The main structures of the egg are the yolk, which is part of the ovum, the albumen, or egg-white, and the shell.

The yolk is a fluid packed with fat globules and enclosed in thin, transparent yolk membranes, with the rest of the embryo sitting on top. The yolk mass is not so uniformly featureless as might be expected from an acquaintance with eggs in the kitchen. In the centre there is a small ball of whitish, protein-rich yolk, called the latebra, about five millimetres across in domestic chickens' eggs. A narrow column of the same material extends to the surface of the yolk where it broadens out into a flat cone, looking rather like a golf tee. This is the nucleus of Pander on which the germinal disc sits.

The surrounding yolk is yellow and consists of tightly-packed globules alternating in light and dark layers, which are probably caused by caretenoid pigments. The light and dark layers are laid down by night and day respectively and can therefore indicate how long it took the yolk to form. In fact the first phase of yolk formation starts months or years before the egg is laid but the bulk of the material is laid down five to nine days (in a chicken egg) before laying, when the yolk expands from 10 to 35 mm in diameter.

Yolk is rich in fats and proteins. It provides the food required by the developing embryo and also feeds the newly hatched bird until its alimentary system takes over. The size of the yolk differs between species. Precocial species, which leave the nest early, lay proportionately larger eggs with proportionately larger yolks than altricial species, whose young are fed in the nest. As the table below indicates, precocial species have more yolk, which is also richer in fats, to fuel the extra development of the embryos within the eggs and then to sustain the chicks until they can find their own food.

The albumen forms between about 55 and 70% of the mass of a new-laid egg and is composed almost entirely of the viscous protein albumen and water. There are layers of thicker and thinner albumen, as can be seen when a chicken's egg is broken into a pan. Surrounding the yolk there is a thick, viscous layer that is drawn out into twisted 'chalazae' that lie along the long axis of the egg. The chalazae lie in a mass of thick albumen that extends to the ends of the egg where protein fibres, called the ligaments are attached to the shell.

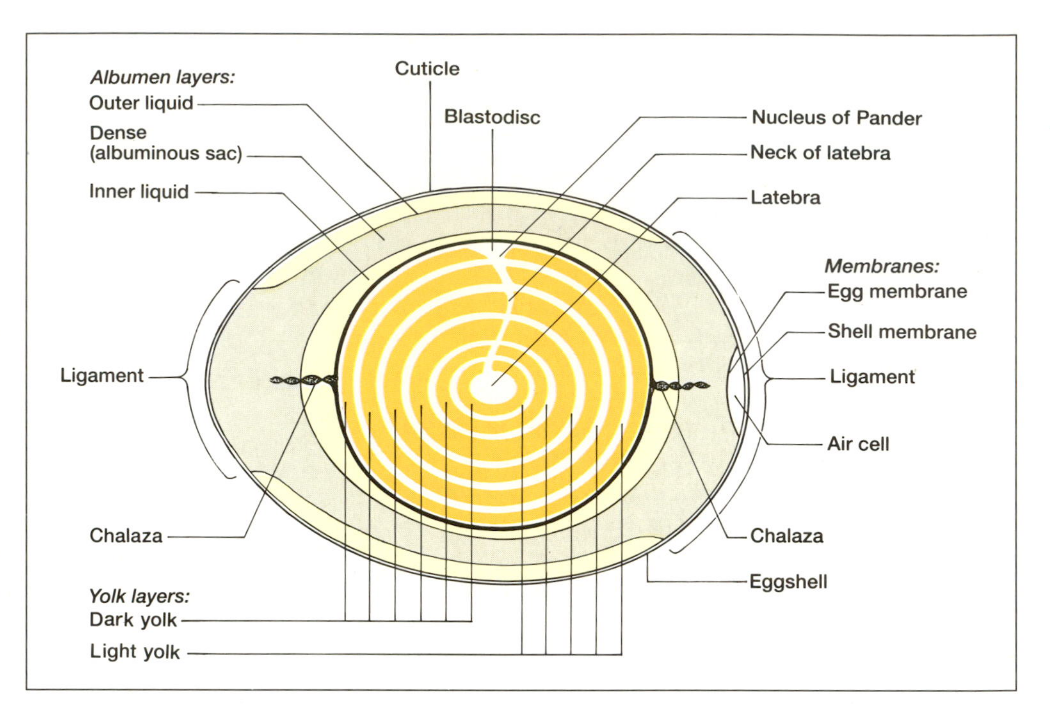


*A section through a new-laid chicken's egg. The shell, albumen and yolk are complex, layered structures that form a life-support system for the embryo, represented at this stage by the blastodisc*

*Yolk, albumen and shell components of eggs. The higher proportion of yolk in precocial birds fuels the greater development of the embryo before hatching and acts as a food reserve after hatching*

| Developmental mode | Yolk (%) | Albumen (%) | Shell (%) |
|---|---|---|---|
| Altricial | 21.8 | 70.4 | 7.8 |
| Intermediate | 29.9 | 62.3 | 7.8 |
| Precocial | 36.6 | 53.7 | 9.7 |

The viscosity and structure of the albumen make a shock-proof packing for the embryo, allowing it to rotate so that the early embryo remains uppermost on the yolk when the egg is turned over. The structure of the protein prevents drying-out and it has bactericidal properties. The albumen also provides some nutrition to the embryo.

The shell is a complex structure made up of three parts: the underlying membranes, the main chalky portion, called the testa, and the external cuticle.

The inner of the two parchment-like shell membranes is attached to the albumen and the outer is attached to the testa. Immediately after laying, the membranes separate at the blunt end of the egg to form the air space, which seems to act as a condenser for saving water. The testa is a complex structure consisting of an organic matrix (3% of the weight) in which crystalline calcareous deposits are laid down. It provides the main strength of the egg and its thickness is generally proportional to the size of the egg.

The shell surface is penetrated by large numbers of pores (diameter 50 μm) which lead into pore channels that allow the exchange of gases with the egg's interior: the total cross-sectional area of the pores is proportional to the weight of the egg and hence its metabolic requirements. The cuticle imparts the characteristic texture of an egg – glossy for woodpeckers and tinamous and chalky for penguins. It adds strength to the shell and also acts as a barrier against bacteria.

## Embryonic development

The germinal disc, the early embryo in a newly-laid chicken egg, can be seen as a white spot, 3–4 millimetres across with a clear centre, lying on the upper surface of the yolk. It consists of two layers of cells, the ectoderm above and endoderm below, separated by a fluid-filled space. Within a few hours of laying some of the ectoderm cells spread into the fluid to form a third layer, the mesoderm. These three layers are the precursors of all the organs in the adult bird. The ectoderm forms the skin and nervous system, the mesoderm forms the muscles, heart, skeleton and kidneys, and the endoderm gives rise to the stomach, liver and lungs. Development can pause at this stage, as happens in species where the clutch is completed before incubation starts.

The first sign of organ formation is the appearance of a groove running along the embryonic disc whose sides then fold over and join to make a tube. This is the neural tube which becomes the spinal cord and thickens at one end to make the brain. On each side of the neural tubes a row of blocks appears. These become part of the voluntary musculature associated with the backbone but include the eye muscles.

To cope with its demand for nutrition from the yolk and for the passage of gases and wastes, the growing embryo creates four membranes. One grows over the yolk to become the yolk sac. A network of tiny capillaries appears, links up with the newly formed heart and begins to transport nutrients from the yolk. Another forms the allantois. This is a membraneous bag that receives waste products from the kidneys and is eventually left behind in the

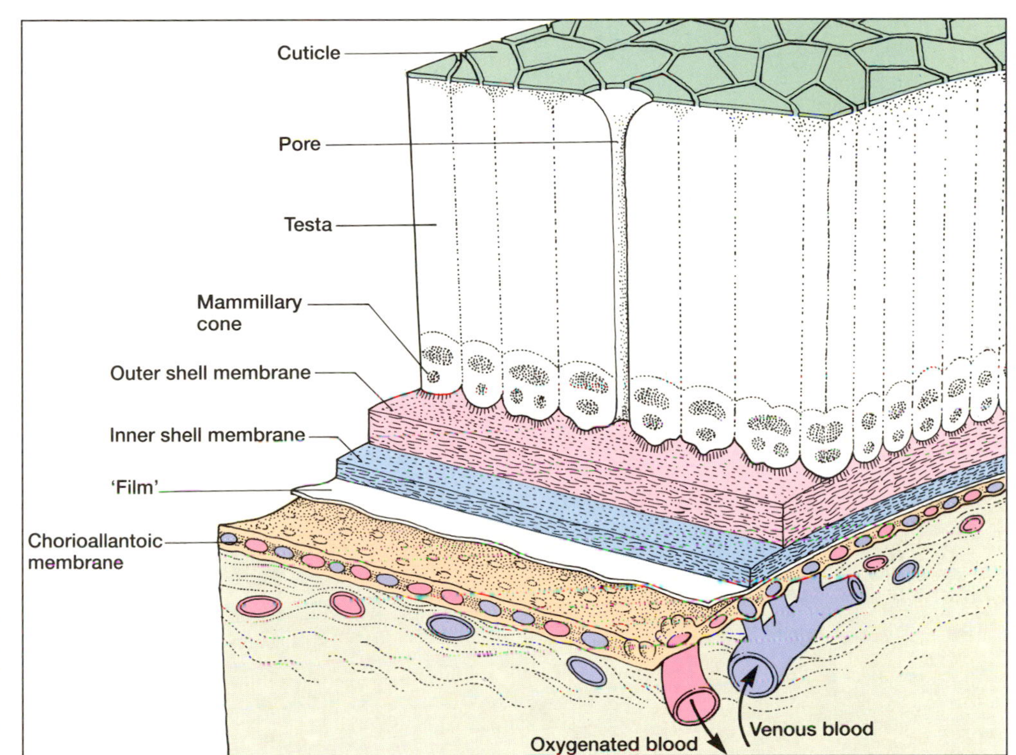


*The internal structure of the egg of the domestic chicken. The complex structure of the shell protects the embryo as well as allowing it to breathe. The chorio-allantoic membrane forms a kind of lung through which blood flows from the embryo. It is pressed against the smell membranes so that gases can be exchanged through the pores*

empty shell as a bag of solid uric acid crystals. The outside wall of the allantois fuses with the third membrane, the chorion, and presses against the inside of the shell to form a 'lung' to effect ventilation. The fourth membrane, the amnion, forms a fluid-filled bag around the embryo to act as a protective cushion.

By the first three days of incubation, the chicken embryo has formed the rudiments of its organs, including two pairs of buds which will become the limbs, and by six days most of the organs are taking shape. The most obvious are the enormous pigmented eyes. At ten days, the chicken embryo is recognisable as a bird with a beak, wings, feet, complete with a skeleton of cartilage, and tiny spots on the skin that are the beginnings of the feathers.

At this stage, the embryo is lying across the axis of the egg but it now turns so that it lies along the axis, turned on its left side and with the head tucked between its legs. For the remainder of its time in the egg the embryo grows steadily and some of the organs start to function. Not all organs will be completed before hatching. The degree of completion depends on whether the chick is altricial or precocial, and the reproductive organs will not be finalized until maturity. However, the embryo is already becoming responsive to its environment and is able to make limited movements.

About seven days before hatching the embryo swallows the remaining egg fluids and shifts its position, bringing its head from between its legs and tucking it under its right wing. The remains of the yolk are drawn up into the body and the navel closes over them.

The egg loses weight during incubation through water evaporating through the shell. Rate of loss is related to the fresh weight of the egg and also to the relative humidity of the air surrounding it. Water loss during incubation averages 16% of fresh egg weight.

Just before hatching, the embryo's metabolism is too high for it to lose carbon dioxide through the shell pores at a sufficient rate. The beak is thrust into the airspace and air-breathing starts. The lungs expand and blood flow is directed to them while the chorio-allantois circulation is shut off. Carbon dioxide continues to build up until the tip of the beak, protected by the egg-tooth, or caruncle, hammers a hole through the shell, which is already weakened by withdrawal of calcium and other elements for the formation of the embryonic skeleton. This process is 'pipping' and the chick is now ready to hatch.

RWB

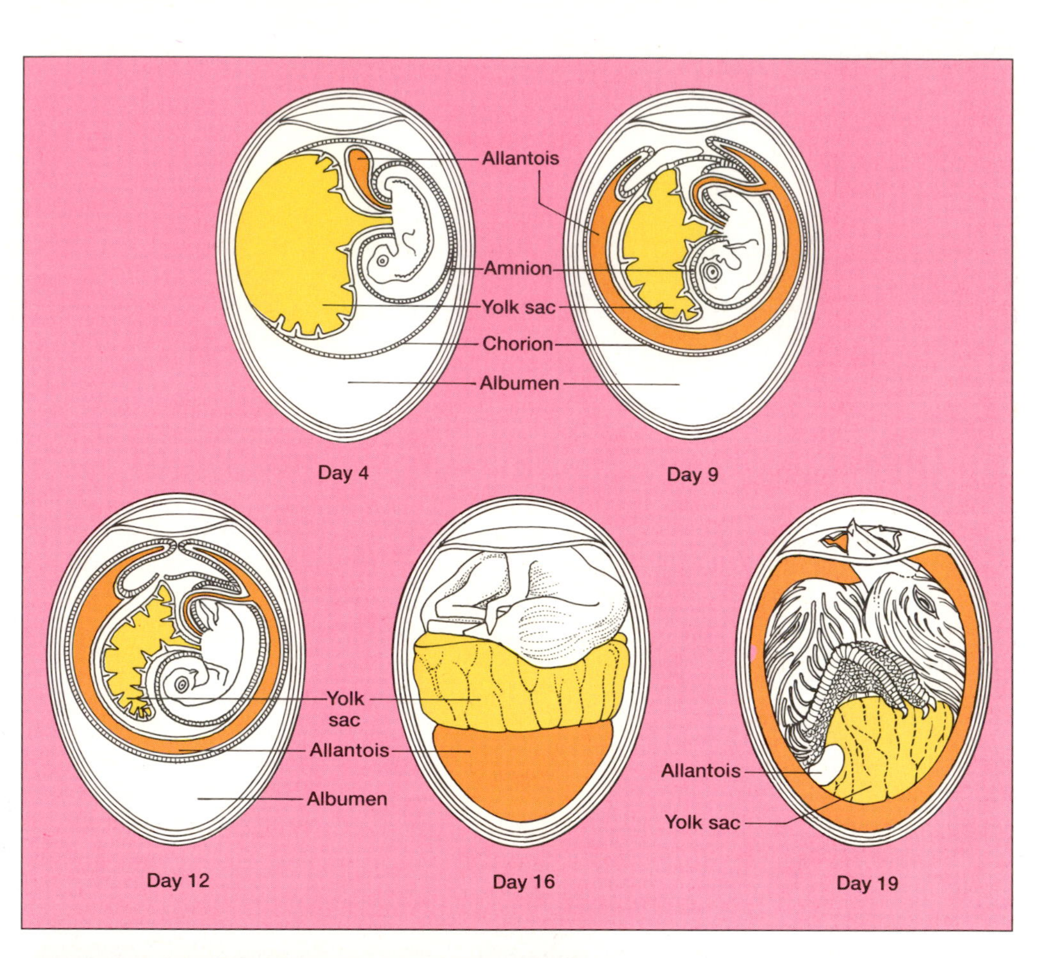


*Above. The embryo and its development. By Day 4, the embryo is enclosed in the amnion; its large head and eyes have formed, the main blood vessels have developed, and the heart has been beating actively for a full day. By Day 9, albumen and yolk are shrinking as the embryo withdraws food by way of tiny vitelline veins in the yolk sac. By Day 12, the extremities are developing and the embryo is capable of convulsive movement. Down begins to form by Day 16, each feather protected by a sticky film. At Day 19, the chick has used all the yolk and albumen except that of the yolk sac attached to the abdomen, and its feet and toes are well-developed. The chick will hatch in two days*

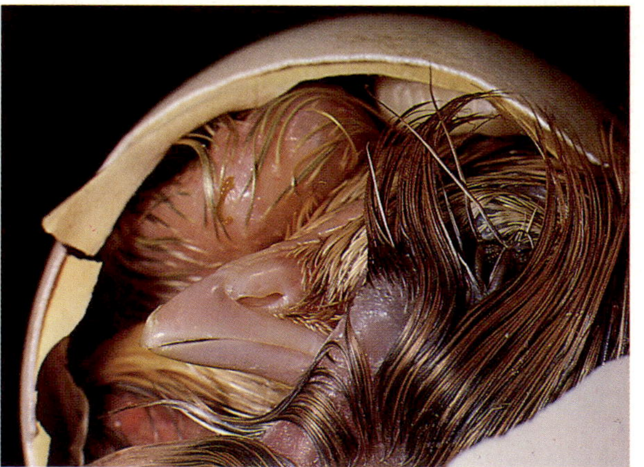

*Above left. This chick of the domestic chicken, the descendant of the Red Jungle Fowl (Gallus gallus), is poised to hatch, around 21 days after laying*

# The nervous system and senses

## The nervous system

The functions of the nervous system are far from simple and to execute them involves an extreme complexity of anatomical structure and physiology. The nervous system does three things: (1) it provides the means for constant gathering of information both about the rapidly changing state of the bird's body and about the equally rapidly changing world in which the bird lives, and the bird's position within it; (2) it analyzes and integrates this information, stores it in the form of memory, and compares it with already stored (learned) information; finally, (3) it uses this processed information to give commands to the internal organs and muscles whose activities generate the bird's external behaviour.

While stressing that the bird's nervous system always functions as an integrated whole, it is convenient for descriptive purposes to subdivide it into two major components, the central nervous system and the peripheral nervous system. In turn the peripheral system may be further subdivided into a sensory system and a motor system. Each division is concerned with one of the three main functions described above and can also be identified with particular groups of anatomical structures.

It is the sensory system which gathers information via the specialized sensory receptors and conveys this in a coded form through the sensory nerve fibres to the central nervous system. The central nervous system consists of the brain and the spinal cord and it is in these structures that information is analyzed and processed (and in which learning takes place). From the central nervous system commands are sent out through the motor nerves to muscles and the endocrine glands.

The properties of the sense organs are crucial in determining what information is available to the central nervous system for the control of behaviour. In turn the central nervous system, especially the brain, determines the complexity of information that can be dealt with, the degree to which past experience (learning) affects behaviour, and the subtlety of control which can be exerted over the body.

### *The brain*

The brains of all birds share a common design in which the integration and control of specific functions are located in certain identifiable structures or regions. Differences are found between brains of different species but these amount only to differences in the absolute size of the brain and to differences in the relative sizes of the individual components. These size differences reflect the relative importance of particular activities or sensory systems in the life of the individual species. As the table below shows, relatively large olfactory lobes are found in species such as kiwis, shearwaters, petrels and the Turkey Vulture, in which the sense of smell is known to be of importance in locating food.

Particularly large and important structures in all avian brains are the cerebellum, cerebral hemispheres and optic lobes (Diagram (*a*)). The cerebellum consists of a single large structure divided on its surface into 10 lobules. The overall function of the cerebellum is to control all aspects of posture and balance, and it receives information from both the limbs and the balance-sensitive organs of the inner ear. The cerebellum has been likened to an 'automatic pilot' which maintains stability and direction during flight.

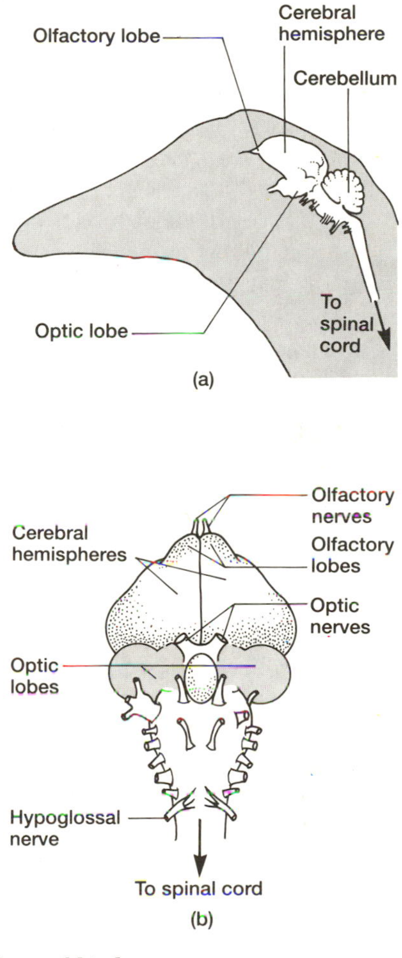


*The relative size of the olfactory lobe in certain orders of birds*

| Order | Size of olfactory lobe as a percentage of the cerebral hemisphere |
|---|---|
| Apterygiformes (kiwis) | 33 |
| Procellariiformes (petrels and allies) | 30 |
| Caprimulgiformes (nightjars and allies) | 25 |
| Apodiformes (swifts and hummingbirds) | 19 |
| Columbiformes (pigeons) | 17 |
| Piciformes (woodpeckers and allies) | 9 |
| Psittaciformes (parrots) | 6 |
| Passeriformes (perching birds) | 5 |

Note that tree-dwelling species tend to have the smallest olfactory lobes

The cerebral hemispheres are a pair of structures whose almost smooth surface belies a complexity of internal structure and differentiated function. It is to the cerebral hemispheres that much sensory information is directly relayed and then analyzed, and where learned information is stored and compared with current sensory input. The cerebral hemispheres send motor nerves either directly to bodily organs and muscles or to the spinal cord.

The optic lobes are particularly large, reflecting the importance of vision in the daily life of birds. These lobes receive sensory information directly from the eyes, but send motor nerves to both eye and neck muscles. They are thus responsible for the control of movements which permit the visual tracking of moving objects.

*The sensory and motor nerves*
The nerve fibres of the sensory and motor systems enter and leave the brain in 12 paired bundles of cranial nerves, each labelled according to its origins or destination within a particular sense organ or structure in the head region of the bird (previous page, Diagram (*b*)). For example, the olfactory and optic nerves are entirely sensory and gather information from the bird's olfactory apparatus and the eyes respectively, while the hypoglossal nerve provides the motor pathway which controls the muscles of the syrinx and is therefore responsible for the control of bird songs and calls.

In a similar way the spinal nerves enter and leave the spinal cord in pairs. They pass through openings between individual vertebrae of the spinal column in which the spinal cord is housed. Because of the great variation across species in the number of vertebrae, the total number of pairs of spinal nerves varies. There are, for example, 38 pairs in the pigeon and 51 pairs in the Ostrich. The bundles of nerve fibres are labelled according to the region of the spinal column from which they emerge and, as in the case of the cranial nerves, the size of the bundle (number of fibres) reflects the importance of particular muscle groups or organs in the bird's lifestyle. For example, in strong flying birds the number of fibres is much greater than average in the brachial plexus from where fibres run to the muscles of the wing, while in running birds the spinal nerves are greatly enlarged in the lumbosacral plexus from where fibres run to the legs.

## The senses

The senses may be divided into three categories according to whether they provide information about events and conditions which occur either, (1) in the world remote from the bird (the remote senses include vision, hearing, smell, balance, magnetoreception and baroreception); (2) at the body surface (body surface senses include touch, taste, temperature); or (3) within the body (monitoring the position of limbs, internal temperature, etc.). Information from all three of these different types of senses is essential for a bird's healthy survival and daily activity. The actual sensory receptors involved in all of these senses are extremely small and always occur in large numbers.

The remote senses, unlike the body surface senses, rely on complicated sense organs (eyes, ears, nasal cavity) which channel information from the outside world to the sensory receptors. The receptors themselves are arranged in elaborate ways within the organ. The structure of the organ, such as the shape of the eye or the

*The brain and spinal cord. Side and bottom views of the brain and main cranial nerves are shown in (a) and (b), previous page. In (c) below, a top view of the brain and spinal cord, the principal cranial nerves are shown and the main spinal nerves, including the brachial and lumbosacral plexus, are indicated. The brains of four different bird species are shown for comparison in (d).*

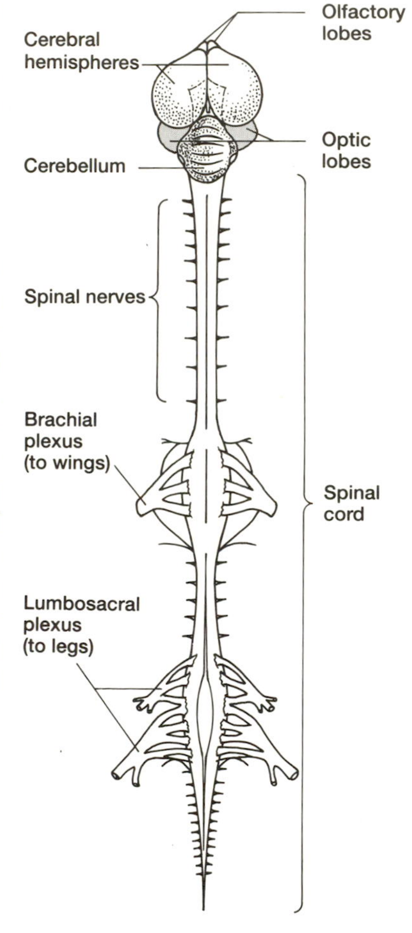

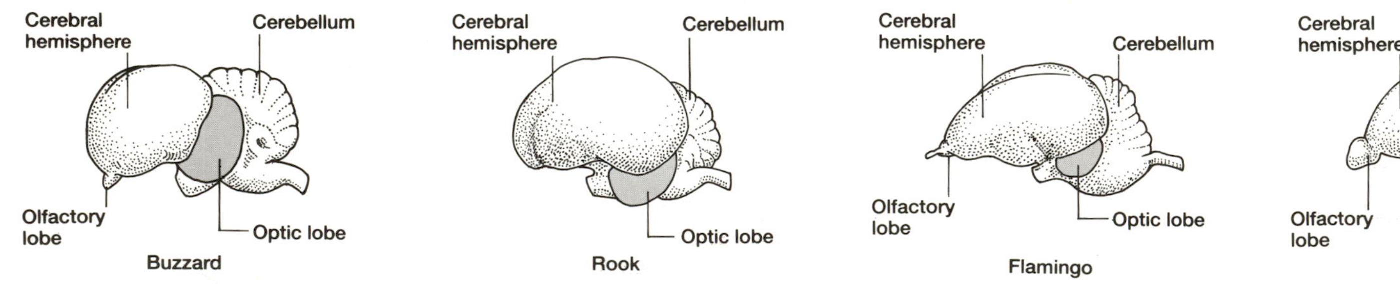


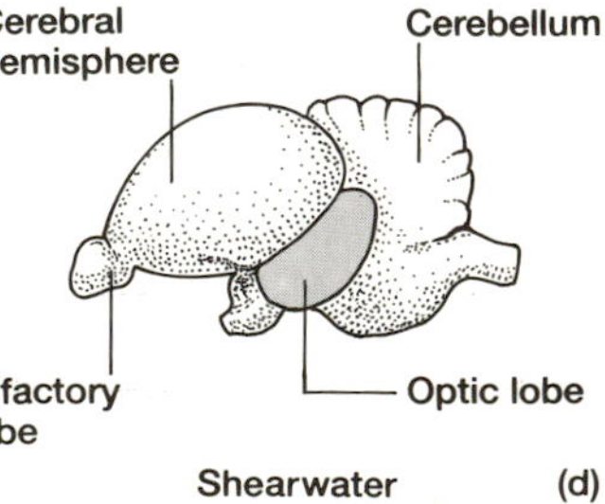

position and size of the ear openings, will itself influence the way information about the world reaches the receptors and hence influence the bird's sensory capacities.

The receptors of the body surface senses tend to be spread throughout the body and are not influenced by an elaborate organ. However, individual receptors are often densest in key parts of the body, as is the case with the concentration of taste and touch receptors in and around the mouth and bill.

### *Eyes and vision*

The bird eye is similar in basic structure to that found in all classes of vertebrate. It has a simple optical system, consisting of a cornea and a lens, which projects an image of the world onto the retina at the back of the eye. The retina contains many millions of sensory receptors, arranged in a complex layer. These receptors are of two main types, rods (for night vision) and cones (for daylight vision). Up to six different types of cones may occur in any one retina. The role of the rods and cone receptors is to convert the pattern of light in the image into coded information which is sent to the brain. It is primarily in the brain that this information is decoded and analyzed to render such details as colour, brightness, the position of objects, and speed of movement through the environment.

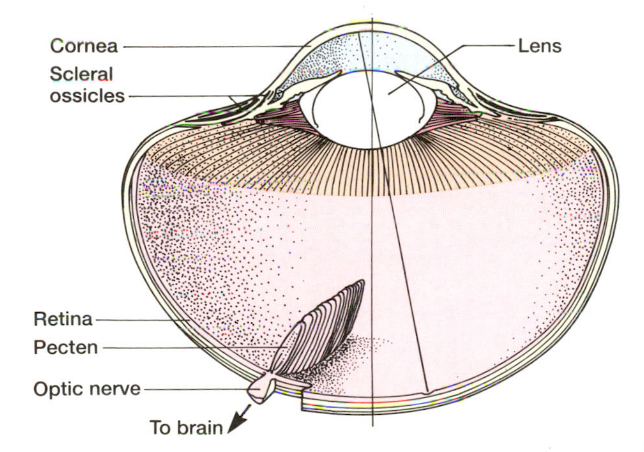


*Left. The eye structure of a domestic chicken, shown in horizontal section*

The eyes of birds differ from those of mammals in three main ways:

1. In no bird species are the eyes globular in shape as in the mammals; this is probably a result of demands for weight reduction. The various flat and tubular shapes are maintained by a ring of bones, the scleral ossicles.
2. Nutrients and oxygen are supplied for the maintenance of the retinal receptors by blood vessels lying on the retinal surface in mammals. In birds, however, this function is carried out by a structure called the pecten. This structure protrudes from the retinal surface into the chamber of the eye and nutrients diffuse passively from its pleated surface to the retina. The development of the pecten differs markedly between species and in some it extends from the retina nearly as far as the lens. There are, however, no agreed explanations to account for the diversity in the shape and size of pectens found between species.
3. In mammals, the optical system adjusts its focus for objects at different distances by muscles which change the shape, and hence the refractive power, of the lens. Birds also have this mechanism but in at least the pigeon and chicken there is an additional mechanism which also alters the shape of the cornea. This mechanism probably employs the ring of muscles which lie between the margin of the cornea and the scleral ossicles. When the muscles contract, the diameter of the cornea at its margin is reduced. This increases the curvature of the corneal surface and hence its refractive power.

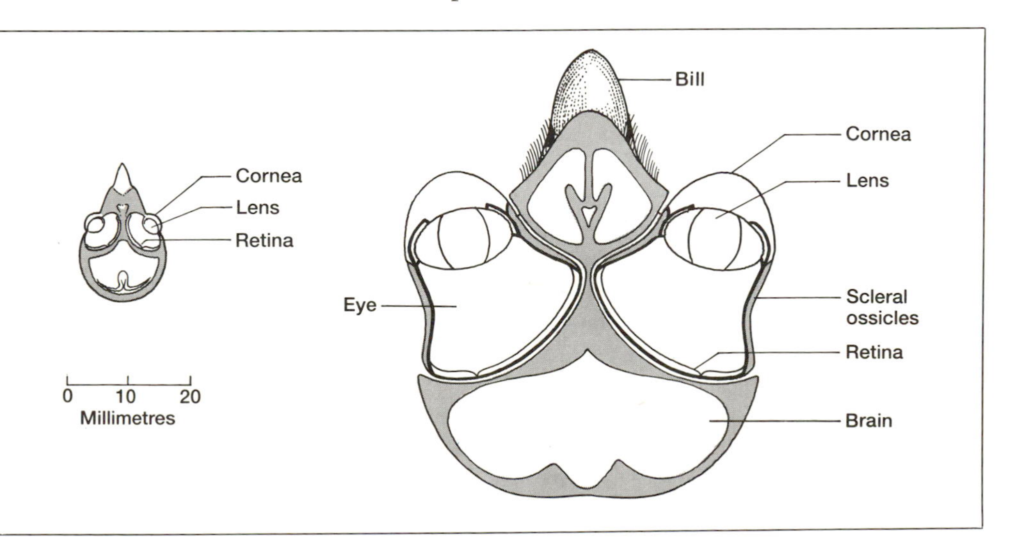


*Below. Heads of, right, the Eagle Owl (Bubo bubo) and, left, the Blue Tit (Parus caeruleus), shown for comparison in horizontal section. Notice the difference in size, shape and position of the eyes in the skull*

*Visual fields* There are marked differences between bird species in the position of the eyes within the skull and the extent to which eye movements are possible. These differences in eye geometry give rise to important differences in the extent of the birds' visual fields and the way that these can be altered as birds scan their surroundings. For example, the eyes of the Mallard, like those of many birds, are set on the side of the head. As a consequence, this duck gains complete visual coverage of the hemisphere above and behind the head, and is troubled by no blind spot from which a predator can approach unnoticed.

On the other hand, the Tawny Owl, like many predatory birds, has more frontally placed eyes, giving it a larger frontal binocular portion to its visual field but an extensive blind area behind the head.

In many birds the eyes are highly mobile and each eye can be moved independently. Such eye movements can alter considerably the part of the world that can be scanned. Whilst foraging, the European Starling can converge its eyes forward and down to look binocularly about its bill tip, or it can swing the eyes backwards and up to scan the whole of the area above its head for predators.

*Colour vision and ultraviolet sensitivity* Investigations of the colour vision and visual receptors suggest that it is safe to conclude that most, if not all, birds have colour vision. Birds in which colour vision has been unequivocally demonstrated include such diverse species as penguins, pigeons, ducks, passerines, owls and hummingbirds.

There is good evidence that some bird species (e.g. the pigeon, Mallard, some passerines and the Black-chinned Hummingbird) can see into the ultraviolet part of the spectrum. This means that birds can appreciate colours that are not visible to mammals, including man. It has been speculated that such sensitivity aids hummingbirds to detect flowers, which often strongly reflect ultraviolet light. The waxy bloom on many fruits also reflects ultraviolet light more strongly than the surrounding leaves. Hence ultraviolet sensitivity could assist fruit-eating birds to detect their food. Ultraviolet light penetrates thin cloud cover and it has been suggested that this could be used to determine the position of the sun through cloud cover, as an aid to navigation.

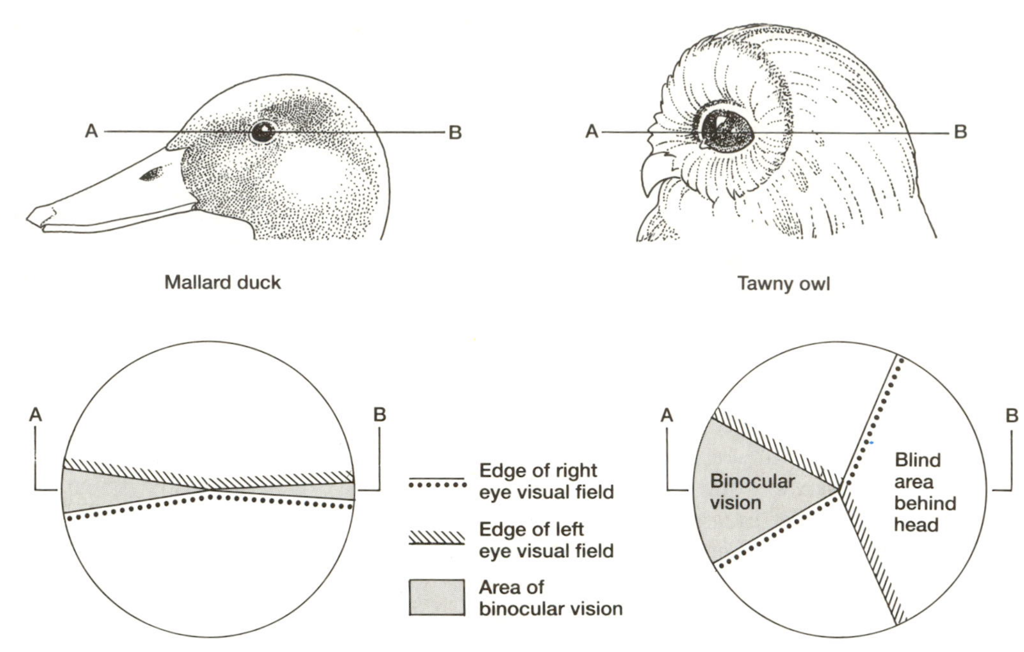


*Visual Fields. In these side views of the heads of the Mallard (Anas platyrhynchos) and the Tawny Owl (Strix aluco), visual fields are shown in a plane corresponding to the lines A–B. The beaks point towards the left. Notice that in the Mallard the visual field of each eye is over 180 degrees wide, providing overlap which gives the bird binocular vision both in front and behind the head. In the Tawny Owl the visual fields are much smaller and the eyes are more frontally facing, with the result that the owl has a larger frontal binocular field but a substantial blind area behind its head. In the Mallard, the binocular field in fact stretches over the top of the head to give complete visual coverage of the hemisphere above*

*Polarization sensitivity* The ability to detect the plane of polarization of light has been demonstrated in the pigeon and in some passerine species. However, the mechanism is not understood and the validity of the early demonstrations has been questioned. Polarization detection is known in the compound eyes of some invertebrates and in some fish but has not been described satisfactorily in mammals. As in the detection of ultraviolet light, the value of this ability may be as an aid to detecting the position of the sun, though this time when the sky is partially cloud covered.

*Resolution of detail* The ability of birds to resolve spatial detail (visual acuity) has been the subject of much speculation, often based upon anecdotal field observations. Older writers assumed that the visual acuity of raptorial birds far exceeds that of man but it is only in recent years that definitive data have become available. These have shown the visual performance of these birds to be more modest than once supposed. While the Wedge-tailed Eagle has an acuity about two and a half times better than that of man, acuity in two species of falcon, the American Kestrel and the Australian Brown Falcon, has been found to be almost identical to that of man at similar light levels. However, these birds outperform the visual acuity of the pigeon and of passerine species; both have poorer acuity than humans.

*Absolute sensitivity* Detailed analysis of the eye structure and of the visual performance in owls has revealed that owls are about 100 times more sensitive than birds such as the pigeon, which are active during the day and roost at night. However, the difference in sensitivity between human and owl eyes is only about two and a half times and this can be accounted for by differences in the brightness of the image produced in the two types of eye. (In photographic terms the owl eye has a lower f-number than that of man.) Comparison of owl sensitivity with the actual amount of light available in the natural environment at night indicates that this level of sensitivity cannot account entirely for the nocturnal behaviour of these birds, and it has been proposed that activity at night also requires reliance on hearing and on specific behavioural adaptations, including the 'perch and pounce' prey capture technique and familiarity with landmarks within the bird's territory. In other words, a woodland owl may depend on knowing where the obstacles and perches are because it perhaps cannot always see them at night.

## *Ears: hearing and balance*

As in other vertebrates, the avian ear contains structures which mediate two senses: hearing and balance. Hearing is the detection of sound vibrations, while balance involves the detection of the earth's gravity. Functionally the ear is divided into three parts: outer, middle and inner ears. All parts are involved in hearing while balance is mediated by the semicircular canals of the inner ear only.

*Outer ears and sound localization* The outer ear is a short tube which opens just behind or below the eye. In most birds it functions simply to funnel sounds towards the ear drum. The opening is usually covered by feathers, but it is visible in bald-headed birds such as some of the vultures.

In most birds the two ear openings are placed symmetrically on either side of the head but in some owl species the ear openings are, in contrast, asymmetric in both size and position. In addition, in many owls there are flaps of skin both in front of and behind the ear openings and these can be moved by special muscles. Also, the whole of the head is surrounded by a facial ruff of dense feathers which reflect sounds towards the ear openings. (These facial ruff feathers are hidden from view by the soft feathers of the facial disc which do not reflect or obstruct sounds.) These special features of owl outer ears enhance the accuracy with which these birds can locate the source of a sound, but probably do not improve the bird's absolute sensitivity to sounds. It has been shown that the Barn Owl can locate and catch live prey in total darkness using sound cues alone. There is little evidence on the accuracy with which birds that lack these outer ear structures can locate sounds, but it seems likely that their performance is considerably worse than that of the owls. Hearing in birds such as passerines, however, is probably used more to detect the presence of a sound source rather than determine its exact location.

*Above. In the majority of birds the ear openings, lying just behind and below the eye, are hidden by feathers, the ear coverts. The ear coverts of the Cactus-Wren (Campylorhynchus brunneicapillus) of the southern United States and northern Mexico are dusky grey and white. But in a number of scavenging birds, such as the African White-headed Vulture (Aegypius occipitalis), head feathering is sparse and the ear opening is visible. Such sparse feathering probably makes it easier for vultures to keep their heads clean*

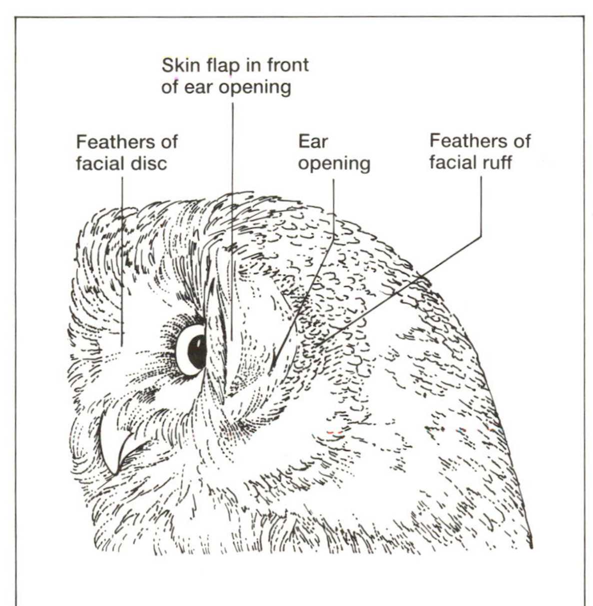


*Left. The outer ears of owls. This head of Tengmalm's Owl (Aegolius funereus) shows the feathers parted at the edge of the facial ruff and the flap of skin in front of the opening.*

*The middle and inner ears* The middle ear of birds, like that of mammals, functions as a device for changing the vibrations of air received at the ear drum into vibrations of the fluid which fills the cochlea of the inner ear. It is the vibrations of this fluid which are detected by the ear's sensory receptors, not the vibrations of the air directly. In the bird ear this transfer of the ear drum's vibrations is achieved by a single bone, the columella, rather than by three bones as in mammals.

The inner ear differs from that of mammals. The cochlea in birds consists of a short, slightly curved tube, whereas in mammals the tube is coiled in upon itself like a clock spring. This coiling has allowed the tube to become considerably longer in mammals, a fact which probably accounts for the observation that hearing in birds is restricted to a much lower frequency range than is common in mammals.

*Frequency range and absolute sensitivity* Bird songs often seem composed of high frequency sounds (see Chapter 10). However, all of these sounds fall well within the range of normal human hearing. In birds, sensitivity to sounds of different frequencies is not uniform and all birds so far studied are most sensitive to sounds within the frequency range between 1 and 5 kHz (1 kHz = 1000 vibrations per second), which corresponds approximately with the top two octaves of a piano. In this frequency region the absolute sensitivity (the minimum intensity of sounds which can be heard) of birds is very similar to that of man. Above these frequencies sensitivity to sounds falls dramatically and the upper limit of hearing in birds is at about 10 kHz (one octave above the highest note of a piano). In young people hearing extends to about 18 kHz and hearing in mammals such as the rodents and bats extends to 90 kHz and 120 kHz respectively.

At the other end of the scale there is evidence that the pigeon can detect very low frequency sounds below the range of human hearing (infrasound). However, the means by which pigeons detect infrasound is not known and the bird may not in fact 'hear' these sounds; they may be detected via touch receptors in other parts of the body, perhaps the legs.

*Discrimination between sounds* In view of the apparent complexity of bird songs it is perhaps surprising to find that within the frequency range 1–5 kHz birds do not outperform man in their sensitivity to changes in frequency, duration or intensity of simple pure tones. However, it may be that a bird's auditory system is especially tuned to complex sound signals, perhaps even specific to its species. Therefore it may be only when complex sound signals are used that birds can outperform man. It was at one time thought that birds exceeded humans in their resolution of time differences between sounds, the idea being that birds could resolve individual notes in a complex song better than man. However, recent investigations suggest that the bird ear may not be specialized for such high speed resolution after all.

*Echolocation* The Cave Swiftlets and the Oilbird are famed for nesting and roosting in the total darkness of caves. Both species have been shown to employ active sonar or echolocation to assist them in navigating within the caves. The birds produce pulses of sound (heard as clicks) and are able to detect the presence of objects by analyzing the echoes which return.

Unlike the mammals which employ echolocation, such as bats and porpoises, these birds are restricted to the use of low frequency sounds. The frequency of the Oilbird's broad band clicks is between 1 and 12 kHz with most sound energy concentrated between 2 and 4 kHz, where the bird's hearing is most acute. As a consequence of using low frequency pulses, Oilbirds are unable to detect small objects. They cannot reliably detect objects less than about 20 cm diameter, a level of performance actually matched by blind humans who also employ echolocation, using clicks of the tongue or taps, as a sound source. The Cave Swiftlets may perform better than this but their ability to detect objects certainly does not match that of bats or porpoises.

## *Nasal cavity and the sense of smell*

The nasal cavity in birds lies within the skull at the base of the upper bill. It contains many elaborate folds of the underlying bone; these serve to increase the surface area of the mucous

*The nasal cavity. The skull of a chicken is shown in (a), with a detail (b) to show the path taken by air between its entry through the nostrils and exit towards the respiratory system. On the page opposite, vertical sections (c) through the skull in the planes A–B and C–D show the elaborate way in which the bones are formed within the cavity to increase the surface area of the mucous membrane where the olfactory receptors are situated. Sections through planes C–D in three other bird species are compared in (d)*

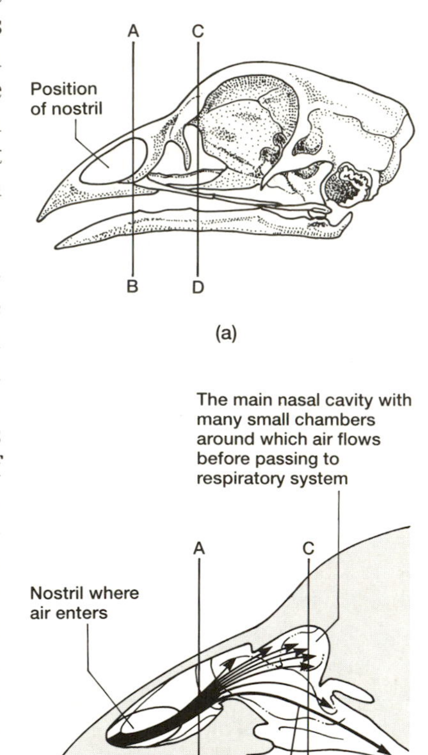


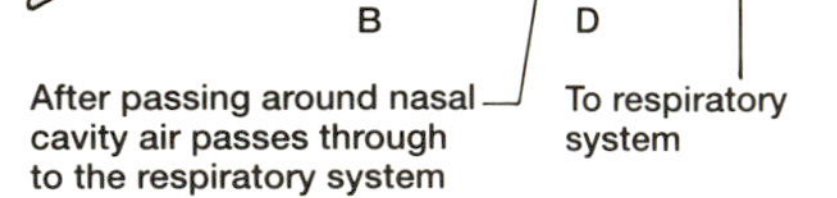

membrane where the olfactory (smell) receptors are situated. The cavity is so designed that all air breathed in through the nostrils is directed over these sensory surfaces. The size, position, and shape of this organ, and the part of the brain which subserves it, the olfactory lobes, vary greatly between species.

Although use of smell in the daily life of a bird varies markedly between species, both laboratory and field studies have demonstrated the importance of smell in species ranging from pigeons to vultures. Some findings appear quite exceptional. For example, Black-footed Albatrosses can be attracted by odour from a distance of at least 30 kilometres to bacon fat poured on the ocean surface. The albatrosses are members of the Procellariiformes and these birds are noted for their large olfactory lobes. Similarly, the Snow Petrel can find hidden pieces of raw fish by smell, fulmars and Sooty Shearwaters are attracted upwind to fish oil, and Leach's Petrels possibly locate their individual nest sites by olfactory cues.

The most important use of olfaction in food finding recorded to date are in the kiwis, Oilbird and Turkey Vulture. Kiwis, which are nocturnal and flightless, detect buried earthworms by smell. Oilbirds are also nocturnal, and feed exclusively on ripe fruits. The Turkey Vulture is able to locate carrion by smell, though this ability is probably not shared by all vultures.

## Magnetoreception

There is good evidence that birds can detect the earth's magnetic field and use it to provide a compass direction when migrating (Chapter 7). Despite these intriguing results the sensory basis of magnetoreception remains unknown. Crystals of magnetite have been found in parts of the pigeon's head and neck and these could form part of a sensory system but no sensory receptors or nerve connections to the brain have been discovered. Further studies have failed to find such deposits of magnetite and alternative theories suggesting that the eye, ears or pineal gland may in some way be involved in magnetoreception have been proposed. The experimental evidence that is available suggests that magnetoreception is closely linked with the visual system.

## Baroreception

It has been shown experimentally that the pigeon can detect changes in barometric pressure. On some occasions these changes could be very small, equivalent to a pressure difference between the floor and ceiling of a room. However, the speed of change is important and it should be noted that humans can detect quite small pressure changes if these are fairly rapid, as for example when rapidly climbing a hill in a car or even travelling in a lift. The mechanism involved in the pigeon is not known but pressure changes can be perceived by humans through the discomfort felt in the ears caused by momentary imbalance in air pressure on either side of the ear drum.

## Touch and temperature

The two simple terms 'touch' and 'temperature' obscure a very complicated set of sensory responses involving at least nine different types of receptors. There are three types of touch or mechanical receptors; these respond either to the acceleration, velocity or amplitude of an object as it touches the skin. There are separate 'cold' and 'warm' temperature receptors and four different types of receptors which respond only to specific types of noxious stimuli. All of these receptors are found at various places on the skin and within the mouth and hence give information about events at the body's surface. The same receptor types within the body give information about the position of limbs and the conditions of internal organs. Knowledge about the working of most of these different receptor types has been gained from just two bird species, the chicken and pigeon, but it is assumed that these receptor types are found in most, if not all, birds.

The most important mechanical receptors are the Herbst corpuscles. These are widely distributed and occur not only throughout the skin, but also along the large leg and wing bones, tendons, and muscles, and near large blood vessels. It is believed that they fulfil a primary role in monitoring functions within the bird's body as well as responding to stimuli in contact with the skin and detecting movements of the feathers.

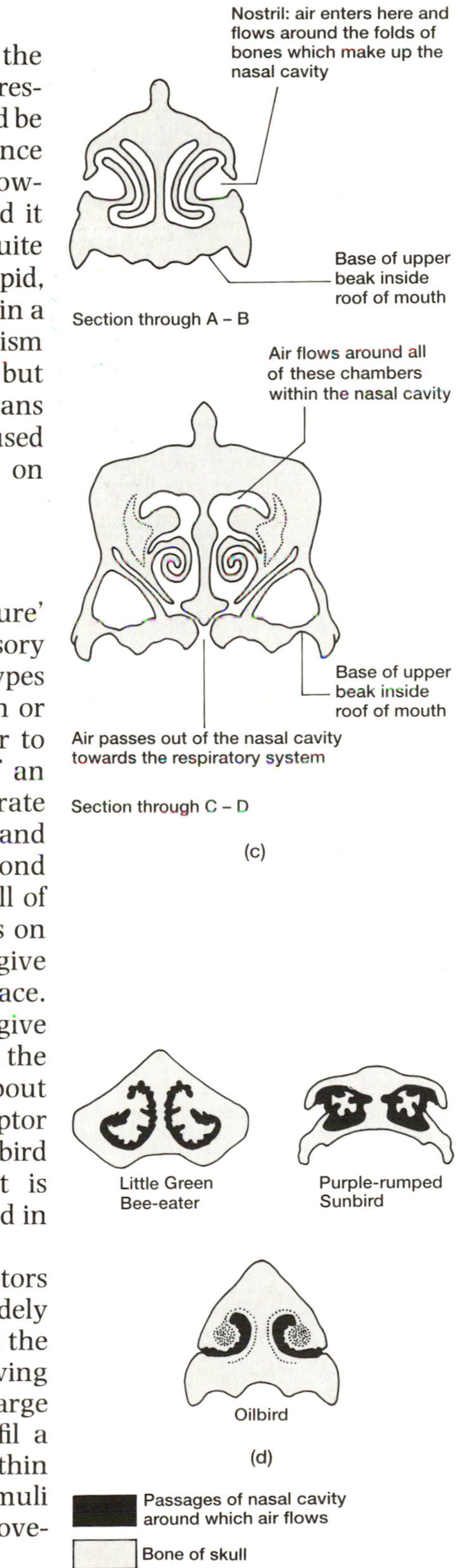

A large number of Herbst corpuscles occur in the beak. Here their number and location are related to the way in which the beak is used during feeding. For example, long-billed shorebirds such as curlews or woodcock, which use their bills for probing in soft mud, have large numbers of Herbst corpuscles near the bill tip. Similar concentrations of these receptors occur inside the tip of the upper bill in ducks and geese, in the so-called bill-tip organ. In granivorous song birds, especially the finches (Fringillidae) which use their beak to crush or cut seeds open, Herbst corpuscles are located at exactly those places in the beak which are involved in seed opening. The long tongues of woodpeckers, which are protruded from the beak to catch prey, are particularly well endowed with Herbst corpuscles. Touch sensitivity is therefore extremely important in the detection and manipulation of prey and food items. The American Wood Stork is capable of efficiently catching live fish, even when blindfolded, using touch sensitivity in its bill alone, and it is thought that the skimmers (Charadriiformes) detect their prey through touch sensitivity as they forage in flight with the lower bill just below the water surface.

## Taste

The need to avoid harmful foods is self evident. Taste provides the most reliable cue to the quality of food and so it is not surprising that most birds have an acute sense of taste. Through taste birds can acquire knowledge about the nutrient value of different foods; this manifests itself in definite food preferences. These can be investigated to learn how birds classify different tastes.

The sense of taste, which is in fact the analysis of chemical composition, is no less sophisticated than the other senses. It requires specialized sensory receptors placed in strategic positions inside the bird's mouth. The number and kinds of taste receptors, and their positions within the mouth, differ between species. As in mammals, birds are found to be sensitive to four main tastes: sweet, salt, sour (acid) and bitter, but sensitivity to these is not shared equally among all birds. For example, sensitivity to salts and acids decreases in the order pigeon, Mallard, chicken, while the Herring Gull and the chicken seem to be indifferent to substances which to humans taste extremely bitter.

That taste can be extremely important to birds when foraging and selecting food items is revealed by a demonstration that Mallards, which had detected food items buried beneath sand using tactile cues from their bill tip organ, could apparently discriminate between palatable and non-palatable foods on the basis of the taste of the food held within the beak. Similarly, shorebirds which probe into sand for food, such as the Dunlin and the Red Knot were able to discriminate between jars filled with sand containing 'taste' and 'no taste'. The 'taste' was supplied by worms which had been removed before the experiment, while 'no taste' was the same sand which had been washed.

The places in which Mallards position food items in their bills prior to ingestion has been shown to coincide with the location of taste receptors. Carrion and insect eating birds probably depend heavily upon taste to determine the palatability of their food, which can often be extremely toxic due to putrefication or poisons contained within the insect's body as a defence. The Carrion Crow has a great abundance of taste receptors in its mouth and is very sensitive to sour substances. Blue Jays are known to be extremely sensitive to the sour taste of plant-derived toxins (e.g. cardiac glycosides, similar to digitalis) stored in the bodies of butterflies and caterpillars.

GRM

# 3 MOVEMENT

## Flight

Flight is the characteristic adaptation of birds, and because of the physical problems of moving in air it is also one of the most demanding adaptations found in nature. Active flapping flight has evolved in four animal groups – insects, pterosaurs (now extinct), birds and bats – and all of these show extreme morphological, physiological and behavioural specializations associated with aerial locomotion. Large wing surfaces support the animal's weight and provide propulsion in flight. The energy requirements of flight are also high. Lungs and blood vessels are enlarged to transport large quantities of oxygen to the flight muscles, which represent a considerable proportion of the body mass. And flying animals must be highly active if they are to find sufficient food to support their enhanced metabolism.

All of these factors have been a major influence on birds, and it is probably justified to claim that no aspect of bird biology is completely unaffected by flight. In understanding birds, it is thus essential to understand how they have responded to the potential of flight, and also to consider how their adoption of flight has constrained their adaptation and radiation.

### Types of flight

Compared to other animal classes, birds show a remarkable homogeneity of form, primarily because the wings of all flying birds must be large to generate sufficient aerodynamic force. Variation in the head and jaws and in the hind legs of birds reflects specialization for different ecological niches, and the same is true for the wings. Wings vary in shape and size, and also to some extent in the distribution of feathers, and in qualities such as surface texture and the shape of the cross section. The wings of all flying birds function aerodynamically in the same way, and variations in the wings can be traced to different types of flight, allowing the bird to exploit its environment in the most appropriate way for its mode of life.

The characteristic flight mode is active flapping, in which the bird flies straight and level. This is a very efficient means of covering distance, although it can require extremely high power output (work per unit of time). All flying birds can use flapping flight, and it is usually essential for take-off and for controlling the flight path. Gliding, on extended wings and without wingstrokes, costs a minimum of energy, and is particularly attractive for large birds. But the disadvantage of gliding is that the bird must lose height unless it can use rising air currents to maintain or gain altitude.

One of the main advantages of flight is its speed. Flapping and gliding are fast compared with walking and running on land, and very fast compared with swimming in water. The fastest land mammal, the cheetah, can achieve 18 body lengths per second, but in flight a swift can achieve 67, a chaffinch 72, and a European Starling 80 body lengths. This is comparable with supersonic aircraft, which can reach about 100 lengths per second. The fastest flying birds can reach air speeds in excess of 20 m/s, and even small birds typically travel at 10–15 m/s. High speed also permits long distance flights. Small passerine birds typically migrate 1000 km non-stop in about 24 hours, a feat which in terms of either endurance or speed is far beyond the capacity of terrestrial animals of comparable size.

### How birds fly

Much of our understanding of the mechanics of flight in birds is based on analogy with aircraft. There is considerable justification in this approach since both aim to travel through the air, and the wings of both operate in the same way. In steady level flight a bird and an aircraft must generate forces which support the weight against gravity and which provide propulsive thrust against the friction and other forces which comprise drag.

#### *Aerofoil action*

The mechanism of force generation by the wings is known as aerofoil action. The aerofoil is the shape of the cross-section of the wings: it has a rounded leading edge and a sharp trailing edge, and an asymmetric profile so that air flowing over the upper surface has further to travel than air passing underneath, and so flows faster. Bernoulli's principle states that, in a moving fluid, pressure falls when speed rises. The wing therefore experiences reduced air pressure above the aerofoil, and increased pressure beneath. This pressure difference gives rise to a lift force, acting transverse to the direction of movement. The magnitude of the lift depends on two factors: the speed of flight, and the degree of assymetry in the aerofoil profile (often described as the angle of incidence, or the angle between the mid-line of the aerofoil and the direction of air flow past the aerofoil). A flat plate set at a small angle to the air flow will act as an aerofoil (experiment, carefully, by holding your hand from the window of a moving car). But the flat plate is relatively inefficient, and the rounded leading edge of the aerofoil gives it a streamlined shape which improves performance markedly by reducing drag. So also does the use of camber, whereby the cross-section is not flat, but is curved downwards. Extreme camber enables the wing to be most efficient in slow flight, and is found in birds such as pheasants and grouse for which a rapid take-off is essential for survival.

The main limit on aerofoil action is stall. If the angle of incidence (or the camber) becomes too great, or the speed of the air becomes too low, air ceases to flow smoothly over the upper surface, and the aerofoil cannot generate lift. This can be disastrous to aircraft, but is less critical to birds, for an animal is usually sufficiently agile to be able to control its flight even after an accidental stall. Indeed, many birds stall deliberately as a means of initiating manoeuvres, and in flapping flight in some species the wing effectively stalls during every wingbeat.

The lift force generated by the aerofoil is the result of pressure differences between top and bottom surfaces. But the aerofoil also causes movements in the air flowing past and behind it. To provide a flow of momentum matching the upwards lift force, air must flow downwards in the region behind the wing, and it is forced to do this by trailing vortices shed behind the wing tip; these form the wake of the wing. A vortex is

*How birds fly. The aerofoil below left represents the streamlined, cambered cross-section of the bird wing. It deflects the airflow passing over it downwards, and the wing experiences the resulting force as a pressure distribution over the surface of the aerofoil. This force is termed lift. There is also a drag force due to friction on the wing surface*

*Below, top. When a wing generates lift it also generates a system of trailing vortices behind the wingtips. The strength of these vortices is determined by the magnitude of the lift force*

*Below, bottom. In an aircraft the vertical force balances, and lift is equal to weight. Separate engines generate a thrust to overcome drag. A bird has no separate engine. In gliding, flight weight is used to counteract drag, and the bird must descend relative to the air. A bird flaps its wings so that lift has a horizontal, forward component which balances drag*

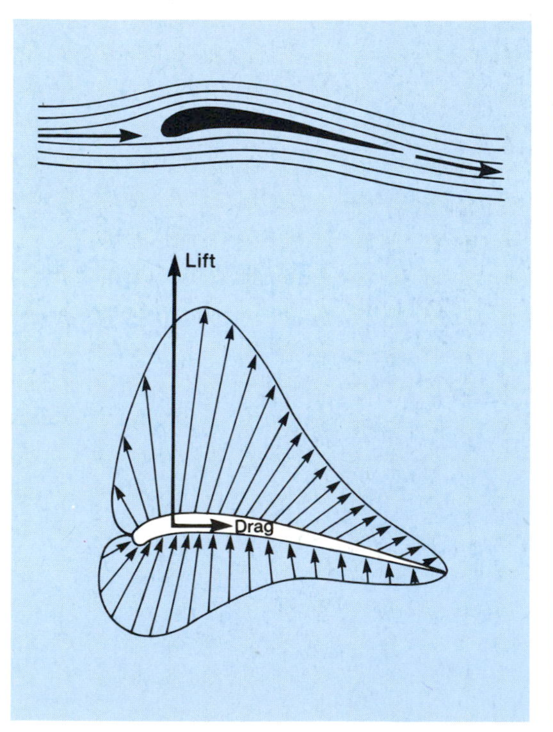


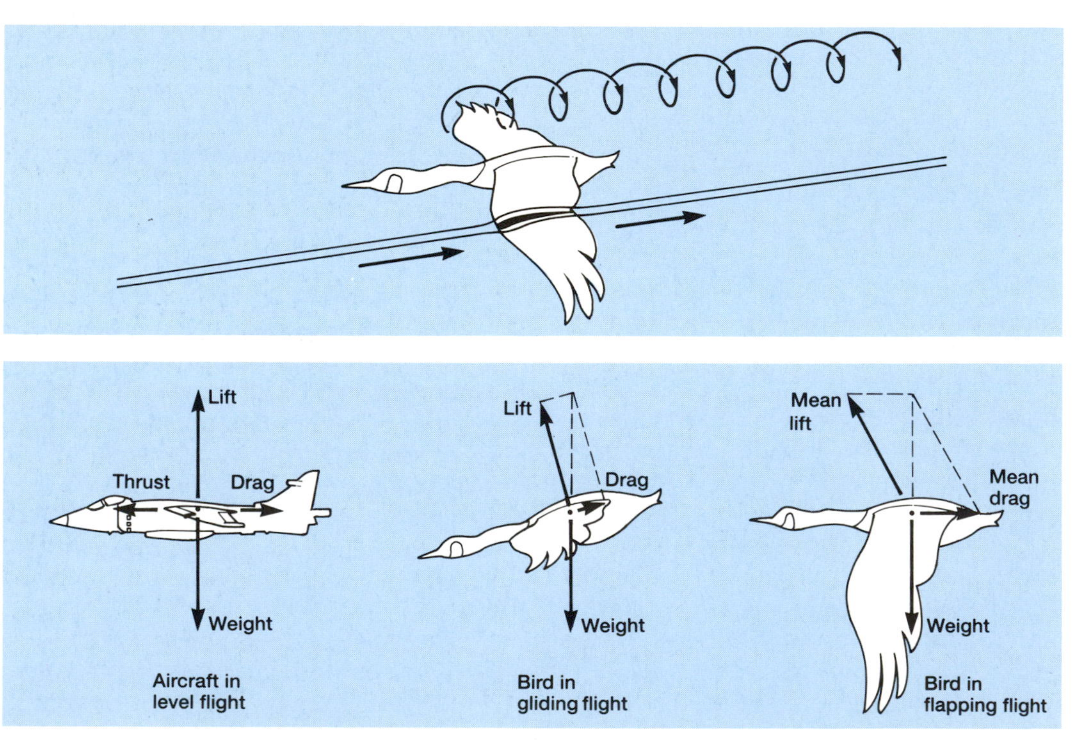

an intense air rotation – examples of natural vortices are tornadoes and dust-devils in air, and whirlpools and 'maelstroms' in water – and vortices encourage air around them to rotate about the vortex. The trailing vortices in the wake are an inescapable consequence of force generation by the aerofoil, for without them the wing would generate no lift. Vortices are often not visible, but in the right atmospheric conditions they can be seen as condensation or 'vapour trails' behind aircraft wings.

### *Drag*

Lift acts at right angles to the direction of movement of the aerofoil. This is not the only force on the wing, however, for energy must be expended to overcome friction on the wing and body surfaces, and to generate the vortices in the wake. These energy losses are experienced as drag forces, and conventionally drag is viewed as the sum of several components: frictional drag of the wings is termed profile drag, friction on the body is parasite drag, and the vortex force is called induced drag. All of these forces act parallel to the direction of movement of the wing. The magnitude of the components of drag vary with the size and shape of the wing, and also with flight speed; with a well-designed streamlined aerofoil, drag is typically less than one-tenth of lift in normal flight.

In steady, level flight, lift and weight, thrust and drag must be in equilibrium. Lift acts vertically and can balance weight, and a horizontal forward thrust force is needed to counter drag. Aircraft generate lift by aerofoil action on the wings, and use engines to generate thrust independently. It is at this stage that the analogy between birds and aircraft breaks down, however. Birds have no equivalent to a jet engine, and must provide both thrust and weight support by the lift from the wings alone. It is this which gives flapping flight its special character.

### *Mechanics of gliding flight*

First, consider the situation where the bird does not flap its wings, but instead glides. This is a very common type of flight, for as well as being used by large birds, gliding is also found in a

*In flapping flight (below left) the magnitude and direction of lift vary throughout the wingbeat, but are greatest during the downstroke*

*Below right. Two patterns of vortex wake have been observed in flapping flight. In slow flight, and in birds with relatively short wings, vorticity is generated only during the downstroke, and the wake is a series of vortex rings. No lift is produced in the upstroke. In the faster, cruising flight of long-winged birds lift is generated throughout the wingbeat and the vortex wake is an undulating pair of line vortices*

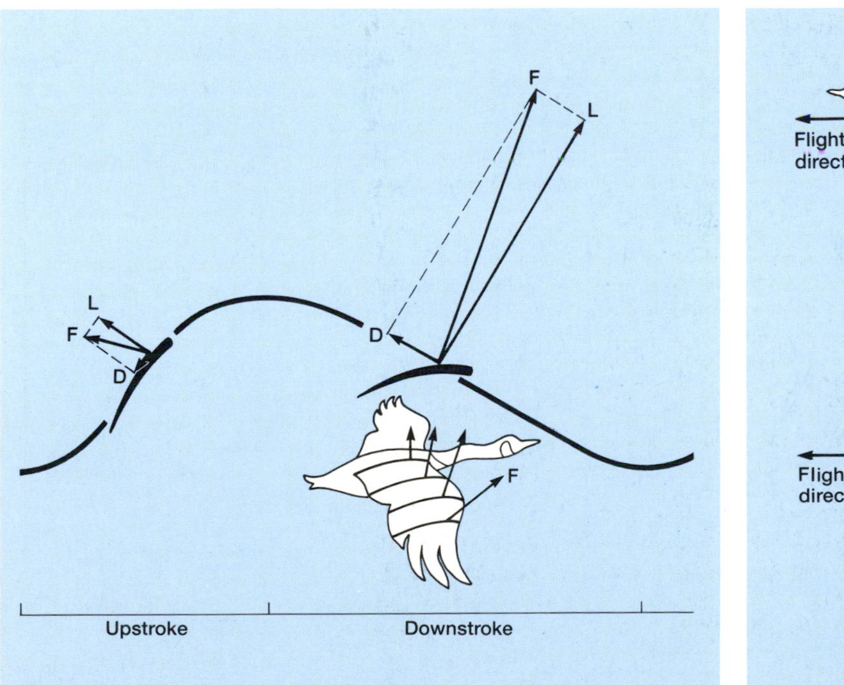


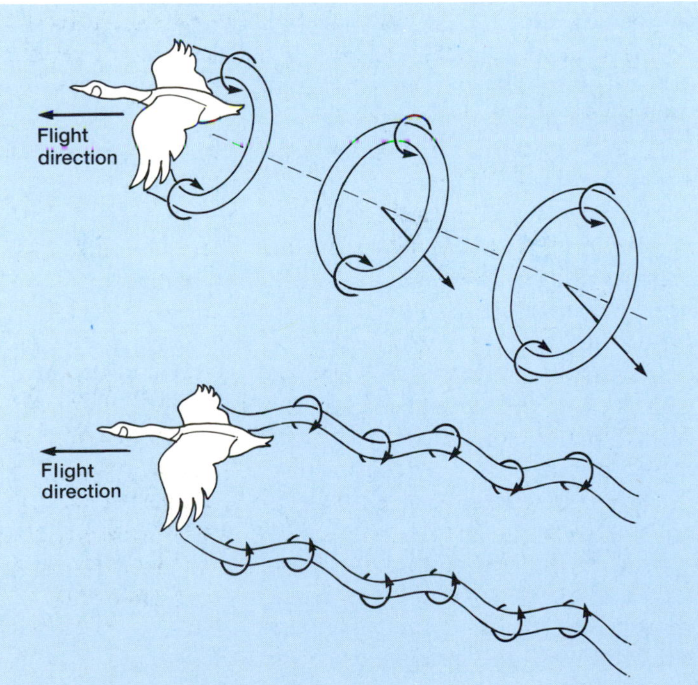

wide range of other animals (including squirrels, marsupials, lizards, frogs, and even a snake). Only three forces act on the animal: lift, drag and weight. The animal can travel in a steady descending glide if weight acts in part to balance drag; for this reason gliders must lose height. The angle of the glide is proportional to the ratio between drag and lift, and since lift much exceeds drag, glide angles are generally shallow.

*Mechanics of flapping flight*

Gliding is a relatively simple mode of flight since the bird can keep its wings stationary relative to the body. The bird descends because it has no means of generating a horizontal forwards force to balance drag. The wings are flapped to provide the thrust which can convert a descending glide into level horizontal flight. It is false to think of flapping flight as a way of generating a vertical force to keep in the air.

The wingbeat in flapping flight is complex, and varies considerably in geometry according to flight speed and to the design of the wings. It can be divided into two phases: during the downstroke the wings move downwards relative to the body, and forwards relative to the air; in the upstroke they are brought back and up, and are often twisted and flexed so that the wingtip moves close to the body. The greater part of the aerodynamic forces are generated during the downstroke, for in this phase the lift points forwards and upwards, providing both weight support and thrust.

Lift and drag vary periodically during the wingbeat, but providing the bird has suitable wingbeat geometry it can balance mean lift, drag and weight in equilibrium. The bird can then fly straight and level.

While the downstroke is responsible for the main propulsive forces, the action of the upstroke is more variable: in some circumstances it generates considerable lift, yet otherwise there is very little useful force. The pattern of vortices in bird wakes has been instructive in showing how upstroke force generation varies, for as explained above, the presence of wake vortices is diagnostic of lift generation by aerofoil action. In slow flight, and in cruising flight in birds with small or short wings, the upstroke gives no force, for the bird could only generate useful lift at the expense of substantial induced drag. The wings accelerate and decelerate rapidly at the beginning and end of the downstroke, and the wake consists of a series of ring vortices each generated by a single downstroke. In longer-winged birds the wake vortices are less intense, and wake structure and wingbeat kinematics vary with flight speed. In cruising flight the wing only deforms slightly during the upstroke; the lift thus generated is used primarily as weight support rather than thrust; the vortex wake consists of a pair of continuous undulating vortices trailing from the wingtips.

The rate at which the wings are flapped depends on the size of bird. It is up to 80 wingbeats per second in the smallest hummingbirds, typically about 15 in small passerines and about two in some slow-flapping owls and herons.

*The wingbeat kinematics of a pigeon in slow flight, at about 3 m/s. During the upstroke the wing is flexed and the wingtip moves close to the body*

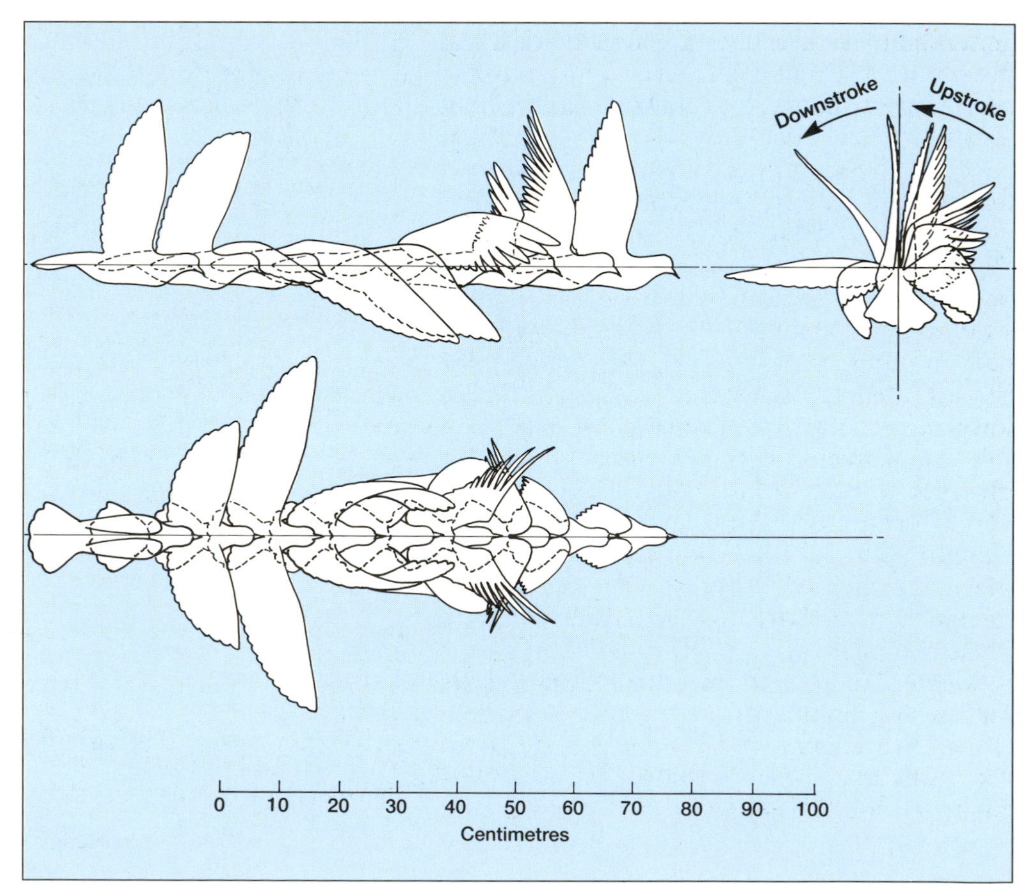

## *Energy consumption in flight*

A flying bird must do work with its flight muscles to move the wing to generate the components of lift which provide propulsive thrust. The rate at which this work is done is the mechanical power required to fly, and is calculated as the sum of the rates of doing work against parasite, profile and induced drags. Flight muscles produce heat as well as mechanical work when they contract, and the total metabolic power consumed in flight reflects this heat loss: metabolic power is estimated to be approximately four times mechanical power. The remainder of the energy must be dissipated as heat by respiration or by convection from the body surface.

Like wingbeat kinematics, mechanical power varies with flight speed. The friction drags rise with speed (approximately as the cube of speed), while induced drag falls as speed increases (as 1/speed), because at higher speeds forward momentum contributes to wake vortex generation. As a result the total mechanical power curve is U-shaped, and there is a range of speeds over which energy consumption in flight is least. Very slow, or very fast, flight is energetically much more demanding, and may for many species be impossible.

The physiological energy demands of flight can be measured in a number of ways: birds can be flown in wind tunnels with masks to measure oxygen consumption; and change in total body mass measured during a long flight can be related to energy since the fuel value of fat is known. Recently it has even become possible to use radioactively labelled isotopes to determine changes in water balance, and hence to estimate oxygen uptake. All of these techniques are fraught with difficulties, but have nevertheless given considerable insight into the energy demands of activity in birds. The available measurements indicate the enormous energy requirements of flight: metabolic energy rates for cruising flight in the range 7–10 watts for a 0.1 kg bird compare with an average 4 watts for walking and running in terrestrial mammals of the same size, and 0.2 watts for swimming fish.

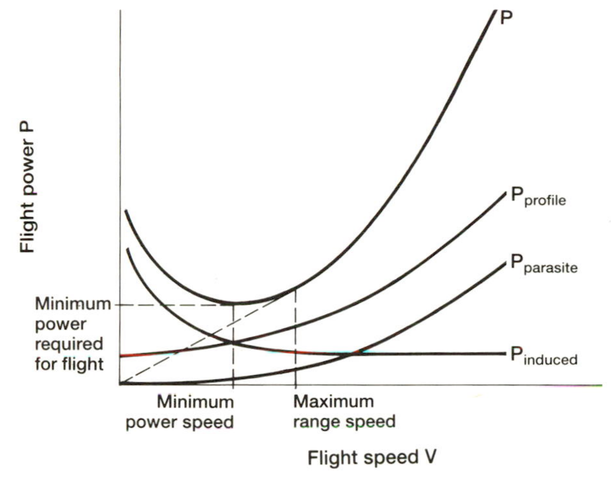


*Far left. The power required to fly. Power, P, depends on flight speed V, and is determined by the sum of the parasite, profile and induced drags. Least power is required to stay aloft neither at very slow nor at very fast speeds but at an intermediate velocity, the minimum power speed. This latter speed is less than the maximum distance for each unit of energy expended*

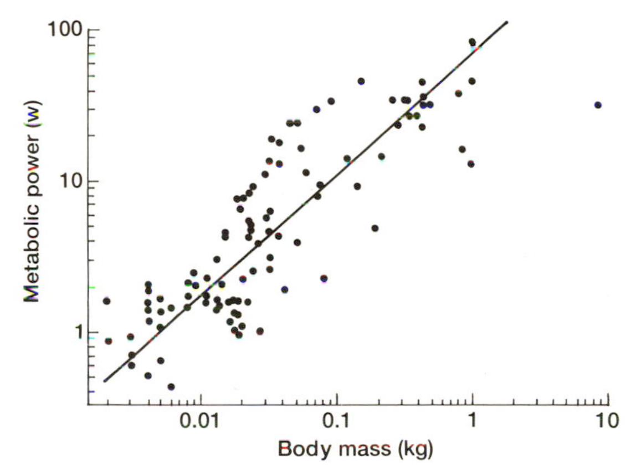


*Power requirements for flight increase with a bird's weight*

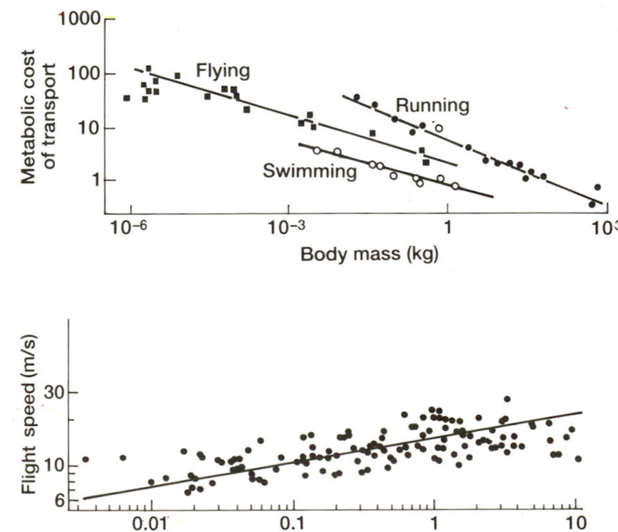


*Cost of transport (the expenditure of effort involved in moving a unit mass a unit distance) declines with body mass. Flying is more economical than walking or running because of the high speeds involved, but less economical than underwater swimming. The single open circle among the running animals is the cost for a surface-swimming duck. Notice that 'flying animals' include insects*

*Flight speed also increases with body mass, because larger birds must fly faster to generate enough lift on their wings*

The penalty of flight is its high energy cost, but its advantage is in speed, for as we have already seen, flying animals travel several times faster than animals on land or in water. The performance of a mode of locomotion is usually measured by *cost of transport*, which is defined as the energy required to carry unit weight of the animal through unit distance (it is computed as the ratio of power to the product of speed and weight). Cost of transport is approximately constant in flying animals, declining slightly with mass, and is markedly lower than for terrestrial movement because of the high speeds involved. Swimming is more economical still than flying, because although speed is low, power is also small: this explains in part why most fish do not maintain body temperature above that of their environment, while high body temperatures are essential for the high levels of activity of birds and mammals.

### *Flight speeds*

The shape of the power curve allows us to define characteristic speeds of sustained flight. At the *minimum power speed* the power is least, and for a given amount of fuel the bird can remain in the air for the longest time. This speed should, for instance, be used during night flights by swifts, or in foraging by birds such as harriers which quarter the ground searching for prey. The *maximum range speed* is slightly higher: at this speed the power/speed ratio (or equivalently the cost of transport: see above) is minimum. Flight at this speed allows the bird to maximize flight range for given energy, and is most appropriate during migration or other sustained flights.

Comparison of measured bird flight speeds can be misleading for several reasons: birds vary speed according to their ecological goal, and adjust air speed in wind to maintain the appropriate ground speed. Moreover, measured speeds are notoriously inaccurate. A graph of measured cruising flight speed against body mass is shown previous page, and some sample speed measurements in the table opposite: larger birds tend to fly faster, but at any mass there is a wide range of speeds, reflecting the range of wing designs and ecological adaptations in birds of comparable size.

The fastest record in steady flight is for the Eider duck (21 m/s); the record for speed is probably held by the Peregrine Falcon, which has been reported to exceed 70 m/s in its dives. Surprisingly, and contrary to popular belief, swifts are among the slowest birds (5–8 m/s). Slow flight is essential for aerial feeding insectivorous birds since their prey travels slowly: swifts appear fast to the observer because they enjoy diving and wheeling in the air close to buildings.

## Styles of flight

Birds show a wide range of adaptations to flight. The different styles of flight are associated with different aerodynamic specializations, and in each case optimization of aerodynamic and ecological performance demands distinct patterns of wing design. Two parameters, aspect ratio and wing loading, are used to quantify the size and shape of the wings:

*aspect ratio* is a measure of the shape of the wings, and is defined as the ratio of the breadth (or span) of the wings to the mean chord (or length of each aerofoil section). This quantity is calculated as wingspan squared divided by wing area. The average value for birds is approximately 7, with a range from around 4.5 in some Galliformes and tinamous and up to 15–20 in albatrosses.

*wing loading*, defined as the ratio of weight to wing area, describes the size of the wings. Unlike aspect ratio, which is dimensionless, wing loading is a dimensional quantity and its value depends on the units of measurement. It also varies with size, tending to be larger in larger birds. In units of newtons per square metre ($N/m^2$) it ranges from 18 in a small hummingbird to 230 in the Whooper Swan.

These two quantities are widely used both in aircraft engineering and in studies of bird flight, and they have well-understood interpretations. Wing loading is the ratio of a force (weight) to

*Wing dimensions. Body mass, aspect ratio, wing loading and observed flight speeds are given for a sample of typical birds. Speeds, where shown, have been obtained mainly from radar measurements, and are cruise or migration speeds in light winds or still air. They are probably representative of the typical flight speeds adopted by these birds*

| | Body mass (kg) | Aspect ratio | Wing loading ($N/m^2$) | Flight speed (m/s) |
|---|---|---|---|---|
| Red-winged Tinamou | 0.82 | 6.9 | 123 | |
| Red-throated Diver | 0.96 | 12.2 | 106 | 17 |
| Great-crested Grebe | 0.89 | 10.6 | 144 | |
| Wandering Albatross | 8.7 | 15.0 | 140 | 15 |
| Northern Fulmar | 0.82 | 11.4 | 73 | |
| Wilson's Storm Petrel | 0.038 | 8.0 | 19 | 11 |
| White Pelican | 8.7 | 8.5 | 85 | |
| Brown Pelican | 3.3 | 10.6 | 71 | 12 |
| Northern Gannet | 2.82 | 13.2 | 118 | |
| Common Cormorant | 2.1 | 10.0 | 103 | 14 |
| Magnificent Frigatebird | 1.49 | 14.7 | 40 | 10 |
| Grey Heron | 1.32 | 7.8 | 40 | 12 |
| White Stork | 3.5 | 7.5 | 62 | |
| Marabou Stork | 6.4 | 7.6 | 72 | 15 |
| Greater Flamingo | 2.9 | 7.5 | 87 | |
| Bewick's Swan | 6.2 | 9.2 | 147 | 20 |
| Whooper Swan | 9.1 | 13.9 | 233 | 17 |
| Barnacle Goose | 1.15 | 10.1 | 98 | 19 |
| White-fronted Goose | 1.7 | 10.8 | 92 | 15 |
| Mallard | 1.01 | 9.1 | 113 | 18 |
| Eider | 2.18 | 8.4 | 194 | 21 |
| Andean Condor | 10.2 | 7.5 | 104 | 11 |
| Hooded Vulture | 2.0 | 6.7 | 45 | |
| Rüppell's Griffon Vulture | 7.3 | 6.9 | 88 | 12 |
| Northern Goshawk | 0.91 | 7.0 | 53 | 17 |
| European Sparrowhawk | 0.19 | 6.5 | 28 | 12 |
| Common Buzzard | 1.0 | 5.8 | 33 | |
| Tawny Eagle | 2.0 | 7.1 | 44 | |
| Osprey | 1.1 | 8.9 | 39 | 13 |
| Common Kestrel | 0.2 | 7.9 | 31 | 9 |
| Black Grouse | 1.0 | 5.9 | 85 | |
| Hazel Grouse | 0.28 | 5.9 | 58 | |
| Red-legged Partridge | 0.50 | 4.6 | 76 | |
| Grey Partridge | 0.35 | 6.3 | 94 | 13 |
| Ring-necked Pheasant | 1.2 | 5.5 | 123 | 15 |
| Common Crane | 4.8 | 7.3 | 85 | 19 |
| Moorhen | 0.31 | 6.7 | 69 | |
| Great Bustard | 7.95 | 8.1 | 154 | |
| Oystercatcher | 0.42 | 9.7 | 64 | 14–16 |
| Lapwing | 0.22 | 7.6 | 28 | |

| | Body mass (kg) | Aspect ratio | Wing loading (N/m²) | Flight speed (m/s) |
|---|---|---|---|---|
| Black-tailed Godwit | 0.22 | 7.2 | 35 | |
| Common Sandpiper | 0.050 | 7.2 | 27 | |
| Dunlin | 0.045 | 8.6 | 30 | 13 |
| Avocet | 0.28 | 7.3 | 35 | |
| Arctic Skua | 0.42 | 10.6 | 40 | |
| Herring Gull | 1.0 | 10.0 | 50 | 10–11 |
| Common Tern | 0.12 | 13.2 | 25 | 9–12 |
| Common Guillemot | 1.0 | 9.2 | 174 | |
| Feral Pigeon | 0.35 | 7.2 | 55 | 13 |
| Woodpigeon | 0.46 | 6.6 | 58 | 17 |
| Budgerigar | 0.036 | 7.3 | 34 | |
| Scarlet Macaw | 0.99 | 6.6 | 48 | |
| European Cuckoo | 0.11 | 7.2 | 24 | |
| Scops Owl | 0.05 | 5.7 | 10 | |
| Barn Owl | 0.28 | 6.7 | 19 | |
| Tawny Owl | 0.50 | 5.7 | 27 | |
| Eagle Owl | 2.7 | 5.9 | 43 | |
| European Nightjar | 0.076 | 8.0 | 20 | |
| Common Swift (Roosting) | 0.042 | 10.5 | 29 | 6.5 |
| Common Swift (Migrating) | | | | 11 |
| Giant Hummingbird | 0.020 | 10.8 | 31 | |
| Allen's Hummingbird | 0.0034 | | | 11 |
| Amethyst Woodstar | 0.0023 | 8.1 | 32 | |
| Common Kingfisher | 0.32 | 6.9 | 32 | |
| Green Woodpecker | 0.16 | 5.2 | 38 | |
| Barn Swallow | 0.022 | 8.0 | 16 | 9 |
| House Martin | 0.020 | 7.9 | 15 | |
| Redwing | 0.058 | 7.7 | 35 | 8 |
| European Blackbird | 0.092 | 6.4 | 35 | |
| Great Tit | 0.020 | 4.8 | 16 | |
| Blue Tit | 0.010 | 6.8 | 17 | 8 |
| Chaffinch | 0.022 | 5.9 | 20 | 10–14 |
| House Sparrow | 0.028 | 5.5 | 26 | 8–11 |
| Large Ground Finch | 0.036 | 4.8 | 24 | |
| Goldcrest | 0.006 | 4.8 | 13 | |
| European Starling | 0.076 | 7.2 | 37 | 9–10 |
| Magpie | 0.215 | 5.7 | 35 | |
| Carrion Crow | 0.46 | 6.8 | 37 | 14 |

an area, and is therefore equivalent to a pressure. It is related to the mean pressure force over the wings, and therefore is proportional to the square root of flight speed. Birds with relatively small wings have high wing loadings, and therefore high flight speeds. Wing loading increases with size, and therefore large birds on average fly faster.

Aspect ratio is interpreted as a measure of aerodynamic efficiency. A wing of high aspect ratio has a lower induced drag, and therefore permits shallower glide angles and lower mechanical energy costs in flight. Aerodynamic performance can be improved by increasing aspect ratio by making the wings longer and thinner, as is done in modern sailplanes. However, this is not always advantageous: very long wings must be strong and heavy if they are to avoid the risk of breaking, they can be a hindrance in a cluttered environment, and moreover can limit the flexibility a bird enjoys in its mode of flight. Extreme aspect ratios are found only in albatrosses, which live in a predictable and uncluttered marine environment.

### *Gliding and soaring*

Gliding is the simplest mode of flight, and is very cheap compared to flapping flight since the flight muscles produce only static forces to keep the wings outstretched and horizontal. Physiological measurements indicate that metabolic energy consumption in gliding Herring Gulls is only 2.2 times the energy cost at rest – and this figure is likely to be representative of other birds – while flapping flight may cost 10–15 times resting. But the disadvantage of gliding is that the bird must lose height.

Vultures and birds of prey have glide ratios (speed of forward travel divided by speed of descent) between 10:1 and 15:1; the best avian glide ratio of 23:1 has been measured in the Wandering Albatross. Modern sailplanes can achieve 45:1, but such very shallow glides could be disadvantageous to birds because they can be achieved only at the expense of reduction in manoeuvrability and loss of the ability to flap the wings.

Despite the unavoidable loss of height, gliding is important to many birds. It is particularly so in soaring flight, when it is used to exploit moving air currents. Soaring is used by many large birds when searching for food, during migration, and sometimes during commuting flights. Two main types of soaring have been recognized in birds: thermal soaring makes use of rising currents of warm air (thermals), and static soaring depends on vertical movements of the air as winds are deflected by slopes or waves. So-called dynamic soaring, which depends on variations in horizontal wind speed at different heights above the sea, has often been said to be the means of flight over the sea in albatrosses, but this type of flight has never been demonstrated; albatrosses appear to use static wave soaring.

Gliding and soaring birds tend to be relatively large in size, for two reasons. First, larger birds have problems carrying sufficient fuel and musculature to meet the energy demands of continuous flapping flight, and use gliding as a means of reducing their overall energy cost. Second, for aerodynamic reasons, the optimum glide ratios attainable by many smaller birds are relatively poor, and gliding therefore gives them no overall advantage.

Thermal soaring is typical of large terrestrial birds: vultures, eagles, buzzards and storks circle in thermals to gain height, and then glide cross-country until they reach a new thermal and begin to climb again. Like all gliding birds these species have low wing loading, but they tend to have relatively short, broad wings of low aspect ratio; they also have emarginated and separated wingtip primary feathers: these features are adaptations for take-off from ground or trees and for maximizing turning and climbing performance in weak and narrow thermals, and presumably these benefits outweigh the disadvantage of low aspect ratio.

Static soaring is most evident in marine soaring birds, because in marine habitats wind is often deflected upwards by cliffs or the face of waves. Marine birds (e.g. albatrosses, frigatebirds, gulls) can adopt the long, thin pointed wings with high aspect ratio which maximize glide ratio, and also minimize power consumption in flapping. Frigatebirds are unusual among marine species in their use of thermals over the sea in the trade wind zones.

### Hovering and windhovering

Hovering permits birds to forage in places which would otherwise be difficult to reach: it is confined to relatively small birds, and for example allows small hummingbirds (mass 2–10 g) to suck nectar from flowers which are too weak to perch on. It is defined as flight with zero air speed – the bird cannot use its forward momentum to help to generate lift, and all of the airflows which support the weight must be generated by the beating wings alone – and is therefore very demanding in energy. Pigeons (300 g) can hover briefly, but the largest bird able to hover for sustained periods is the African Pied Kingfisher (100 g), which hovers over lakes when fishing well away from the shore.

Most hovering birds have relatively long wings which minimize induced power, for this is the dominant energy component. Hummingbirds, with relatively short wings, are an exception to this rule, probably because an increase in wing loading allows fast flight and the easier maintenance of feeding territories.

Hummingbirds hover with their wings fully extended during the entire wing stroke. They generate lift during both downstroke and upstroke. The stroke plane is roughly horizontal but the body axis is strongly inclined to the horizontal. The wings describe a figure of eight with symmetrical halfstrokes; the morphological downstroke is actually a forwards stroke while the upstroke is a backstroke. The wings are twisted during the backstroke and the morphological upper side faces downwards with slight inverted camber. The slight thrust produced during the forward stroke equals the slight drag produced during the backstroke. Wingbeat frequency is high, varying with size; 15 strokes per second for the Giant Hummingbird (mass 20 g) and 52 strokes per second for the Ruby-throated Hummingbird (3.7 g). The arc through which each wing moves (the stroke amplitude) is large, usually about 120°.

Windhovering, in which the bird flies into a wind, is more appropriate for larger birds. Its purpose is to permit the bird to remain stationary relative to the ground while scanning for prey, and it is found in a range of raptors including kestrels and other falcons, and some buzzards. Windhovering is also used by terns and kingfishers while fishing. Studies of kestrels have shown that they prefer to windhover in winds of strength around the minimum power speed: windhovering therefore becomes a relatively economic means of flight, but requires considerable agility and control to keep the head sufficiently still to find prey.

*Hovering hummingbird. The two extremes of wing position in hovering are shown*

### Energy-saving flight

The mechanical and energetic demands of flapping flight are such that selection has placed a premium on ways of reducing the costs of flight wherever possible. Evidence of this is seen in the overall light build of birds, and in the close correlations between wing design, behaviour and ecology. Birds also use various specialized styles of flight in which the energy cost is less than in normal level flight.

One of the best-known economic flight styles is *formation flight*, which is used, for example, by geese, swans and cranes. By flying just outside the wingtip vortices created by its neighbours, a bird in a V-formation can exploit the wake upwash generated by the vortices, and theoretically might save as much as 70% of the induced power: at cruising speed this represents about 15% of the total aerodynamic power. Similar advantages may apply to birds in flocks. Flight in groups probably has other advantages, such as reinforcing social hierarchies, and assisting young or novice birds to learn the route.

A more surprising solution to the energy problem is the use of *intermittent flight*. The flight path remains level on average, but the bird intermittently gains and loses height.

*Undulating flight* takes advantage of the low energy costs of gliding, and brief descending glides are interspersed with short bouts of flapping during which height is regained. By flying in this way many birds can reduce flight energy slightly (estimated at between 10 and 15%), since weight support is continuous while thrust is intermittent. By varying the duration of the glide and climb phases birds can also make the best use of local updraughts. This flight mode is used by many birds in a range of situations: it is most significant to heavier birds with larger wings such as pelicans, storks and cranes, for these species can glide most efficiently, and owing to their size have the greatest need to save energy.

By contrast, *bounding flight* is a much more stereotyped behaviour which is confined to smaller birds (up to about 200 g body mass) with low aspect ratio and relatively high wing loading; it is very common in passerines and in some other families such as kingfishers, parrots and woodpeckers. The flight path consists of brief bursts of flapping, alternating with a passive phase in which the wings are folded, during which the bird moves in a ballistic path. Bounding flight does not reduce mechanical energy at normal flight speeds, but reflects instead a compromise between conflicting demands of aerodynamic performance and flight muscle physiology: by varying the duration of flapping the bird controls the mean power output from the flight muscles, and obtains a crude form of 'gear'. This is very important for small passerines, for without a mechanism of this kind they could not accommodate the large changes in body mass that occur prior to migration.

A third flight style which can reduce energy costs is *ground effect*. The induced drag of a wing is greatly diminished when it is near a flat surface. Many birds habitually fly close to water, and ground effect can reduce total mechanical power by as much as one fifth for skimmers foraging a few centimetres above a water surface. Larger aquatic birds such as pelicans and swans (during take-off) benefit from ground effect, but it is not a reliable source of energy saving since a smooth water surface is essential for ground effect to function.

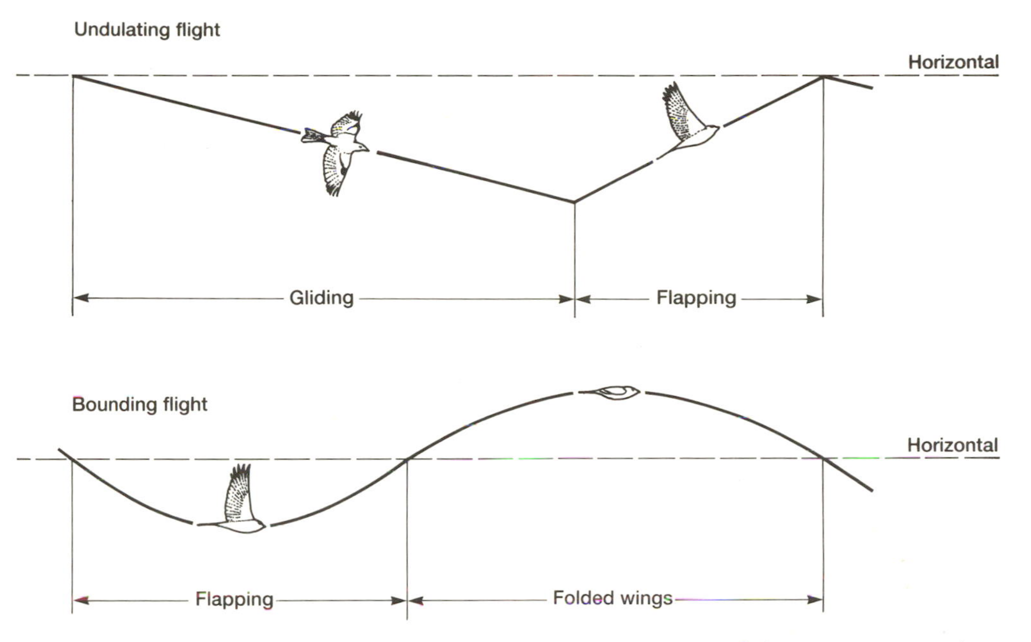


*Intermittent flight can save energy. There are two forms, undulating flight, where flapping is interspersed with gliding, and bounding flight, where flapping is interspersed with passive phases in which the bird follows a ballistic path*

*A blue tit using bounding flight, the wings folded tight against the body*

*Take-off, climbing, and landing*

Birds that take off from water and have high aspect ratio wings and a high wing loading, such as divers, mergansers, geese and swans, have to skitter along the surface to achieve the necessary speed for flight, with the feet contributing the extra vertical force needed to keep the bird out of water while acceleration is accomplished by both feet and wings. These birds have rather economical flight. Red-throated Divers skitter 15–40 m before taking off, with their feet moving synchronously with the wings (two steps, one by each foot, per wingbeat cycle). They take off at a speed of about 10 m/s and reach 18 m/s in horizontal short distance flights. By contrast grouse have very high wing loadings but low aspect ratios and expensive flight. However, their flight muscle mass is large, 22–29% of body mass, which is high compared with 7–19% for the former group. They are therefore able to take off almost vertically.

In ascending flight, work is done against gravity and the bird has to flap. Landing on the other hand is often preceded by a gliding phase. To decrease speed before landing, a bird can climb into the air to use up kinetic energy or air-brake with its wings, legs, and tail. The final touch-down may involve a stall. For steep descents many species also often side-slip and perform half-rolls, or make steep turns as divers sometimes do. Many water birds use their feet as water-brakes by extending their legs forwards with the feet angled (swans, geese) or by dragging them behind in the water (loons or divers, mergansers).

*Left. The largest birds find difficulty in accelerating to their flying speed. They commonly need to jump off a tall structure to benefit from gravity's pull, or run, in the case of the Mute Swan (Cygnus olor) across water*

*Above. A vertical leap gives smaller birds like the Sage Grouse (Centrocercus urophasianus) the airspeed needed for flight*

*Manoeuvrability and control of movements*
Catching prey, avoiding obstacles and taking off and landing require the ability to increase power output and the forces generated on the wings for brief periods to control the flight path. Most types of manoeuvre involve turning: in a turn the bird must direct some of the lift force laterally to prevent sideslip, and must therefore bank its wings. This is one reason why birds foraging among vegetation should have relatively short, broad wings of low aspect ratio: the short wing span avoids collisions with obstacles, while the wing area is maintained to permit slow flight and small turning radius. A much longer, thinner wing is appropriate for aerial hunting birds (insectivores such as swifts and swallows, and predators such as falcons, skuas and frigatebirds), for this enables them to turn rapidly without losing height. With high aspect ratio, agility is improved by reducing wing loading, making the wings narrow and pointed.

Flying birds need high stability but also rapid and sensitive control of wing movements: these are to some extent conflicting demands, but their exceptional sensory performance and ability to manoeuvre rapidly mean that stability in level flight is less important than it would be for an aircraft, for instance. Pitch, rotation about the bird's transverse axis, can best be controlled by a long, sturdy, dorsoventrally flattened tail, with help from fore-and-aft movements of the wings relative to the bird's centre of gravity. Roll, rotation about the bird's longitudinal axis, leads to sideslip which can be stabilized by sweep-back or by dihedral (a V-attitude) of the wings. Roll can also be controlled by differential twisting of the wings, and a quick turn (high rate of roll) can be produced by partial retraction of one wing giving a lift difference between the two wings. Yaw, rotation about the bird's vertical axis, is controlled by the tail or by twisting and flexing of the wings to change their local lift to drag ratio.

## Limits to size in flying birds

The size range of flying birds spans four orders of magnitude, from the smallest hummingbirds with mass about 1.5 g, to a maximum around 15 kg in a number of different groups, including bustards, pelicans, swans, the California Condor, albatrosses and vultures. Although these are all relatively large birds, with a wingspan above 2 m, their mass is relatively modest compared with that of mammals, being little heavier than the average domestic dog. Flightless birds are much heavier: penguins reach 35 kg, and Ostriches 150 kg, while the extinct Elephant Birds of Madagascar and the Moas of New Zealand may have been as heavy as 400 kg.

The limits to size are set by a complex interacting mosaic of mechanical, ecological and physiological constraints. The lower size is constrained primarily by physiology: small warm-blooded animals seem to find it harder to maintain heat and water in balance during high activity, and the heart may be unable to contract sufficiently fast to supply enough oxygen in the blood to the flight muscles. There is no reason to expect the presence of a mechanical lower limit to mass in birds: insects of considerably smaller size are able to fly using broadly similar aerodynamic mechanisms. A number of factors conspire to set an upper limit. As we have seen, the metabolic power consumed in flight increases with body mass. The power available from the flight muscles also increases with mass, but because wingbeat frequency decreases in larger birds, the margin between power needed to fly and power available becomes narrower. Additionally, at larger sizes the risk of damage to muscles, tendons and bones becomes appreciable. There is no sudden limit above which powered flight is impossible, but large birds are more likely to adopt economic or less strenuous modes of flight such as gliding and soaring, formation flight, or to use ground effect. There are also ecological factors which limit size in modern flying birds, although these can only be hinted at: many now extinct birds were considerably heavier than 15 kg, yet fossils indicate they had the morphological specializations associated with flight. The huge Pleistocene condor *Teratornis* from Argentina is estimated to have reached a mass of at least 40 kg, and yet flew like its contemporary relatives in the New World vultures (Cathartidae). We must suppose that

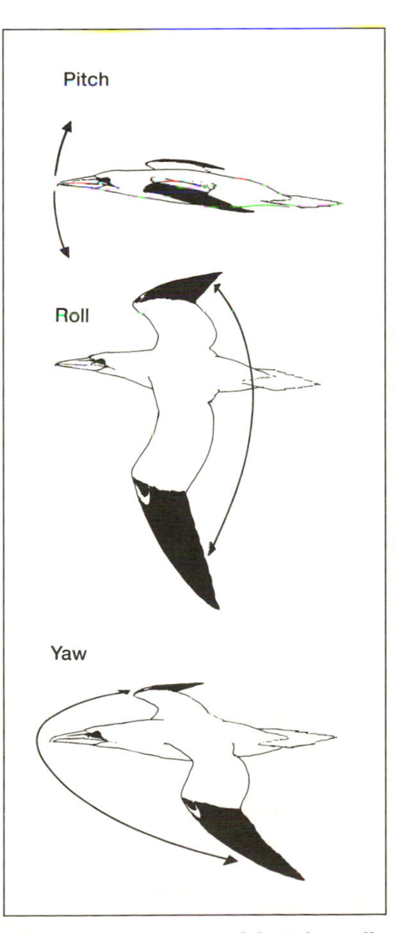


*Flying birds maintain stability about all three axes and so avoid pitching, rolling, and yawing from side to side*

although the mechanical and physiological problems involved in flight in relatively heavy birds can be solved, the ecological constraints of finding roost or nest site and guaranteeing sufficient food are insoluble in today's environment.

## Flightlessness

Flight conveys many advantages to birds; probably the most important are the abilities to forage over a wide area, to migrate over long distances, and to escape terrestrial predators. But flight is also very demanding energetically. Thus in situations when it can give few benefits a bird may find it preferable to dispense with aerial locomotion altogether. Many types of bird do not fly for some portions of the year: for instance, ducks have small wings so that they may fly fast, but when they moult the wing area is reduced so much that flight is impossible for some weeks. True flightlessness, in which all capacity for flight is lost, and usually the wings are reduced, is remarkably common in birds from a wide range of different taxa: the Galápagos Cormorant and the Kakapo (a parrot) from New Zealand are examples of birds with close relatives which can fly, but which do not use flight for foraging; in each case a relatively sedentary species has occupied habitats free of predators and this has rendered flight redundant. This trend is particularly marked in rails of the order Gruiformes, in which a number of island species no longer fly. Presumably similar trends were associated with the loss of flight in the ancestors of Ostriches and other ratites. Once flight is lost there is no constraint preventing increase in size.

Pressure for size increase is also associated with flightlessness in wing-propelled diving birds. A number of species, of which the best known are the auks, use their wings both for flight in air and for swimming on or under water. This compromise is feasible only over a limited size range, and the largest auks gave up flight as they increased size. Similar trends also occurred in the ancestors of penguins, in which flight has been completely lost, and in some steamer ducks in the Falkland Islands.

UMN and JMVR

## Terrestrial locomotion

Humans walk to go slowly and run to go faster. Similarly, many birds change gait as they increase speed. Their normal slow gait is a walk in which (as in human walking) they move their feet alternately, setting each down before the other is lifted. The fast gait may be a run or a hop, and some small birds use hopping as their only gait. In running, the feet move alternately (as in walking) but each is set down after the other has been lifted: there are stages in the stride when both feet are off the ground simultaneously. Gamebirds (Galliformes) and other mainly ground-living birds use running as their fast gait. In hopping, the feet move together and are set down side by side. This gait is characteristic of small birds that spend much of their time in trees, such as finches. Another fast gait, the asymmetrical hop, is intermediate between running and normal (symmetrical) hopping. The feet are not set down alternately at equal intervals as in running, nor are they set down precisely side by side: instead, one foot is set down slightly after the other. Many corvids (members of the crow family) do this, for example magpies.

In one of the few measurements of maximum running speeds of birds, Bobwhite quail were found to be able to run for 20 minutes at 0.7 metres per second and to sprint for 30 seconds at 2 metres per second. Flying birds generally seem to be rather slow runners, compared to similar-sized mammals, presumably because

*Standing 2.5 m tall and weighing up to 150 kg, Ostriches (Struthio camelus) can attain running speeds of 60 km per hour*

they have a smaller proportion of leg muscle in their bodies (in consequence of having big wing muscles. However, Ostriches have large leg muscles (and rudimentary wing muscles) and are exceedingly fast. A biologist drove alongside a running ostrich on the Mara plains in Kenya with the speedometer reading 60 kilometres per hour (17 metres per second or 38 mph), which is a speed fast enough to win most horse races.

Physiologists have measured the energy cost of walking and running, by measuring how much oxygen animals extract from the air. The cost is generally about the same for birds and mammals of equal weight, travelling at the same speed, but is unusually high for geese and penguins, which have a waddling style of walking.

Many birds bob their heads backwards and forwards as they run. A hint about the possible function of this behaviour comes from observations of pigeons which were trained to walk on a moving belt so as to remain stationary relative to the laboratory. They bobbed their heads when walking on firm ground, but not on the moving belt. Head bobbing keeps the eyes stationary relative to the ground for part of the stride and may make it easier for the bird to detect moving objects.

RMcNA

## Methods of swimming and diving

Around 390 species of birds in 9 orders swim regularly, either swimming on the water's surface, or diving from the surface to swim underwater or plunging into the water from a height. Most aquatic birds such as ducks float on the surface and row or paddle with their feet. Drag, the resistance to forward movement, changes very little up to a certain swimming speed (approximately 0.5 metres per second for Tufted Ducks), but then increases in an exponential fashion with increased speed. Thus, any increase in speed above 0.5 metres per second will require a large increase in energy expenditure and it is unlikely that water birds exceed their optimum swimming speed for more than a few seconds. Ducks also paddle with their feet when diving or swimming under water. Eiders and Surf Scoters use their wings as well as their feet, but auks and penguins use their wings alone. For ducks and auks, the adaptations for aquatic locomotion have compromised their ability to walk and fly respectively. Penguins, of course, have lost the ability to fly altogether. Their wings have developed as flippers and are used as hydrofoils: penguins fly underwater. They are well streamlined and can swim under water at very high speeds. King Penguins have been filmed in a large aquarium swimming at 3.4 metres per second.

Although penguins can float and swim at the surface, they undertake long sea journeys under water where the drag on the body is less than at the surface. Periodically they have to surface to ventilate their lungs. As they approach the surface the drag on the body increases and it has been suggested that leaping clear of the water (porpoising) is energetically less costly than surfacing although this may only be so at very high speeds (more than 2.5 metres per second for the Adélie Penguin).

Like many other species of bird, penguins dive for food and may spend a large proportion of their time under water. As the table overleaf indicates, the duration of individual dives may vary from a mean of approximately 20 seconds in Tufted Ducks diving to a depth of 2 metres, to approximately 5 minutes in Emperor Penguins. The longest and deepest recorded dives for Emperor Penguins measure, remarkably, over 15 minutes and 265 metres respectively.

*Swimming penguins must frequently surface to breathe. However, the reason why they periodically porpoise or leap clear of the water is uncertain. The tactic may reduce the energetic cost of swimming for these Adélie Penguins (Pygoscelis adeliae) approaching the Antarctic ice*

*Dive durations and depths of some naturally diving birds*

| Species | Mean duration (seconds) | Maximum duration (seconds) | Maximum depth (metres) |
|---|---|---|---|
| Tufted Duck | 20 | 46 | 6 |
| Western Grebe | 30 | 63 | — |
| Great Northern Diver | 47 | 62 | 60 |
| Imperial Cormorant | 68 | 95 | 25 |
| Guillemot | 77 | 202 | 180 |
| Gentoo Penguin | 128 | 189 | 100 |
| Emperor Penguin | 300 | 900 | 265 |

During the majority of dives, birds use oxygen stored in the body at a rate similar to that when they are exercising at a similar level in air. This oxygen is then replaced and $CO_2$ expelled during the short period at the surface and the bird dives again. When Tufted Ducks dive they have to overcome their positive buoyancy and paddle hard both to reach and to remain at the bottom. Oxygen is consumed at a rate similar to that when they are swimming at the surface at maximum sustainable velocity. Penguins, on the other hand, are close to neutral buoyancy and more efficient at underwater locomotion than ducks. Their oxygen consumption during diving may be little higher than resting values. The oxygen storage capacity of the body in aquatic birds is greater than in their terrestrial relatives, owing to greater blood volume and a larger respiratory system.

On the few occasions when a bird remains submerged for much longer than usual (perhaps when foraging under ice), or is unexpectedly and temporarily unable to surface (perhaps if disoriented under ice), there is a noticeable reduction in heart rate which indicates that blood flow, and therefore oxygen supply, to most parts of the body (maybe even to the active locomotory muscles) is reduced. The brain and heart may be the only parts of the body receiving sufficient oxygen. The rest of the animal may have to produce energy in the absence of sufficient oxygen, which means that lactic acid is produced. After such dives, when this emergency 'oxygen conserving' response is invoked, the animals usually remain at the surface to recover for a considerable period before the next dive.

PJB

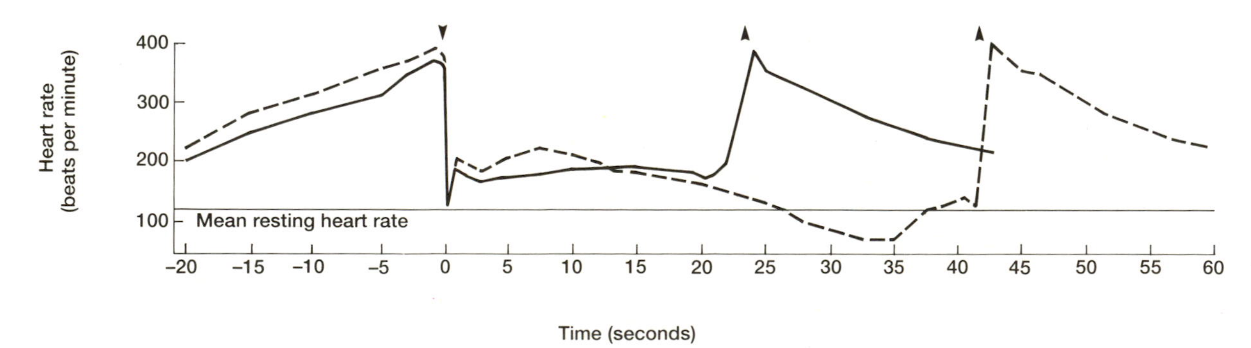


*Average heart rate before, during and after normal vertical feeding dives (continuous line) and during long dives under a pond covered at the surface to simulate ice (dashed line). The downward arrow indicates the point of submersion and the upward arrows points of surfacing. During normal feeding dives, heart rate remains substantially above resting heart rate, whereas towards the end of the longer dives it falls*

# 4 BIRDS, ANCIENT AND MODERN

## Fossil birds

### Archaeopteryx

Students of the history of birds are fortunate to have the beginning of bird evolution revealed by fossil remains. One of the few certainties about the origin of birds is that they evolved from reptiles; and the seven almost complete specimens of *Archaeopteryx* from the Lithographic limestone of the Late Jurassic period in Bavaria found over the last century or so provide a linking form with a mixture of reptilian and avian characteristics. These specimens provide the only clues to the probable type of reptile that gave rise to birds: an event that occurred in the Jurassic (190–136 million years ago) or possibly as early as the Triassic (225–190 million years ago).

Birds appear to be derived from archosaurs, one of the dominant reptile groups of the Triassic, and the one that gave rise to the dinosaurs, pterosaurs and crocodiles. Early suggestions of a proto-avian ancestor centred around arboreal, lizard-like forms. Later it was suggested that there were striking similarities of shape between *Archaeopteryx* and some of the small bipedal dinosaurs.

The most likely ancestors of birds appear to be a branch of the saurischian dinosaurs, the coelurosaurs. These were carnivorous and usually bipedal, with a fairly upright posture and with forelimbs that were not used in ordinary locomotion and which could have become modified for other purposes. The coelurosaurs occurred from the Triassic to the Cretaceous, and produced a number of medium-sized to small forms, some little larger than the domestic fowl.

Uncertainty concerning avian origin and affinities still persists. The full range of smaller reptiles of the Triassic and Jurassic is incompletely known; and birds display some specializations of structure that do not allow a ready association with specific reptilian taxa.

It was fortunate for avian systematists that the discovery of *Archaeopteryx* in 1861 provided a base-line from which the subsequent evolution and adaptive radiation of birds could be deduced. The discovery was also of value for its evidence of one class of vertebrates evolving from another. Some dissent came from Christian creationists who had accepted biblical statements on animal origins as absolute and unquestionable, and therefore found the existence of such an animal troublesome.

However, a new and unexpected attack on the validity of *Archaeopteryx* has been mounted recently. The astronomer, Fred Hoyle, and the mathematician, N. C. Wickramasinghe, have proposed a 'star-dust' hypothesis for the initiation of major evolutionary changes, claiming that the latter occurred following entry of cosmic dust into the earth's atmosphere. For the purposes of their hypothesis they wished the evolution of birds to have occurred at the end of the Cretaceous period. They appeared, however, unaware of the considerable range of avian fossil material from the Cretaceous period, and in addition seemed at first to assume that there was only a single specimen of *Archaeopteryx*, and that by discrediting this they could justify their views.

*Archaeopteryx, the earliest known bird, is preserved in sedimentary limestone of the Early Cretaceous deposited in southern Germany around 140 million years ago*

The seven known individuals of *Archaeopteryx* are preserved in the soft sediments of swamps in what is now Bavaria. These sedimentary limestone deposits were used in the last century for lithographic stone. When the slabs were carefully split for this purpose some cleaved along the plane of deposition, revealing not only the almost complete skeleton of one of these animals, but also the impressions of flight feathers like those of recent birds, forming complete wings and also bordering the tail. These can be best seen on the most complete skeletons, one in London, the other in Berlin, collected in 1861 and 1877 respectively.

Ignoring the avian characters of the skeletons, the cosmic dust theorists claimed that these were just reptiles, that the feather impressions had been added later deliberately to deceive, and that there had been subsequent collusion by museum scientists to maintain this deceit until the present day. In a manner surprising in the scientifically trained, they have ignored or brushed aside contrary evidence in pursuing and publishing their thesis.

The following points are particularly relevant. The first discovery of a bird from this Jurassic locality in Bavaria was an impression of a single typical feather. As stated above, there are now seven *Archaeopteryx* specimens; they were found by different individuals and are located in various institutions. In some cases these institutions have fossils that were not recognized as *Archaeopteryx* until much later, but all have evidence of typical feathering. Feathers and bones of other birds have been found in rocks of the Early Cretaceous, 136–100 million years ago, as far apart as Spain and Australia.

Although the cosmic dust theorists have dismissed the flightless aquatic toothed birds of the Late Cretaceous as swimming reptiles, there appears to be no way in which one can invalidate the co-extant ichthyornithids and hesperornithids. *Ichthyornis* species had a skeleton in which the general structure of wings, pelvic girdle, deeply-keeled breastbone, and legs was almost indistinguishable from that in recent birds. In addition, some bones found in the Late Cretaceous appear referable to the subclass that contains the living birds. The hypothesis that birds, as a class, evolved suddenly after the end of the Cretaceous appears not only untenable but absurd.

*Archaeopteryx* shows a mixture of reptilian and avian characters. It was about 40 cm long, about a third of this being tail. It resembles reptiles in that the front of the skull is a tapering snout rather than a bird-bill, and has small teeth; there are many ribs; the forearm digits are not fused as in a bird's wing and have claws; the pelvis is small and the tail long.

Avian characters include metatarsal bones of the lower leg that are fused to some degree, resulting in a leg structure like that of recent birds. In addition, the toes are reduced to the typical three pointing forward, with a small, short opposing hind toe. The pectoral girdle has a fused furculum or wishbone, larger and stouter than in recent birds. The scapulae lie on the back, near the spine, and their forward ends are attached to large broad coracoids. It has been argued that there are adequate attachments here for flight muscles, but in addition a small bone has recently been identified as the missing sternum or breastbone, short and narrow but showing some flanges for muscle attachment. Although the tail is long it appears inflexible and seems to have functioned as a single long steering rod, bordered on either side with a row of feathers that meet at the tail and create an elongated homologue to the modern avian tail.

The most important bird-like feature is the evidence of feathers; and some systematists claim that it is only the possession of feathers that wholly separates birds from other groups of animals.

## Origin of flight

Besides demonstrating the origin of birds from reptiles, *Archaeopteryx* also provides the earliest information on the evolution of flight in birds, which in turn is linked with the possession of feathers. Typically, feathers are highly specialized skin structures that can be regarded as modified scales but are much more complex in structure. Their primary use in birds generally appears to be as an external and erectile

*A reconstruction of Archaeopteryx, showing its characteristic mixture of reptilian and avian features*

covering which allows both effective insulation and temperature regulation.

In this form they occur as a series of small, flexible overlapping structures known as contour feathers, that cover the whole body except for legs and feet, bill and eyes; all but the last of these having a horny sheath. Contour feathers probably evolved as a necessary adjunct to warmbloodedness; the alternative being the fur that occurs on mammals and also appears to have been present on some of the later pterosaurs.

In recent birds it is apparent that contour feathers are modified in shape and size on different parts of the body in various species in order to function more effectively. Since feathers can become enlarged and specialized in function they might be expected to do so on the edges of body structures such as limbs if this conferred some functional advantage; as shown by wing and tail feathers of *Archaeopteryx*.

There has been a suggestion that the original function of wing-feathers in *Archaeopteryx* was to give a greater surface for catching insects with the forearms by clapping them together. Apart from any other functional objections this ignores a very important point. Feathers are fine structures in constant danger of abrasion and there is nothing to suggest that those of *Archaeopteryx* differed in this respect. They would be unlikely to last long if used in this way. Another point not taken into account in this hypothesis is the presence of similar feathers bordering the tail where they could hardly have been used for flycatching. In addition, the vane asymmetry and curvature of *Archaeopteryx's* flight feathers would have greatly aided controlled flight, and so provide further evidence that *Archaeopteryx* could fly.

Even if *Archaeopteryx* was flightless, the presence of these large feathers on wings and tail would have had an important function during movement. If raised and spread they would have imparted a general lift to the body, particularly if the creature was running bipedally on a fairly level surface.

Although the presence of the wing and tail feathers of *Archaeopteryx* or any other very early bird could be explained by linking them with movement of some kind, a stronger reason is needed for the evolution of flight. General convenience in moving around would not seem to provide sufficiently strong selective pressures for the evolution of flight. Instead, since recent birds provide evidence of a rapid loss of the ability to fly when living in places where predators are absent (e.g. oceanic islands), one can reasonably argue that birds primarily evolved and retained flight in order to evade or avoid predators.

It was at first suggested that *Archaeopteryx* originated from a tree-climbing, lizard-like reptile that evolved extensions from the limbs in order to glide from tree to tree. There are present-day examples of this habit in various animal groups, including amphibians, reptiles and mammals, usually involving flanges or wing-like extensions of the skin. The latter usually extend to the hindlimbs.

In contrast, *Archaeopteryx* has such structures limited to forelimbs and tail, while the hindlimbs are free of encumbrance and appear adapted for running, with fused metatarsals and forward-pointing toes. The opposing hind-toe does not appear well-developed for grasping or perching.

There are still supporters of the view that *Archaeopteryx* represents an evolutionary stage at which flight still required a launch from an elevated point in a tree; and recently this has been coupled with assertions that the sharp claws of the wings must have been used for tree-climbing.

However, it could be argued that wing-claws are more likely to be long and sharp through disuse than from constant use; and structures that become non-functional during evolution may persist for some time if their presence is not detrimental. Were they functional, then by analogy with recent birds it is possible that wing-claws could have been used for fighting, or nest-defence, or used for clambering by flightless young as are the secondarily-evolved wing-claws of the Hoatzin or some rails. In addition, a habit of climbing would have led to modifications of muscles and bones of forelimbs that would probably have made their use in flight less likely.

More cogently, the need to protect feathers from abrasion and damage would make climbing rather unlikely. Much of the behaviour and posture of birds has evolved to minimize plumage damage. Even so, feathers become abraded and are constantly replaced during the lifetime of birds. The single feather that was the first fossil evidence of *Archaeopteryx* is as likely to have been moulted as to have been lost at death.

The evolutionary stem of which *Archaeopteryx* was a part could have evolved flight in response to predators, beginning as terrestrial, running animals. The ability to launch oneself and travel through the air, however poorly, is an effective way to baffle pursuers, as various animals from grasshoppers to flying fishes can demonstrate. Since the technique is increasingly effective as greater distance is placed between pursuer and prey, there is strong selective pressure for travelling the greatest possible distance and this might provide the impetus for the evolution of more efficient flight.

The skeleton of *Archaeopteryx* has the bone-structure necessary for flight. The exceptionally large furcula and coracoids could have compensated to some extent for the very small sternum. In the wing itself the main difference from modern birds was the absence of fusion of digits to provide a more rigid structure. The feather impressions of the fossils show the typical division between primary and secondary flight feathers, and the asymmetry of feather shape characteristic of the wings of recent birds. The tail would have provided additional lift and functioned as a rudder in steering.

The skeleton of *Archaeopteryx* is, as mentioned, a mosaic of reptilian and early avian characters. Although the latter make it the earliest-known bird, some of the specialized, derived characters that it shows appear to differ from those of later birds, notably aspects of the structure of the skull and fusion of the leg bones. *Archaeopteryx* is likely to have been a divergent side-branch of the main evolutionary stem, and is usually separated in the subclass of 'ancient birds', the Archaeornithes.

## Other fossil birds

For a long time there was a considerable gap, of about 25 million years, in the evidence of evolutionary development between the Archaeornithes of the Late Jurassic and the more typically structured birds to which later groups belong and that began to appear in the Early Cretaceous. However, in recent years an increasing number of specimens have been discovered that appear to relate to a group of bird species that are less derived, and more generalized or primitive in structural characters than the more modern birds.

These finds help to fill this gap in continuity of avian evolution, but do not merely show intermediate stages in development. They had evolved their own peculiarities. To separate them from the more typical birds with which they were to some degree co-extant, they are assigned to the subclass of 'other birds', the Enantiornithes.

At present there is no evidence of the existence of the Enantiornithes prior to the Early Cretaceous, but this might be the chance result of geological deposition. Specimens have been assigned to the subclass from the Early Cretaceous of England, where it is represented by *Wyleyia* of Sussex, and from Spain and possibly eastern Australia. In the Late Cretaceous it includes *Gobipteryx* of Mongolia, *Alexornis* of Mexico and a group of species from Argentina on which the subclass was originally based.

Skeletal material of the Enantiornithes is more like that of recent birds than the bones of *Archaeopteryx*. The metatarsal bones of the hindlimbs and the digits of the wings are more typically bird-like but still less completely fused than in modern birds. The upper part of the coracoid is simpler, less highly adapted to provide attachment surfaces, and the humeral head is less well-developed. The structures suggest a simpler form of flight, perhaps differing from that of recent birds.

The leg bones of the Argentine material indicate considerable adaptation for different locomotion in birds that had wing-spans ranging from about one-third to one metre. They include long slender bones of a kind now typical of running landbirds. Unlike those of recent birds these leg bones show some indication of the three digits usually fused to form the modern tarsometatarsus. Some resemble the lower limb bones of present-day penguins, short and very broad, but with the three bones partly separated. The most extreme are short, broad tarsometatarsi in which the outer of the three parallel bones of which it is composed has been reduced to a thin splint while the inner is enormously enlarged, additional bones showing that the foot is a flat structure dominated by one huge toe with an equally large claw.

During both Early and Late Cretaceous two other evolutionary lines appear to have been present, co-extant with the Enantiornithes. One ended in the Cretaceous, the other persisted as a source of present-day birds. The former was the subclass of 'toothed birds', the Odontornithes; the latter was the recent birds or 'new birds', the Neornithes.

The Odontornithes appear to have been seabirds and there is fossil evidence of at least two divergent groups. In their skeletal structure they closely resembled recent birds, to a point where some systematists have regarded them as specialized early representatives of the latter. However, they show more generalized characters in a number of minor aspects of body structure. The difference that most clearly sets them apart from more recent birds is the presence of small pointed teeth in sockets along the jaws.

The more typical of these birds were the *Ichthyornis* and *Apatornis* species, known only from the Late Cretaceous of North America. They are somewhat similar to modern seabirds such as terns or puffins in the general proportions of their skeletons; with rather short legs, and largish heads with long, tapering bills. The sternum has a deep keel for flight-muscle attachment like that of recent birds. They were obviously capable of sustained flight, although a flatter structure of the upper wing bone, the humerus, indicates that flight may have differed to some degree from that of recent birds.

The other forms of toothed birds are adapted for underwater swimming, and resemble the present-day divers or loons in their general morphology. The body is longer and streamlined, the legs are strong and, as in divers, set well back towards the rear of the body for propulsion, with long toes that would have been lobed or webbed. The neck is long and thin, and the bill tapering and narrow. They would obviously have hunted live prey.

*A reconstruction of Hesperornis, a toothed bird of the Late Cretaceous. Hesperornis was flightless and probably ate fish*

The earliest forms of these toothed swimmers are *Enaliornis* species from the Early Cretaceous of England. In the Late Cretaceous they were replaced by similar species, *Baptornis* in North America and *Neogaeornis* in Chile. These were up to about one metre long, and had become further specialized to a point where the wings were reduced to a few small slender bones too small to be used for flight, and the sternum had lost its keel.

*Hesperornis* and *Coniornis* species of the Late Cretaceous of North America were similar in shape to these others. *Hesperornis regalis* was the largest, reaching a length of about 1.7 m. In these genera the wings had been reduced still further to a single vestigial bone (a reduced humerus) and could not have been apparent as more than a tuft of feathers. All these species adapted for subaquatic movement would have been clumsy or helpless when moving about on land, and must have nested by the water's edge. None appears to have survived into the Tertiary period.

## The origin of modern birds

Inadequate fossil evidence creates considerable uncertainty concerning the period of origin of the recent birds, the Neornithes. Fossils provide only a scattering of bones or bone fragments, from various periods of the Cretaceous, not certainly identified and tentatively assigned to waterbirds such as sulids or flamingos. More material is available from the Late Cretaceous in north-east USA; and has been re-identified recently as early forms of charadrine shorebirds, together with an extinct taxon showing similarities and presumed affinities with both pelecaniform and procellariiform birds. The specimens indicate that birds of this subclass were present, and had already begun to differentiate before the end of the Cretaceous.

There are two alternative hypotheses concerning the evolution of recent birds within the Cretaceous and Palaeocene periods. The first hypothesis is based on rates of evolution, and of genetic change as shown by DNA–DNA hybridization studies. The hypothesis appears to fit such information as is already available. It suggests that the earliest divisions within recent birds began over 100 million years ago in the Cretaceous Period.

It is suggested that the principal adaptive radiation that produced the main orders of recent birds occurred in the mid-Cretaceous. The earliest branching is likely to have been that of the ratites, and of the waterfowl and the gamebirds, diverging from other recent birds. The divisions that produced the subsequent non-passerine families occurred for the most part during the latter half of the Cretaceous but with some probably delayed until the earliest

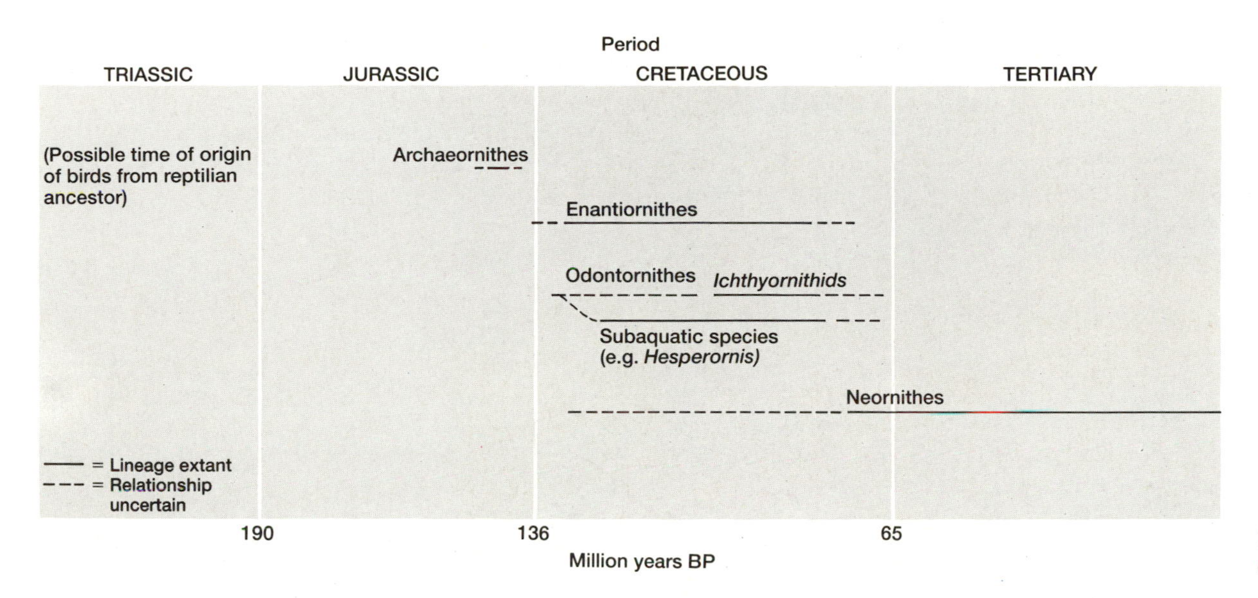


*The periods at which the ancestral groups of birds appear in the fossil record*

part of the Tertiary – the Palaeocene or Eocene.

Towards the end of the Cretaceous the dinosaurs, pterosaurs, most of the larger reptiles with the exception of crocodiles and turtles, and also many of the larger marine invertebrates, died out. Various explanations have been put forward for this, ranging from rapid climatic changes to cosmic disasters. Present information suggests that the main types of present-day birds came through this period without loss. The disappearance of the Enantiornithes and Odontornithes might have been due to competition with the results of an increasing adaptive radiation of recent birds rather than environmental factors. If some major environmental disaster did occur at the end of the Cretaceous period, then the survival of birds and mammals suggests that for terrestrial creatures living in the open the ability to regulate body temperature may have been an important factor in survival.

The alternative evolutionary hypothesis for recent birds proposes that it was not until the Odontornithes and Enantiornithes and the other groups of animals had died out at the end of the Cretaceous that the recent birds were able to diversify into their greatest variety of forms.

Even with the first hypothesis it can also be suggested that, as in the mammals, the loss of Cretaceous animal groups might have made possible the rapid evolution of birds at, or just after, this period. Certainly, even by the early Tertiary most of the major divisions of our present avifauna appear to have produced representatives.

In addition, a number of taxa that are now extinct, such as the bony-toothed birds and the giant Diatrymids, were also present. In the details of their bone structure some of these showed a mixture of characters present in other taxa, suggesting that they may have been linking forms, now lost, that might have filled gaps between some of our more disparate avian orders and families.

In general the Tertiary avifauna appears to have been similar to that of the present-day, but some now-extinct groups evolved, sometimes serially, to occupy major ecological niches. The seas offered opportunity, and giant, gliding bony-toothed birds, the Osteodontornithidae, were present until the Pliocene. The Pelecaniformes, and later the auks, produced penguin-like underwater hunters. Huge cursorial, flightless birds occurred in a number of epochs – in Europe and the Americas, Diatrymids in the Paleocene, Phorusrachids in the Oligocene to Pleistocene; in Australia, Dromornithids in the Miocene to Pleistocene; in Malagasy and Africa, Elephant Birds, Aepyornithids, from ?Miocene to recent; and in New Zealand, Moas from Pleistocene or earlier to recent times.

Although most non-passerine families seem to have differentiated by the early Tertiary, the songbirds were fairly late in their big adaptive radiation. In the earlier Eocene only a few were present, and at first the vacant niches for small insect- and seed-eating birds appear to have been filled by species of non-passerine families, smaller than any existing in those families at present.

The main radiation of passerines producing the present array of families seems to have occurred later in the Eocene, 36–45 million years ago. Another active period of adaptative radiation which may have produced many of our present genera of both non-passerine and passerine birds was in the Miocene at around 20 million years ago. Following this, the avifauna of the world probably decreased during the climatic fluctuations of the Late Pliocene and the Pleistocene, although isolated island populations and populations temporarily cut off by climatic change showed bursts of speciation.

CJOH

*Weighing up to 450 kg, the extinct elephant birds of Madagascar (including Pachyornis elephantopus, shown here) were among the largest birds ever to walk the earth. Aepyornis persisted until comparatively recently. Egg-shell fragments only 1000 years old have been radiocarbon dated*

## How orders of birds are distinguished

Having described fossil birds we now turn to modern birds. The primary aim of the discussion will be to give a sketch of the physical characteristics and habits of the various groups of birds. The division of birds into these groups is not arbitrary. It results from the mighty labours of taxonomists whose usual aim is to classify together birds with a shared evolutionary ancestry.

The starting point for many taxonomic endeavours is also the most familiar, the species. Designating species is a problem that is more complex than it appears at first sight. These problems will be surveyed in the section 'What is a species?'.

Under the system of nomenclature invented by the Swede Carl von Linné (Linnaeus: see p. 294) every species has a generic name and a specific epithet. Thus the Mallard belongs to the genus *Anas* and bears the specific epithet *platyrhynchos* which, incidentally, describes its flat duck bill. Although there may be many species in a single genus, and species in different genera may have the same specific epithet, no two zoological species may have the same generic and specific names. The species category has a biological definition (see below), but all higher categories in the Linnaean hierarchical system, including genus, family, order, class, and so forth, are arbitrary designations of genealogical relationship. The orders of birds, however, each constitute lineages rather than aggregates of unrelated species; but any order (or other taxonomic category) of birds is not necessarily the same age as any other order.

Most species are placed within this Linnaean hierarchical framework, but if their relationship to a particular group is uncertain, they are designated in the classificatory scheme as *incertae sedis* (of uncertain taxonomic status).

The world's 8800 living bird species are classified in approximately 27 orders. This classification has persisted in surprisingly stable form since the end of the nineteenth

**How the taxonomic hierarchy applies to the Green-headed Tanager**

| | |
|---|---|
| Kingdom | *Animalia (Animal)* |
| Phylum | *Chordata* |
| Sub-phylum | *Vertebrata (Vertebrates)* |
| Class | *Aves (Birds)* |
| Sub-class | *Neornithes* |
| Infra-class | *Neoaves* |
| Order | *Passeriformes (Passerines)* |
| Sub-order | *Passeri (Oscines)* |
| Family | *Fringillidae (Finches and relatives)* |
| Sub-family | *Emberizinae* |
| Tribe | *Thraupini* |
| Genus | *Tangara* |
| Sub-genus | — |
| Species | *seledon* |
| Sub-species | — |

*Above. The Green-headed Tanager (Tangara seledon) is endemic to the woodland of south-east Brazil and north-east Argentina*

century, and is based mostly on the work of two German anatomists, Maximilian Fürbringer and Hans Gadow. These scientists attempted to gather information from as many anatomical systems as possible using a large sample of avian groups.

The quality of optical equipment available at the time did not permit the detail of description possible today, but despite these limits the work remains highly useful. The Fürbringer–Gadow system employs about 40 anatomical characters, which are of varying degrees of utility in identifying genealogical relationships among birds. Although the anatomical variation exhibited by these characters is often continuous, the variants were usually designated as discrete types. 'Diagnoses' of the orders (enumerations of the main features that characterized the group) could then conveniently refer to these anatomical types.

In this way, the structure of the palate originally received four designations: dromaeognathous, desmognathous, schizognathous, and aegithognathous, depending on the size, shape, and orientation of the constituent bones of the palate. More types were added to this scheme by subsequent authors, as finer levels of detail were discovered. The result is that palatal structure is now one of the primary bases for the division between ratites such as the Ostrich and the rheas, which are dromaeognathus, and other birds. The other indicator is the presence or absence of a sternal keel.

The presence of various thigh muscles formed the basis for a classification by Alfred Henry Garrod that was incorporated into the Fürbringer–Gadow system. These muscles included the caudofemoralis, iliofemoralis, and flexor cruris lateralis (these are the names currently used for these muscles, not the ones used by Garrod). Later workers added other hindlimb muscles to Garrod's scheme. Shared presence or absence of these muscles by different groups of birds were thought to reflect degree of relationship.

The arrangement of the tendons of the muscles flexing the toes was also considered to be of importance in identifying higher-level relationships among birds. These tendons exhibit different patterns of fusion and insertion among birds.

Other anatomical features that have been used to designate ordinal status include the aspects of pterylosis (feather arrangements),

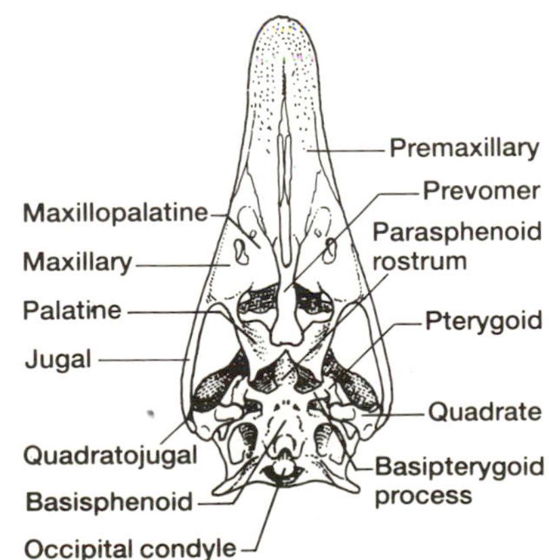


*Darwin's champion, T.H. Huxley (1825–95), divided birds into four groups with different palate structures. In the dromaeognathous condition (above), characteristic of tinamous, the prevomers extend far back*

presence or absence of an oil gland, shape of numerous elements of the skeletal system, morphology of the muscles and cartilaginous elements of the syrinx (vocal apparatus), intestinal convolutions, and arrangement of the carotid arteries.

The primary potential flaw in this system is that it is based on similarity without regard to the hierarchical level of that similarity. Similarity of form may be of three kinds: (1) it may be independently derived in unrelated lineages; (2) it may be retained from a more or less distant common ancestor; (3) it may be derived in the immediate common ancestor. Although the scheme described here attempts to eliminate taxonomic groups based on independent origins of similar features, as (1), it does not necessarily distinguish between primitive (2) and derived (3) similarity. Thus, although feathers are a derived characteristic of birds, it would not make sense to group ostriches and sparrows just because they all have feathers. Recent authors have attempted to bring greater rigour to classificatory systems by testing the validity of the characters on which they are based. To date few groups have been subjected to tests; however, the few tests we do have indicate that the old Fürbringer–Gadow system retains at least some validity at the ordinal level. Research at lower taxonomic levels continues to yield surprises, however, and although the older work has great value, it would be a mistake to assume that all taxonomic problems were solved in the last century! MCM

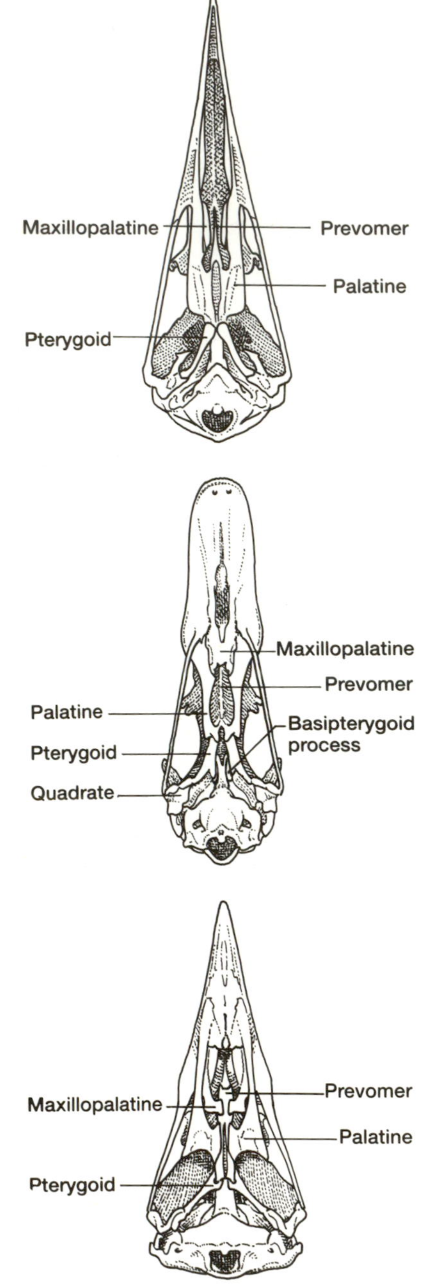


*In the palate of schizognathous birds such as gulls and gamebirds (top), the prevomers are completely fused and variable in size. In desmognathous birds such as ducks, herons and pelicans (centre), the prevomers are fused or absent. Finally, in the aegithognathous birds, the passerines and swifts (bottom), there are large fused prevomers that are truncated anteriorly.*

## A FAMILY TREE OF THE ORDERS OF BIRDS BASED ON DNA–DNA HYBRIDIZATION

The problems of how to distinguish the orders of birds and how to describe their inter-relationships have traditionally been tackled by examining anatomical features. Now modern molecular biology is increasingly being brought to bear on these problems. The most useful technique in this research has proved to be DNA–DNA hybridization.

The genetic material, deoxyribonucleic acid (DNA), is a molecule composed of two strands. When DNA is heated the two strands separate, only to rejoin in their original configuration on cooling. If the DNA of two species is mixed, the DNA again separates on heating. However, when cooled, the strands from the two species congeal together and hybridize. The two strands in hybrid DNA are less closely bound chemically. Therefore when re-heated they separate or 'melt' at a lower temperature than the DNA of a single species. If the two species are closely related, and their DNA therefore rather similar, the weakening of the chemical bonds will be slight, and the lowering of the melting point will also be slight. If the species are distantly related, the lowering of melting point will be greater. Thus does the extent of melting point depression give information on relationships.

Generally DNA–DNA hybridization confirms the picture established by traditional taxonomy. Where the two approaches yield different answers it is not possible (in the absence of a more detailed fossil record) to say which answer is 'right'. In our account of the orders of birds, traditional taxonomy has been followed since this will make it easier for the reader to tie together information in this encyclopedia with accounts in other bird books. But the family tree opposite, showing the relationships of the orders of birds and approximately when they evolved, is based on DNA–DNA hybridization, especially the work of Charles Sibley.

MdeLB

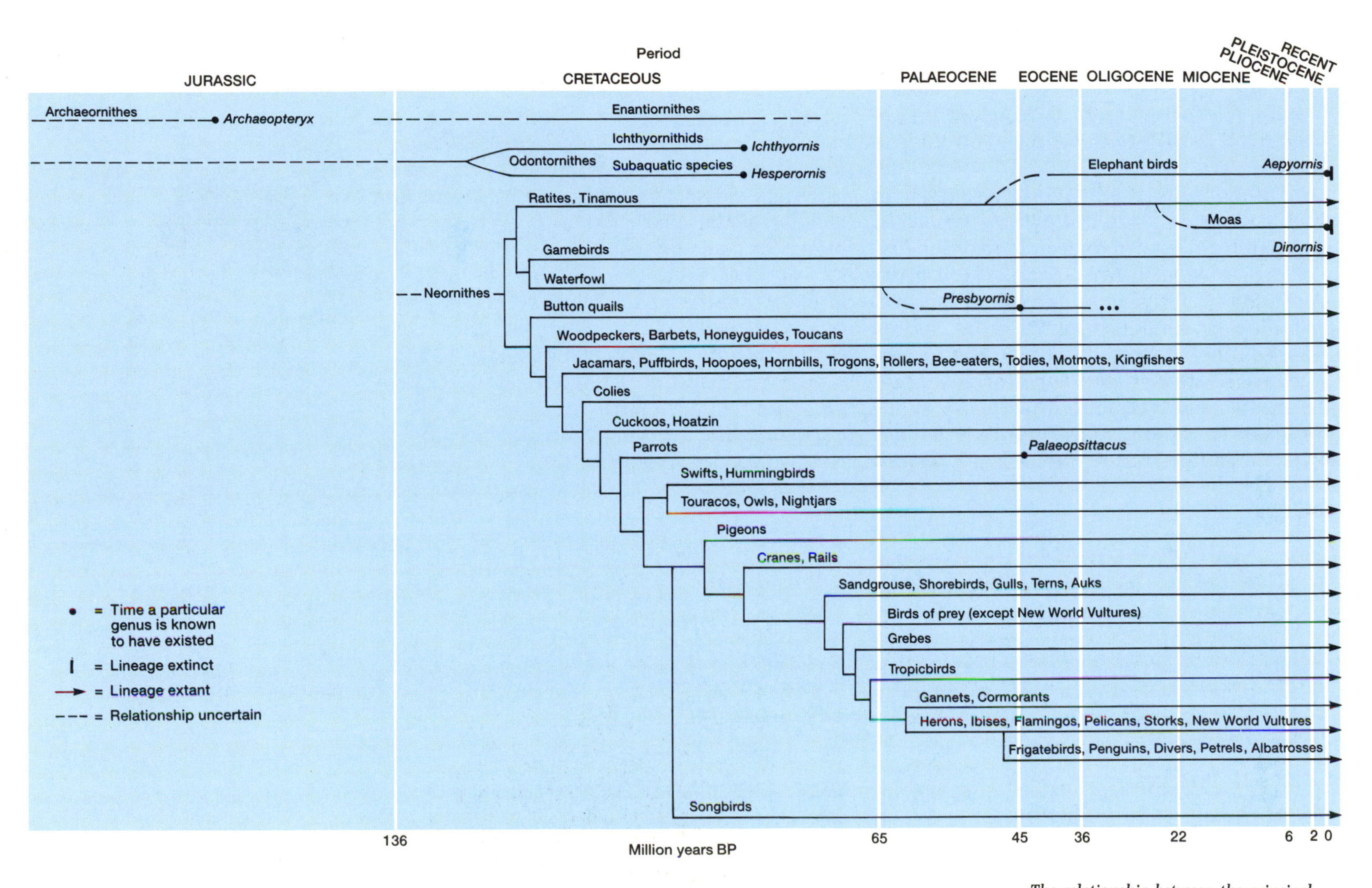


*The relationship between the principal bird groups as suggested by recent DNA hydridization studies. Vertical lines represent the approximate time of evolutionary separation. It can be seen, for example, that the lineage leading to frigate birds and penguins diverged from the heron/ibis lineage about 48 million years ago*

## What is a species?

Blue Tits breed with other Blue Tits, and not with Great Tits. This everyday observation, that the members of a species interbreed with each other and not with the members of other species, is an illustration of the biological species concept. This has been the predominant species concept in ornithology, as well as in other zoological disciplines, for the last half century.

Biological species are reproductive communities, defined as groups of interbreeding populations that are reproductively isolated from other such groups. Species are prevented from interbreeding by isolating mechanisms, factors that prevent gene exchange. These may be behavioural signals that identify an individual as a member of a certain species, or any aspect of external morphology; such factors ensure that errors in mate recognition do not occur, and are termed pre-mating isolating mechanisms. In some cases these mechanisms may be insufficient to prevent mating, but frequently post-mating isolating mechanisms will take over. (For example, foreign sperm may be unable to survive in the oviduct of a female of a different species, or the hybrids may be sterile, a situation sometimes called 'genetic death'.)

In birds, the features that prevent interbreeding are often quite obvious. For example, Ostriches and sparrows have diverged sufficiently since their common origin that there is no question of these species ever exchanging genes. Among more closely related species, plumage colour and pattern, song, and other behavioural displays are most important. For example, two American grebes (*Aechmophorus occidentalis* and *A. clarkii*), long regarded as colour variants of the Western Grebe, have recently been taxonomically split into two species. It was observed that the two forms, which appeared to differ only in the amount of black pigment on the face and in bill colour, rarely formed mixed pairs. Careful study of the vocalizations revealed subtle but important differences.

Among the passerines, song is a critical factor in maintaining the integrity of species, because males use song both to attract mates and to defend territories. Many songbirds have rather variable songs, but it appears that certain components of the song are nevertheless species-specific. It is these components that identify the bird as a potential mate to females of his species, whilst the variable aspects of the song permit individual recognition. Neighbouring males learn to recognize their neighbour's songs and cease to respond to them so long as territorial boundaries are honoured.

In many species both appearance and song are important cues to species recognition. Male Red-winged Blackbirds may lose their territory to another male if their red epaulets are artificially blackened – apparently the red epaulets serve as a signal to other males of the species to 'keep out'. In some species plumage appears to be unimportant, with song being the primary means of recognition. The New World flycatchers include numerous complexes of species that are extremely similar in appearance ('sibling species'). Experiments with tape recordings of song and stuffed specimens have shown that crested flycatchers will respond to recordings only of their own species' song, regardless of the identity of the stuffed bird perched near the speaker.

When species are living in the same place (sympatric), their status as cohesive reproductive units, distinct from others, can generally be determined by direct observation. It is rarely so easy to determine the status of those living in

*Above. The American Clark's Grebe (Aechmophorus clarkii), also called the Mexican Grebe (left), was until recently thought to be a race of the Western Grebe (A. occidentalis) (right). It is now regarded as a separate species. Both birds share the same general range, though Clark's Grebe is less common in the north of the continent*

*Left. Red epaulets raised, a male Red-winged Blackbird (Agelaius phoeniceus) signals other males to 'keep out'. Experimentally deprived of such ownership symbols, a male is liable to lose his territory*

different geographic areas (allopatric). If two similar forms live in different parts of a continent and do not overlap geographically, it is not known whether they could interbreed if they came into contact. In such cases, species status can only be inferred by analysis of behavioural and morphological or other characteristics. Extensive similarity in these features is used as an indicator of the potential to interbreed and produce viable, fertile offspring. Thus, in the case of the California and Florida populations of Scrub Jay the potential to interbreed has been assumed on the basis of plumage similarities.

Species are thought to arise through a process, the *allopatric model of speciation*, which is initiated when an ancestral population becomes split, either through a physical change in the environment such as the emergence of a mountain range, the breakup of a land mass, or the movement of a glacier; or through dispersal of some members of the original population into a new geographic area. Isolation of the splintered populations, with no genetic exchange, allows genetic and morphological differentiation to proceed unchecked, and after a period of time (perhaps many thousands of years) the populations may be quite different from each other. If the populations come into contact with each other again secondarily, the differences between them may be enhanced as the new reproductive isolating mechanisms are reinforced; differences that prevent members of the groups from wasting reproductive effort through hybridization will be selected for, as will differences that allow the groups to avoid competing with each other for food and other resources.

When reproductive barriers are complete, so is the process of speciation. Sometimes these barriers will be complete by the time of secondary contact, in other cases the barriers will be complete only after a period of contact, and in some cases the barriers may break down. The latter is thought to be true of the North American flickers which interbreed freely in their zone of overlap in the Great Plains.

Many biological species have broad distributions, occupying parts of a continent or parts of several continents. In most cases such species are not morphologically uniform throughout their range, but vary, presumably in response to local environmental conditions. These local forms were formerly considered to be different species, and indeed the practice of describing such forms as separate species continued into the first decades of this century. With intensified field research, however, it was discovered that many of these 'species' are actually parts of a reproductive continuum, connected over the expanse of continents by interbreeding such that their morphological (and genetic) distinctness is smoothed. This regular exchange of genetic material is termed *gene flow*, and the graded sequence of differences that gene flow creates in a species over a large geographic area is called a *cline*. It is possible for the populations at opposite ends of a cline to be very different from each other, but because they are connected by interbreeding forms they are nevertheless regarded as parts of the same biological species.

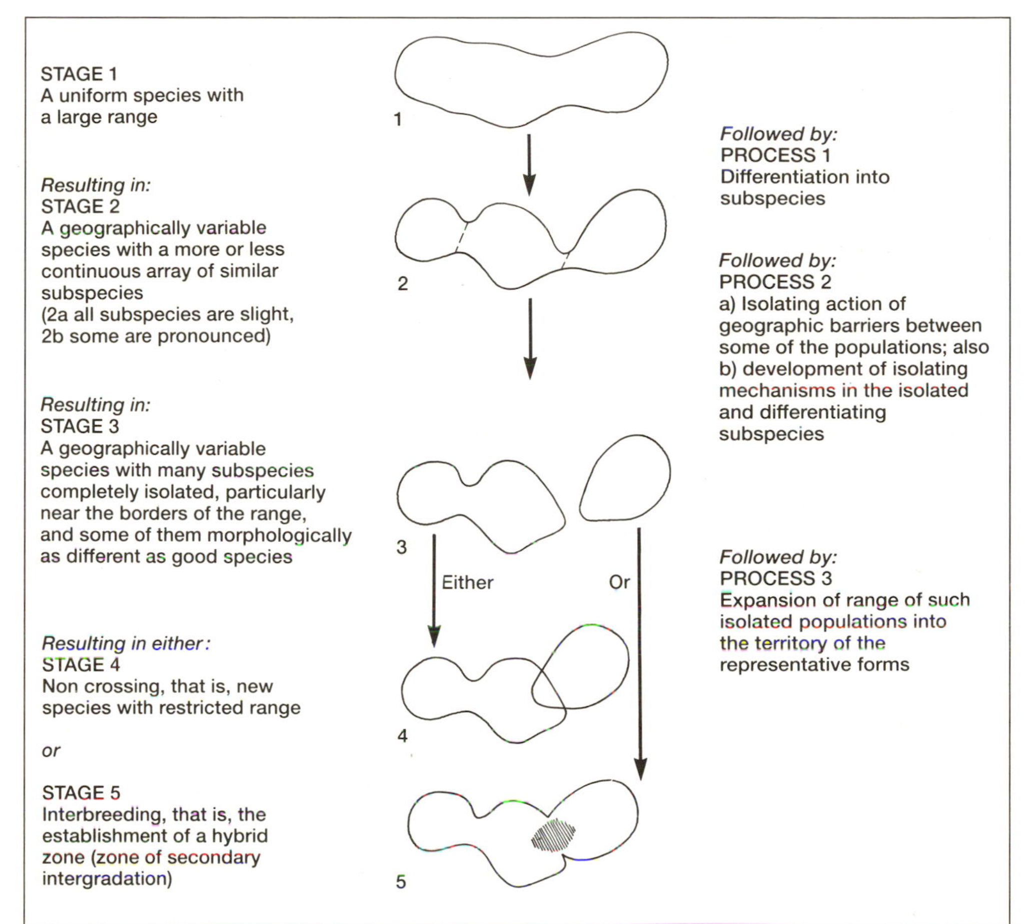


*The genetic impact on bird populations of geographical isolation, the so-called allopatric model of speciation*

An intriguing aspect of geographic variation is the phenomenon of *ring species*, in which a series of races form what amounts to a geographic loop. Neighbouring forms on the loop interbreed but the forms at either end of the loop are reproductively isolated, despite occurring in the same region. Thus the Herring Gull of the North Atlantic is connected to the west by a series of slightly different races that occupy the cool northern countries. The end form of the series is the Lesser Black-backed Gull which does not interbreed with the Herring Gull where the two co-exist in the north-east Atlantic. A similar phenomenon occurs with the Great Tits that encircle the Himalayan massif.

When a population or group of populations is distinct from other members of its species, however, it is often considered to be a race or subspecies of its parent form.

Species vary in their tendency to differentiate geographically; for example, there are at present as many as 31 races described for the North American Song Sparrow (*Melospiza melodia* spp.), and none for the White-throated Sparrow (*Zonotrichia albicollis*). To be sure, taxonomists differ in their tendencies to recognize differences as indicative of subspecies status; for example, Harry Oberholser, who would be termed a 'splitter', described 10 races of yellowthroat in the early part of this century, of which only one (*Geothlypis trichas*) is currently accepted. For the most part, however, the days of taxonomic excess are over in ornithology, and 'lumpers', who are disinclined to recognize separate races, now predominate.

Geographic variation may follow certain patterns associated with patterns in climate. These patterns are sufficiently regular that they have been termed 'ecological rules', although there are many exceptions and the causal mechanism is often unknown. Bergman's Rule states that races of warm-blooded vertebrate

*Herring Gulls and allies belonging to the Larus argentatus group form a ring species. Various sub-species developed in Pleistocene refuges and then spread, as indicated by the arrows. Today's forms are differentiated by mantle colour (ranging from pale grey to near black as indicated) and by leg colour (ranging from yellow to pink as indicated by the outline colour for each form's range). In north-western Europe the Herring Gull, Larus argentatus (pale grey back, pink legs), has probably spread from the west and now overlaps with the Lesser Black-backed Gull, Larus fuscus (dark grey back, orange legs), though the two species do not interbreed*

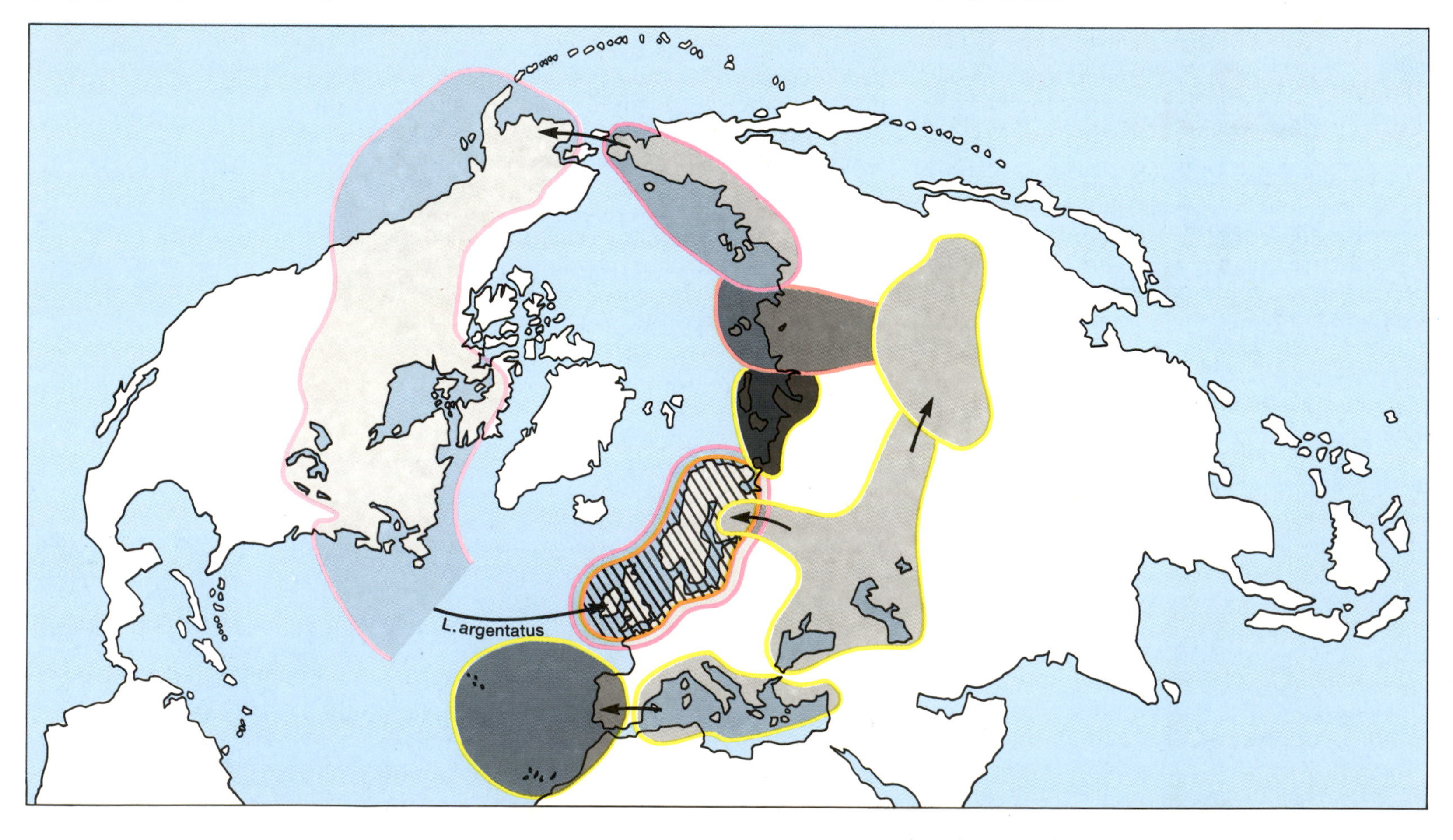

species (not just birds) tend to be larger in cooler climates than races of the same species in warm areas, presumably because the larger forms have a smaller surface area, relative to their body volume, than smaller forms and hence heat loss is reduced. Allen's Rule, which is more usefully applied to mammals, states that appendages through which heat loss may occur tend to be smaller in cool climates. According to Gloger's Rule races in cool, dry climates are more lightly pigmented than races from warm humid climates. The reasons for this are unknown, although this rule appears to hold better than the other two.

Geographic variation is often very complex, and although it has long been assumed to be due to selective adaptation to local conditions, this assumption is rarely tested. A simple transplant experiment has demonstrated that the differences between subspecies are not always entirely genetic. Eggs were exchanged between northern and southern Florida populations of Red-winged Blackbirds and between populations in Colorado and Minnesota, and the hatchlings were raised by foster parents. These populations differed in beak and other body dimensions, but the transplanted birds approached the average size of their foster population, rather than their natal population, in these body dimensions. This demonstrates a significant environmental component to the differences between these two geographic races, and indicates a need for more research on the heritability of population differences.

Although the biological species concept is widely accepted in ornithology, at least in its theoretical foundation, it is by no means universally so. In recent years the philosophical validity and utility of the biological species concept has been called into question. An important criticism of the concept is that judging the status of allopatric forms is highly subjective because it depends upon the individual researcher's assessment of features that could serve as isolating mechanisms. The Florida and California Scrub Jays are considered the same species, yet they have very different breeding systems: the Florida form is a cooperative breeder, where offspring usually stay on their natal territory for a period of time

*Left. The North American Song Sparrow (Melospiza melodia spp.) has split into 31 geographic races which vary, among other features, in breast and crown pattern. Races include Melospiza melodia heermani from California, M.m. fallax from the southwestern United States, M.m. melodia from the southeast, and M.m. atlanticus from the east*

and help their parents to raise additional broods. Why are these not considered separate species? There are significant, genetically based behavioural differences between eastern and western Marsh Wrens; why should these not be split?

An alternative to the biological species concept is a phylogenetic species concept that considers species to be 'the smallest diagnosable cluster of individuals within which there is a parental pattern of ancestry and descent'. In other words, species are recognizable lineages. 'Recognizable' means that one or more behavioural, morphological, physiological, or genetic characteristics identify a group of individuals as a lineage, a group composed of closest relatives. Whereas biological species are defined by interbreeding, for phylogenetic species interbreeding is merely one criterion for species status. Although the members of a species breed together, interbreeding may also be a primitively retained trait. In the case of the North American flickers, the red-shafted and yellow-shafted forms would be considered separate phylogenetic species. Although they interbreed in a narrow zone of overlap, the parental forms are nevertheless separate lineages that maintain their genetic and morphological integrity.

Unlike the biological species concept, recognition of phylogenetic species cannot be made by inference, it must be the result of careful population-level analyses of genetic and morphological variation. This increased rigour in application of species concepts can only lead to increased knowledge, not just of species relationships and evolutionary patterns, but of population biology in general. This view is not widely held among ornithologists at the moment, but it is beginning to receive considerable attention.

MCM

*The aptly named Yellow-shafted Flicker (Colaptes auratus auratus) lives east of the North American Rockies; to the west it is replaced by the Red-shafted Flicker (C.a. cafer). The two forms interbreed in a narrow overlap zone on the Great Plains*

## HOLOTYPES

When a description of a new species is published, one specimen is usually designated the 'type specimen', or *holotype*, for that species. All other specimens collected at that locality at the same time are designated *paratypes*. If no holotype is chosen, then every specimen in the 'type series' is known as a *syntype*. One of these syntypes may later be chosen as 'the type' for that species, in which case it is called a *lectotype*. If the holotype and paratypes or lectotype and syntypes are all somehow destroyed, a new type may later be designated, the *neotype*. This practice of type designation is traditional in taxonomy, but its utility is questionable; types may be atypical, or variation may be such that no single specimen is typical.

Thus, in 1977 a new species of wood-wren, *Henicorhina leucoptera*, was described from Peru. The holotype of this Bar-winged Wood-wren is American Museum of Natural History, no 812091, and the description is based upon that specimen, collected in the Cordillera del Condor, above San José de Lourdes, dept. Cajamarca, Peru, 2200 m, 18 June 1975 by John Fitzpatrick. A total of 13 individuals was collected at the type locality during two visits to the area; however, only those specimens collected with the holotype are considered paratypes.

MCM

*The Bar-winged Wood-wren (Henicorhina leucoptera), a species named in 1977*

## A survey of modern birds

The following account of the various orders of birds gives details of the families into which the order is divided, and the number of living species in each family. Sometimes this number is qualified by 'about'. This arises because, as explained in the section 'What is a species?', there is often some doubt about the validity of particular species. Therefore the number of species in a family sometimes remains uncertain. Another problem is that one or two new bird species are still being described every year!

For each family, details of the size, habitat and distribution are provided. For flying birds the length given is the distance from the tip of the bill to the tip of the tail of a bird laid on a flat surface. For non-flying families the height of the standing bird is given. Under geographical distribution the following obvious abbreviations are commonly used; N(North), E(East), S(South), W(West) and C(Central).

The account of the order continues with a general description of the birds concerned. The food eaten is specified and the breeding habits outlined. Mention is made of any specially interesting habits.

### Ratites

Ratites are large flightless birds belonging to six separate orders, two of which are extinct. Because the relationship of ratites to each other is clearly closer than their relationship to other birds, they are discussed together here. All living species are black, grey, brown or white.

#### OSTRICHES RECOGNIZE THEIR OWN EGGS

In the Ostrich, incubation duties are shared by male and female. Since the sex ratio among breeding adults is around 1.4 females for every male, there are hapless females who cannot secure the exclusive help of a male. Consequently, females lay in communal nests. Several (1–6) hens, the so-called 'minor' hens, lay up to 13 eggs each in the ground nest of the 'major' hen. She, along with the male, incubates the clutch that may contain up to 30 white unpatterned eggs.

While incubating, the major hen can properly cover only about 20 eggs. These are mostly but not exclusively her own. The extra eggs laid by the minor hens are pushed to the periphery of the nest where they are much more vulnerable to predation by, for example, hyaenas. The ornithologist who discovered this discrimination, Brian Bertram of King's College, Cambridge, could not determine what clues the major hen used to recognize her own eggs and retain them in the prime, central position.

The value to the major hen of accepting and incubating some of the minor hens' eggs is that they 'dilute' her own. If a jackal arrives at the nest and steals just one egg, there is at least a chance it will not belong to the major hen – as it certainly would if she had laid all the eggs. Despite the discrimination against their eggs, the minor hens also receive some benefit from this system. A proportion of their eggs are incubated, a better outcome than failing absolutely to nest because of the scarcity of males.

MdeLB

## Order *Struthioniformes*

**Family** Struthionidae • Ostrich • One species • Largest living bird 2.5 m tall/150 kg.
Four subspecies live in open country in Africa; another, Arabian, subspecies became extinct early in the twentieth century.

## Order *Rheiformes*

**Family** Rheidae • Rheas • 2 species • 1.5 m tall/up to 45 kg.
Open country from Brazil to tip of S America.

## Order *Casuariiformes*

**Families** Casuariidae • Cassowaries • 3 species • 1.5 m tall/up to 140 kg.
Forests of N Australia, New Guinea and adjacent Indonesian islands.

Dromaiidae • Emu • One living species • 1.5 m tall/up to 70 kg.
Drier open country of S and W Australia. Also several extinct sub-fossil forms.

## Order *Dinornithiformes*

**Families** Dinornithidae and Anomalopterygidae • Moas • About 22 species • All extinct.
From chicken size to 3 m tall.
Confined to New Zealand and most exterminated by early man about 500 years ago. One species may have survived into the early nineteenth century.

## Order *Apterygiformes*

**Family** Apterygidae • Kiwis • 3 species • From 50 cm/1.2 kg (Little Spotted Kiwi) to 85 cm/3.5 kg (Brown Kiwi).
Forest and scrub of New Zealand.

## Order *Aepyornithiformes*

**Family** Aepyornithidae • Elephant Birds • At least 11 species • All extinct • The thick shelled eggs held up to two gallons and the birds may have weighed up to 500 kg.
Madagascar and southern Africa, where exterminated by man about 1000 years ago.

The ratites probably evolved from a common flying ancestor. The Ostrich and rheas still have prominent wings, the cassowaries and emus have wing remnants but the kiwis' wings are barely detectable. The kiwis are probably the nearest relatives of the extinct moas.

In some respects the Ostrich is the most distinct anatomically. It has a two-toed (3rd and 4th) foot while the other living ratites have three prominent toes (2nd, 3rd and 4th).

Most ratites are long-necked birds that can spot and often outrun predators in the open country where they live. The cassowaries are somewhat different; a shorter neck and stockier build are suited to pushing through dense tropical forest. Finally, the squat kiwis mostly live in dense undergrowth.

Although flightless, ratites can well defend themselves with their bills, feet and powerful legs. Cassowaries have the most dangerous weapon in the form of one toenail, the innermost, which may reach over a foot in length. It becomes pointed and sharp, and kicking cassowaries kill more New Guineans than any other animals. A cassowary kick literally slices open the victim.

*Diet.* With a broad, slightly flattened bill, most ratites are basically vegetarians, although all will eat almost anything they can secure. The digestive tract is duly long, especially in the Ostrich and rheas. In addition, the rheas have two well developed caecae in which bacterial digestion of cellulose occurs. Chicks are insectivorous when small, and slowly become more vegetarian with age.

Kiwis are specialized exceptions to this. They mostly feed at night, using a long bill with nostrils at its tip to probe for earthworms, their main food.

*Reproduction.* The social structure of ratites is quite similar. In all except the Ostrich, where incubation is shared by male and female, the males do all incubation and rearing of downy chicks. Several females lay eggs for each male. The system has been most thoroughly studied in rheas where each dominant male may display to and attract up to a dozen females, who will proceed to lay eggs for him in a nest scrape he makes. The female lays an egg every second day for up to 10–14 days before shifting to the next dominant male. Over the course of two months each female may lay up to 30 eggs in the nests of four to eight different males.

Males incubate the eggs alone and then care for the chicks. Synchronized hatching appears to be advantageous to the rhea chicks. Since the eggs may be laid several days apart, the presence of other eggs in the nest seems to slow down the process of incubation by up to 3 to 5 days for the first egg and speed up the development of the last eggs laid by as much as 3 or 4 days. To do this, communication between eggs is required and experiments have demonstrated that communication by voice between developing chicks inside the eggs does occur.

Clutch sizes and incubation periods are as follows: rheas, 15–30 yellow or green eggs, 35–40 days; emus, 10–20 dark green eggs, about 56 days; cassowaries, 4–8 light green eggs, about 50 days.

Kiwis are again exceptional. One or two very large white eggs (about 25% of the female's weight; 60% of the egg is yolk, the highest proportion in any bird) are incubated by the male alone for 65–85 days. Chicks are active and fully feathered at hatching. It is possible parents feed the chicks in the first weeks after hatching; thereafter chicks feed themselves.

DB

### DID RATITES EVOLVE FROM FLYING BIRDS?

Ratites are an ancient group of birds, at least 80 million years old. They probably evolved from a flying bird resembling modern tinamous, a group to which the ratites are certainly related.

In becoming flightless, ratites lost the keeled sternum to which the flight muscles attach in flying birds. However, they retain other skeletal features that are associated with the ability to fly: fused wing bones; presence of an alula (bastard wing) on the wing; and pygostyle for anchoring tail feathers. In addition the structure of the ratite cerebellum, the part of the brain responsible for in-flight coordination, suggests a flying past.

Ostriches and rheas have retained prominent wings which are now used in courtship display. Cassowaries and emus have lost all but remnants of their wings. In cassowaries these consist of several long feather shafts which appear like long rounded fingernails. The reasonably well developed wings of Ostriches and rheas, along with behaviours like the 'broken wing display' which draws predators away from the nest, further indicate that these birds have probably evolved from flying ancestors.

The large numbers of flightless island birds, especially rails, is extra evidence that the loss of flight can occur quite rapidly in evolutionary terms.

DB

### DID RATITES EVOLVE ONLY ONCE, OR ARE THEY SURVIVORS OF SEVERAL LINEAGES?

The current and historic geographic distribution of ratites in the southern hemisphere has resulted in much debate on their origins.

One theory assumes that all ratites derived from some ancient bipedal dinosaur or tinamou-like ancestor while the southern continents were still attached. The suggestion is that after the break-up of Gondwanaland into Africa, Australia, South America and Antarctica occurred, between 130 and 80 million years ago, the common ratite ancestor that was present on all continents then evolved independently on each continental mass. In South America, Australia, New Zealand, Madagascar and New Guinea, there were probably no large mammalian predators during this period. The ratites evolved and increased in size as a defence against the smaller, slower mammals that had developed in those areas. Without large cats or canids, ratites were able to compete in South America and the Australasian region. In North America and Eurasia none was able to cope with the large mammalian predators and the more rigorous climate.

This theory does not fully explain the Ostrich in Africa. Perhaps the Ostrich (and the Elephant Birds in Madagascar) evolved before the large mammalian predators arrived in Africa. Or perhaps the Ostrich evolved alongside those predators, depending for its survival on its large size and speed.

A second theory suggests that all the ratite groups evolved from separate stocks on each continental mass to fill similar habitats. The evidence of similarities in anatomy and behaviour seems to refute this idea.

DB

## Order *Tinamiformes*

**Family** Tinamidae • Tinamous • 45 species • Size from 20 cm/450 g to 50 cm/2300 g.

Forest, scrub and grassland of mainland America from S Mexico to Patagonia.

The tinamous are ground-dwelling birds that fly only rarely. Instead, they escape predators either by running or by relying on the camouflage of their barred or speckled brown plumage. In general shape they somewhat resemble gamebirds, to which they are quite unrelated. Females are usually larger than males. Calls are loud, flute-like, often polysyllabic whistles.

Although their flesh is strangely translucent, tinamous are delicious to eat.

Several skeletal features (skull, especially palate, bill covering, pelvis) indicate the close relationship of the tinamous and the ratites. This suggestion is strengthened by similarities in the egg-white proteins of the two groups.

*Diet.* This is mostly seeds and fruit, supplemented by insects and other small animals.

*Reproduction.* Polygyny and polyandry are apparently widespread. The most common pattern seems to be for one or more females to lay a clutch for a male, and then proceed to deposit eggs in another nest for another male. Completed clutches in the sparsely lined ground nests contain between one and twelve wonderfully glossy unpatterned eggs of various hues; turquoise, pale green or pale chocolate brown according to species.

Incubation, by the male alone, lasts 19–20 days. He also tends the downy, precocial young.

## Order *Sphenisciformes*

**Family** Spheniscidae • Penguins • 16 species • Size from 30 cm/1 kg (Little Blue Penguin) to 110 cm/up to 40 kg (Emperor Penguin).

Primarily seabirds of cooler southern hemisphere waters, penguins occur around Antarctica, at sub-Antarctic islands, New Zealand and southern Australia, southern Africa, and along the west coast of S America north to Peru and Galápagos. The range extends further north in S Africa and S America than elsewhere because of cold currents, the Benguela and Humboldt respectively, sweeping up from the south.

Penguins are flightless. On land they stand upright, walk with a shuffling gait and occasionally slide on their bellies. In their more familiar element, the sea, the birds' adaptations are revealed. Their legs are set at the rear to serve as a rudder in conjunction with the tail. The stiff flippers, which cannot be folded and lack flight feathers, propel penguins through the water at sustainable speeds of 5–10 km/h. The plumage pattern, dark bluish-grey above and white below, may make it difficult for both predators and prey to spot penguins since, seen from below, they are pale as the sky and, seen from above, dark as the sea depths.

Penguins, densely covered with three layers of short feathers, are unusual among birds in virtually lacking bare areas between feather tracts. Instead, the feathers grow more or less uniformly over the entire body. They and a well-defined fat layer provide insulation. Heat loss is also controlled by a counter-current exchange system in the flippers and legs. Thanks to this system, blood flowing to the extremities transfers its heat to returning blood, thereby ensuring that heat remains within the body. So efficient is Emperor Penguin heat retention that the bird's metabolism speeds up to generate extra heat only when the temperature drops below −10°C. Conversely, on warm summer days polar species lose heat by panting, eating snow to provide the necessary water. Species of warmer latitudes shed heat on land from their large flippers and bare facial areas.

*Diet.* Fish (taken especially by inshore penguin species), squid, crustacea. Krill (euphausiid crustacea) are particularly important in the diet of Antarctic species such as the Chinstrap and the Macaroni Penguins. In all species the tongue and palate are spine-covered, the better to grip slippery prey.

*Reproduction.* Usually social, often breeding in enormous colonies on offshore islands. Some species are burrow nesting. Mates often recognize each other by call. Parents and chicks also recognize each other, using calls and location as the main recognition cues. Such vocal recognition is probably most important where chicks congregate in protective creches.

There are normally two (greenish) white eggs, but only one in the genus *Aptenodytes* (King and Emperor) because during incubation that egg rests on the uptilted feet and is warmed by a covering fold of abdominal skin. Incubation takes 33–62 days. Both sexes feed the downy young by regurgitation. Most species breed during southern spring and summer. Two breeding cycles merit further comment. In the Antarctic, Emperor Penguin laying occurs in May. Males incubate alone for the two dark mid-winter months when air temperature falls to −60°C. They lose up to 45% of body weight in a fast which, including courtship, lasts nearly four months. Chicks depart only two-thirds grown in December, allowing parents adequate time to moult and build up body reserves before the next season. The King Penguin of the sub-Antarctic islands has a breeding cycle lasting rather over a year. Consequently, successful pairs rear two chicks in three years, an arrangement unique among birds.

*Special adaptations.* During the two- to six-week moult penguins are restricted to land (or ice). Fattening for this fasting period usually occurs immediately after breeding.

The unusually dense bones of penguins aid diving. Emperor Penguins have been recorded diving for up to 15 minutes to 265 m, King Penguins to 240 m. However the krill-eating Chinstrap Penguin usually dives no deeper than 45 m and the inshore-feeding Gentoo no deeper than 20 m, in dives lasting 0.5–1.5 minutes. During diving the heart rate drops in an Adélie Penguin from 100 to 20 beats per minute. This reduces oxygen consumption and so prolongs the dive. But it is not known how deep-diving penguins avoid the 'bends' (the painful and dangerous formation of nitrogen bubbles in the blood of human divers who decompress too rapidly).

## Order *Gaviiformes*

**Family** Gaviidae • Divers or loons • 4 species • Size from 60 cm/1.5 kg (Red-throated Diver) to 85 cm/5 kg (White-billed Diver).

Waterbirds breeding beside lakes and ponds in arctic and subarctic latitudes of N America, Greenland, Iceland, and Eurasia between extremes of about 83 degrees N (Red-throated in Greenland and Ellesmere Island) and 41 degrees N (Great Northern in Great Lakes region of N America). Generally migratory, wintering south of breeding range on sea; N Pacific, N Atlantic, Mediterranean, Black Sea.

The upperparts of these aquatic predators are variously black, grey or brown with bold patterns in summer plumage while the underparts are always white. The legs of divers are exceptionally specialized for swimming, being flattened and set so far back that the birds are unable to walk properly.

*Diet.* Mostly fish caught crosswise in the bill; otherwise frogs, crustacea, aquatic insects, etc. Prey is usually caught in the top 10 m of water, but dives to 80 m have been recorded.

*Reproduction.* Normally two olive or brown eggs, incubated by both sexes for about four weeks. The downy chicks leave the nest within a couple of days and may be brooded on the parent's back. They become independent after about 60 days.

The nests are usually sited on a lake shore or on an islet within the preferred clear oxygen-rich lakes. Pairs are territorial except sometimes Red-throated Divers nest in loose groups. A possible reason for this contrast is that this species often nests on a small pond but flies to a larger water body to feed, while the other three species feed on the home lake during the breeding season.

The territorial calls include very loud shrieks, croaks and whistles; this may even be the basis for the expression 'crazy as a loon'.

## Order *Podicipediformes*

**Family** Podicipedidae • Grebes • 20 species • Size from 25 cm/115 g (Least Grebe) to 48 cm/1400 g (Great Crested Grebe).

Waterbirds of freshwater lakes and marshes in all continents except Antarctica. Absent from deserts, extreme N Eurasia and N America. Some migratory movement to sea in winter.

In non-breeding plumage the grebes are grey, brown or black above, white below. At breeding many species acquire colourful tufts on the head and/or vivid necks. Although superficially similar to the divers or loons, the grebes have maintained a distinct pedigree for 70 million years. However, they are similarly well adapted to an aquatic life; the tarsi are elliptical, the tail reduced and the feet lobed (not webbed as in divers). Also an aquatic adaptation is the dense plumage of around 20 000 feathers. Stripped from the birds, the breast pelts, called grebe fur, provided muffs for European ladies and saddle blankets for South American Indians. Three species are flightless and now confined to a single lake system; Short-winged Grebe of Lake Titicaca (Peru/Bolivia), Junín Grebe of Lake Junín (Peru) and giant Pied-billed Grebe (probably extinct) of Lake Atitlán (Guatemala).

*Diet.* Aquatic animals, both fish and invertebrates. Uniquely among grebes the Western Grebe spears fish in the manner of darters. Perhaps to protect delicate intestines from fish bones, grebes eat their own feathers.

Grebes provide one of the best examples of 'character displacement', the contentious phenomenon in which competition may cause two related species to be more different at sites where they live together than at sites where one species lives in the absence of the other. For example, over much of South America the Silver Grebe is the only resident grebe. It has a similar bill over most of its 3000 km range. The exception occurs at Lake Junín. There, the only home of the large-billed Junín Grebes, the Silver Grebes have smaller bills. The smaller-billed Silver Grebes catch more small prey, mostly insects, while the Junín Grebes concentrate on larger prey, fish. And not only do grebe species divide up prey according to their bill size. Even within species smaller-billed individuals go for small items while their larger-billed conspecifics aim for larger morsels.

*Reproduction.* Grebes breed either colonially

or solitarily, nesting on floating, but anchored vegetation. Two to nine eggs are laid and incubated for 20–30 days by both sexes. Uniquely, the Hooded Grebe lays two eggs but only takes one chick from the nest, invariably abandoning the other egg. The small precocial young are carried aboard their parents and fed by them for 6 to 12 weeks.

The courtship of grebes, the subject of pioneer animal behaviour studies, is notably elaborate. Sometimes male and female reverse roles, even mounting each other.

## Order *Procellariiformes*

**Families** Diomedeidae • Albatrosses • 14 species.
Procellariidae • Gadfly petrels, shearwaters, fulmars and allies • About 55 species.
Hydrobatidae • Storm petrels (Mother Carey's chickens) • 20 species.
Pelecanoididae • Diving petrels • 4 species.

The order of birds with the greatest size range, from the 14 cm/25 g Least Storm Petrel to the 125 cm/12 kg Royal Albatross. The Procellariiformes range the seas worldwide, generally coming to land only to breed, either on continental coasts or, more commonly, on offshore islands. Transequatorial migrations are undertaken by many species breeding in high latitudes.

Variously patterned in black, grey, brown and white, the Procellariiformes are *par excellence* the wide-ranging seabirds of the open oceans. Also known as the 'tube noses' on account of paired tubular nostrils surmounting the bill, the birds have a characteristic musty oily smell due to stomach oil which, along with oil from the preen or uropygial gland, is spread on the plumage. It has been suggested that the tubular nostrils serve as a conduit allowing controlled application of stomach oil.

*Diet.* A variety of marine food – fish, squid, plankton – caught in several ways. Albatrosses and gadfly petrels often catch squid at the surface by night, shearwaters and diving petrels dive for fish and plankton respectively, prions filter copepods from the surface, and storm petrels, pattering on the water in the manner of St. Peter (hence the name), collect a miscellany of planktonic titbits. Fishery waste is scavenged by many species and the giant petrels of the Southern Ocean are active predators, notably of penguins.

The olfactory lobes of petrels' brains are unusually large. There is evidence that the birds can smell their way to food at sea but it is not known whether this is helped by the tubular structure of the nostrils.

*Reproduction.* One large white egg (up to 38% of the female's weight in the Grey-backed Storm Petrel) is incubated by both sexes in alternate shifts lasting from a few days up to three weeks. Total incubation time is between 40 (smaller storm petrels) and 80 (larger albatrosses) days. The downy young is fed by regurgitation by both parents on a diet that may include stomach oils. With a high energy to weight ratio these substances are economical to transport over large oceanic distances. The fattened young may outweigh their parents before they are deserted to complete their development while starving. Thus, when ready to fly some 2 (smaller storm petrels) to 9 (larger albatrosses) months after hatching the single chick departs for the sea alone.

Petrel pairs, usually faithful from year to year, often nest in large dense colonies. In the case of some burrow-nesting species active at the colony at night, the essential recognition between mates is achieved by voice.

The breeding cycle of the largest albatrosses, such as the Wandering Albatross (wingspan 3–3.5 m), takes so long (11 months from laying to fledging of young) that successful pairs only breed every second year. The intervening year is used for moulting while another pair occupies the breeding territory.

Young shearwaters are sometimes harvested by man from their colonial breeding grounds. For example, the Tristan Islanders of the South Atlantic visit nearby Nightingale Island, home to over half the world's Great Shearwaters. On Nightingale, where some two million pairs cram into 200 hectares, the loss of tens of thousands of chicks, which are boiled for their fat, does not adversely affect the population.

## Order *Pelecaniformes*

**Families** Pelecanidae • Pelicans • 7 species • About 1.5 m long and weighing 2.5–15 kg. Freshwater and coasts on all continents. Generally in warm climates.

Sulidae • Gannets and boobies • 9 species • Size from 70 cm/0.9 kg (Red-footed Booby) to 95 cm/3.1 kg (Atlantic Gannet).
Gannets (3 species) occur in N Atlantic and temperate seas off southern Africa and Australasia. Boobies occur in tropical oceans.

Phaethontidae • Tropicbirds • 3 species • About 1 m/400 g.
Tropical oceans.

Phalacrocoracidae • Cormorants • 29 species • From 45 cm/1 kg to 100 cm/5 kg. Freshwaters and coasts worldwide. Few species at high latitude. Galápagos Cormorant is flightless.

Fregatidae • Frigatebirds • 5 species • About 90 cm/1 kg.
Tropical oceans.

Anhingidae • Darters • 4 species • About 90 cm, weighing 1–2 kg.
Freshwater at lower latitudes of N and S America, Africa, Asia and Australasia.

Although comprising six distinct families of aquatic birds the pelicans and their allies are united in being the only birds that have all four toes connected by webs (totipalmate). All except tropicbirds have very small nostrils that sometimes lack an external opening, for example in the gannets which need to prevent the entry of water when diving head first into the sea from 30 m. All have a gular or throat pouch, most developed in pelicans where it serves to strain fish from water, and in male frigatebirds.

The pelicans, gannets and boobies are various combinations of black, grey, white and brown, the tropicbirds white with black markings, and the other three families dark (sometimes greenish) brown or black with white markings.

Cormorants, darters and frigatebirds have somewhat permeable plumage. Cormorants and darters therefore dry out, wings spread-eagled, after swimming. Frigatebirds rarely get wet, soaring instead over tropical oceans on wings with a particularly low wing loading (about 3 kg/sq m; cf. 5–6 kg/sq m in gulls).

*Diet.* Mostly fish and, in the tropical oceanic species, also squid. Gannets, boobies, tropicbirds and the Brown Pelican plunge from often a great height into the sea to seize prey. All the pelicans scoop water and prey into their capacious gular pouch. When the water, which initially can weigh more than the bird itself, is drained off, the prey remains to be swallowed. Frigatebirds are notorious for their piratical habits, chasing other birds (e.g. boobies) until they disgorge and then catching the vomit in mid-air. Nevertheless most prey, for example flying fish plucked from the sea surface, are probably self-caught. Cormorants and darters (snakebirds) largely capture fish underwater after a dive from the surface. With a peculiar hairy stomach lining (to cope with fish bones?) darters impale their quarry dagger-style on the bill, the strike of which is accelerated by a unique hinge mechanism between the eighth and ninth neck vertebrae.

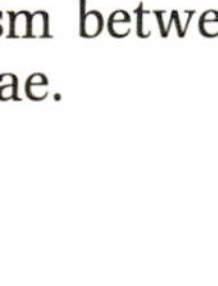

*Reproduction.* Generally colonial. One to six chalky white eggs are laid in a crude nest (tropicbirds build no nest). Incubation by both sexes takes 3–8 weeks. Gannets and boobies incubate the eggs by covering them with their feet. The young, often blind and naked at hatching, are altricial and fed by regurgitation. Growth is particularly slow in frigatebirds, which may spend up to six months in the nest and then remain dependent on the parents for several more.

*Special feature.* In the Orient, cormorants have long been used to help men catch fish.

## Order *Ciconiiformes*

**Families** Ardeidae • Herons and bitterns • 60 species • 30–140 cm long.
Shores and marshes worldwide except northern America and Eurasia.

Scopidae • Hammerhead • One species • 50 cm.
Wetlands, including those that are wooded, of SW Arabia and sub-Saharan Africa.

Balaenicipitidae • Whale-headed Stork • One species • 120 cm.
Swamps of C and E Africa.

Ciconiidae • Storks • 17 species • 75–150 cm/2–9 kg.
Open, often wet country worldwide, excluding most of N America and northern Eurasia.

Threskiornithidae • Spoonbills and ibises • 31 species • 50–110 cm long.
Shores and marshes worldwide, S of approx. 45 degrees N.

Phoenicopteridae • Flamingos • 4 or 5 species • 90–130 cm/2–3 kg.
Saline lakes of high pH throughout tropical and warm temperate zones.

The Ciconiiformes are long-legged, often long-necked birds adapted for wading and capturing animal food. The sexes are alike, or at least very similar. The largest family, the herons and bitterns (Ardeidae), have a particularly long sixth neck vertebra which causes the neck to adopt an S-shape. An aid to heron classification is the characteristic powder down, a type of feather that is never shed. Instead it grows continuously and frays into a powder that is used as a dressing to protect the plumage against fish slime. Storks, generally bulkier than herons, fly with the neck extended, as do the spoonbills and ibises. Another feature of most spoonbills and ibises is the lack of face feathers. The flamingos, the only group with webbed feet, are sometimes placed in a separate order, the Phoenicopteriformes, and sometimes allied with the geese and ducks in the Anseriformes. They gather in enormous congregations; for example, some 2.5 million Lesser Flamingos dwell in East Africa's Rift Valley.

*Diet.* All manner of animal food, particularly aquatic forms – fish, amphibia, crustacea. However, the largest storks, the three species of marabou of southern Asia and tropical Africa, are vulture-like scavengers. A number of species, for example the White Spoonbill and the woodstorks, probably detect food more by touch than by sight, snapping shut their probing bills within 25 milliseconds of feeling prey. Like whales, flamingos are filter-feeders. Water is drawn into the upside-down bill and then forced out by the fleshy tongue past bristle filters. These sieve off invertebrates and/or planktonic algae which are rich in carotenoids, the pigments that colour the flamingo's brilliant plumage. Indeed, captive adults do not breed successfully without an abundance of these pigments, which are nowadays usually supplied by the addition of synthetic canthaxanthin to the food and not by the former method of supplementing the diet with carrots and peppers.

*Reproduction.* Many species are colonial. Excluding flamingos, breeding habits are somewhat uniform. Some three to six white, blue or brown eggs are incubated by both sexes for 18–30 days on a crude platform nest (though Hammerheads build a domed nest one metre across). The downy altricial young are reared by both parents, the nestling period lasting up to 18 weeks in large storks. Flamingos lay a single red-yolked white egg which is incubated for four weeks. The young is fed, until fledging (about 10 weeks), on flamingo milk. Secreted by the parent's oesophagus, the milk has 8% protein and 15% fat, but no carbohydrate. Even when young flamingos have left the nest and gathered in creches, the parents apparently feed exclusively their own youngster, presumably recognizing its call.

*Conservation.* Many species have suffered at the hands of man. Flamingo tongues were once relished as a delicacy by Roman Emperors. Similarly, in the last century millions of egrets were slaughtered so that their plumes might adorn hats. The ensuing outcry led to the formation in Britain of the Royal Society for the Protection of Birds (1889) and in the United States of the National Audubon Society (1886). The Sacred Ibis (*Threskiornis aethiopicus*), venerated by the ancient Egyptians, has been absent from that country since the first half of the last century. Today, however, the main threat to the Ciconiiformes comes not from direct hunting but from habitat destruction, especially wetland drainage.

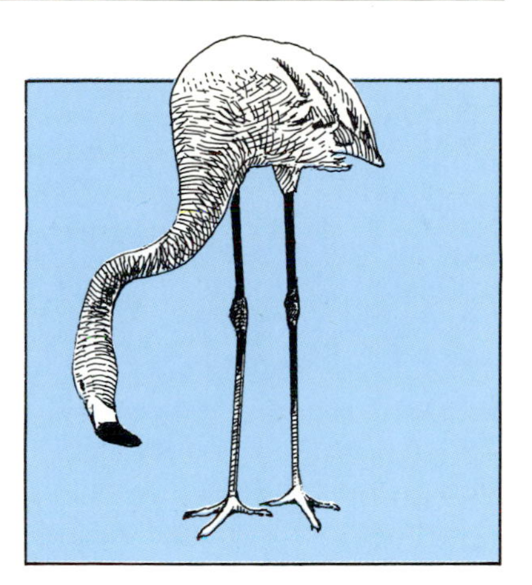

## Order *Anseriformes*

**Families** Anatidae • Ducks, geese and swans • 147 species • Size from 30 cm/220 g (White Pygmy Goose) to 150 cm/15 kg (Trumpeter Swan).
Salt and freshwater wetlands worldwide, except Antarctica.
Anhimidae • Screamers • 3 species • Around 80 cm/3 kg.
South American wetlands.

Although wading and swimming birds, the three screamer species have the general shape of a turkey and the powerful voice of a goose; hence the name. Their beak is short, their legs and feet long and barely webbed and their wings armed with two spurs on the leading edge. As with penguins the feathers grow uniformly over the body, and not in tracts. Unlike all other birds except *Archaeopteryx*, the ribs of screamers lack the uncinate processes which strengthen the rib cage. The bones are unusually well pneumatized, and the sexes very similar.

The general long-necked, web-footed outline of geese, ducks and swans is well known. All species except the Magpie Goose of Australasia and some South American sheldgeese of the genus *Chloephaga* undergo a three or four week period of flightlessness due to the simultaneous shedding of the flight feathers. (Two of the three species of steamer duck *Tachyeres*, South American marine diving ducks, are flightless throughout the year.) Sexes are alike or very unlike. In the latter case the brighter males may take on a drabber plumage, called the eclipse plumage, when moulting the flight feathers.

*Diet.* Screamers mostly eat aquatic plants. The Anatidae feed by diving, dabbling or grazing. The diet is equally varied, ranging from the largely piscivorous mergansers (*Mergus* spp) to the wholly vegetarian geese. Unlike mammals, the vegetarian species lack gut-dwelling bacteria capable of digesting cellulose. Consequently, the birds obtain nourishment only from the plant cell contents which are released by the grinding action of the stone-filled gizzard, but not from the cellulose cell walls that are passed undigested in the faeces.

*Reproduction.* Screamers are monogamous. They lay two to six brownish or greenish-white eggs. After incubation by both sexes for about 6 weeks, the young leave the nest quickly but are fed by the parents.

Swans and geese may pair for life and are similar in plumage. In contrast many ducks pair for a single breeding season only and the sexes are dramatically different in plumage. (The relationship between plumage and mating systems is discussed on pp. 262–3.) Four to fourteen white or pale green or blue eggs are laid in a nest often lined with the female's own down, and incubated by female or male or both. Except for the Magpie Goose, the precocial young feed themselves under the watchful eye of male or female or both. The Black-headed Duck is an obligate nest parasite, always laying in other birds' nests. Other species lay some of their eggs in other nests.

*Relations with humans.* Wildfowl have long been hunted, mostly without undue detriment to the population. It is, for example, open to question as to whether hunting was responsible for the 1875 extinction of the Labrador Duck. The history of wildfowl domestication is also long. The eastern Greylag Goose entered captivity over 4000 years ago and the Mallard over 2000 years ago. The Muscovy Duck was already domesticated in South America in pre-Columbian times. Eiderdown was collected for bedding from Eider nests, particularly in Iceland. Mute Swans in Britain were, and still are, Crown property except where 'royalties' (or ownership rights) are granted to certain companies which mark their swans' bills during the ceremony of swan-upping.

## Order *Falconiformes*

**Families** Cathartidae • New World vultures • 7 species • Up to 120 cm/14 kg (Andean Condor).
Open country and forests from S Canada to southernmost S America.

Sagittariidae • Secretary-bird • One species • About 130 cm/4 kg.
Savannas of sub-Saharan Africa.

Pandionidae • Osprey • One species • About 60 cm/1.5 kg.
Coasts, lakes and rivers worldwide.

Falconidae • Falcons and caracaras • 60 species • Size from 15 cm/45 g (Falconets) to 60 cm/2 kg (Gyr Falcon).
Worldwide except Antarctica.

Accipitridae • Kites, Old World vultures, harriers, hawks, eagles, buzzards, etc. • About 217 species • Size from 30 cm/100 g (small Accipiters) to 140 cm/8 kg (large vultures).
Worldwide except Antarctica.

The characteristic hooked bill, fleshy cere at the base of the bill, strong talons and carnivorous habits of the birds of prey, or raptors, are familiar. The New World vultures, the family that includes the Andean and Californian Condors, share several features with Old World vultures; more or less unfeathered head, a ruff of feathers at the base of the neck, generally black or brown plumage, perhaps with white patches. Nevertheless, New World vultures possess features which justify their position in a separate family. They are voiceless. The nasal septum, which separates the two nostrils, is perforated. Several New World vulture species live in forest where their unusually well-developed sense of smell is a useful aid in carrion detection. So named because its long crest was supposed to resemble the bunch of quill pens used by a secretary in pre-keyboard days, the Secretary-bird is a long-legged diurnal bird of prey, specializing in catching rodents, insects and snakes on the open plains of Africa. The strong, sharp-clawed feet of the Osprey are well adapted to catching its prey, fish. The toes bear spiny gripping studs and the outer toe can be moved to face backwards, giving the two forward, two back arrangement of owls. The falcons and caracaras are generally agile predators, most species relying more on live prey than carrion. Seen against the light the inside of their eggs appears buff while it appears green in the eggs of the Accipitridae. Another point of difference is that the falcons defaecate below the perch while the hawks can squirt a jet of faeces over several feet. The Accipitridae, the largest bird of prey family, includes a diverse array of birds, for example carrion-eating vultures, ponderous eagles, agile bird-catching Accipiters and harriers whose well-developed ears aid prey location as the birds quarter grassland.

In birds of prey the sexes are more or less similar in plumage although females are often larger. In general, species feeding on carrion show no sexual dimorphism. As the prey becomes more agile so the degree to which the female outweighs the male increases, a trend that reaches its peak in the bird-catching Accipiters. A female sparrowhawk weighs almost twice as much as her mate.

*Diet.* The hooked bills and strongly-clawed feet of birds of prey indicate their carnivorous habits. All manner of prey – carrion, snakes, insects, fish, birds, mammals – is eaten. Some species have remarkable feeding habits. The Lammergeier, a vulture of Eurasian mountain ranges, opens bones by dropping them from a height onto rocks. The bird then eats the marrow of the smashed bone. In addition to scavenging, Egyptian Vultures throw stones at Ostrich eggs in order to break them open. Other species have a very specialized diet. The Everglades Kite feeds solely on a single species of freshwater snail (*Pomacea*, the apple snail) while the Honey Buzzard rips open wasp nests to obtain the grubs. In a flurry of dusk activity the Bat Hawk catches its daily bat ration.

*Reproduction.* One to six immaculate or boldly marked eggs are laid in a nest that is often bulky and may either be built by the birds themselves or appropriated from other birds; the latter is particularly the habit of falcons. Incubation, lasting almost two months in the condors, is by both sexes (e.g. vultures) or by the female alone (e.g. Osprey). If the latter the male provides his mate with food at or near the nest. The downy young are tended by both parents until fledging, at five months in the larger eagles and vultures. Even after fledging young eagles, not yet proficient at hunting, may depend on their parents for food for several more months. Brood reduction is frequent in birds of prey. In some species younger chicks simply die of starvation. In nine others (e.g. Harpy Eagle, Lesser Spotted Eagle) the elder chick invariably kills its younger sibling.

Most raptors are monogamous. However, polygyny is frequent among harriers and polyandry known from Galápagos and Harris' Hawks. Most species are territorial but some nest in colonies and then hunt solitarily, while others, including the insectivorous Red-footed Falcon and the griffon vultures, both nest and forage gregariously.

## Order *Galliformes*

**Families** Megapodiidae • Megapodes • 9 species • Size from 26 cm/1 kg to 65 cm/8 kg. Primary forest to scrub in the East Indies, Malaysia, New Guinea, Australia and W Pacific Islands.

Cracidae • Guans, curassows and chachalacas • 42 species • Size from 50 cm/500 g (chachalacas) to 95 cm/4.8 kg (Great Curassow). Americas, from S Texas to N Argentina.

Tetraonidae • Grouse • 16 species • Size from 30 cm/300 g (Black-breasted Hazel Grouse to 90 cm/6.5 kg (male Capercaillie). Forest, plains and tundra of northern Eurasia and America.

Phasianidae • Pheasants, quail and partridge • About 180 species • Size from 14 cm/45 g (quails) to around 2 m/5 kg (Indian Peacock: the length includes 1.5 m of 'tail', actually enlarged tail coverts).
All habitats except wetland. Worldwide except northern Eurasia and southern S America.

Numididae • Guineafowl • 7 species • Around 50 cm/1.5 kg.
Open and forest habitats of sub-Saharan Africa, with an isolated population (Helmeted Guineafowl) in Morocco.

Meleagrididae • Turkeys • 2 species • Around 1 m long and, in the wild, up to 9 kg (males).
More or less open American woodland, from S Canada to Mexico (Common Turkey) and in Guatemala, Belize and S Mexico (Ocellated Turkey).

The Galliformes, or gamebirds, have short rounded wings ill-adapted for sustained flight; only certain quail undertake extensive migrations. The fowl-like gamebirds generally live on the ground supported by large heavy feet that have three front toes and one shorter hind toe. Exceptions are the tree-dwelling cracids and the megapodes. Both these groups have exceptionally large feet where the hind toe is the same size as the fore toes. Another cracid peculiarity is the trachea, elongated in the males of some species into a loop between the pectoral muscles and the skin. This probably serves to amplify the call. The grouse are characterized by partly feathered tarsi and feathered nostrils. In the Rock Ptarmigan the feathering extends to the feet, an adaptation to the birds' snowy habitat. Many species in the largest family, the Phasianidae, have leg spurs. This family includes the relatively drab quail of the New and Old Worlds, the partridges, and the pheasants and peacocks which are among the most spectacular of all birds. Included too is the Junglefowl of South-East Asian forest. This stock probably provided today's domestic fowl. It may have been domestication by Bronze Age people around 4000 BC and was certainly in captivity in India by 3200 BC. Guineafowl also have a long history of domestication, dating back to the Greeks and Romans who called them Numidian birds; hence the modern scientific family name, Numididae.

Because of its prominent place on the dinner table the Common Turkey needs little introduction. Although less abundant than formerly, when an estimated 10 million gobbled across N America, the species' future in the wild has been secured over recent years by active management programmes.

*Food.* Usually omnivorous, taking vegetable matter – seeds, leaves, fruits, buds – and invertebrate prey. Chicks include a particularly high proportion of insects in their diet. Generally ground-feeding, except for the tree-dwelling cracids.

*Reproduction.* The reproductive habits of the megapodes are unparalled among birds. The eggs, white or chalky brown, are laid in mounds or burrows but not incubated by the parents. Instead the necessary heat comes from the sun, from fermenting vegetation or from volcanic activity. When the young hatch they tunnel out of the nest mound or burrow and head into the bush where they immediately fend for themselves. They can fly within 24 hours of hatching.

In the Solomon Islands the nest pits of the Common Scrub Fowl are dug in volcanically-heated soil which provides an incubation temperature of 34°C. Beyond selecting the site, the adults do nothing to regulate clutch temperature.

In contrast, the well studied monogamous Mallee Fowl of southern Australia actively regulates the temperature of its nest mound, a structure used year after year. The male digs a hole about a metre across and a metre deep, fills it with vegetation and covers the whole with sandy soil. From September to January the female lays eggs (total 5–35) at intervals of several days into chambers in the vegetation made by the male. Early in the incubation period most heat comes from fermentation of the vegetation. Excess heat is released when the male uncovers the mound. Later, in mid-summer, most heat comes from the sun so soil is added if overheating is imminent. The eggs hatch very asynchronously after 40–90 days of development at around 34°C.

In the other five families nests are simple, usually on the ground. However, cracids and tragopans (Phasianidae) nest in trees. Eggs are white in cracids. In other families the eggs are buff to olive, either immaculate or spotted. Up to twenty eggs are laid (e.g. by Grey Partridges); thus gamebirds produce among the largest clutches of any birds. In Red-legged Partridges double clutching is known. The female lays one clutch to be incubated by the male and then immediately lays another to incubate herself; over thirty eggs may be laid in all. Except in some Phasianidae and megapodes, incubation, lasting from 16 days in smaller quails to 5 weeks in larger cracids, is by the female alone.

Chicks are universally precocial, able to feed themselves from shortly after hatching. Young are usually tended by the female alone. The emancipation of males from raising young has, through the action of sexual selection, contributed in many species to the evolution of spectacular male plumage. Variation in mating system and plumage dimorphism is especially evident in grouse. In some species (e.g. the Rock Ptarmigan) the sexes differ little and are monogamous. In others (e.g. Sage Grouse and Black Grouse) the males, substantially larger than females, gather in leks to attract females and mate, but play no other role in rearing the young.

## Order *Gruiformes*

This diverse order of ground-feeding birds, that are also usually ground-nesting, includes 12 families. It is the order with the highest percentage of recently extinct or currently endangered species. For example six of the 15 crane species are currently endangered. Sexes are generally alike, with both sharing nesting duties. An exception is the bustards where the duller females undertake all incubation and rearing of young.

**1. Family** Mesitornithidae • Mesites • 3 species • Approx. 30 cm.
Dull brown birds of Madagascar forest and scrub, mesites are poorly known. With a diet of fruit and insects they lay one to three eggs on a platform nest one or two metres above the ground. Young probably precocial.

**2. Family** Turnicidae • Buttonquails or hemipodes • 16 species • Around 15 cm.
Resembling true quails (Phasianidae), the delicately patterned buttonquails dwell in grassland and scrub in sub-Saharan Africa, the Mediterranean area and thence east to China and the Philippines; also Australia. With a diet of seeds and insects the male alone incubates the clutch (usually of four eggs) which hatches in about 13 days, an exceptionally quick period for a precocial species.

**3. Family** Pedionomidae • Plains Wanderer (also known as Collared Hemipode) • One species • Around 12 cm.
Australia. Resembles buttonquails in all respects described above except incubation lasts longer, 23 days.

**4. Family** Gruidae • Cranes • 15 species • Size from 0.9 m/3 kg to 1.8 m/10 kg.
All continents except S America and Antarctica. Cranes are long-legged, long-necked birds that are all migratory. They are omnivorous birds that engage in spectacular displays enhanced by loud calls emanating from an unusual

trachea convoluted like a trumpet's coil. Cranes usually resort to wetlands to breed, laying one to three eggs, variable in colour. Incubation lasts around 30 days and the young are precocial.

**5. Family** Aramidae • Limpkin • One species • 65 cm/1 kg.
The Limpkin is usually found in swamps from the south-eastern USA to C Argentina. So named for its limping gait, the Limpkin is a specialist feeder on freshwater snails. Four to eight buff eggs are incubated in approximately 20 days. Though precocial the young are fed snails by their parents.

**6. Family** Psophiidae • Trumpeters • 3 species • 50 cm/1–1.5 kg.
Grey chicken-sized forest dwellers of northern S America, omnivorous trumpeters live in loud (hence the name) groups. Nesting habits are poorly known. About seven eggs probably laid in tree holes.

**7. Family** Rallidae • Rails • 129 species • Size from 18 cm/30 g (Baillon's Crake) to 60 cm/3 kg (Takahe).
Distribution is worldwide, including a remarkable number of remote islands where endemic species have now lost the power of flight (e.g. the world's smallest flightless bird, the Inaccessible Island Rail). At least 15 forms have vanished in the past century. Rails, usually of subdued hues, live rather surreptitiously at ground or water level. Most are laterally compressed and have short rounded wings; 10 or 11 primaries is the norm. They fall into three groups, the long-billed rails, the stubbier-billed crakes and the fully aquatic coots with lobed feet.
Nests are usually well concealed in low dense vegetation. Two to sixteen brown eggs are incubated for 20–30 days. Young are precocial. They may be cared for by young of an earlier brood, in addition to both parents.

**8. Family** Heliornithidae • Finfoots • 3 species • 30–60 cm.
Despite the scattered distribution on overhung forest streams of tropical America, tropical Africa and SE Asia, the three finfoot species are rather similar, indicating a previously wider distribution. All have a long neck, swim with brightly-coloured lobed toes and bear a long stiff tail. The diet includes frogs, crustacea etc. Two to six speckled cream eggs are laid on a nest one metre above the water. Incubation lasts 11–14 days. The male Sungrebe of tropical America has special underwing folds into which the young can be slotted whether he is swimming or flying.

**9. Family** Rhynochetidae • Kagu.
One species, about 60 cm long, is endemic to New Caledonia. The grey forest-dwelling Kagu is restricted to this western Pacific island's central mountains and is threatened by forest destruction (nickel mining) and introduced mammals. Kagus have an omnivorous diet, eating especially snails. A single speckled buff egg is incubated for 35 days. Young are precocial.

**10. Family** Eurypygidae • Sunbittern.
One species, about 45 cm long, lives along forested watercourses from S Mexico south to central Brazil. The bird, named from the bright orange flashes seen on the outspread primaries, is omnivorous. Two buff eggs laid in a bulky nest are incubated for about 27 days. The downy young are altricial.

**11. Family** Cariamidae • Seriemas • 2 species • About 70–80 cm.
Resembling small cranes, the seriemas live on the grassy plains of central S America. Although omnivorous they are, like the Secretary-bird, adept at catching snakes. They lay two vaguely marked pale buff eggs (incubation about 25 days) in a nest on or close to the ground, and tend the young in the nest.

**12. Family** Otididae • Bustards • 22 species • Size from 40 cm/0.6 kg to 120 cm/18 kg, which places the largest species, the Great Bustard and the Kori Bustard, among the heaviest flying birds.
Bustards, tall striding birds, dwell on the plains, savannas and semidesert of Africa, southern Eurasia and Australia, where they are omnivorous. Often polygynous, the larger males indulge in spectacular highly visible displays. The larger species generally parade on the ground while the smaller ones leap or fly repeatedly into the air. Up to six reddish or olive eggs are laid on the ground. Incubation (3–4 weeks) and caring for the precocial young is the hen's work. Several species are threatened by hunting and agricultural encroachment.

## Order *Charadriiformes*

Although superficially rather diverse the order's 17 families are united by anatomical similarities in palatal bones, voice box and leg tendons. They split into three groups; the 12 families of waders or shorebirds, the four families represented by the skuas, gulls, terns and skimmers and finally the single auk family.

### Sub-order *Charadrii*

**Families** Jacanidae • Jacanas or lily-trotters • 7 species • From 15–30 cm and from 40–230 g.
Tropical marshes and riverbanks of the Americas, Africa, SE Asia and N Australia.

Rostratulidae • Painted snipe • 2 species • About 25 cm/100 g.
Swamps of S America, Africa, SE Asia and Australia.

Haematopodidae • Oystercatchers • 6 species • Around 40 cm/500 g.
Mostly coastlines of all major continents, but not at highest latitudes.

Charadriidae • Plovers and lapwings • 62 species • From 15 cm/35 g to 40 cm/300 g.
All types of open country worldwide.

Scolopacidae • Sandpipers • 81 species • From 13 cm/18 g to 65 cm/1000 g.
Open countryside. Most species breed in temperate, boreal or Arctic regions of N hemisphere and migrate to estuaries and wetlands of more equable latitudes.

Recurvirostridae • Avocets and stilts • 7 species • Around 40 cm/200–300 g.
Fresh to saline waters of all major continents.

Phalaropodidae • Phalaropes • 3 species, all with lobed feet • Around 18 cm/50 g.
Breed beside N Eurasian and American ponds. Winter on freshwater or open ocean.

Dromadidae • Crab Plover • One species • 35 cm/300 g.
Indian Ocean coastlines.
Burhinidae • Stonecurlews or thick-knees • 9 species • Around 40–50 cm/500 g.
Generally open, often arid country of major continents except N America.
Glareolidae • Pratincoles and coursers • 17 species • Around 25 cm/80 g.
Open scrubby country of southern Eurasia, Africa and Australia.
Thinocoridae • Seed snipe • 4 species • Size from 19 cm/60 g (Least Seed Snipe) to 30 cm/300 g (Rufous-bellied Seed Snipe).
Open Andean country from N Ecuador southward.
Chionididae • Sheathbills • 2 species • About 40 cm/500 g.
Antarctic peninsula and sub-Antarctic islands.

Most members of the Charadrii are recognizable as shorebirds, long-legged often long-billed inhabitants of beaches, mudflats and open spaces. The exceptions are:

1. the jacanas which resemble moorhens and trot on aquatic vegetation aided by very long toes (their 'toespan' is about 15 cm).
2. the fork-tailed pratincoles which pursue aerial insects.
3. the quail-like seed snipe.
4. the sheathbills, white scavengers of southern shores.

*Diet.* Most take a variety of animal food picked from the ground or from the water (swimming phalaropes) or hauled from below the ground surface using the lengthy bill. Crab Plovers and seed snipe eat crabs and seeds respectively while the sheathbills are opportunist scavengers, particularly active around penguin colonies.

*Reproduction.* Mating systems vary from polygyny through monogamy to polyandry. Two to six (but four is an especially common clutch) mottled eggs are laid on the ground or in a crude nest (jacanas, sheathbills). The Crab Plover is unusual, laying a single white egg at the end of a two-metre tunnel dug in sand. Incubation, lasting up to a month in the larger species, is undertaken by both sexes, by the female alone or by the male alone (some sandpipers, plovers, jacanas, and all seed snipe and phalaropes). Where one sex incubates it is generally the duller and is courted by the more showy bird of the opposite sex.

Downy and often precocial young are cared for by the sex(es) that incubated them. The dowitchers *Limnodromus* spp. (Scolopacidae) are the exception; females incubate, then males tend the young.

## Sub-order *Lari*

**Families** Stercorariidae • Skuas and jaegers • 6 species • About 50 cm and weighing 1.5 kg (*Catharacta* spp.) or 250–800 g (*Stercorarius* spp.).
Breed in cool temperate or polar zones of both N and S hemispheres. Winter at sea worldwide.
Laridae • Gulls • 45 species • Size from 25 cm/90 g (Little Gull) to 75 cm/2 kg (Great Black-backed Gull).
Worldwide but especially northern coasts.
Sternidae • Terns and noddies • 42 species • Size from 23 cm/50 g (Little Tern) to 50 cm/700 g (Caspian Tern).
Worldwide, mainly coasts.

Rynchopidae • Skimmers • 3 species • About 40 cm and 250 g (females) or 350 g (males).
Coasts and large rivers of tropical Africa, SE Asia, eastern N America and much of C and S America.

The Stercorariidae include the three bulky *Catharacta* species, of which the Great Skua is the most familiar to northern hemisphere bird-watchers, and the three lighter more agile *Stercorarius* species, also known as jaegers. All are brown with white wing flashes and, in contrast to the next three families, the females are slightly larger than males. Gulls and terns are familiar long-winged birds that combine white, grey, and black plumage. The three skimmer species, black above and white or pale below, have uniquely among birds an eye pupil that is not round. Instead it closes to a narrow cat-like vertical slit. The lower mandible is markedly longer than the upper.

*Diet.* Skuas and gulls have a catholic diet – fish, seabird eggs, human refuse. The larger species are frequently predatory and skuas chase other seabirds to make them disgorge. Many terns eat fish, small squid and crustacea caught by plunging into sea or freshwater, but the marsh terns take insects and amphibia. The skimmers' feeding trick involves flying low over the water with the lower mandible knifing through the surface layer. When the mandible hits prey, primarily fish, the jaws snap shut. The shock of the impact is contained by extra strong neck muscles.

*Reproduction.* Monogamy is normal. The nest is often an unlined ground scrape but sometimes a more substantial lined structure. Noddies nest in trees and on ledges. Fairy Terns make no nest; they may lay the single egg directly onto the fork of two branches. Clutches of one to three blotched green or brown eggs normal, or of three to four, and occasionally five eggs in skimmers. Downy young fed by parents. Fledging period 7 weeks in larger species.

## Sub-order *Alcae*

**Family** Alcidae • Auks • 22 living species • Size from 16 cm/90 g (Least Auklet) to 45 cm/1 kg (Common Guillemot or Murre) and 5 kg (Great Auk, extinct 1844).
Coastal waters and oceans of N hemisphere. Rarely south of 25 degrees N.

As is common among birds that swim and then dive to catch underwater prey, most auks are dark above, pale below. Although the often-bright feet are webbed, underwater propulsion is mainly achieved by beating the wings. Consequently, wing area represents a compromise. The wing must be large enough to keep the birds airborne but small enough to beat underwater. Because the result of this compromise is a small wing, the wings are beaten fast and the wing loading of auks is unusually high for a bird, about 16 kg/m². The largest species, the extinct Great Auk, became flightless like penguins. Several species (e.g. Atlantic Puffin) have bright bills during the breeding season. The bill plates are shed during moult.

*Diet.* Fish and crustacea are caught by diving underwater to depths of up to 180 m (Common Guillemot) and 60 m (Atlantic Puffin).

*Reproduction.* Some species breed solitarily but most breed in very dense colonies, up to 70 pairs per square metre in the Common Guillemot. Nest sites, usually coastal, include cliff ledges, crevices and burrows. Eggs are one or two, white or buff or shades of green-blue, more or less marked. Incubation, usually by both sexes, lasts for 29–42 days. The downy chick stays in the nest for a variable period, 2–50 days. At one end of this spectrum occupied by the majority of species (e.g. puffins) the chick is fed at the nest site until it is nearly full grown and can then depart and feed itself at sea. At the other end are four Pacific murrelet species (e.g. Xantus's Murrelet) that take their two young to sea within two days of hatching. The Razorbill and the two guillemot (murre) species are intermediate. The father accompanies the single young to sea when it is one quarter grown at about three weeks old.

## Order *Columbiformes*

**Families** Pteroclididae • Sandgrouse • 16 species • Around 30–40 cm and 200–400 g. Arid lands of Africa, southern Europe and SW and C Asia.

Raphidae • Dodo and solitaires • 3 species (all extinct) • Around 1 m/(10 kg?). One species endemic to forests of each of the three Mascarene Islands, Mauritius, Réunion and Rodrigues.

Columbidae • Pigeons and doves • About 300 species • Size from 15 cm/30 g (Diamond Dove) to 80 cm/2.4 kg (Victoria Crowned Pigeon). Woodland; fewer species in more open country. Distributed worldwide except Antarctica and higher northern latitudes.

The sandgrouse share with pigeons a thick heavy plumage and certain skeletal features. In egg pattern and feather structure they more resemble the Charadriiformes and are sometimes placed in their own order, the Pteroclidiformes. They are quite unrelated to grouse (Tetraonidae). Sandgrouse are strictly terrestrial and are marked in blacks, greys and browns to provide camouflage in their open habitat. Males are more boldly patterned than females.

The dumpy flightless Dodo and solitaires almost certainly evolved from flying pigeon-like ancestors. Man's hunting, introduced predators (cats, pigs etc.) and forest fires probably all hastened their demise; the Dodo of Mauritius became extinct about 1665, the Réunion Solitaire around 1715 and the Rodrigues Solitaire around 1760.

Pigeons are easily identified by their small head, with bare cere, short neck, soft dense plumage and cooing call. Incidentally, there is no significant difference between pigeons and doves; the larger species tend to be called pigeons while the smaller ones are termed doves. Grey and brown plumages are most common but some tropical species, especially fruit pigeons, are brighter. Sexes are usually similar.

*Diet.* Sandgrouse subsist on dry seeds. This thirst-promoting diet must be supplemented by water so sandgrouse regularly (about every other day) fly up to 30 km to water holes. Drinking flocks can number thousands.

Dodos and solitaires were probably vegetarian, as are today's pigeons. Some species of pigeon eat green leaves, others seeds or fruit. The stomach of some fruit-eating specialists rubs the nutritious pulp or pericarp off the seed, which is then coughed up. Pigeons can drink by sucking up water, without needing to tilt back the head.

*Reproduction.* Sandgrouse are monogamous. Brown-blotched eggs, usually three, are laid in a rudimentary ground scrape. Incubation takes 21–31 days, the female taking the day shift, the male doing night duty. Downy precocial young feed themselves but water is provided principally by the male. At the drinking hole he shuffles to introduce water into the sponge-like grip of the modified belly feathers. After carrying this water over distances as great as 30 km the male stands erect so his chicks can drink from the central belly groove of his plumage. The fledging period is 4–5 weeks.

The Rodrigues Solitaire apparently laid a single egg and chicks gathered in creches. The other species' breeding habits are unknown.

Pigeons are monogamous. They lay one or more commonly two rather small white eggs on a twig platform. Incubation lasts 2–4 weeks, the male taking the day shift, the female doing the night duty. The altricial nestlings are initially fed entirely on pigeon milk, a secretion of the crop of both parents. The milk has the consistency of cream cheese, with an approximate composition of water 75%, protein 15%, fat 9% and minerals 1%. Regurgitated items from the parents' diet are later added to milk in the chick's meals. Fledging of the so-called squabs takes under 2 weeks in the smaller species, up to 3 months in the larger.

## Order *Psittaciformes*

**Family** Psittacidae • Parrots • About 330 species • Length 9–100 cm. All major landmasses (except Antarctica) south from about 20 degrees N.

Containing such birds as lories (Australasia and Oceania), cockatoos (Australasia), lovebirds (Africa), Budgerigars (Australia) and amazons and macaws (C and S America), the parrot family is united by several characters. Attached to the skull by a flexible joint, the short strong curved upper mandible fits over the shorter lower mandible. The tongue and jaw musculature are unusually complex and often powerful; some species can crack Brazil nuts. Parrots, typically arboreal, have strong zygodactyl (two toes forward, two back) feet. These can be used to manipulate food during eating. Indeed few birds surpass parrots in 'manual' dexterity. Their nearest relatives are pigeons and cuckoos.

Feathers of parrots tend to be sparse, hard and glossy with powder downs intermingled. Although many parrots are more or less green, some species sport vivid colour combinations of red, yellow and blue. The Eclectus Parrot of New Guinea and Australia is unusual in that the female (red and blue) is more brilliantly coloured than the male (basically green).

Although principally tropical, sedentary and reaching their greatest diversity in Australia and S America, parrots were formerly more widespread. There is, for example, a fossil parrot from the lower Miocene (approx. 20 million years ago) of France. More recently the extinct Carolina Parakeet, last seen in the Florida Everglades in the 1920s, ranged north to North Dakota and New York.

*Diet.* Most parrots are vegetarians, eating fruits, nuts and grain, although some species include a proportion of insects in the diet. The pygmy parrots of New Guinea add lichens and fungi to the above fare. The Kea of New Zealand has developed the unsavoury habit of gouging fat and muscle from dying sheep. The lories and lorikeets are nectar specialists, using their elongate tuft-tipped tongue to mop up nectar and pollen.

*Reproduction.* The nest is an unlined hole in a tree, inside a termite nest or on the ground. Several pairs of the South American parakeet club together to build a communal stick nest in a tree. Within the structure each pair has its own nest.

Eggs are white; from one (large macaws) to twelve (small species). Incubation, generally by the female, but sometimes by both sexes, lasts 16–35 days. The altricial young, initially naked, are tended by both sexes and fed by regurgitation. Larger species may take over two months to fledge.

Although many species are highly sociable, parrots are usually monogamous. An exception is the flightless and threatened Kakapo of New Zealand. This species is the heaviest parrot (average weight of males 2.1 kg and females 1.3 kg). The booming call of males attracts females by night to the lek. After mating there the females assume total responsibility for the incubation and rearing of young.

*Special features.* Although certain parrot species, for example the Galah of Australia, have become so numerous as to rank as agricultural pests, some 30 species are threatened. The principal reasons are destruction of forest habitat and their popularity as easily tamed cage birds. This popularity stems partly from their bright colours and partly from their ability to mimic the human voice. Although flocks of wild parrots are noisy, they are not known to mimic other species and so it is not clear why parrots have the ability to 'talk'. The African Grey Parrot in particular excels as a mimic.

## Order *Cuculiformes*

**Families** Musophagidae • Turacos • 22 species • Size from 35 cm/230 g to 75 cm/1 kg. Forested zones of sub-Saharan Africa.
Cuculidae • Cuckoos, anis, ground cuckoos (incl. Roadrunner) and coucals • About 128 species • Size from 18 cm/30 g to 65 cm/700 g. Range worldwide excluding highest latitudes and some oceanic islands.
Opisthocomidae • Hoatzin • One species • 60 cm/800 g. Wooded riverbanks in the Atlantic drainage of northern S America.

Allied to cuckoos by similarities in foot structure and egg-white protein, the arboreal turacos are medium-sized, often brightly coloured birds. The feathers of the head and breast commonly lack binding barbules, and consequently seem hairy. Although green is a common enough colour in birds, it is usually due to the interaction of two or more pigments, or of pigment and feather structure. Only in the noisy sociable turacos is the colour due to a green pigment, called turacoverdin. The red colour often seen in turaco wings is due to another pigment, turacin. If an isolated red feather is used to stir water, the copper-based turacin leaches out to stain the water – but this does not happen when a living turaco is drenched in a rainstorm!

The cuckoos and allies are generally brown, grey or black with little difference between the sexes. Most are arboreal and solitary but some are terrestrial (e.g. Roadrunner) or social (e.g. group-living anis). The family includes the European Cuckoo whose disyllabic call needs no description; and in fact many other species have similar simple calls.

The Hoatzin, whose general appearance resembles a shaggy guan, poses a taxonomic puzzle, but at least its egg-white proteins suggest an affinity with cuckoos.

*Diet.* The turacos are frugivorous, although their vernacular name of plantain-eater is misleading for they eat neither plantains nor bananas.

Most cuckoos are insectivorous, many specializing on noxious hairy caterpillars shunned by other birds. Snails, fruit and small vertebrates are also eaten.

The Hoatzin eats green leaves which, prior to digestion, are crushed by the horny walls of the enlarged muscular crop and not, as is usual in birds, by the gizzard. Because of the crop's enlargement the sternum and flight muscles are small and powers of flight correspondingly feeble.

*Reproduction.* Turacos are probably monogamous. They lay two to three white or pale eggs in a nest placed 5–20 m up in a tree or bush. Incubation by both sexes lasts around 3 weeks. The downy young, reared on regurgitated fruit and insects, remain in the nest for only some 10 days before scrambling into the nest tree well ahead of fledging at 4 weeks.

Within the cuckoo family all 47 members of the sub-family Cuculinae and three species in the Neomorphinae are brood parasites that rely on other species to raise their young. The several females of an ani group lay about six eggs each in a shared communal nest. The outcome is a very large clutch, up to 29 eggs. Other species have more conventional habits. Two to six eggs are incubated on a crude platform nest by both sexes for about 2 weeks and the altricial young are raised by both parents.

Hoatzin nests, usually over water, support two to three eggs which are incubated for 4 weeks and are much prized as human food. Both parents, plus up to half a dozen helpers, feed the young by regurgitating green leaves, an exceptional diet for a growing chick. Within a few days of hatching the chick leaves the nest to explore the surrounding branches, where progress is aided by two remarkable large claws on the wings. If danger threatens, the chick drops into the water below, dives if necessary, and then swims to the nearest branch before climbing back to the vicinity of the nest. Claws and swimming ability are lost by fledging.

*Special feature.* Roadrunners drop their body temperature when the air temperature falls sharply, for example at night. To warm up again they expose dark pigmented skin on the back to the sun's dawn rays.

## Order *Strigiformes*

**Families** Strigidae • (Typical) owls • 124 species • Size from 12 cm/40 g to 70 cm/4 kg. All habitats worldwide except Antarctica and some remote islands.

Tytonidae • Barn owls • 10 species • Size from 23 cm/180 g to 50 cm/1300 g. All habitats on most major land masses but absent from extreme north, C Asia and New Zealand. The (Common) Barn Owl is perhaps the world's most widely distributed species.

Owls are familiar and instantly recognized by their upright stance, generally brown plumage (desert species are paler), hooked bill and sharp talons, with a reversible outer toe. But the forward-facing eyes are the most striking feature. They are virtually fixed in the eye sockets and peer out of a facial disc. Because the eyes are fixed the neck is very flexible.

Differences between the typical owls and barn owls are slight. The facial disc of the latter is heart-shaped rather than circular, there are minor differences in bone structure and the barn owl middle toe has a serrated comb on the claw.

*Diet.* Owls catch mammals, birds, insects, frogs, crustacea and fish, but rarely eat carrion. Smaller prey are swallowed whole and the indigestible parts such as bones, feathers or beetle carapaces regurgitated in pellets which can be analysed.

At least 80 of the 134 owl species catch their prey by night, either by pouncing from a perch or by slowly quartering the terrain. The energetic cost of this latter hunting technique is reduced because the owl's flight is buoyant due to the bird's low wing loading.

Although the binocular vision of owls certainly facilitates nocturnal hunting, the absolute visual threshold of the Tawny Owl, reflecting the darkest conditions in which it can usefully see, is only two to three times superior to the threshold in humans. Hearing is probably equally important in prey detection. Exceptionally large ears border the facial disc which in fact may serve to channel sound into those ears. Precise location of sounds is aided by the fact that the ears are asymmetrically placed on the skull. Detecting prey sounds is also helped by the unusually quiet flight of owls, a consequence of their soft plumage. This carries the additional advantage that the prey cannot hear the owl's approach. Interestingly the fishing owls do not have soft plumage, perhaps because the prey is unlikely to hear the approaching owl.

*Reproduction.* Owls are monogamous and commonly territorial. The species-specific calls play a role in maintaining territories. Nests may be burrows, holes in trees, old nests of eagles etc; little material is added. Up to twelve white eggs are laid. In some species the number varies according to the food supply; in years of plenty a large clutch, in lean years breeding is foregone. Lasting up to 5 weeks, incubation is by both sexes or by the female alone, in which case the male, often the smaller sex (see also birds of prey), may feed his mate on the nest. The downy nestlings (fledging period up to 8 weeks) are tended by both sexes. Asynchronous hatching is common so, in times of food shortage, the smaller younger owlets die and are eaten by their bigger siblings.

## Order *Caprimulgiformes*

**Families** Caprimulgidae • Nightjars or goatsuckers • 70 species • Size around 25 cm and 50–100 g.
More or less open habitats (occasionally forest) in temperate and tropical regions.

Podargidae • Frogmouths • 12 species • About 30–50 cm.
Tropical rainforest of SE Asia and Australia.

Aegothelidae • Owlet-nightjars • 8 species • Around 25 cm and 50–100 g.
Dense brush or more open woodland of Australia and islands of extreme SE Asia.

Nyctibiidae • Potoos • 5 species • 25–50 cm.
Forest or semiforest from S Mexico to S Brazil and Paraguay.

Steatornithidae • Oilbird • One species of 45 cm/400 g.
Forested country from central Peru north to Venezuela; also Trinidad.

All species have long wings, feeble feet, a short bill and a gaping mouth. As befits birds that are primarily nocturnal or crepuscular the nightjars and allies are dressed in sombre grey or brown plumage, often delicately patterned with streaks or bars. By day they roost quietly, relying on superlative camouflage. Nightjars may roost on the ground, or lengthwise along branches, a habit shared with frogmouths. Owlet-frogmouths perch more conventionally across branches. Potoos sit atop stumps. Oilbirds spend the day in their breeding caves.

The order's scientific designation and the name goatsucker stem from the ancient and incorrect belief that nightjars used their large mouths to milk goats.

*Diet.* The Oilbird is apparently the only nocturnal fruit-eating bird. A well-developed sense of smell may be important in locating ripe fruit. Other species are primarily insectivorous; moths are especially prominent in the diet. Prey is caught during aerial chases (nightjars), perhaps on the ground (owlet-nightjars), by pouncing to the ground (frogmouths) or by brief flycatcher-like sallies (potoos).

*Reproduction.* Details are poorly known because of the birds' cryptic solitary habits; only the Oilbird is colonial. The nest is sparse or non-existent in ground-nesting nightjars. Oilbirds make a more substantial nest by regurgitating fruit onto a cave ledge. One to four white or buff more or less blotched eggs are laid. Incubation, lasting up to 35 days in Oilbirds and larger potoos, is typically undertaken by both sexes. Nestlings are usually fed by both parents on insects held in the bill, or on regurgitated fruit in the case of the Oilbird. The fledging period normally lasts up to 50 days, but again the Oilbird is exceptional. The young Oilbird reaches a maximum weight 50% above the parental weight around its 70th day and then loses weight as it completes development in the remaining 30 or so days. Fat young Oilbirds were harvested to yield cooking and lighting oil; hence the name.

Most species are probably monogamous. However, many nightjars are sexually dimorphic and some males (e.g. Standard-winged Nightjar) sport the spectacular ornaments characteristic of polygynous species.

*Special features.* The Poorwill of western N America is the only bird known to hibernate. Arizona Indians long knew of the phenomenon which only came to ornithologists' attention in 1947 when it was learnt that these nightjars may creep into canyon crevices to sleep through the winter at a body temperature of 18 degrees C, well below the normal 39 degrees.

Oilbirds use echolocation to detect obstacles in the dark caves where they breed and roost. Resonating with the birds' clicks and cries a cave containing thousands of nesting Oilbirds is one of nature's marvels.

## Order *Apodiformes*

**Families** Apodidae • Swifts • About 80 species • Size from 10 cm/8 g (Glossy Swift) to 25 cm/175 g (White-naped Swift).
Worldwide except higher latitudes and remoter islands. Temperate species migratory.

Hemiprocnidae • Crested swifts • 3 species • Around 20 cm/100 g.
Open woodland of SE Asia.

Trochilidae • Hummingbirds • About 320 species • Size from 6 cm/2 g (Bee Hummingbird, the world's smallest bird) to 22 cm/20 g (Giant Hummingbird).
All habitats in the Americas from southern Alaska to Patagonia, but most species 10 degrees N to 25 degrees S.

Their scientific name implies that swifts and hummingbirds lack feet. Though this is incorrect, they do have very short legs. Additionally the wings are pointed and long because, although the humerus is short, the 'hand' is lengthy.

Contrary to appearances, the swifts do not fly exceptionally fast nor are they related to swallows and martins. Generally brown, black or blue (with similar sexes) the swifts are perhaps the most aerial of all birds. While on the wing they mate, collect nest material, drink and even, in some species, pass the night at high altitude.

Hummingbirds, frequently clad in dazzling iridescent plumage (males normally brighter), are renowned for their flying achievements. They fly at about 12 m/s. Depending on species the wings are beaten 15–78 times per second causing the humming noise after which the group is named. The flight muscles are accordingly very well developed, amounting to 25–30% of body weight (the average among birds is about 15%). Since lift is provided by both up and down strokes, powered by the pectoral and supracoroncoid muscles respectively, these muscles are of roughly equal size in hummingbirds. In contrast in a 'normal' bird like a tit, the pectoral muscle is 15 times larger.

*Diet.* Aided by a wide gape swifts consume small aerial insects and spiders. Over 1000 insects can be delivered to a young Common Swift when a parent regurgitates a single bolus of food. It is not known whether swifts primarily snap up individual insects or locate a swarm and then repeatedly pass through it, mouths agape.

The crested swifts are not quite so aerial as the true swifts. Nevertheless, they feed mostly on flying insects caught at dusk and dawn.

Hummingbirds are nectar specialists, supplementing their diet with insects caught in flight or snatched from leaves or spiders' webs. Nectar is taken either while hovering or when the hummingbird perches in front of a flower.

There may be a close correspondence between the shape of the bill of a hummingbird species and the shape of the flowers on which that species specializes. For example, the extraordinary Swordbill with a 10 cm bill uses it to extract nectar from a passionflower with an 11 cm corolla. Some species can defend clusters of flowers from other intruding hummingbirds; others depend on visiting flowers scattered over a wide area, a strategy called trap-lining.

*Reproduction.* Swifts (including crested swifts) often nest colonially in sites allowing ready aerial access, for example inside chimneys and caves or under the overhang of a waterfall. The nest is glued in position by the birds' saliva secreted by salivary glands which, in both sexes, enlarge during the breeding season. Certain cave-nesting swiftlets of SE Asia make their nests entirely of saliva; such nests provide top grade birds' nest soup. One to six pure white eggs are laid and incubated by both sexes for 17–28 days. The nestling period is variable (up to 3 months). When food is scarce nestling development slows, and they become torpid and cool.

Many trap-lining male hummingbirds (those that visit widely scattered, undefended nectar flowers at intervals) assemble at leks for mating. Indeed, mating is the only contribution of most male hummingbirds to breeding. The nest, built of plant material and spiders' webs and lichen-camouflaged, is saddled on a twig or suspended from a leaf tip. Normally the female alone incubates the two white eggs for 14–23 days (depending on species) and raises the nestlings on a regurgitated mix of nectar and insects (fledging period 18–38 days).

*Special feature.* Some cave-dwelling swiftlets navigate within their dark breeding caves by echolocation. The frequency (pitch) of their audible clicks is similar to those of the Oilbird.

## Order *Coliiformes*

**Family** Coliidae • Mousebirds or colies • 6 species • Size *c.* 30 cm (half is the tail)/50 g. Wooded and bushy country of sub-Saharan Africa, excluding Madagascar.

Placed in the single genus *Colius* the non-migratory mousebirds are clad in drab brown or grey soft plumage which is poorly waterproof. The 10-feathered tail is long and sharply graduated with stiff shafts. Unusually the outer toes, first and fourth, are reversible and so can be used forwards or backwards. The acrobatic mousebirds commonly feed and sleep upside down. Mousebirds are highly social, habitually gathering by day in flocks up to 20 strong and often sleeping in tightly-packed clusters. At least one species, the Speckled Mousebird, becomes torpid at night. The name mousebird derives from the ability to scurry rodent-like through trees and along the ground.

*Diet.* The stubby bill, fleshy about the nostrils, is used for stripping buds, leaves and fruits. Mousebirds are therefore agricultural pests. Insects are also eaten.

*Reproduction.* Two to seven, usually three, eggs are laid. The eggs may be white or buff, immaculate or streaked brown. Incubation by both sexes, in an open cup nest placed in a bush, takes 12–14 days. The altricial young, fed regurgitated food by both parents, fly some 18 days after hatching.

## Order *Trogoniformes*

**Family** Trogonidae • Trogons • About 35 species, mostly around 30 cm long.
Forest and woodland of America (21 species), sub-Saharan Africa (3 species) and the Oriental region from India to Philippines (11 species). Mostly restricted to tropical latitudes but may extend altitudinally to temperate zone.

The beautiful trogons share several peculiarities, despite living on three continents. Their feet have two toes in front and two behind but, uniquely in trogons, it is the first and second toes that have shifted to the rear. The skin is particularly flimsy and the feathers inclined to fall out when museum skins are prepared. The bill is short and the eye surrounded by brightly coloured bare skin. There is therefore no doubt of the trogons' common ancestry, but where these birds evolved is uncertain. A possible fossil trogon from the Oligocene of France suggests a wider distribution formerly.

Males are usually metallic green or brown above. Below, they are variously red, blue, green or yellow. Females are drabber. A resident of montane rainforest from Mexico south to Panama is the Resplendent Quetzal. Adorned by 60 cm tail plumes the bird was sacred to the Mayas and Aztecs and became incorporated in the pictorial representation of their god Quetzalcoatl.

*Diet.* Trogons spend much time perched quietly on mid-level branches whence they sally forth on the wing to pluck fruits and to snatch invertebrates from the air or from foliage.

*Reproduction.* Two to four immaculate eggs of various colours are incubated by both sexes for 17–19 days. Nestlings, initially naked, are tended by both parents for around 17 days before fledging from the nest cavity. Many species, especially in the New World, excavate their own nests in rotting wood or in the papery nests of wasps which are eaten, along with the larvae, in the process.

## Order *Coraciiformes*

The nine coraciiform families include brightly coloured land birds, mostly living in tropical or sub-tropical climes, especially the eastern hemisphere. All members have the three front toes joined for at least part of their length. All nest in cavities, digging holes in, for example, earth banks or rotten trees. Although some kingfishers lay up to ten eggs, clutches of two to seven are the norm. The eggs are white or pale and rarely marked. Except for downy young cuckoo-rollers, hoopoes and woodhoopoes the young are born blind and naked. All are cared for in the nest.

**1. Family** Alcedinidae • Kingfishers • 87 species • Size from 10 cm/8 g (African Dwarf Kingfisher) to 46 cm/500 g (Laughing Kookaburra or Jackass).
Worldwide except highest northern latitudes and some oceanic islands.

Frequently brilliantly coloured, the dagger-billed kingfishers have a partly joined second and third toe and live in a variety of habitats – mangrove, rainforest interior, savanna and, of course, the familiar waterside habitats.

*Diet.* As variable as habitat. Fish are caught by diving; arthropods and small vertebrates are caught by sitting and pouncing down to the forest floor. Reptiles are common prey for kookaburras.

*Reproduction.* Incubation is by both sexes for 18–22 days and the nestling period is 20–30 days. Helping behaviour is known in the Laughing Kookaburra and the African Pied Kingfisher.

**2. Family** Todidae • Todies • 5 species • About 11 cm.

The five West Indian tody species – one species each on Cuba, Jamaica and Puerto Rico and two on Hispaniola – are bright little forest insectivores. During the three-week incubation period in a burrow that is often right-angled, todies are rather inattentive. Each adult may spend less than one-quarter of its time brooding. Fledging period around 19 days.

**3. Family** Momotidae • Motmots • 8 species • 16–50 cm.
Tropical forest and more open country from Mexico south to NE Argentina and Paraguay.

The brilliantly blue-green or brown motmots perch languidly and then pounce onto forest arthropods to snatch them in a characteristically serrated bill. In most species the vanes of the two longest central tail feathers are missing towards the tip. A racket-tail is then attached by the exposed shaft. Both sexes dig the nesting burrow and incubate for 15–22 days. The fledging period is 3–5 weeks.

**4. Family** Meropidae • Bee-eaters • 24 species • Size from 17 cm/15 g (Little Bee-eater) to 35 cm/80 g Blue-bearded Bee-eater).
Forest and open country of Africa, southern Eurasia and Australia.

Among the Old World's most colourful birds (sexes similar), bee-eaters usually have elongate central tail feathers. The longish pointed bill is used as the birds sally forth to snap up dragonflies, flying bees and other venomous Hymenoptera. The bee-eater then removes the sting by rubbing and beating its prey on a perch before swallowing.
*Reproduction*. Forest bee-eaters and smaller open country species are both solitary nesters. Other species are colonial; some embankment colonies can house over 10 000 Rosy Bee-eaters. In such colonies helping behaviour often occurs, usually between related birds. For example, up to five helpers can assist at nests of the East African White-fronted Bee-eater. Egg-dumping has also been recorded. Incubation and fledging last around three and four weeks respectively.

**5. Family** Leptosomatidae • Cuckoo-roller • One species, 43 cm long, living in woodland of Madagascar and the Comoros Islands.

The little-studied Cuckoo-roller, the only coraciid with powder down, probably evolved from an early roller invasion of Madagascar. The species lives on large insects and chameleons. These are fed by the male to the female when she alone incubates.

**6. Family** Coraciidae • Rollers • 16 species (including five species of ground-roller of Madagascar, sometimes separated in the family Brachypteraciidae) • 25–45 cm.
Forest and open country of Africa, southern Eurasia and Australia. Most species migratory.

The wide-billed large-headed rollers are often mainly blue. They get their name from a rolling, rocking, courtship flight. Food is mainly large arthropods, either pounced upon or caught in flight. Incubation and fledging periods are 18 and 25–30 days respectively.

**7. Family** Upupidae • Hoopoe • One species (28 cm/70 g) which lives in open country in Africa and Eurasia south of about 60 degrees N.

Known from ancient Egyptian hieroglyphics, the fawn Hoopoe is split into about nine subspecies, all of which reveal striking black-and-white wing stripes in flight. Probing with its decurved bill, the Hoopoe extracts ground-dwelling invertebrates. Incubation by the female alone lasts 17 days and the nestling period about 4 weeks, during which time no faeces are removed from the nest. Hoopoes use stinking preen-gland secretion for defence.

**8. Family** Phoeniculidae • Woodhoopoes • 6 species • 22–45 cm.
Forest of sub-Saharan Africa, excluding Madagascar. Fossils known from Europe.

With long tails and decurved scimitar bills the woodhoopoes are black and often glossy. They feed primarily by bark-probing. As with Hoopoes the preen gland produces a foul defensive secretion. Several species live in groups. Where this happens group members help the single breeding pair by feeding the brooding female (clutch around three green/blue eggs; incubation 17 days) and the young (fledging period about 4 weeks).

**9. Family** Bucerotidae • Hornbills • 45 species • Size from 40 cm/85 g (*Tockus* spp.) to 1.6 m/4 kg (Helmeted Hornbill).
Forest, savanna and semi-desert of sub-Saharan Africa and SE Asia (to New Guinea).

Uniquely among birds, the first two neck vertebrae (atlas and axis) of hornbills, the largest coraciids, are fused. Mainly brown, black or white, the hornbills are quite unrelated to their New World ecological counterparts, the toucans. Their name is due to a bony casque mounted on the bill but the exact function of this structure is not known.

Other features are long eyelashes, bare and often bright skin around the eyes, fairly pronounced sexual dimorphism and flight that is noisy and whooshing because the short underwing coverts do not cover the flight quills.

*Diet.* Hornbills are omnivorous. Forest species concentrate on fruit. Ground hornbills are carnivorous.

*Reproduction.* This is remarkable. Except for the two ground hornbills, the female begins sealing the entrance of the nest hole with mud. She then enters and, but for a narrow slit, seals herself within with faeces and food remains, with or without male assistance. The sealed-in hen lays the clutch (one to six white eggs) and, in most genera, moults all her flight feathers, becoming temporarily flightless. Incubation lasts 25–40 days and the young fledge in 45–85 days. The female, once more able to fly, breaks out of the nest in the middle of the nestling period or alongside the fledging chicks. This latter feature varies with species. During the female's imprisonment she is fed by her mate. In the African *Bucorvus*, members of the group, not just the male, bring food to the nest.

## Order *Piciformes*

Woodpeckers and their allies possess zygodactylous feet where two toes point forward and two back. The leg muscles and tendons used to work these feet are also distinctive. Excluding some temperate zone woodpeckers, all species are non-migratory. All lay pure white eggs in holes from which usually hatch naked blind altricial young (jacamar young have long white down).

**1. Family** Galbulidae • Jacamars • 15 species • 15–30 cm.
Forest, secondary growth and savanna from S Mexico to S Brazil.

Often brilliant green, the long-billed jacamars are viewed as the New World's ecological equivalent of the Old World bee-eaters. Like the latter they perch quietly before pursuing flying insects, but they are not bee specialists. Both sexes dig the nesting burrow, incubate the two to four eggs for about three weeks and raise the young to fledging, also in about 3 weeks.

**2. Family** Bucconidae • Puffbirds • 30 species • 15–30 cm.
Forest, secondary growth and savanna from S Mexico to S Brazil.

Often dullish brown and grey, the large-headed puffbirds sit with neck feathers puffed out; hence the name. They feed like jacamars. Two or three, rarely four eggs are laid. The incubation period is unknown. The White-whiskered Puffbird pair share out incubation in an unusual manner. The male sits from early afternoon through to the following dawn when the female takes over for her single 5–8 hour shift. The fledging period is about 3 weeks.

**3. Family** Capitonidae • Barbets • About 76 species • Size from 9 cm (tinkerbirds) to 35 cm (Great Barbet).
Forest, secondary growth and savanna of tropical America, sub-Saharan Africa and SE Asia, east to Borneo. This fragmented tropical distribution parallels that of trogons.

The plump gaudy barbets have a robust conical bill surrounded by bristles, which give a bearded effect. Calls of Old World species have a metallic ringing quality, leading to such nicknames as tinkerbirds. Barbets eat fruit and insects. Incubation of two to four eggs lasts about 2 weeks. Fledging periods are 3–5 weeks. In some species the juveniles of a first brood help feed the nestlings of a second.

**4. Family** Indicatoridae • Honeyguides • 15 species • Size from 10 cm/10 g to 20 cm/55 g (Greater Honeyguide).
Forest and open woodland of sub-Saharan Africa, Himalayas and extreme SE Asia.

Uniquely among non-passerines, the honeyguides have only nine primaries. The drab brown, grey or olive plumage of honeyguides belies their singular habits. Confusingly, the birds eat little or no honey. Instead, in addition to insects, they eat wax from bees' nests. The digestion of wax may be aided by symbiotic bacteria or by special digestive enzymes. At least two species, the Greater and the Scaly-throated Honeyguide, obtain their wax ration by guiding. They give a chattering call which attracts the attention of a large mammal, commonly a human or a honeybadger. The bird then flies towards a known bee nest in short stages. Led to the nest by the honeyguide, the follower opens it up to obtain the honey, leaving a wax reward accessible to the honeyguide.

Most honeyguides are brood parasites, laying thick-shelled eggs in the nests of hole-nesters like woodpeckers and barbets. Incubation period is short at 12–13 days. The young parasites kill the host young and fledge after 38–40 days. Only the polygynous Himalayan species, the Indian Honeyguide, rears its own young.

**5. Family** Ramphastidae • Toucans • Around 40 species • From 15–66 cm • Toco Toucan is largest.
Forest and sparser woodland from S Mexico to N Argentina.

Characterized by bright colours and bare skin around the eye, the toucan's most noticeable feature is a large, gaudy but surprisingly lightweight bill. Used for plucking difficult-to-reach fruits, the bill is strengthened by a honeycomb network of bony fibres within the horny outer sheath. In addition to fruit, toucans eat insects and prey on birds' nestlings, lizards etc.

Toucans commonly roost in tree holes. Due to an unusual muscular and bony attachment the tail can be folded over the bird's back so that the tip lies beside the head and space is saved.

Incubation of the two to four eggs by both sexes lasts around 15 days. Young have particularly well-developed heel pads for shuffling about in the hard nest. These are lost by fledging at 40–50 days. Co-operative breeding has been recorded in Collared Aracari.

**6. Family** Picidae • Woodpeckers, piculets and wrynecks • About 200 species • Size from 8 cm/8 g (Scaled Piculet) to 55 cm/550 g (Imperial Woodpecker).
Most dry land habitats; worldwide except northernmost latitudes and remoter islands.

The true woodpeckers (Sub-family Picinae; 169 species) are supremely adapted for life on tree trunks. The skull is thick-walled to withstand the shocks involved in pecking into wood. The tongue is often long, barbed, supported by elongate hyoid bones and made sticky by salivary gland secretions, all adaptations allowing the extraction of insects from narrow crevices. Also, the tail is stiffened to prop the bird as it ascends tree trunks. Piculets (Sub-family Picumnae; 29 species) lack the stiffened tail, as do the ant-eating wrynecks (Sub-family Jynginae; 2 species) which perch crosswise on branches and do not excavate their own nesting holes.

Most woodpeckers and piculets eat gleaned arthropods and fruit. The sapsuckers lap sap that oozes from holes they bore in bark. The flickers feed on ants on the ground while the Acorn Woodpecker stores acorns in bark holes in the autumn for later retrieval and consumption. A typical Acorn Woodpecker stores around 400 acorns per season. However, one woodpecker group (groups rarely exceed ten birds) stored 220 kilograms of acorns in a watertank near Flagstaff, Arizona.

Woodpecker courtship is preceded by drumming, in which both males and females strike the bill, at around 20 blows per second, on a dead branch or similar which acts as a sounding board.

The hole nest is either re-used from a previous year or freshly excavated by both sexes. Ten thousand wood chips have been counted under a Black Woodpecker nest hole. Incubation of the two to twelve eggs lasts 9–20 days. The sexes incubate alternately by day but the male alone takes the night shift. Fledging periods last up to a month in larger species.

## Order *Passeriformes*

Around 5200 species, well over half of all birds, belong to the order Passeriformes, the perching birds or passerines. The order includes the most familiar garden birds – tits, chickadees, robins and sparrows – as well as other species found in virtually all land habitats. No passerine is a true water bird; although the dippers (Cinclidae) come close. Most passerines are small or medium-sized birds. The largest species are the ravens (e.g. *Corvus corax*, about 1.5 kg) in the family Corvidae and the Australian lyrebirds (males weigh about 1.15 kg) in the family Menuridae.

The perching feet of passerines have four toes, which are unwebbed. The strong hind toe, a key to the order's perching ability, is never reversible. Additional passerine features are nine or ten primary feathers and distinctive sperm. All these characteristics are shared by the two basic divisions of the passerines, the sub-Oscines and the Oscines.

Virtually all the passerines of North America and Eurasia belong to the sub-order Oscines, also known as the song birds. This is a somewhat confusing term since anyone who has heard the marvellous fluty calls of the antbirds (see below) of Amazonia could justifiably argue that these birds too should be called song birds. However, there is no argument over the fact that the songbirds include remarkable songsters, including species which duet and mimic other birds. The sounds are produced by five to eight syrinx muscles.

In contrast, the so-called sub-Oscine passerines have four or fewer pairs of syrinx muscles. These birds, in the sub-orders Eurylaimi, Menurae and Tyranni, are concentrated on the continents of the southern hemisphere, with a particularly rich radiation in South America. The reason for this radiation of the Tyranni seems to be as follows.

As part of the Cretaceous fragmentation of Gondwanaland, South America split from Africa some 110 million years ago, with primitive passerine stock aboard. For the next 100 million years, until it collided with North America in the Pliocene, South America was not connected to other land. The isolated primitive passerine stock evolved into two principal groups, the manakins (Pipridae), cotingas (Cotingidae) and tyrant flycatchers (Tyrannidae) on the one hand, the woodcreepers (Dendrocolaptidae), ovenbirds (Furnariidae) and antbirds (Formicariidae) on the other. It is the wealth of species in these groups that makes South America, home to over 3000 bird species, the world's most bird-rich continent. While the evolutionary radiation of the Tyranni was underway, the Oscines radiated in Eurasia and North America. Most ornithologists consider that the Oscines remain among the more rapidly evolving. For that reason they can be thought of as the most advanced, which is also the reason why they are customarily placed at the end of any list of bird orders.

Because of rapid passerine evolution, the boundaries between families or sub-families are often less than clearcut. For example a European birdwatcher can easily distinguish a thrush (Turdinae) from a flycatcher (Muscicapinae) from a warbler (Sylviinae). His colleague watching south Asian birds will find these groups much less clearly distinct. Similar doubt hangs over relationships among the families with nine primaries, including the Emberizidae, Icteridae, Parulidae and Fringillidae.

Being small, most passerines are dependent on high energy foods. In practice this usually means seeds or insects or some judicious mix of the two. Nevertheless, certain groups specialize on alternative diets. The sunbirds (Nectariniidae), often considered the Old World counterparts of the New World hummingbirds, consume nectar. The hook-billed shrikes (Laniidae) impale prey on thorns – hence the name 'butcherbird' – and the resulting gibbet may sport mice and small lizards alongside large insects.

*Diet.* The members of several passerine families have a more or less frugivorous diet. Tanagers (Thraupinae) provide an example of unspecialized frugivores. The specialist frugivores, such as manakins, cotingas and birds of paradise (Paradisaeidae), are generally confined to tropical latitudes where a year-round fruit supply is assured. At the extreme of specialization, the birds eat only fruit, concentrating on those (e.g. laurels) rich in protein and fat.

Because a day's rations can be secured in a small fraction (approximately one tenth in the Bearded Bellbird) of the daylight hours, it is no surprise that some male frugivores have devoted the remaining time to increasing their reproductive success. With the aid of spectacular plumage (e.g. birds of paradise), loud calls (e.g. bellbirds, Cotingidae) or communal dances (e.g. manakins), male frugivores undertake some of the most remarkable of all avian displays in their efforts to attract females. These promiscuous male frugivores do not help the female rear the young since she can find all the food needed by herself.

*Reproduction.* Considering the number of species involved, the breeding habits of passerines are remarkably uniform. Monogamy is the norm (Chapter 10), although the male frugivores described above are a notable exception. Most species construct their own cup-shaped or domed nest, the work being undertaken by one or both members of the pair.

Only some 16 species are brood parasites. Eggs are more or less heavily marked and of various colours. Clutch sizes range from one to about fifteen, and are generally larger in hole-nesters. Incubation lasts up to about 28 days in the large species.

The lyrebirds have an exceptionally long, 50-day, incubation period because the female, the sole incubator, leaves the nest for between three and six hours each morning, during which time the egg cools and development slows.

Young passerines hatch blind, naked and helpless. One or both parents raise these altricial young to fledging in a period ranging from about 8 to 45 days.

## Sub-order *Eurylaimi*

**Family** Eurylaimidae • Broadbills • 14 species.
Tropical forest of C Africa and SE Asia.

## Sub-order *Menurae*

**Families** Menuridae • Lyrebirds • 2 species.
More or less wet forest of SE Australia.
Atrichornithidae • Scrub-birds • 2 species.
Forest edge; W and E Australia.

## Sub-order *Tyranni*

**Families** Furnariidae • Ovenbirds • About 220 species.
All habitats; C America to extreme S.
Dendrocolaptidae • Woodcreepers • 48 species.
Forest, woodland; N Mexico to C Argentina.
Formicariidae • Antbirds • About 230 species.
Wooded habitats; C Mexico to C Argentina.
Tyrannidae • Tyrant flycatchers • About 375 species.
All habitats; Americas except extreme N.
Pittidae • Pittas • 29 species.
Tropical or sub-tropical forests; Africa, SE Asia and Australia.
Pipridae • Manakins • About 53 species.
Tropical forests from S Mexico to Paraguay.
Cotingidae • Cotingas • 65 species.
Forests from SE Mexico to SE Brazil.
Conopophagidae • Gnateaters • 9 species.
Forest undergrowth; eastern tropical S America.
Rhinocryptidae • Tapaculos • 29 species.
Undergrowth; Costa Rica to southern S America.
Oxyruncidae • Sharpbill • One species.
Humid forest; Scattered pockets from Costa Rica to SE Brazil.
Phytotomidae • Plantcutters • 3 species.
Brushland; S America (S of 5 degrees S).
Xenicidae • New Zealand wrens • 4 species.
Forest and scrub; New Zealand.
Philepittidae • Sunbird asities • 4 species.
Madagascar forest.

Sub-order *Oscines*

**Families** Hirundinidae • Swallows and martins • 74 species.
Mainly open habitats; worldwide.
Alaudidae • Larks • 75 species.
Open country; all major continents.
Motacillidae • Wagtails and pipits • 54 species.
Open country; worldwide.
Pycnonotidae • Bulbuls • About 120 species.
Forest and scrub of Africa and S Asia, east to Borneo.
Laniidae • Shrikes • 69 species.
Mostly semi-open country; northern America, Africa, Eurasia.
Campephagidae • Cuckoo-shrikes • 72 species.
Forests; Africa, S and E Asia, Australia.
Irenidae • Leafbirds • 14 species.
Forests of S and E Asia.
Prionopidae • Helmet shrikes • 9 species.
Wooded savanna of sub-Saharan Africa.
Vangidae (incl. Hyposittidae) • Vanga shrikes • 13 species.
Forest and brushland; Madagascar and Comoros.
Bombycillidae • Waxwings and Silky flycatchers • 8 species.
Forests of N Eurasia and America S to *c.* 10 degrees N.
Dulidae • Palmchat • One species.
Open woodland of Hispaniola.
Cinclidae • Dippers • 5 species.
Swift streams; N Africa, Eurasia and western America.
Troglodytidae • Wrens • About 60 species.
All habitats; extreme NW Africa, Eurasia, most species in N and S America.
Mimidae • Mockingbirds • 30 species.
Mostly scrub; Americas (Great Lakes southwards).
Prunellidae • Accentors • 13 species.
Brush or open country; N Africa, Eurasia, except S Asian peninsulas.
Muscicapidae
Subfamilies:
Turdinae • Thrushes • 305 species.
All habitats; worldwide.
Timaliinae • Babblers • 252 species.
All habitats; Africa, S Asia, Australasia, Oregon to Baja California.
Sylviinae • Old World warblers • About 350 species.
All habitats; worldwide excluding highest latitudes.
Muscicapinae • Old World flycatchers • About 155 species.
More wooded habitats; Old World, Australasia.
Malurinae • Fairy-wrens • 26 species.
All habitats; Australia and New Guinea.
Paradoxornithinae • Parrotbills • 19 species.
Reeds, dense grass; C Asia extending W into Europe.
Monarchinae • Monarch flycatchers • 133 species.
Tropical woodland of sub-Saharan Africa, S Asia and Australasia.

Orthonychinae • Logrunners • About 20 species.
Forest to scrub; SE Asia and Australia.
Acanthizinae • Australian warblers • About 65 species.
All habitats in SE Asia and Australasia.
Rhipidurinae • Fantail flycatchers • 39 species.
Woodland; SE Asia and Australasia.
Pachycephalinae • Thickheads • 46 species.
Rainforest to scrub; SE Asia and Australasia.
Paridae • Tits • 46 species.
Woodland; Africa, Eurasia and N America (S to Mexico).
Aegithalidae • Long-tailed tits • 7 species.
Woodland; Eurasia, western N America (Brit. Columbia to Guatemala).
Remizidae • Penduline tits • 10 species.
Open country and reedbeds; Africa, Eurasia and N America (S to Mexico).
Sittidae (including Neosittidae sittellas) • Nuthatches • 21 species.
Woodland, rocky outcrops; N America, Eurasia (not desert), extreme N Africa, Australia.
Climacteridae • Australasian treecreepers • 8 species.
Forests of Australia and New Guinea.
Certhiidae • Holarctic treecreepers • 5 species.
Forests; Eurasia, Africa, N America.
Rhabdornithidae • Philippine treecreepers • 2 species.
Forests; Philippines.
Zosteropidae • White-eyes • 85 species.
Canopy, forest edge, bushes; sub-Saharan Africa, S and E Asia, Australasia.
Dicaeidae • Flowerpeckers • 50 species.
Woodland; SE Asia, Australasia.
Pardalotidae • Pardalotes or diamond birds • 5 species.
Forest; Australia.
Nectariniidae • Sunbirds and spiderhunters • 116 species.
All dryland habitats; Old World tropics.
Meliphagidae • Honeyeaters • 169 species.
Mostly woodlands; Bali, Australasia, Oceania N to Hawaii, southern Africa (sugarbirds).
Ephthianuridae • Australian chats • 5 species.
Dry or wet open country; Australia.
Emberizidae
Sub-families:
Emberizinae • Old World buntings and New World sparrows • 281 species.
All habitats (but rarely dense forest); worldwide except SE Asia and Australasia.
Catamblyrhynchinae • Plush-capped Finch • One species.
Bamboo thickets; Andes, from Colombia to Bolivia.
Thraupinae • Tanagers and honeycreepers • 233 species.
All wooded habitats; S Canada to C Argentina.
Cardinalinae • Cardinal grosbeaks • 37 species.
Wooded habitats; Canada to C Argentina.
Tersininae • Swallow Tanager • One species.
E Panama to C Argentina.

Parulidae • Wood warblers • 119 species.
Forest and scrub; Alaska to N Argentina.
Vireonidae • Vireos and pepper-shrikes • 43 species.
Forest and scrub; Alaska to N Argentina.
Icteridae • American blackbirds (including cowbirds, caciques, troupials and American orioles) • 94 species.
All habitats throughout the Americas.
Fringillidae
Sub-families:
Fringillinae • Fringilline finches • 3 species.
Woodland; Eurasia and Canary Islands.
Carduelinae • Cardueline finches • About 122 species.
All habitats; virtually worldwide except Australia.
Drepanidinae • Hawaiian honeycreepers (or finches) • 23 species.
Mostly rainforest; Hawaiian Islands.
Estrildidae • Waxbills • 124 species.
All habitats; sub-Saharan Africa, S Asia, Australia.
Ploceidae
Sub-families:
Ploceinae • True weavers • 95 species.
All habitats; mostly Africa, east in S Asia to Java and SW China.
Viduinae • Parasitic viduine weavers and whydahs (widow birds) • About 10 species.
Savanna of sub-Saharan Africa.
Bubalornithinae • Buffalo weavers • 3 species.
Scrubland of sub-Saharan Africa.
Passerinae • Sparrow weavers and sparrows • 37 species.
Mostly open habitats; Africa and Eurasia; introduced elsewhere.
Sturnidae • Starlings • About 106 species.
All habitats; Africa, Eurasia to NE Australia and Oceania E to Tuamoto.
Introduced N America, New Zealand and S Australia.
Oriolidae • Orioles and figbirds • 28 species.
Forest and woodland; Africa, Eurasia (not NE) and Australia.
Dicruridae • Drongos • 20 species.
Wood and scrubland; sub-Saharan Africa, S Asia, N and E Australia.
Callaeidae • New Zealand wattlebirds • 3 species.
Forest; New Zealand.
Grallinidae • Magpie larks • 2 species.
Open woodland; Australia and adjacent islands.
Corcoracidae • Australian mudnesters • 2 species.
Open woodland; E Australia.
Artamidae • Wood swallows • 10 species.
Scrub to forest clearings; SE Asia and Australia.
Cracticidae • Bell magpies • 9 species.
Woodland and grassland; New Guinea and Australia.
Ptilonorhynchidae • Bowerbirds • 18 species.
Forest, scrub and grassland; New Guinea and Australia.
Paradisaeidae • Birds of paradise • 43 species.
Mostly wet forest; Moluccas, New Guinea and NE Australia.
Corvidae • Crows, magpies and jays • 113 species.
All habitats; worldwide.

MdeLB

## Food

The ability to fly has been the dominant influence on avian adaptation, and the energy needed for flight has meant that birds tend to ingest and digest quantities of food of high nutritive value. Thus, although birds now exploit almost everything that lives upon, over or just beneath the earth's surface or in the shallow layers of its waters, they are, and have always been, predominantly animal-eating.

Most bird species eat arthropods, especially insects. Many of these insects are caught on the vegetation that they themselves consume, and this close association between plants and birds over millions of years has presumably given rise to the eating of seeds and berries, and to the sucking of nectar from blossom. Today the seeds of many plants are bird-dispersed and, in the tropics, their flowers are bird-pollinated.

### Animal-eating birds

All types of animal are taken. Annelid worms are eaten by waders; squid and cuttlefish by penguins and albatrosses; snails by the Song Thrush and the Limpkin; and brine shrimps by the Greater Flamingo.

Fish are food for the Osprey, for mergansers, for pelicans and for the Black Skimmer. They catch their slippery prey in different ways: the Osprey with its feet, the undersides of which have a spiny roughened surface; the pelican in a scoop-like pouch suspended beneath its bill; the Goosander between the saw-like edges of its upper and lower jaw; and the skimmer with a bill of unique construction – the lower mandible is flattened like a knife blade and projects some way beyond the upper. The skimmer feeds in shallow water containing high concentrations of fish, and flies a straight path with the lower bill immersed and the mouth open. When the edge of the mandible strikes prey, the head doubles under the bird's body and the bill snaps shut. The fish is drawn out of the water as the skimmer lifts its head, and is swallowed in flight.

Lizards and frogs are taken by herons and by the Roadrunner, snakes by the Secretary-bird, and mammals and birds by owls, falcons and hawks. The heaviest of the eagles, the Harpy Eagle, feeds largely on monkeys. Carrion is food for crows, gulls and vultures, and bone marrow is included in the diet of the Lammergeier. Many of the frigate birds are pirates, harassing boobies in order to force them to disgorge their meals of fish, and all the marine skuas are kleptoparasites, making a living by stealing food from other sea birds. True parasitism is very rare, but there is a Galápagos finch (*Geospiza difficilis*) that preys upon boobies by pecking at the soft quills of emerging feathers, and consuming the oozing blood.

Arthropods, notably insects, provide sustenance for a greater number and variety of birds than any other plant or animal food. Swifts, nightjars and flycatchers take them on the wing; woodpeckers search with sticky tongues for larvae under bark; and leaf-eating caterpillars are collected by and for many young temperate-zone birds. Beeswax is a curious food of animal origin taken by the honeyguides.

*Using its uniquely extended lower jaw, a Black Skimmer (Rhynchops niger) ploughs the water surface in the hope of snapping up a tasty morsel*

## Eaters of non-flowering plants

Among birds utilizing the lower plants are pygmy parrots that feed mainly on fungi; the ptarmigan that eats lichens; and breeding geese taking arctic mosses and liverworts during the summer.

Brent Geese (Brant) and sheathbills browse seaweeds; Capercaillie the resinous buds and needles of conifers; and Lesser Flamingos subsist on tiny blue-green algae that they filter from the soda lakes of East Africa. The flamingo's bill is characteristic and peculiar in that the lower mandible is large and trough-like and the upper one small and lid-like; the bird holds this bill upside-down, taking mud and water in along the whole of the gape with the enlarged tongue acting as a piston. Coarse particles are kept out by stiff excluder bristles along the edges of the jaw and tongue, and water is expelled three or four times a second past other finer filter devices that trap the algal filaments.

## Eaters of flowering plants

Flowering plants support a greater variety of birds. For instance, many ducks, geese and swans take the starchy roots and tubers of water plants; leaves make up the diet of the grazing geese and the New Zealand Ground Parrot (Kakapo); and flower petals are eaten by the Palm Chat. Since almost all birds lack the ability to digest cellulose (the substance that makes up the cell walls of leaves and stems) grass and herbage are unusual foods and, where they are taken, need to be consumed in large amounts in order to produce the required level of nutrition.

A couple of specialized vegetarian habits have arisen from insect-eating. Small woodpeckers called sap-suckers drink the sap that oozes from holes that they drill in the tender bark of trees such as aspen, and devour also any insects attracted to the sugary fluid. Similarly, nearly one-fifth of the world's birds at one time or another take another sugary liquid, the nectar produced by flowers; these include the hummingbirds, sugarbirds, sunbirds and flowerpeckers. All are small and structurally rather similar, but they are not closely related,

so nectar-feeding must have evolved more than once, probably by ancestral forms searching for insects among flowers. Nectar is mainly a source of carbohydrates, and these birds all take insects or fruit as well. In the tropics, many flowering plants are adapted for bird-pollination since there is a relative scarcity of flower-visiting insects, and here the birds are modified so that their bill and tongues suit the blossoms that they visit.

## Berry and seed-eaters

The most widely utilized plant foods are seeds, nuts, berries and fruits, doubtless a reflection of their excellent food value. Thistle seeds are a significant item in the diet of the goldfinch; the oily kernel of the palm nut is eaten by macaws and by the Palm-nut Vulture; and acorns are eaten by the jay. The jay and the oak tree have evolved a symbiotic relationship in which the bird stores acorns to use later at times when they would be otherwise unavailable. However, many of these buried acorns are never recovered by the jays so that they germinate. By sacrificing some seeds to jays, an oak benefits from others dispersed and deposited in good germination sites.

A similar symbiotic relationship exists in western North America between another member of the crow family, Clark's Nutcracker, and the seeds of the piñon pine.

## Specialists and omnivores

The above survey of food habits suggests that birds are dietary specialists; in fact, true conservatives are the exception, although examples include the Palm-nut Vulture, whose range, in East and Central Africa, closely matches that of palm plantations, and the Lesser Flamingo which is almost totally dependent on the filamentous blue-green alga *Spirulina*. Obviously, those with an idiosyncratic diet are vulnerable to fluctuations in their food supply, and a number of the world's threatened birds (for example the Everglades Kite) come into this category.

Most birds are catholic in their tastes, taking a mixture of items, and stable populations are composed of omnivorous individuals that can adjust their habits when one type of seed, berry or insect, for instance, becomes scarce or another common. Many of the most successful species have been those living in proximity to man, such as the House Sparrow and the European Starling, both of which have flourished since the transition from a hunter-gathering economy to settled agriculture around 12 000 years ago.

## Seasonal differences in diet

For the majority of birds, the food supply is far from constant. Daily variations are caused by the cycle of day and night and the ebb and flow of the tide. Longer-term changes are controlled by the climate; in the temperate zones these are regular, but in the tropics they are associated with the more unpredictable arrival of the rains. At higher latitudes, birds typically respond to scarcity by migrating, others overcome fluctuations simply by altering their diet with the season.

To some extent dietary change depends on food availability, but preferences are governed by certain changes to the bird's internal chemistry. Migrants that lay down fat prior to departure must include high-calorie foods in their diets in order to do so; indeed, any species that suffers a regular seasonal food shortage increases reserves of body-fat *before* that famine occurs.

Many birds also change their diet with age, and protein-rich foods such as insects are a common food for rapidly growing youngsters which are vegetarians as adults. Juveniles of ground-nesting birds may find these animal items for themselves; others, that hatch with their eyes closed and remain in a nest for some days, will be supplied by their parents, who may not collect the same food for their offspring that they are eating themselves. Pigeons and flamingos of both sexes and the male Emperor Penguin are extraordinary in feeding their single or twin chicks on a proteinaceous secretion from the cells lining the crop (or oesophagus in flamingos). This secretion is similar in composition to mammalian milk and is produced under the influence of the same prolactin hormone.

*Upper left. Camouflaged by variegated plumage in summer and white plumage to match the snow in winter, the Rock Ptarmigan (Lagopus mutus) includes much vegetable matter in its diet*

*Upper right. Flamingos use a muscular tongue to push water out of the mouth past bristles which catch food particles, rather in the manner of baleen whales*

*Below. An Anna's Hummingbird (Calypte anna) visits a fuchsia flower in the mountains of California and receives a dusting of pollen (the yellow specks on the throat) which will perhaps fertilize the next plant it visits*

## Modifications of the bill and tongue for feeding

In a few species, the food taken by the two sexes will differ because the pair are structurally dissimilar, an adaptation that probably allows them to avoid competing with one another; thus the female sparrowhawk catches larger prey than her more lightly-built mate. The most extreme example of this sexual differentiation was found in the extinct Huia where each sex employed a different technique to obtain insects from decaying wood – the male chiselled out grubs while the female's long, curved, more flexible beak was used as a probe.

Modifications for food-catching and preparation are mostly confined to the bill and hindlimbs since for over 100 million years the forelimbs of most birds have been adapted for flight. The wide range of adaptations of the beak and the tongue for different diets has provided a classic subject for evolutionary investigation.

The generalized, although not the most primitive, bill may be considered to be that of the omnivorous crow. It is straight, pointed and roughly triangular in section, and is found in similar form in gulls, gannets, woodpeckers, bitterns, the Secretary-bird, penguins and the male Huia. The fruit-eating thrushes have similar pointed beaks and a wide gape that enables them to swallow berries whole. Deepening, shortening and down-curving produces the bill of the carnivorous owls, falcons and hawks, and that of the vegetarian parrots. A stout, short, arched bill is found in grazers such as geese and wigeon and also in the seed-eating finches. A fine slender bill will usually belong to insect or tick-feeders such as warblers or oxpeckers and, with lengthening, becomes that of the nectar-sucking hummingbirds and flowerpeckers. Other long but much stronger bills are those of the earth-probing waders and kiwis.

The 'teeth' that hold slippery fish between the jaws of the Goosander have already been mentioned. Vertical flattening, with similar saw-like serrations, is found in the bill of the frugivorous toucans. The snail-eating Limpkin has a flattened beak with a gap between the mandibles where the snail shell is lodged before being broken. Horizontal widening and laminations along the edges of the mandibles are characteristic of the beaks of dabbling ducks, particularly those of the plankton-sieving shovelers, spoonbills and flamingos. A very broad, short bill with a wide gape may belong to an insect-catching nightjar or swift that hunts on the wing.

The tongue tends to conform to the shape of the beak and is subject to analogous modifications. In the tongues of nectar-feeders, peculiar curling and fraying of the edges with a

*This female Northern Goshawk (Accipiter gentilis) dwarfs her mate (left), in a manner characteristic of the bird-hunting hawks*

*Tools for the task. A bird's bill greatly influences its feeding habits and can vary enormously in form; different species are efficient, for instance, at catching flies, sipping nectar, dabbling for tiny seeds, or tearing at carrion. Shown here, from the top, are the bills of a Parula Warbler (Parula americana), Sparkling Violetear Hummingbird (Colibri coruscans), Common Snipe (Gallinago gallinago), Great Tinamou (Tinamus major), Common Shoveller (Anas clypeata), Turkey Vulture (Cathartes aura), and Ivory-billed Woodpecker (Carnpephilus principalis)*

division at the tip leads to a fine double tubular structure up which nectar is drawn by capillary action. The tongue of honey-eaters is cut like a tiny brush at the end to enable syrup to be lapped. Those of penguins and mergansers have backward-directed spines on the upper surface which give a good grip on fish. The shovelers and flamingos have fringed tongues which, together with the laminate bill edges, strain minute food organisms from water. The long, mobile tongue of the woodpecker probes for insects in wood and catches them with the aid of a sticky coating of saliva. A few species that bolt large items of food whole, such as the pelicans, have only rudimentary tongues.

## How birds chew

A loss of teeth is common to modern birds and is related to a need to reduce excess weight at the head-end of an efficient flying machine. Since rapid and fairly constant absorption of finely macerated food is important for the maintenance of their high metabolic rate, birds have evolved an excellent chewing mechanism – the gizzard – that is situated near their centre of gravity between the wings. Birds tend to take in large quantities of food rapidly, store it in bulk in a crop and gradually grind the supply in a muscular gizzard with the aid of ingested small stones and grit, the size of which depends upon the coarseness of the food.

## Pellet formation

To prevent overloading this fast-moving system with indigestible or only slowly digesting items, many birds rid themselves of unwanted components of their meals in casts or pellets regurgitated through the mouth. The pellet may be composed of bones, beaks, claws, shells and husks, and is normally wrapped for easy passage in softer items such as fur, wool, feathers and plant fibres. Pellet-formation is typical of birds of prey but also occurs in albatrosses, which eject the beaks of squids; kingfishers, which cast the bones of fish; insectivores, which eject the hard exoskeleton; and some fruit-eaters, which regurgitate the larger, harder seeds.

## Drinking

Drinking is necessary only if the food does not provide enough water to balance any loss that occurs by excretion or evaporation through the lungs – birds do not sweat and do not normally produce liquid urine. The frequency of drinking depends on the type of food and the bird's ecology; most seed-eaters need to drink at least once a day and may take in about 10% of their body weight, but some small birds of arid regions, such as the Zebra Finch and Budgerigar, can survive for many months on dry seeds without drinking or losing body weight. Presumably they can adequately supplement low dietary moisture with water produced internally through metabolism.

The method of drinking varies. 'Sipping and tilting' is the commonest – the bill is dipped in the water, a sip taken and the head raised so that water trickles down the throat by gravity. Swifts sip from the surface of a pond while on the wing and sandgrouse suck in a large mouthful before raising the head. Pigeons, doves and a few others immerse the bill and suck, not lifting the head until their crop is full.

Marine species have particular problems with drinking in that they take in salt with their food or drinking water. Small species are more sensitive to salt stress than larger ones and most marine ducks, for instance, are quite big. The salt is eliminated by the excretion of a concentrated solution through glands that have been inherited from the birds' reptilian ancestors and are situated just beneath the skin above the eyes. A duct from each gland leads into the nose from which saline fluid flows down the bill and is removed when the bird shakes its head.

## Food selection by smell and taste

How do young birds discover what is edible and what is not? The first appearance of pecking in those that hatch with their eyes open is not dependent on learning; they may peck at a number of objects that contrast with the background and accuracy improves with experience. Their sensory responses will determine whether the peck is repeated.

*Uniquely among birds, the nocturnal kiwis (Apteryx australis) of New Zealand have nostrils at the tip of the bill. They find food using a sense of smell which is more acute than in many other species*

In the food selection of most species, a sense of smell seems unimportant – none of the brightly-coloured bird-pollinated flowers has a scent, nor have bird-distributed berries, unlike the blooms that appeal to insects or the sizeable ripe fruits that attract mammals. However, large olfactory organs are present in the brain of the flightless kiwis, and experiments with captive individuals show that they can find buried food by its odour. Some New World vultures also locate carrion by its smell. Sensory corpuscles are common on the bill, especially of ducks and waders, and on the tongue of woodpeckers; these detect vibrations when the bird is searching 'blind' for living prey.

In contrast to mammals, only a few taste buds are present – 350 in the mouth of the parrot as opposed to about 17 000 in the rabbit. Most birds can detect salt, sour and bitter as accurately as we can, but many seem relatively indifferent to sweetness. Hummingbirds, on the other hand, can discriminate between a number of sugars, completely rejecting saccharin. Wild birds are known to avoid certain insects (usually striped red or yellow and black) apparently because they are distasteful, and it is supposed that these avian predators are the agents in the evolution of insect mimicry whereby edible ones develop the warning coloration of those that are unpalatable or poisonous. The implications are that the taste of their food is of considerable importance; at first, its flavour may convey no information to the naive bird who will accept a range of items. If it eats and then vomits, it will associate the taste of the food being expelled through the mouth with an uncomfortable 'gut reaction'. Originally a gourmand, the bird is now a gourmet, and taste will thereafter 'jog the memory'.

Once conditioned, the bird judges potential food by its flavour, but obviously the next, more efficient, stage is to recognize prey by sight alone before it needs to be caught and tested in the mouth.

## Feeding by sight

In the food selection of most birds the visual sense is paramount. The eyes are large and vision is acute. Young Herring Gulls, with no experience of food, will peck at models that are pointed, low, moving and red. Thus they hatch with preferences that match the parent's beak from which they will obtain food and which has a red spot near the tip. Many other juvenile gulls and terns will peck preferentially at red; so will the chicks of the Oystercatcher, the Moorhen and the Common Coot, although, unlike the other two, the bill of the adult coot is white. Young pheasants, goslings and ducklings, that obtain their own food from the start, almost all prefer green.

The preferences of some adult birds can be guessed from a study of the colours of flowers that they pollinate and berries they distribute; over 80% of the former are red or orange, while red, orange or black berries predominate over any others. Conspicuousness or contrast can overide the importance of colour in feeding (a fact suggested by the common occurrence of shiny black berries) and shape and movement also affect the likelihood of the object being pecked.

## Other aspects of food selection

It is obvious that birds, unlike mammals, are relatively immune to the toxic alkaloids in the berries, such as yew and nightshade, that they eat in quantity. The European Cuckoo must be resistant to the irritant histamines in the hairs of the caterpillars that form a large part of its diet, while vultures can eat putrid carrion that would poison many other animals.

Parental example is an additional factor that affects food selection in the individual. Only the megapodes, that leave their eggs to incubate in rotting vegetation, are independent of such influences. Ducks and geese lead their young to suitable feeding grounds; most game birds pick up and drop morsels of food to bring it to their chicks' attention; young coots and moorhens follow their parents and are fed from the bill, sometimes even obtaining food from the young of the previous brood.

In numerous other species, the parents bring food to the nest. In parasitic species, at least, the importance of this early experience seems to be slight, since cuckoos and honeyguides are not given by their foster-parents the diet that makes them unique as adults – hairy caterpillars on the one hand and beeswax on the other.

The eventual diet that the young bird learns to select will be governed by the food available, by sense organs and behaviour patterns, and by body structures such as bills and feet. Thus titmice, but few other species, learn to open milk bottles because they have the necessary 'hammering and tearing' motor patterns and can cling with their centre of gravity sideways of their toes; the size of seeds taken by finches is determined in large part by the size of the beak – each species learning by trial and error to take the most suitable. The ability of some birds to use the foot with the bill in feeding is perfected by practice; it has, in part, a structural basis in that most birds that do it have short legs.

JK

*A Blue Tit (Parus caeruleus) opens a milk bottle to drink the cream. This habit was first noticed in Britain in 1929 and has since spread widely*

## TOOL-USING

Sometimes, there is no physical characteristic that leads us to expect a particular diet. For example, there is nothing to indicate that the Song Thrush is alone among the British thrushes in possessing the capacity to learn to crack the shells of snails using a rock as an anvil; its close relative, the blackbird, does not employ the thrush's technique. Egyptian Vultures learn to throw stones at Ostrich eggs in order to break them so that they may eat the contents. Another classic example of a diet that has no apparent physical foundation is provided by a finch *Camarhynchus pallidus* of the Galápagos Islands, where there is no woodpecker. This bird probes with a cactus spine into the bark crevices of trees, thereby disturbing insects beyond the reach of its short tongue; as the insect appears, the thorn is discarded and the food seized and devoured. A few herons, for instance the Green-backed Heron in North America, have learnt to bait fish with bread, like true fishermen. Tool-using of this type is a rarity but gives a further indication of the great diversity of birds' feeding behaviour.

JK

*The Green-backed Heron (Butorides virescens) casts a miscellany of titbits upon the water to attract fish within range of its dagger-like bill. Birds in Florida have even been seen to use popcorn as bait*

### BIRDS AND BERRIES

Small berries are adapted for dispersal by birds in a relationship of mutual advantage similar to that of flower-pollination. The plant puts resources into an edible coating around its seeds and, in exchange, the avian visitor eats the nutritious pulp and ejects the hard seeds in droppings or pellets at a distance from the parent plant, where they germinate better for having been through a digestive tract.

The interaction is one in which both plants and fruit-eaters have been modified during evolution; thus bills and guts are adapted for efficient swallowing and digestion of berries, and the plants produce fruits that match in size the birds' gapes and are rich either in carbohydrates or in proteins and oils. The eye-catching, bright red or shiny black carbohydrate-rich berries are food for many opportunist thrushes, crows and pigeons that rely on the fruit for energy but obtain their protein from invertebrates. On the other hand, a few tropical plants, such as the olive and the laurels, have large seeds covered by a thin layer of dull-coloured flesh that is highly nutritious and can supply the entire diet of specialist frugivores like the cotingas.

Some berries also contain alkaloids that deter mammals which are not helpful distributors. The spread of plants that suit them and produce fruits in a sequence through the season or the whole year is obviously promoted by berry-eating birds that can thereby adjust and improve their own habitats. However, this co-evolved system is open to exploitation by 'cheaters' – that is, by seed-predators like finches that digest and destroy the seeds as well as the fruit that contains them.

JK

## Optimal foraging

### Choosing the best feeding strategy: optimal foraging

Being a bird is an expensive business. Living, growing, behaving and reproducing consume energy and nutrients, resources that have to be replaced as food from the outside world. In most cases, food means the bodies or bodily products of other organisms. Birds are thus obliged to be predators of one sort or another even though 'predation' may simply mean eating plants or filtering organic detritus from a pond.

Food is the source of energy and nutrients on which a bird depends. However, finding, consuming and digesting it cost energy and nutrients. These activities also cost time which could be spent doing other important things like looking for mates or sleeping. To carry on surviving and reproducing, birds and other animals have to ensure that the amount of energy and nutrients they take in from a predatory act exceeds the amount they have to spend on the act, and that they could not have done better by spending the same amount of time feeding in some other way.

The decisions birds make as predators – where to look for food, what to eat, how long to search before moving on and so on – can thus be thought of as analogous to those of a business executive. Any business transaction incurs a cost in terms of initial investment but reaps benefits in the form of a rate of return. Maximizing returns on investment makes for a successful business; failing to maximize them may mean going to the wall. The 'economics' of predation work in the same way except that the measures of cost and benefit are no longer dollars and Deutschmarks but increases and decreases in the probability of reproducing successfully in the future. The extent to which an animal maximizes the ratio of nutritional returns to time and energy investment in its predatory decisions determines the likelihood that it will pass on the genes that have predisposed it to decide in that way. Natural selection, therefore, acts as a natural optimizing agent, honing the decisions that animals make so that, within the context of the alternative courses of action open to them, they maximize their behavioural efficiency.

This simple argument forms the basis of a powerful approach to studying predatory behaviour. If it is right then, in principle, all we need to know to predict what an animal will do in any give situation are the costs and benefits implicit in the choices open to it. The choice reaping the greatest benefit relative to its cost – the optimal choice – is the one the animal should have been selected to make.

Of course, it is never quite that easy. Phrasing the argument as we have done above makes a lot of important assumptions all of which affect critically our predictions about optimal choices – for instance, assumptions about the currency the animal is attempting to maximize (e.g. is it trying to maximize rate of energy intake or rate of acquiring particular nutrients such as calcium?), assumptions about the costs involved in alternative options (e.g. time and energy costs, lost opportunity costs, the risks involved) and many more. Getting all these right is a considerable task and leaves plenty of scope for error. Nevertheless, a welter of evidence from laboratory experiments and field observations over the last decade indicates that the optimality approach to decision-making in animals is a powerful and revealing one.

Birds have been at the forefront of optimal foraging studies for a number of reasons, notably their observability in the field and the suitability of many smaller species for behavioural experiments in the laboratory. Many of the best and most quoted examples of optimization in the context of predation therefore come from birds.

## Where to eat and what to eat

We can consider the wide range of foraging decisions that birds have to make as falling into two broad groups: those to do with choosing *what* to eat and those to do with choosing *where* to feed and how long to feed there. Since most natural food supplies tend to be neither uniformly nor even randomly distributed through time and space but occur in discrete clumps (such as bunches of berries, insects in rotting tree stumps or concentrations of earthworms under cowpats), models predicting where to feed assume that what is optimized is the choice of and time spent in *patches* of food. Optimality models thus divide into '*prey*' models and '*patch*' models, though the two share some basic mathematical characteristics.

### *Testing 'prey' models*

*Crows and whelks: choosing the most profitable prey.* Along the west coast of Canada, Northwestern Crows commonly forage along the shoreline for molluscs, especially whelks. There is a problem with eating whelks, however: they have a tough shell which the crow cannot break with its bill. To get the flesh, therefore, crows pick up whelks, fly to a nearby rock and drop the whelks from the air so that the shells smash on the rock. The act of flying up to drop the whelk (which often has to be repeated several times) costs time and energy, costs which have to be offset against the eventual nutritional returns.

In a study of foraging crows, Reto Zach noticed that they took mainly the largest whelks on the shore and tended to drop them from a height of about five metres. Did this preference for size and height maximize returns on energy investment for the crows? A series of simple experiments which involved dropping whelks of various sizes from different heights suggested it did. First, larger whelks were easier to break from any given height. Second, the

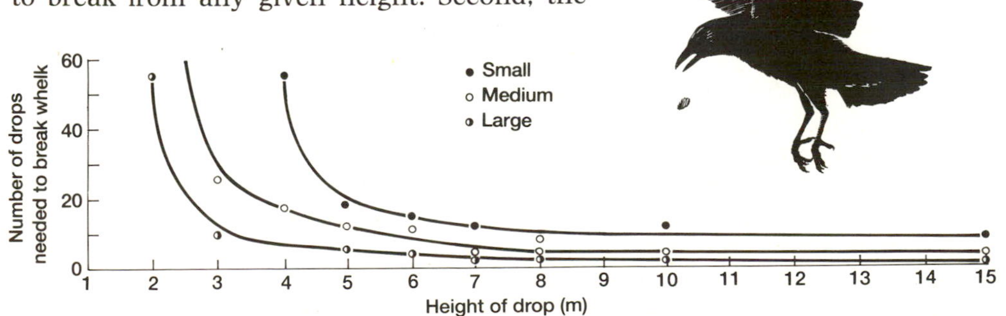


*How Northwestern Crows (Corvus caurinus) deal with whelks. Large whelks are not only more rewarding as food, but require fewer drops from any given height to break the shell*

total amount of vertical flight (number of drops × height of each drop) needed to break a large whelk was minimized by choosing a drop height of just over five metres. More drops were needed to break the shell at heights below five metres and breakage wasn't improved much by flying any higher. Crows thus appeared to choose prey and drop heights that minimized the amount of upward flight required.

*Birds and worms: optimal diet breadth.* As demonstrated with the whelks, natural food items do not come in neat, standard-sized packages like tins on a supermarket shelf. They vary in, among other things, nutritional content, ease of capture and the time it takes to eat them. If a bird is to maximize its net returns it should go for the most profitable (most nutritious net of the costs of catching and eating) prey as the crows appear to when feeding on whelks. However, if profitable prey are scarce, searching for them will cost a lot of time so that returns over an entire foraging trip are low. Under these conditions it may pay to include a few less profitable items in the diet. The optimal policy then becomes: 'always take the most profitable prey when encountered but only take unprofitable prey if they increase the rate of return on the foraging trip as a whole'.

John Krebs and co-workers tested this simple optimal diet model in an ingenious experiment using captive Great Tits. Great Tits are small, easy-to-train, insectivorous birds with a high energy demand and Krebs's experiment was based on the assumption that tits would forage so as to maximize their net rate of energy intake. Birds were offered large, profitable and small, unprofitable pieces of mealworm on a conveyor belt so that their encounter rate with each size could be controlled precisely. When encounter rates with profitable prey were low birds did best by taking any worm they came across and thus should have fed unselectively. When they were high, however, they did best by taking only the large worms and should have ignored all small worms no matter how often they encountered them since taking any small worm would have cost time that could have been spent taking the next large one.

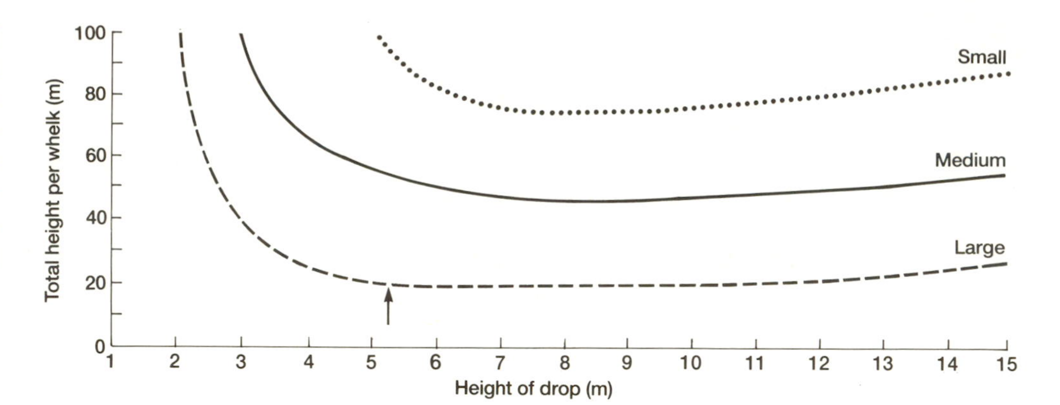


*Drops of about 5 m (arrow) minimize the amount of upward flight Northwestern Crows need to break open large whelks*

The results bore out the predictions remarkably closely except that tits still tended to take a few unprofitable worms when they should have ignored them. Later experiments suggested that this was due to discrimination errors by the birds who were not infallible in telling large worms from small. This does not mean the tits in Krebs's experiment were failing to optimize – they may simply have been doing the best they could in the face of their limited discrimination abilities.

Strong supporting evidence for the optimal diet breadth model has also come from the field. Studies of Lapwings foraging for earthworms on pasture, for instance, have shown that they take almost exclusively the range of worm sizes predicted on the basis of energy profitability (determined by a wide range of size-specific costs involving difficulties of detection and extraction and loss to thieving gulls) and search time. Furthermore, as the model predicts, whether or not unprofitable worms are taken depends not on their own availability but on that of profitable worms. Similar selectivity has been found in Redshanks foraging for worms on tidal mudflats.

### Testing 'patch' models

*Starlings feeding chicks: how much to collect from a patch.* European Starlings feed their young on various invertebrates such as caterpillars, beetle larvae and earthworms. To collect them, the parent bird must spend time and energy flying out to a suitable food patch, searching for and gathering prey and returning, loaded up, to the nest. How many prey should it collect on each trip to maximize the efficiency with which it delivers food to its nestlings? Since it may have to make up to 400 round trips a day to meet demand, this becomes an important question.

The problem of load size can be explained by a simple graphical model, above right, with time along the horizontal axis and prey load size along the vertical axis. Since the starling has to fly to and from a patch, the total time spent flying can be represented as round-trip travel time on the horizontal axis. When it arrives in a patch, the bird gathers prey quickly, because its bill is unencumbered, so the loading rate curve climbs steeply. As its bill fills up, however, gathering further prey becomes difficult so the loading rate begins to drop off. The starling's problem becomes one of when to give up. If it leaves the patch too soon it spends a lot of time travelling for only a small load. If it leaves too late it wastes time so it should deliver its prey and start again, on the steep part of the curve. The departure time/load size at which they would do best is the point on the loading rate curve where load size divided by travel plus search time (i.e. the net rate of food delivery) is maximized; this can be found very simply by constructing the tangent to the curve from the origin of travel time (line AB in the diagram). Is there any evidence that starlings optimize load size in this way?

In a field experiment, Alex Kacelnik trained parent starlings to collect mealworms from a tray onto which he dropped worms at a diminishing rate to simulate a declining loading curve. By varying the distance of the tray from the birds' nest boxes, Kacelnik used the model described above to predict the load size at which starlings would leave the tray for their nest. The predicted relationship between round-trip flight time and load size is shown for three plausible currencies of food delivery in the graphs on the right. The load sizes actually chosen by the birds are shown as dots. The fit between actual and predicted load sizes is close for all three currencies but it is best when starlings are assumed to be maximizing the rate of food delivery net of the energetic cost of foraging to the parent and of begging to the chicks.

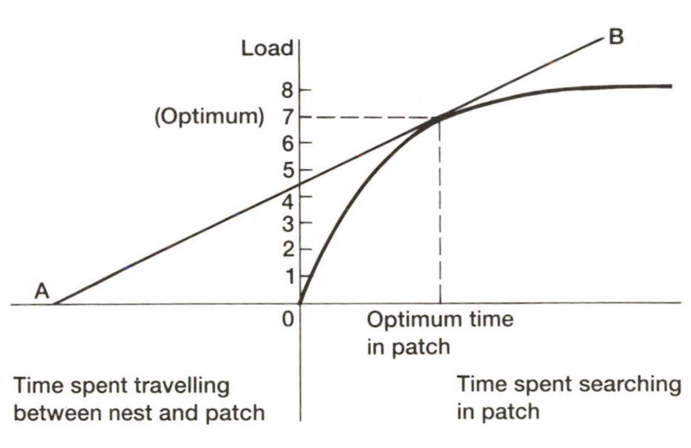


*Juncos and risk-sensitivity: gambling on a bonanza.* The examples so far have assumed that birds forage to maximize some measure of food intake or delivery. In some cases, however, it may be more important to minimize the risk of doing very badly. A small bird trying to survive the winter, for instance, may be hard put to find enough food during the day to live through the long, cold night. Suppose on its last foraging trip before dark the bird needs to find four units of food to avoid starving overnight. It has a choice of visiting one of two patches. One yields four units every visit, the other yields four sometimes but is equally likely to yield eight or nothing at all. Which should the bird visit? Clearly the answer is the first patch. Even though the other patch sometimes yields a bonanza which would provide the bird with a welcome reserve of extra food, it carries the risk of yielding nothing and thus certain death. The first patch yields less each visit, but it is enough to get through. But now suppose the bird needed eight units to survive. Its only option under these circumstances is to choose the second patch because only here does it stand any chance whatsoever of getting what it needs even though it also risks getting nothing. Although the first patch guarantees it something it is now not enough.

This scenario highlights a situation in which a bird should take into account not just the *average* reward rate at different patches but also the *variability* in reward rate – how likely is a patch to yield a bonanza or nothing at all? Such *risk-sensitivity* in foraging decisions have important consequences for a bird's prospects of survival.

*Left. European Starlings delivering food to their chicks can predict the optimal load size with which to return to the nest. The diagram below illustrates the relationship between predicted and actual load sizes for three different currencies of food delivery: A maximizes the rate of delivering energy to the chicks; B maximizes the rate of energy delivery net of the energetic costs of flight by adults and begging chicks; C maximizes the energetic efficiency of food delivery*

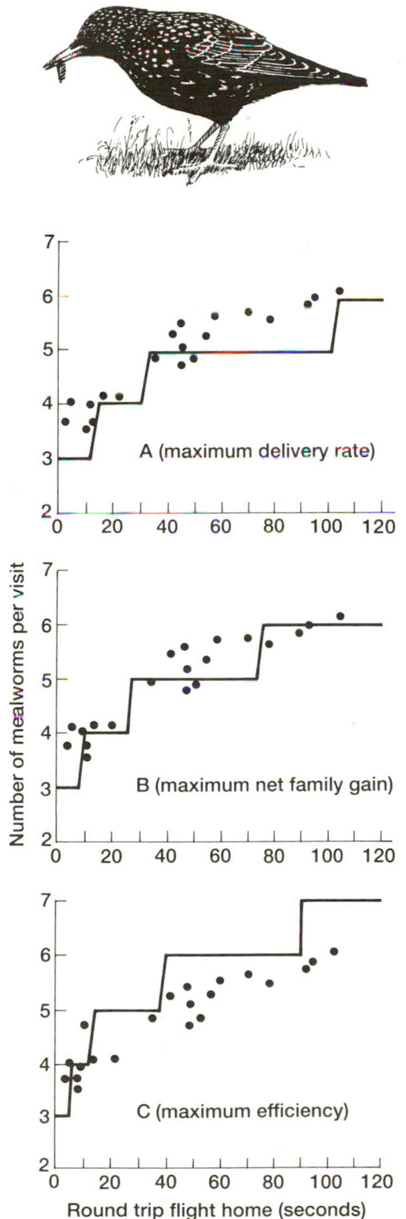

Experimental tests of risk-sensitivity among birds have centred around Tom Caraco's work with Yellow-eyed Juncos and White-crowned Sparrows, small North American buntings. In one experiment with juncos, Caraco offered captive birds a choice between a feeding station that provided two seeds on each visit and a station that provided nothing on half the visits but four seeds on the other half. Birds were deprived of food for different lengths of time before each test so that their food demand varied. As predicted above, birds switched from preferring the constant (two seeds every time) station to preferring the variable station as deprivation time and thus food demand increased.

## CACHING

Risk-sensitive foraging is one way in which some birds cope with shortfalls in their food supply. Caching is another. Caching or food-hoarding effectively provides a means of smoothing the fluctuations in natural food supplies which can lead to shortages, by creating a recoverable store. The major advantage of a food store is that it is the bird's own creation. The quality, density and distribution of food items within it are, in principle, known to the bird in advance of recovery, and much of the costly uncertainty of searching out new sources of food is removed. However, the effectiveness of the store in this respect clearly depends on the bird's ability to remember storage locations and on whether any other predators have stumbled across the store in the meantime and plundered it.

Recent studies suggest that many food-hoarding birds can remember storage sites accurately and for long periods, sometimes up to several weeks. Many members of the parid (tit, chickadee) family are food hoarders, storing seeds, insects and other invertebrates in any convenient small cavity, such as crevices in bark. Field studies suggest that these stores are short term, with Marsh Tits, for instance, recovering stored seeds within four to twelve daylight hours. When experimenters placed extra seeds at different distances from stored seeds, it was the stored seeds that were taken first on 77% of occasions. This implies that storage sites are remembered in the wild. Seeds are not recovered as a result of the bird simply bumping into them by chance. Subsequent laboratory experiments with Marsh Tits have shown that up to 90% of search time during recovery was spent at sites (sectors of trays of moss) where seeds had been stored previously.

To pin down the role of memory more closely, David Sherry and co-workers used the fact that birds like tits, with laterally placed eyes and little binocular vision, cannot perform some discriminations learned with one eye when they are forced to use the other eye, unless the initial discrimination is learned using the binocular portion of the visual field. In Sherry's experiments, Marsh Tits stored seeds while one of their eyes was covered with a small plastic cap. After storing, the cap was either taken off and replaced on the same eye or switched to the eye which had just been used to store. When allowed to use the same eye during seed recovery birds performed as well as if they had been able to use both eyes. When forced to use their 'naive' eye, however, their recovery rate was no better than that expected from chance encounters. Accurate seed recovery thus appears to depend on the acquisition, retention and availability of visual information, implying a crucial role of memory in relocation.

While food-hoarders may be able to remember where they put their food, memory alone does not guarantee it will still be there when they return. For many food-hoarding species, especially those using long-term hoards, stored food is vulnerable to loss from a variety of sources, including decomposition and pilfering. To get round this, hoarders have adopted a variety of tactics for protecting their stores.

Some measures against pilferers are based on the spatial distribution of the store. Scatter-hoarders like tits and nuthatches often rely on secretive hoarding behaviour and cryptic stores to reduce the risk of discovery. The spacing of items in scatter hoards may itself act as a deterrent because increased spacing reduces the likelihood that a potential thief will discover more than a handful of storage sites. Larder-hoarders like Acorn Woodpeckers, however, which stockpile food at particular locations, often have to spend time aggressively defending their stores from intruders.

Anti-theft behaviour may be finely tuned to the immediate risk. Hans Källander, for instance, noted that Rooks scatter-hoarding walnuts were not secretive if the other Rooks around them were also hoarding, but if a non-hoarding individual wandered through, a hoarding bird would pick up the item it was storing and wait until the newcomer had left.

CJB

## Does foraging optimally mean birds are clever?

No. A popular misconception on encountering optimality arguments for the first time is that, to fulfil their predictions, animals must be very clever. After all, it takes sophisticated mathematics and computer programs to make the predictions so it must take mental processes at least as complicated to abide by them. Actually what the maths and models do is to simulate the process of natural selection acting on the consequences for survival and reproduction of alternative courses of action; they say nothing about the mental routes by which animals arrive at their decisions. In fact there is good evidence that animals use remarkably simple behavioural rules to achieve what look like clever decisions – for example, always going for the biggest prey; because bigger prey are usually more profitable, a 'take the biggest' rule means the animal ends up following the optimal policy of taking the most profitable prey as if it was actually counting the calories.

CJB

## Energy requirements

Birds use energy in the metabolic processes concerned with maintaining life (such as respiration and excretion) and in activities such as flying. To meet these requirements they must obtain sufficient energy from their food. Biologists arbitrarily divide energy expenditure into various levels (although in reality it is continuously variable). The energy expenditure of a bird is lowest when the bird is inactive and unfed and when the ambient temperature is high (that is, in the thermoneutral zone). This level of expenditure is known as the basal metabolic rate (BMR), which scientists measure by the amount of oxygen consumed or carbon dioxide produced by the bird during respiration. The basal metabolic rate varies with the weight of the bird; large birds use more energy than small ones in absolute terms but they use relatively less per unit weight. Passerines have a higher basal metabolic rate than non-passerines of the same weight. Primitive birds such as Ostriches and kiwis have notably low basal metabolic rates, and swifts high ones.

*Studying the gases expired by an incubating Wandering Albatross (Diomedia exulans) permits an estimate of its energy requirements. Incubation surprisingly takes only 1.3 times as much energy as resting quietly*

In addition to the basal metabolic rate, energy is required for a variety of other processes and activities. When the ambient temperature is lower than the thermoneutral zone, heat must be produced to maintain the bird's body temperature. The metabolic rate of an unfed, resting bird that is also producing heat is known as the standard metabolic rate (SMR). Of greater importance is the energy expended in activity: the daily energy expenditure of an active bird in the wild is often at a level of three or four times the basal level.

Basal metabolic requirements vary principally with the time of day, geographical location and season. For example, they are highest during that part of the day when the bird is normally active (during daylight hours for diurnal birds and at night for nocturnal ones). This variation occurs even when the bird is inactive.

Superimposed on this is an annual variation: the change from night to day is greater in winter than in summer. Basal metabolic rates of resident tropical species are typically one third lower than those of temperate species. Migrants also have reduced metabolic rates in their tropical wintering ranges. In contrast, exposure to cold enhances basal metabolic rates; and small birds in winter tend to have a high metabolic rate.

In general, free-living birds allocate about half of their daily energy expenditure to maintenance requirements; the rest is available for various activities of which foraging for food itself is particularly demanding of energy.

Small birds may need to eat their weight in food each day because of their high metabolic rate. Thus a Blue Tit weighing 11 g needs 1 kcal per gram of body weight each day in winter, which is equivalent to over 300 small insects, weighing about 10 g in total. Large birds, on the other hand, need relatively less food. A kestrel weighing 220 g needs only a third of a kcal per gram of body weight each day in winter: this means about seven voles, weighing in total 120 g.

Energy requirements vary greatly throughout the year because of changing environmental conditions and because of changes in the activities of the birds themselves. In winter, birds need extra energy to keep warm. They also lay down fat to tide them over the long nights or over periods when food is scarce and difficult to collect. However, unlike mammals, small birds do not put on a lot of fat in winter; a bird such as a Blue Tit may have enough fat for only a night, or a day or two of poor feeding conditions. Because they do not have a long-term store of fat, birds must regulate their time and hence energy budgets over one or a few days and they are therefore greatly affected by the environment. High energy costs force them to change their behaviour; thus male Red-winged Blackbirds engage in less territorial display and spend more time foraging at low temperatures.

Many birds escape adverse weather by migrating between their breeding grounds and more hospitable wintering areas. This itself is very demanding of energy and so migrating birds often lay down a store of fat to use as fuel for the journey. Dunlin, travelling between England and northern Russia, put on some 30 g of fat in the spring at a rate of 0.6 g a day, enough for a journey of two and a half thousand miles.

Energy is often required to defend resources such as a winter feeding territory or a breeding territory. Thus, territorial Hawaiian honeycreepers expend 17% more energy than non-territorial ones.

A lot of energy is put into producing offspring, especially when dependent nestlings or fledglings have to be fed by the parents. It costs little for the male to develop his testes and to

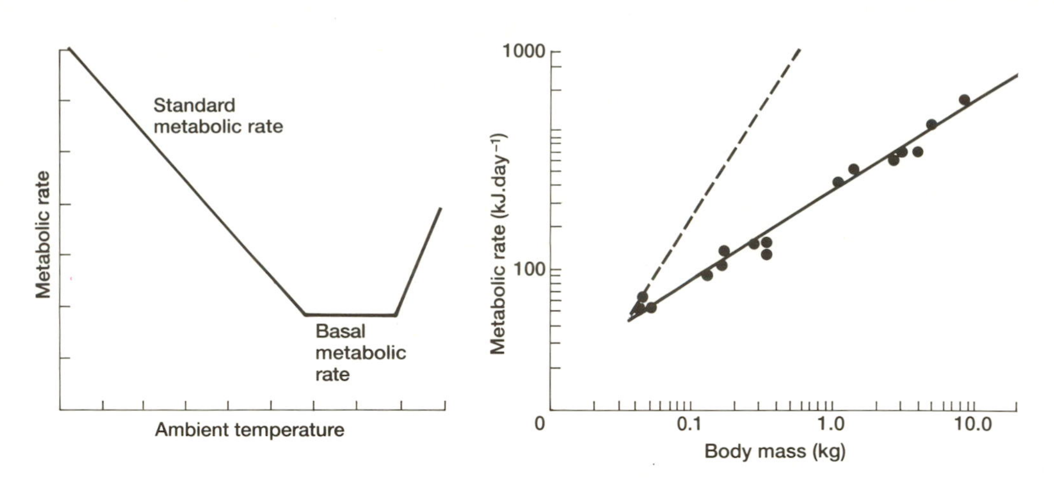


*Above left. A generalized plot of a bird's metabolic rate in relation to temperature. At low temperatures standard metabolic rate is high. It falls as temperature rises until a minimum metabolic rate, the basal metabolic rate, is reached at moderate temperatures. Higher temperatures still precipitate a new rise*

*Above. The relationship between body mass and metabolic rate for various petrels (solid line). The dashed line indicates the expected relationship if species of all body sizes had the same metabolic rate per unit weight, highlighting how large birds have, weight for weight, a lower metabolic rate than smaller species*

produce semen, usually less than 1% of his basal metabolic rate, but at this time he may also be incurring costs in defending a territory and attracting a mate. The female, on the other hand, concentrates her resources on producing eggs. The energy cost of growing ovarian tissue and producing eggs depends mainly on the relative size and energy content of the egg and the size of the final clutch.

The eggs develop over several days so the energy cost is spread somewhat but peak daily requirements can still be very high: about 50% of daily BMR in a passerine such as the European Starling and more than 200% of BMR in the Wood Duck. Males sometimes contribute to the female's energy budget by feeding her at this stage. Even so, females may use up much more energy than their partners at this stage. The daily energy expenditure of female Dippers while laying is 4.3 times BMR compared with 3.2 times BMR for their mates.

Incubating birds must both maintain a constant egg temperature and increase the temperature of eggs that may have cooled while they have been absent from the nest. When the ambient temperature is high the bird's body heat may be sufficient to keep the eggs warm but at low temperatures, when additional heat is needed, the costs of incubation increase. Costs at this stage can be reduced by having a well-insulated nest, or one in a cavity, but they increase with clutch size. The bird may also start incubation at a high body weight so that it has some energy in reserve. The embryos themselves can contribute some heat. In the American Kestrel this amounts to about 20% of the heat required. The metabolic rate of Zebra Finches while incubating is 20–30% higher than that of a non-incubating adult. On the other hand, an incubating bird does not spend much time flying so the daily metabolic rate may be low. Female Dippers during incubation expend energy at 3.2 times BMR – about the same as a territorial male.

Precocial young, which can feed themselves, may require little energy investment apart from that of defence from predators. Altricial nestlings, however, need to be fed by the parents for a number of weeks. Birds generally work at about four times BMR when feeding nestlings, in part depending on the number or mass of nestlings in the brood. However, birds cannot work at more than four times basal metabolism for long, perhaps because sustained higher levels of expenditure would risk depleting body reserves or because food could not be processed fast enough, even if it could be collected. This energy limit may prevent birds from rearing many chicks at any one time. For the first few days of life, altricial nestlings cannot regulate their own body temperature and hence one or both parents must spend time brooding them. They then grow rapidly to a peak weight, often putting on fat at this stage, and so the time and energy spent foraging by the parents greatly increase. Peak energy requirements may, however, be spread out if the nestlings have hatched over several days; energy may also be saved by slow growth, low nestling activity, low metabolic rates or reduced heat requirements, for example in large broods when nestlings huddle.

During moult, energy is needed to replace feathers; in addition the loss of feathers may also reduce insulation and so the bird may need more food to keep warm. Enough food must also be eaten to supply sufficient nutrients such as amino-acids for feather replacement and this may be more than that needed purely to meet energy requirements. However, daily energy expenditures of free-living birds are probably low at this time: for example, about three times BMR for Dippers.

Daily energy requirements depend on how the birds allocate their available time to different activities (their time budgets) and this in turn depends on the stage of the annual cycle. Birds must sometimes devote a lot of time and energy to finding food. In winter, when temperatures are low, days short and food scarce, Blue Tits spend 85% of the daylight hours searching for and catching food; in spring this drops to about 70%. At the other extreme, hummingbirds, which feed on an abundant, localized and energy-rich food may forage for as little as 10% of the day. During the breeding season, a female Dipper that has to spend time incubating eggs has only 15% of the day in which to feed, while one feeding nestlings spends over 60% of the day collecting food for herself and her brood. A

| Kestrels | All percentages of a 24 hour day Winter | | Incubation | | Nestlings | |
|---|---|---|---|---|---|---|
| | M | F | M | F | M | F |
| Roosting | 54 | 54 | 29 | 29 | 25 | 25 |
| Sitting | 21 | 25 | 17 | 4 | 21 | 17 |
| Hunting | 25 | 21 | 46 | 2 | 54 | 45 |
| (Perch-hunt) | 19 | 17 | 29 | 2 | 33 | 33 |
| (Flight-hunt) | 4 | 1 | 13 | 0 | 13 | 8 |
| (Flying/soaring) | 2 | 3 | 4 | 0 | 8 | 4 |
| Incubation/brooding | – | – | 8 | 65 | – | 13 |
| **Dippers** | **Early spring** | | **Incubation** | | **Nestlings** | |
| | M | F | M | F | M | F |
| Roosting | 48 | 49 | 34 | 32 | 22 | 22 |
| Resting | 6 | 8 | 24 | 0 | 42 | 29 |
| Hunting | 42 | 41 | 41 | 10 | 29 | 45 |
| (Foraging) | 28 | 26 | 38 | 10 | 20 | 31 |
| (Diving) | 14 | 15 | 3 | 0 | 9 | 14 |
| Flying | 5 | 2 | 2 | 0 | 8 | 5 |
| Incubating | – | – | – | 58 | – | – |

*Time budgets for kestrels and dippers. For the kestrels (Falconidae), winter birds are paired but have not begun courtship feeding. Nestlings are in the second half of the nestling period. Hunting includes hunting from a perch and hunting from the air (perch-hunt and flight-hunt, plus extra flying) so sitting is additional to perching while foraging. During incubation the male occasionally sits on the eggs but the female undertakes the major part of the incubation.*

*For the dippers (Cinclidae), early spring birds are on a territory but birds within 16 days of producing eggs are not included. Nestlings are in the later part of the nestling period from day 16 onwards. Hunting is either done while walking (foraging) or while underwater (diving) and does not involve flying. Resting includes both preening and singing. Only the females incubate*

lot of time is also usually spent resting, perhaps to keep energy requirements low; resting time may also be a buffer which can be turned over to foraging if, for example, food becomes scarce and more time-consuming to collect.

This variation in activity contributes to a variation in energy requirements of about 30–40% over a year. Daily energy demands are generally highest when the birds are feeding nestlings and low during moult and incubation when the birds are less active. Energy demands can also be high during egg laying and in midwinter. Male kestrels need nearly 100 kcal per day when they are feeding nestlings but less than 70 kcal when they are moulting. Female kestrels, on the other hand, need least energy (55 kcal) when they are incubating eggs. Male Dippers need about 60 kcal during the winter and 70 kcal when feeding nestlings but only about 50 kcal when moulting. Female Dippers need about 70 kcal when laying eggs and about 60 kcal when feeding nestlings but less than 50 kcal when moulting.

Activity, particularly flying time, affects the daily energy budget so much because flight is very demanding of energy. The energy costs of flight average about twelve times that of basal metabolism but they depend on the size of the bird, its aerodynamic properties and its method and speed of flight. Swallows, martins and swifts have exceptionally low flight costs for their size (only about four or five times basal metabolism). This is due to their streamlined shape, long wings with a large wing area, and their extensive use of low-cost gliding flight. In stark contrast, the short flights of European Robins cost 23 times basal metabolism because landing, taking off and flapping are energy expensive. However, robins do not fly very much so the total energy requirements of flight during a typical day are not high. Diving and swimming are also expensive activities; diving costs for Dippers are about six times BMR.

Over a year, birds can consume a considerable amount of energy. A male kestrel, for example, needs over 27 000 kcal a year – the equivalent of more than 2000 voles. A population of birds can thus have a considerable impact on their environment. In some cases this may cause problems for humans. Even small birds such as queleas, each weighing less than 20 g, can be pests when they occur in large numbers. In Africa, there are some 1500 million queleas; a quarter of these may attack grain crops for about 30 days a year. Each bird needs about 18 kcal a day or 2.5 g of seeds; the population as a whole needs 4000 tonnes a day. In addition to the food they eat, some is dropped and wasted and this may be as much as 7 g a day by each bird. Quelea damage costs Africa about $22 million a year, although only a few per cent of the total grain harvest is affected (see Chapter 11).

AKT

# Feather care and moult

## Feather care

Birds spend a significant fraction of each day resting. A proportion of this resting time is devoted to feather care. Since feathers provide birds with their insulation, colour and means of locomotion, they are basic to survival, and feather care is of the utmost importance.

Surprisingly, the mature, functional feather is a light, strong, but dead, structure (see feather structure). As a result feathers cannot be repaired, and damaged feathers survive until they are naturally replaced at the time of moult. Accordingly, all birds attempt to keep their feathers in good condition and devote much time and effort to feather care.

*Like all birds, the Great White Egret (Egretta alba) keeps its plumage in good condition by preening*

### *Preening*

Preening is the most basic activity of feather care. Here the bill is used to restore feather structure, to clean the plumage, to apply preening oils, or simply to rearrange displaced feathers. Preening occurs frequently throughout the day, either as the focus of prolonged activity given over to the behaviour or as a quick response to some immediate discomfort. Most frequently it is performed while the birds are at rest (loafing), but has been known to occur in flight.

Depending on the type of maintenance required, the bill is used in one of two ways. Either the feather is passed through the bill with a 'nibbling' action, or else the bill is 'wiped' across the surface of the feather. In the former, the feather is worked between the mandibles so that each portion of the feather is carefully attended to. This may be important for restoring the relationship between the barbs and barbules, for removing dust particles and stale oils, or for applying new oils and working them into the feather structure. Wiping is probably most important in rearranging feathers, in drying them, or assisting with the application of preening oils.

The preen gland (uropygial gland) is located on the upper surface of the rump just at the point of attachment of the major tail feathers, or rectrices. Although the preen gland is large, it is not easily seen because it is covered by the overlying body (contour) feathers. The gland, which is frequently bi-lobed, is formed of a mass of tubules in which fat cells are formed. It is the decay of these cells that produces the oils used by birds in feather care.

The preening oil leaves the gland through a nipple-like structure. The upper surface of the gland is generally bare, although the 'nipple' usually supports a tuft of modified down feathers. Manipulation of these feathers stimulates release of the preening oil which then

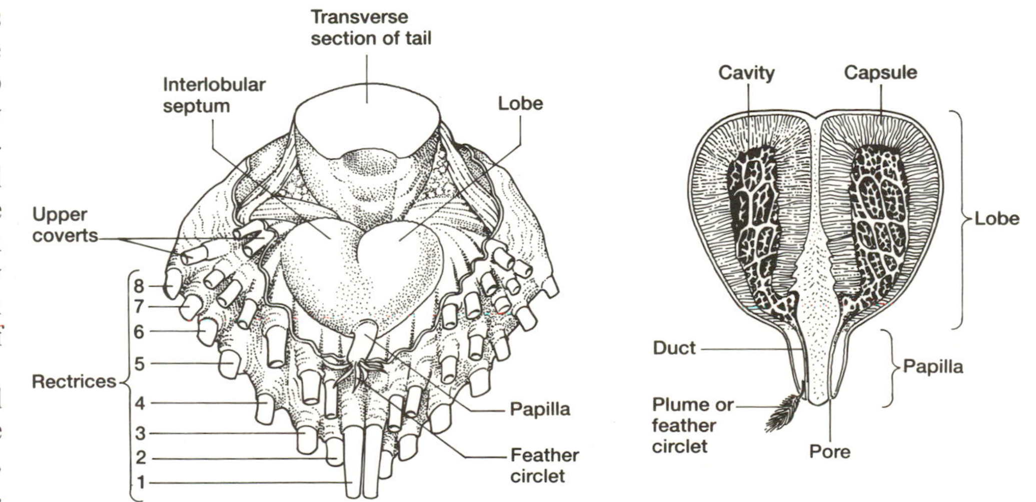


*The preen gland of a domestic chicken showing its location above the base of the tail and (right) its internal structure*

flows out on to the surface of the gland where it becomes available for use. The size of the gland is generally related to the life-style of the bird; those birds that occupy wet habitats have proportionally larger glands, whereas land birds have proportionally smaller glands. There are some birds, however, that totally lack a preen gland: for example, the Ostrich, emus, cassowaries and some doves, parrots and woodpeckers.

Although the oils produced by the preen gland are formed from a mixture of fatty substances, there is considerable intraspecific variation, and even individual variation as to the precise nature of the oils. The size of the preen gland and the nature of the oils produced may also vary with season. For example, in waterfowl, the makeup of the preening oil is different at the time of the eclipse (basic) plumage than it is for the breeding (alternate) plumage. These changes are apparently brought about by the direct influence of specific hormones to which the gland is sensitive.

Preening oil assists waterproofing in birds. This may result from its own oily properties or its ability to help preserve the physical structure and flexibility of the feathers. Its effectiveness as a water repellant, however, shows interspecific variation. For example, although cormorants and anhingas possess preen glands, their flight feathers become waterlogged during diving. As a result, they frequently spread their wings and tails to dry after periods of diving. During preening, oils are applied to the naked parts of the body, including bills and feet. Thus, oils may be essential to maintaining the condition of exposed parts of the skin. Preening oils may even possess antibiotic substances that could help to control bacterial and fungal infections. Finally, it has been suggested that components of the oils, when exposed to sunlight, become converted to vitamin D. The vitamin could then be absorbed through the skin or ingested during preening.

Birds lacking a preen gland compensate in various ways. Ratites' feathers are loosely structured, needing no treatment with preening oil. Some other species use modified feathers, the broken fragments of which provide waterproofing.

Although the bill is the major tool used in preening, not all parts of the body can be reached by it. The head, for example, presents particular problems. Here the feet are essential to grooming but, in addition, the head may be rubbed against the preen gland, other parts of the body, or the perch on which the animal is sitting. In some instances the bird may solicit the assistance of other individuals through a distinctive behaviour in which the head is bowed and the neck presented to the prospective partner. This behaviour, known as allopreening, is complex and its basic function is now debated. Although it may serve in grooming, it seems most likely that its prime function, within a species, is in pair bonding or in signalling social status. Between species, allopreening serves to establish dominance or to assist in social integration, for example, between brood parasites and their potential hosts.

### *Bathing*

Bathing is another behaviour that is used for feather care and also for bodily comfort. Although it is most commonly carried out with water, some birds, such as the pheasants and their allies (Phasianidae), use dust instead.

*Two Hooded Mergansers (Mergus cucullatus) take a bath to cleanse their plumage. Although most bird bathing involves water, some species prefer dust*

Several types of bathing behaviours have been identified. These are: (*a*) stand-out bathing ('splash-bathing') which is carried out while the bird is at the water's edge; (*b*) stand-in bathing, where the bird actually stands or crouches in shallow water; (*c*) in–out bathing, where the bird jumps in and out of shallow or deep water; (*d*) flight-bathing, in which the bird simply dips in and out of the water while in flight; (*e*) plunge-bathing, in which the bird appears to 'dive' from a perch into the water, and out again; (*f*) rain-bathing, in which the birds deliberately expose themselves to rain, either in flight or while stationary; and (*g*) dew-bathing, in which the droplets of dew that accumulate on leaves or grass are used to moisten the feathers. Although birds may use one or a combination of the methods described, most groups of birds predominantly use a characteristic bathing behaviour.

When bathing, birds usually moisten their feathers, without soaking them. This suggests that bathing mainly serves to assist preening, perhaps by making the feathers more pliable, or to ensure an even distribution of preening oil. On the other hand, bathing can result in a complete soaking of the feathers. In this case bathing may be an important method of cleaning the feathers and underlying skin. After such a bath the bird is in a vulnerable condition as its flight efficiency is impaired. Therefore, it needs a nearby and safe place in which to groom itself and to let the feathers dry. Aside from its role in feather and body care, bathing may also provide a mechanism to assist body cooling under hot conditions.

While bathing, the bird typically ruffles its feathers, dips its head and breast into the water and, while fluttering its wings, splashes water up and over its body. These movements form the basis of most bathing behaviour, although additional components, for example, fanning of the tail or tilting the body from side to side, may be added.

Bathing behaviour, like most avian behaviour, is adaptable and birds will use favourable conditions when and where they find them. Whereas most land birds bathe while standing either on the edge of, or in the water, many sea birds bathe while swimming on the surface of deep water. Nevertheless some land birds, for example the Rook, have been observed to bathe while swimming in deep water, while some sea birds such as gulls, use stand-out bathing at the edge of rock pools. Clearly, bathing preference is not always predictable and birds have been known to use a wide variety of water sources, such as the spray from garden sprinklers, the water in stock tanks and even the driving rain of tropical storms, to meet their needs.

The frequency with which bathing occurs varies with individual need, but generally bathing continues with a similar intensity in both winter and summer. The time of day at which bathing occurs, however, varies with season and with species. For example, pigeons tend to bathe in the morning, and House Sparrows in the afternoon. Certain environmental factors also influence bathing. Sunshine is known to stimulate bathing activity whereas increasing wind velocity, increasing cloud cover, and decreasing temperature, tend to suppress it. Despite these generalizations, high latitude species such as penguins and many temperate zone species, for example European Blackbirds and Starlings regularly bathe under freezing temperatures. If free-standing water is not available, snow is commonly used as an alternative. Snow-bathing has been reported from a wide variety of avian families. During the winter, bathing generally occurs under sunny conditions when birds have sufficient time to dry themselves before nightfall.

Dust-bathing is in many ways similar to water-bathing. In the former the bird usually squats or lies down in a sunny, dusty place. Here a body-sized depression is usually formed in which it squats, then using movements that are similar to those of water-bathing, flicks dust on to the various parts of the body. After dust-bathing the bird rises, fluffs its feathers and vigorously shakes itself to remove the excess dust from its plumage. This is usually followed by a bout of preening or scratching.

The role of dust-bathing is not yet fully understood. In most species it is not as common as water-bathing and, with the exception of the pheasants, does not seem to be a substitute for the latter. It appears to be more common in birds of open country but is not restricted to them. There can be little doubt that dust-bathing is a form of comfort behaviour. It may serve a role in removing excess or stale oils, excess water, and the inevitable debris (particles of skin and broken feathers) that accumulates in the plumage. Additionally, dust-bathing may assist in removing parasites such as fleas, lice, and mites from the plumage.

Like water-bathing, dust-bathing shows considerable variation as to its frequency and timing. It appears to be more seasonal than water-bathing and is more common in summer than in winter. This could simply be a consequence of dry conditions being more common in summer, though one might argue that snow-bathing is the winter equivalent of dust-bathing. Dust-bathing is also influenced by the weather. Sunshine and warm temperatures encourage dust-bathing whereas wind and rainfall depress the activity.

*Sunning*

Sunning is another behaviour associated with feather care. It is often referred to as sun-bathing but because it does not share the same functions as true bathing it seems more appropriate to refer to the behaviour as sunning. For example, sunning is not only associated with feather care but is also important in temperature control. Thus sunning can be divided into two types: (*a*) sun-exposure, in which it appears the plumage and skin benefit from direct exposure to the sun's rays; and (*b*) sun-basking, in which the bird seems to benefit mainly from the sun as a heat source. In both cases the bird frequently adopts a characteristic posture which makes this behaviour easy to recognize.

Sun-exposure is widespread amongst birds. It usually occurs in some protected location which is, nevertheless, exposed to full sunlight. Usually it occurs during the mid-day period when the sun is at its highest and its hottest. At this time the bird selects its sunning location then frequently settles on the ground, fluffs its feathers and extends its wings and tail so that the back is fully exposed to the sun. On other occasions the flight feathers may not be fully extended but the body feathers are erected. Sunning birds often appear to be overheated as they frequently pant and lift the wings away from the body, presumably to dissipate heat. Overheating is not tolerated for long and birds frequently have to retreat to shade where preening generally follows.

Sun-exposure probably has several functions associated with feather care. First, it appears that the sun's heat is invaluable in restoring and maintaining the shape of the major flight feathers. Second, the interaction of sunlight with the preening oils may be important to the synthesis of vitamin D. Third, the warmth of the sun may help in the spread of preening oil on the feathers, or in the softening of old oils. Fourth, the sun's rays alter feather colouration even after the feather has completed growth. Fifth, the sun's rays may have a directly beneficial effect in maintaining skin condition. Finally, the sun's rays may increase the activity of fleas, mites, and lice which may in turn make them more easily detected for removal, or may encourage them to disperse to areas where they can be reached during preening.

By contrast to sun-exposure, sun-basking appears to be a mechanism for influencing body temperature. Under cool conditions such exposure to the sun's warmth could be a mechanism for reducing energy demand and may be of particular advantage to small birds in temperate latitudes. However, sun-basking is not restricted to temperate zone species. Unlike sun-exposure, sun-basking is not restricted to the mid-day hours, and in temperate regions also appears to be more common in the non-summer seasons. Although the postures assumed are often similar, the different timing of this behaviour helps reinforce the view that its main value is in providing body warmth.

*A juvenile European Blackbird is here spreadeagled in the characteristic posture of a sunning bird. Sunning is believed to help maintain feather condition*

### *Anting*

Anting is a puzzling and bizarre aspect of avian behaviour. Although it has been frequently observed, its function is still debated. However, because sun-exposure and anting frequently occur as complementary behaviours, and often involve similar postures, both may have a common origin.

During anting the bird usually squats on the ground in an area where ants are common. The feathers are fluffed, the wings and tail are spread, and the ants are allowed to move freely through the plumage and over the skin. Alternatively, the bird may adopt a preening posture and selectively pick up ants which are either rubbed across the plumage or are placed in specific areas.

It is suggested that birds exploit the formic acid produced by many ants to remove stale preening oil from the feathers. Additionally, it may be that the formic acid may help to ease the comfort associated with moult and the discomfort associated with moult and the growth of new feathers. Support for the latter however, is meagre and the coincidence of anting with moult is thought to be a seasonal, rather than physiological, effect. A third suggestion is that anting is important to avian health. For example, formic acid may act as an effective insecticide which could help to kill or detach parasites from the skin and feathers. However, the birds do not always select ants that produce formic acid (Formicidae); they have also been seen to 'ant' with harvest ants (Myricinae), millipedes, and even worms. These observations raise doubts about the purely insecticidal function of anting, and so several other suggestions have been made. Some ants, for example, produce complex glandular secretions that cover the body, and have antibiotic properties that are effective against bacterial and fungal diseases.

This could also help to explain the often noted relationship between anting behaviour and periods of high humidity, a time when fungal diseases are common. Alternatively, it may be that ants become more active under humid conditions, and that the birds may be responding to increased activity by the ants. It has even been suggested that anting is an autoerotic act.

### *Headscratching*

Scratching, in one form or another, is a common part of the comfort behaviours of birds. It is a natural response to minor irritation on the body. However, it is also an important component of preening behaviour where it serves to realign the feathers, much as combing would do. Headscratching, however, seems to be performed in a highly specific way. Consequently, birds are divided into two groups depending upon whether they scratch their heads by (*a*) the direct method (in which the head is scratched under the wing), or (*b*) the indirect method (in which the head is scratched over the wing). It was originally thought that this behaviour was rigidly fixed and thus could be used as a tool in classification. Recent evidence suggests, however, that it may be less rigid than previously thought. For example, the Black and White Warbler and the Hairy Woodpecker use both types of headscratching behaviour. Even within families the pattern is not always consistent. For example, in 40 species of North American wood warblers, 31 species are indirect scratchers, seven are direct scratchers, one uses both (see above) and one remains undetermined. In general, it appears that the direct method is the most primitive and is characteristic of ground dwelling species, whereas the indirect method is the more recently evolved behaviour and is characteristic of the tree dwelling species. Those species that show both methods are characteristically climbers (woodpeckers, woodcreepers, creepers, and nuthatches) with distinctive leg and pelvic anatomy that makes possible both types of headscratching. Apart from feather care, headscratching may be important during moult for dislodging and removing feathers.

## Moult

Feathers are subject to damage and wear. Thus, if plumage effectiveness is to be maintained, birds must be able to replace damaged and worn out feathers. Feather replacement occurs on a regular basis in the life of all birds and is known as moult.

Natural moult occurs when the feather follicle is stimulated to growth. As the new feather begins to form the old feather is gradually pushed out of its socket. The new feathers emerge from the follicle as pin-like structures which gradually break open to reveal the form of the new feather. Once the feather has achieved its mature shape and size it becomes hardened with a tough proteinaceous material (keratin), blood is gradually withdrawn from the feather shaft and the follicle ceases its activity until the next period of moult. Although active growth in the follicle is normally restricted to the moulting period, growth can be stimulated through plucking of feathers or through some traumatic event which can cause a spontaneous dropping of feathers known as 'shock moult'. Whatever the stimulus, moult is a complex process that involves both the dropping and replacement of feathers.

### *Control of moult*

What stimulates the natural moult cycle is not well understood. However, its seasonal occurrence is generally timed to ensure survival and to reduce conflict with other vital events in the annual cycle, such as breeding and migration. These processes are regulated by a complex interaction between environmental information and the physiology of the bird. In temperate regions and high latitudes, the seasonal changes in daylength appear to have the most profound influence on the timing of the major events of the annual cycle. By contrast, in tropical regions where daylength varies very little during the course of the year, or in desert areas where strict adherence to changes in daylength could be fatal, other factors such as the seasonal occurrence of rain, or the sudden appearance of new plant growth, may provide the necessary environmental stimuli. Whatever the case, the various events of the annual cycle, including moult, are programmed to occur at the most appropriate times of the year.

The internal processes that regulate moult are not well understood, even though they have been the focus of much experimental work. The results, however, clearly suggest that the hormonal system plays a fundamental role. However, which hormones are the most important, and how they activate moult, remain a puzzle. In addition, the stimulus for moult in one group of birds may be quite different to that in another. In general, it appears that the reproductive hormones (e.g. oestrogen and testosterone) tend to suppress moult, and so a decrease in their concentration appears to be necessary for moult to occur. But whether these hormones have a direct effect on the follicle, or whether their impact is in turn mediated by the gonad-stimulating hormones produced in the pituitary gland (e.g. follicle-stimulating hormone or luteinizing hormone), is not clear. Likewise, the role of some of the accessory hormones involved in reproduction (e.g. prolactin necessary for incubation) is still open to question. Evidence strongly suggests that the thyroid gland is fundamental to moult and that increased production of the thyroid hormone (thyroxine) will stimulate moult. But again, it is not clear whether the thyroid hormones actually initiate moult or whether their activity is simply essential to maintaining moult once it has begun. Clearly, many questions remain about the controlling mechanisms of moult.

Once begun, moult is generally an orderly and predictable process. As moult proceeds the feathers are shed in a controlled, progressive manner. Because the flight feathers are generally the largest feathers on the body, and are also essentially for flight, they are usually replaced gradually and take the longest to complete moult. Thus seasonal moult tends to begin with the shedding of the first flight feathers and ends when the last one is fully grown. The pattern in which these feathers are moulted is specific and characteristic of the various families of birds. In the interim, all the feathers of the body are shed and replaced, once again in a specific and characteristic manner.

In some birds flight is impaired during moult. This most commonly results from the rapid shedding of all the flight feathers at one time. Species that are characterized by such flightlessness during moult usually inhabit areas where they can escape from predators. Thus many water birds (e.g. divers, grebes, waterfowl, and shearwaters) shed all their flight feathers simultaneously, are flightless for four or five weeks, and spend that time in the comparative safety of open water (lakes, sea) or offshore islands. By contrast, species which rely on a more typical sequential moult may become flightless when too many feathers are lost at one time. This is characteristic of some arctic breeding passerines where moult has to be compressed into a space of weeks between the end of breeding and the onset of deteriorating environmental conditions. The most bizarre example of flightlessness during moult occurs in the females of some hornbill species which moult all their flight feathers while sealed in their nesting cavities.

### *Periodicity and patterns of moult*

The great majority of birds undergo one annual moult which normally follows the breeding season and, depending on the species, may take from four to eight weeks for completion. Generally the larger the individual, the longer the duration of the moult, for example two to three years in eagles and cranes. However, this pattern is not standard for all birds. Among waders (shorebirds) and flycatchers, moult may start on the breeding grounds, be suspended during migration and then be resumed when the birds reach their winter destinations. Such a pattern ensures that the energetic demands of migration do not compete with those of moult, while at the same time flight efficiency is little affected.

In other instances two moults may occur during the year. Such moults are characteristic of species that have distinctive breeding and non-breeding plumages. Thus some waterfowl have a short-lived 'eclipse' plumage that appears following the breeding season, while some songbirds, for example the American Goldfinch, alternate between a brightly coloured breeding plumage and a dull winter plumage. In these cases only one of the moults, usually the post-breeding (pre-basic) moult, is complete (i.e. all the feathers are replaced). The second moult, usually the pre-breeding (pre-alternate) moult, does not include the flight feathers and is thus described as being partial or incomplete.

Incomplete moults are also characteristic of species that have a third moult during the year. For example, some ptarmigans develop a mottled plumage in the autumn as an intermediary stage between the brown summer, and white winter, plumages. Elsewhere, many of the large species of seabirds, such as albatrosses, have a single partial moult of the body plumage each year so that a minimum of two years is required before the entire plumage is renewed. Likewise, species that live in non-seasonal environments, such as the Budgerigar of Australia, often have a series of incomplete moults that occur during the course of the year. Clearly, no general moulting pattern applies to all birds, but the pattern adopted by each species is adapted to its needs.

*An immature White-bellied Sea Eagle (Haliaeetus leucogaster) moulting primary feathers (nearer the wing-tip) and secondary feathers (nearer the body) at the same time, a pattern common to most birds*

Breeding plumages are not always acquired through moult. In many instances the natural wear of the feathers removes their extremities so that some underlying and previously hidden feature becomes apparent. This is commonly seen in songbirds where wear of the feathers is essential to the unmasking of the breeding plumage. For example, the appearance of the red breast of the Redpoll, the black 'bib' of the House Sparrow and the non-spotted, steely plumage of the European Starling, all depend on the gradual wearing away of the soft, lightly coloured tips of the feathers in question. In such cases wear has been exploited as an energy saving device. This permits the bird to change its appearance without the costly demands of an additional moult. Generally, the feathers that exploit such wear are tipped by soft, light coloured extensions of the barbs. By contrast, where resistance to wear is at a premium the feathers are usually black in colour and are heavily impregnated with pigments (melanins) that are resistant to abrasion. This explains why many species have dark flight feathers and why the wings of white birds, such as gulls, are frequently tipped with black.

### Age variation and polymorphism

Not only does moult vary with the season it also varies frequently with the age of the bird. Many of the larger birds, such as seabirds and birds of prey, take several years to achieve sexual maturity and, as a result, do not achieve a fully adult plumage until years after hatching.

Surprisingly, however, the males of many songbird species also show variations in plumage with age. In some cases the differences are subtle, for example retention of juvenile underwing coverts in many North American blackbirds (Icterinae), but in other cases are quite obvious, for example plumages of male Northern Orioles. Where the differences are obvious the immature plumages frequently resemble those of the less brightly coloured females, or even the juvenile plumage. In this case sexual maturity is not the regulating factor, because these birds are capable of breeding when one year old and such dull coloured males often do breed successfully.

Four hypotheses currently exist to explain this phenomenon: (1) the sexual selection or cryptic hypothesis, which suggests that the drab males have a greater likelihood of survival than their brightly coloured counterparts; (2) the female mimicry hypothesis, which suggests the dull coloured males gain advantages, in addition to survival, through resembling females; (3) the winter adaptation hypothesis, which suggests that a plumage adapted for winter conditions continues to provide an advantage to its holder; and (4) the status signalling hypothesis, which suggests that the dull plumage signals subordination which can, in turn, reduce aggression from dominant individuals. Unfortunately, these hypotheses remain somewhat ambiguous and the problem of delayed plumage maturation is not yet clearly resolved.

Similarly, the issue of plumage differences between the sexes (plumage dimorphism) has also stimulated much discussion. Generally it is presumed that the males are more brightly coloured in order to attract an appropriate mate, and that the brighter the male, the greater his success (sexual selection theory). Alternatively, it may be that the female benefits from a drab plumage because this provides a form of camouflage which makes her less obvious to predators (predation theory). This would be particularly advantageous to females that are entirely responsible for incubation and brooding, as is common with many of the true finches (Fringillidae). Again, the issue remains unresolved (see Chapter 10 for a more detailed discussion).

### Non-plumage moult

In addition to feathers, the scales of the legs and feet, and special developments of the bill, must be replaced on a regular basis. Because the basic origin and structure of feathers and scales are similar, it is not surprising that moult of the scales is similar to that of the feather. However, the covering of the bill (the rhamphotheca) is an extension of the skin and is a living, growing, tissue. As a result bill colour, and even shape, frequently undergo dramatic seasonal changes, that, in harmony with the plumage, can indicate breeding condition.

In some birds, special ornamental structures are developed on a seasonal basis. In puffins, for example, highly coloured and laterally compressed 'sheaths' are developed during the

*Above. Many birds alter in appearance as they age. The transformation is particularly striking in gulls – California Gulls (Larus californicus) in this case – from brown juvenile to grey and white adult*

*Opposite. During the summer breeding season when sandeels are brought to the burrow-dwelling chick, the bill and facial ornaments of the adult Atlantic Common Puffin (Fratercula arctica) are at their most gaudy. The horny outer layers are shed to attain the drabber winter condition*

*Opposite below. As a resting Herring Gull (left) sinks towards sleep, its head drops (centre). In deep sleep, the bill is tucked under the breast feathers (right)*

breeding season, but are shed when the breeding season is over. Other groups of birds, such as hornbills and pelicans, similarly shed seasonal embellishments of the rhamphotheca.

Clearly, though moult is most commonly associated with plumage changes it is a process that can affect all the epidermal structures produced by birds. Despite its regular occurrence and its importance in the life of birds, many aspects of moult remain poorly understood.

ALAM

## Sleep and roosting

All birds sleep, and all birds roost. These two terms are often used interchangeably, but in fact they have rather different meanings. When ornithologists say a bird has 'gone to roost', they generally mean that it has moved to the place where it will sleep. Once at the roosting-site, however, a bird may engage in a variety of behaviours other than sleeping. For example, it may preen, it may rest without sleeping, or, in social species, it may spend time in various interactions with its fellows. Thus sleep is only one of several behaviours associated with roosting, though it is probably the most important one.

Sleep differs from rest in that it involves profound alterations in posture, alertness, heart rate, muscle tone and brain activity. In birds, as in mammals, there are two distinct states of sleep, termed 'quiet' and 'active' sleep, or 'shallow' and 'deep' sleep. In birds these sleep states and rest can often be distinguished by postural and behavioural changes. For example, as a Herring Gull sinks from rest into a deep sleep it first of all drops its head onto its breast, and then tucks its bill under its scapular feathers. The respiration frequency and heart beat slow and become less regular, the body temperature falls slightly, and the muscles relax so that sometimes a slight jerking of the head is evident. The bird becomes less sensitive to disturbances around it, becomes more difficult to arouse, and the activity of the brain also changes.

This is measured by attaching electrodes to the scalp and recording on an electroencephalogram (EEG) the electrical pulses created by the activity of nerve cells in the brain. Such recordings demonstrate a shift from 'slow' waves in quiet sleep to 'fast' waves in deep sleep. The exact significance of this change is not known, but it shows that a sleeping bird's brain is very active, and moreover, active in a way different from the waking state, and even different between the two types of sleep.

In humans, brain activity observed during deep sleep is known to be associated with dreaming, but no one knows if birds dream. In both birds and mammals all of these details of sleep behaviour and physiology have up to now only been measured in laboratories, where the subjects can be monitored with sensitive recording machines such as EEGs. Almost nothing is known about these details in free-living birds.

One peculiarity of avian sleep that can be readily observed and recorded in the natural state is eye-blinking. Even while in deep sleep birds open their eyes at regular intervals. There are some records of continuous eyelid closure lasting several hours, but usually a sleeping bird opens its eyes every few minutes or even every few seconds (and so 'peeking' might be a better term than 'blinking'). During a blink, the eye (birds often open just one eye) remains open for several seconds while the bird apparently surveys its surroundings. The eye is then closed again, but if there is some sort of disturbance it remains open. Careful observations have shown that the rate of eye blinking is slower in deep sleep. Also, in several species males blink more and sleep less deeply than females, and the rate of blinking may vary with factors such as the perceived safety of the sleeping site. Birds blink less, for example, when sleeping in groups than when alone, and more if a predator has recently been spotted in the vicinity.

It is also interesting to note that in many species, the eyelid is coloured to contrast with the head plumage. A sleeping Mallard drake often appears at first glance to be awake, because the closed white eyelid, conspicuous against the dark head plumage, gives the appearance of a wide-open eye, but a closer look reveals that this is an illusion. The eye itself is actually dark. The significance of the contrasting colouration is unknown, but one could speculate that it provides a measure of safety, by making a sleeping and vulnerable duck appear vigilant.

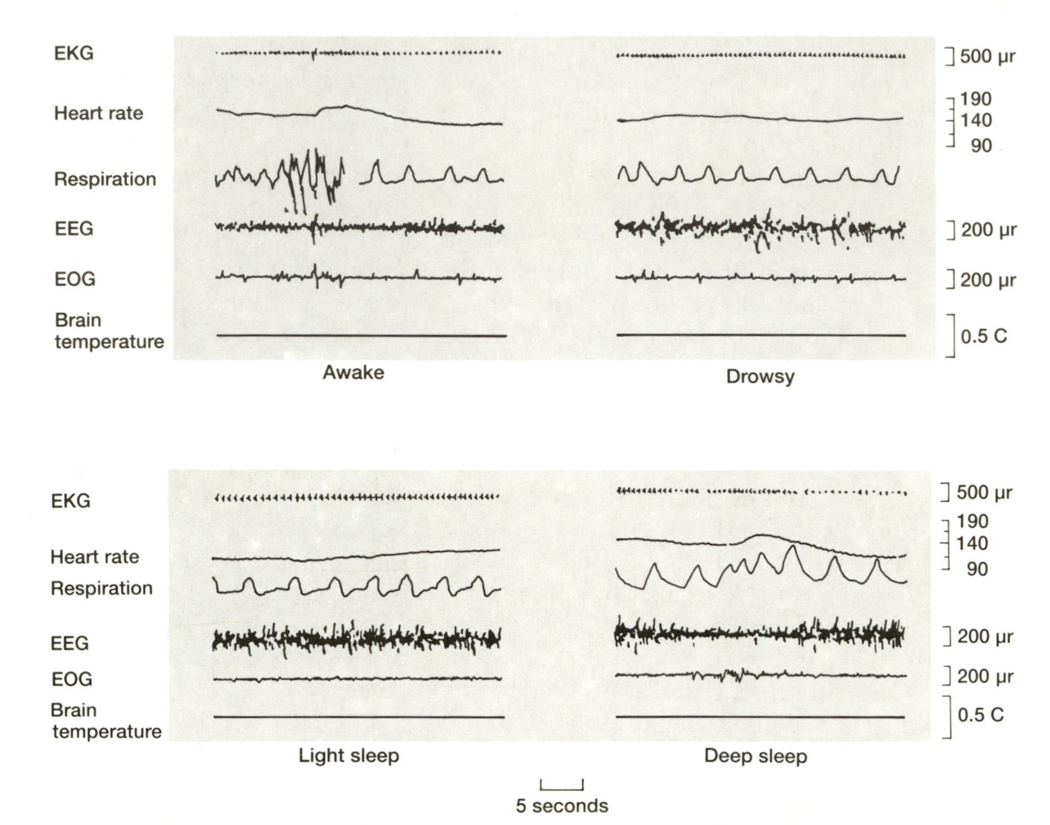


*Changes in the heart rate, brain activity and eye-blinking of a pigeon as it slips from wakefulness into sleep*
*EKG = electrocardiogram*
*EEG = electroencephalogram*
*EOG = electrooculogram*

Some birds are able to fall into torpor, which might be thought of as a particularly deep sleep. An animal in torpor has slowed its metabolism by letting the body temperature fall to just a few degrees above ambient, it is completely immobilized, and requires a long period to warm itself up before it can resume normal activity.

Torpor has been observed in swifts and hummingbirds, and in the nestlings of some seabirds, all of which are subject to occasional brief, but unpredictable and acute, shortages of food due to cold weather or storms. The ability to fall into torpor, thus drastically reducing energy expenditure, has evolved as a way of surviving these inclement periods.

It has been alleged that some birds sleep while flying, particularly the extremely aerial species such as swifts, swallows and some terns. Given the observation that birds usually sleep by alternating short periods of eye closure with brief 'peeks', this seems just possible. However, a close examination of the evidence reveals that it is entirely circumstantial. For example, House Martins were seen descending from a great height early in the morning, and the observer concluded that they must have slept on the wing. As is the case for so many aspects of the sleep behaviour of wild birds, not enough is known to draw a firm conclusion, but the evidence does not seem strong enough to support this claim, particularly since swifts and martins can often be observed sleeping in a more conventional manner.

For birds that roost solitarily or in small groups, roosting sites are usually carefully chosen with safety and shelter in mind. When not breeding, so that they need not be on or near the nest, most birds roost deep in tangled undergrowth, or in a crack, tree cavity, or crevice, or even a burrow in the snow. Many, such as owls, rely on camouflage to conceal themselves. Camouflage is also extremely important in open country where many species such as sandgrouse roost solitarily on the ground. However, other species, such as geese, roost in large groups that cannot be concealed. For many of these birds the same roost sites are often used year after year. Such roost sites are usually well placed with respect to food, water and safety, but the most important thing about them may be that they are traditional, and so individuals can be sure of meeting many conspecifics there.

The temperature inside a large roost may be warmer than outside, and the wind speed lower. In Rooks, Jackdaws and Redshank, three species in which it has been studied, it was found that these energy-efficient places are taken by dominant individuals, who force juvenile and sub-dominant individuals to the edge of the roost where the exposure to the wind is greater and the temperature lower. For these birds, it is apparent that the energy savings due to the improved microclimate is more than offset by the energy costs of flying the great distances that some individuals do to join these roosts. It seems there must be

*Poised for sleep. Black Vultures (Coragyps atratus) watch the sun go down from their roosting site, Cardon Cactus in the Sonora Desert of the south-western United States*

another advantage, likely related to some social factor. One such advantage may be that birds can learn about the location of difficult-to-find food sources by watching the behaviour of their roostmates. Experiments and observations on, among others, Red-billed Queleas and Black Vultures support this 'information centre' idea.

The timing of roost entry and departure usually shows a strong relationship with light levels, though the exact relationship may change seasonally. For example, many songbirds enter and depart their roosts just at civil twilight (sun 6° below horizon) during much of the non-breeding season, but as the breeding season advances they awaken earlier. Close studies of passerines such as the Great Tit have shown that, during the breeding season, the female roosts in the nest cavity, and the male nearby. The male escorts the female to the nest entrance at dusk, and returns again before dawn to awaken her with a low whistle. His attentiveness is an example of mate guarding. Outside the breeding season he is not so attentive, but the pair roost near one another on their territory. Their roost entry and departure times remain closely tied to dusk and dawn. In many songbird species the male often spends time singing just after awakening and before roosting, even in winter. This song is a territorial signal warning other birds to keep out.

Other species also spend much of their roosting time engaged in various social activities. Outside the breeding season, Rooks and European Starlings, for example, gather into smaller 'pre-roosts' up to several hours before entering the main roost. These pre-roosts are very noisy and conspicuous, and the birds are busy with social interactions, sometimes engaging in spectacular aerial aerobatics involving thousands of birds in tight flocks. As darkness falls, birds leave these pre-roosts and fly silently into the large main roost. Other gregarious species such as geese gather into 'post-roosts' after leaving the main roost at dawn. They preen and drink at these gatherings, but much of their time is spent in interactions with other individuals.

Most birds spend about eight hours out of every 24 hour cycle asleep, though there is great variability between species in the amount of sleep, and even within species there may be a large seasonal variation. Some species distribute their sleep time around the clock (European Pochards), but most sleep principally either during the night (Tufted Duck) or during the day (owls, Common Eider), and spend all or most of this sleep time at their roost. As this comparison shows, sleep habits vary greatly even within families of closely related birds. In the continuous daylight of the Arctic summer, birds seem to sleep rather little, but show an inactive period at around midnight, when the sun is lowest in the sky. In general, species living at high latitudes (with short summer days) sleep less than species at low latitudes. This relationship also appears to hold within species. For example, Common Eiders sleep three to five hours in the High Arctic, but over seven in Norway. In a statistical comparison, daylength (or interchangeably, latitude) seems to be the single best variable that predicts how much a bird will sleep. Nevertheless, the relationship between the daily events in a bird's life and the timing of sleep are not well understood. Does harder work, for example while provisioning nestlings, lead to a requirement for more sleep? In some experiments extra food has been supplied to birds, thereby reducing the work load, with the result that the amount of sleep increases, but not very much. Experiments such as these suggest that sleeping behaviour is rather inflexible, but the great variability observed between and within species suggests exactly the opposite!

Of all the mysteries surrounding sleep perhaps none is greater than the basic question of why animals sleep at all. Everyone has experienced tiredness, and the most obvious explanation is that sleep is required while the body and nervous system recharge or recover in some fashion. The recovery might be related to the elimination of some metabolite or waste product which cannot be processed while the brain is awake. The nature of this substance, if it exists at all, is not known. In support of this hypothesis is the phenomenon, common to birds and mammals including humans, that continued sleep deprivation leads to an increasing need for sleep. Longer and deeper sleep follows sleep deprivation, though all the lost sleep may not be made up. An alternative version of this recovery idea holds that it is necessary for the body to shut down while the brain processes all the information collected in the course of a normal day's activity. The apparent intensity and complexity of brain function during sleep may be explained by this idea. However, the recovery-type hypotheses have no explanation for the established fact that some birds sleep a lot and others rather little, within as well as between species.

An alternative theory of the evolution of sleep holds that sleep is a mechanism for immobilizing animals in a relatively safe roost site at times of the daily cycle when it is not safe or not worthwhile to be active. The energy-saving aspects of sleep, such as the slowing of the metabolic rate and the decrease in body temperature seem in line with this idea, but it is otherwise unsatisfactory in that it provides no good explanation for the high level of brain activity during sleep. Currently, scientists are not able to say which, if any, of these theories is correct. However, ornithologists are in the process of taking sleep research out of the laboratory and into the field where, hopefully, some answers may soon be found. RCY

## Diseases and endoparasites

Wild birds suffer a multitude of infections caused by bacteria, viruses or parasites and may also be exposed to toxic materials of natural or man-made origin. The role played by infectious disease in regulating wild bird populations is unclear but it is likely that nutrition plays an important role in the regulation of numbers, with disease usually serving as a secondary factor. Nevertheless, disease may be the primary cause of death in certain circumstances. For example, serious outbreaks of disease with many deaths can be due to a specific infectious agent acting in unusual circumstances such as crowding or stress due to severe weather or migration.

Some diseases are common to domestic poultry and wild birds and the latter may serve as reservoirs of disease-producing agents for domestic species. An example is Newcastle Disease (ND), a highly contagious and frequently fatal disease of domestic birds. A wide range of wild species have been reported as naturally infected with the virus of Newcastle Disease (NDV). The role of wild birds in the introduction of ND to a country and its subsequent spread therein is dealt with extensively

*Daily sleeping times for a selection of wild birds. The data are arranged by latitude (°N)*

| Species | Latitude | Sleeping time (hours) | No. of daylight hours | When sleep occurs |
|---|---|---|---|---|
| Eider | 78 | 4 | 24 | By day |
| American Robin | 69 | 4 | 24 | By night |
| European Starling | 69 | 1 | 24 | By night |
| House Martin | 68 | 3 | 24 | By night |
| Great Tit | 52 | 8 | 16 | By night |
| European Pochard | 49 | 13 | 16 | All hours |
| Tufted Duck | 49 | 13 | 14 | By night |
| Wild Turkey | 39 | 4 | 12 | By night |
| Anna's Hummingbird | 33 | 11 | 11 | By night |
| Pintail | 30 | 8 | 10 | By day |
| Galápagos Penguin | Equator | 13 | 12 | By night |

in books on poultry diseases. Conversely, wild species may acquire domestic infections when they frequent farm yards or poultry rearing establishments. Since wild birds are neither vaccinated against the diseases of domestic species nor fed continuous medication to prevent disease they may be unduly susceptible to diseases transmitted from domestic birds.

Many infectious agents occur in birds but often they are not associated with ill health. This is particularly so with internal parasites (endoparasites). Many hundreds of species have been described from birds but only when they occur in large numbers does ill health become evident.

*Left. Wild birds, for example Collared Doves (Streptopelia decaocto), may harbour diseases dangerous to domestic chickens, and inadvertently introduce them to poultry runs*

## Bacterial diseases

*Avian cholera* is usually an acute sporadic disease caused by *Pasteurella multocida* and a wide range of domestic and wild birds are susceptible. Particularly, disease may occur in waterfowl on migration or in their wintering grounds and severe outbreaks of disease involving thousands of waterfowl (coots, Whistling Swans, geese, ducks, etc) have been reported from western USA and elsewhere. Natural methods of transmission of the infection include the food, by invertebrates or by inhalation. Predation and scavenging on diseased carcases is also potentially important.

*Salmonellosis* has been reported from a wide range of birds, but gulls, because of their scavenging habits, are frequently affected. The most common bacterial species concerned is *Salmonella typhimurium* and the disease affects birds in the same way as other animals, producing diarrhoea and, in severe cases, death. There is evidence that wild birds may be the source of *Salmonella* infections in domestic livestock, such as cattle on feed lots. Outbreaks of salmonellosis have also been ascribed to the build-up of infection in feeding areas such as bird tables in the winter. Infection can be transmitted in the droppings.

*Tuberculosis*. The majority of species in which avian tuberculosis has been diagnosed are either gregarious or carrion feeders. In the United Kingdom, gulls have been reported to show a particularly high infection rate, Woodpigeons have been found commonly infected but others such as starlings, rooks and jackdaws are also affected. Tuberculosis is a chronic disease, infection occurring most usually by ingestion. The bowel is the primary site of infection, followed by the liver, lungs and spleen, where yellowish, cheesy, lesions are found. In wild birds the causative organism is the avian tuberculosis bacillus, but the human tuberculosis bacillus has been found in parrots.

*Psittacosis* or *ornithosis*, often known as parrot fever or chlamydiosis, after the name of the causative agent *Chlamydia*, is characteristically a disease of psittacine birds and of humans. However, it is now recognized in domestic poultry, feral pigeons in cities, and a range of wild birds as well as domestic and wild mammals. The lungs are chiefly affected and pneumonia is a common sign. Other organs affected include the liver and central nervous system.

Other bacterial infections found in wild birds include (i) *yersiniosis* of pigeons, sparrows, caged songbirds; (ii) *erysipelas*, occurring in domestic and wild birds and mammals, is of particular significance in bird parks and zoological gardens; and (iii) *mycoplasmosis* which occurs particularly in raptors and psittacines. Various kinds of mycoplasma occur and they produce different clinical signs, some cause respiratory tract infections similar to ornithosis while others produce disease of the bones and joints (*Mycoplasma synoviae*).

*Although the explosive increase of some gull populations may have been nourished by abundant supplies of garbage, this food source is not without its risks. It may, for instance, harbour botulism, a disease as fatal to birds as to humans*

## Fungal diseases

*Aspergillosis*, due to infection with the mould *Aspergillus fumigatus*, causes disease of the lungs, air sacs and liver. Many species may be infected but waterfowl and gamebirds seem particularly susceptible especially when they are kept in semi-domesticated conditions. Penguins in captivity are very prone to infection.

A further troublesome mycosis is thrush, or sour crop, caused by *Candida albicans* and seen in domestic poultry, game birds and psittacines. Young birds are particularly susceptible.

## Viral diseases

*Newcastle Disease*. While the primary species affected and the source of infection is the domestic chicken, wild birds are susceptible and may spread the virus. Movement of psittacines and caged wild caught birds in commercial trade has been a major source of introduction of a highly pathogenic strain in the USA.

Newcastle Disease is caused by an avian paramyxovirus of serotype 1 (PMV-1), and a closely related virus is responsible for disease in racing pigeons and other members of the Columbidae family. It has been previously assumed that, during outbreaks of ND, pigeons were affected as a result of contact with diseased poultry. While this can occur, the disease of racing pigeons which swept Europe in 1983 was due to a pigeon mutant variant of PMV-1 which did not occur in domestic poultry. As the mutant is adapted to pigeons, it might be thought it would occur in wild pigeons but this has not so far been observed.

*Duck Plague* or *Duck Virus Enteritis* is an acute disease of ducks, geese and swans. It particularly occurs when wildfowl occur at high density, either in the wild or under captive conditions (for example at duck farms on Long Island, USA).

*Fowl pox*. This virus disease occurs in many bird species but, interestingly, not in wild waterfowl. Two forms of the disease occur. The skin form causes wart-like growths on unfeathered parts of the body, the limbs, eyelids, comb and wattles. The form affecting the mucous membranes of the mouth and upper respiratory tract was at one time called 'avian diphtheria'.

*Puffinosis* is a disease of seabirds, originally reported from shearwaters (especially Manx Shearwaters). It is characterized by blisters on the webs of the feet, and mainly affects and kills young birds as they leave their nesting burrows at fledging. The cause of the disease is not known.

*Fledgling Manx Shearwaters (Puffinus puffinus) are the prime victims of puffinosis, which is characterized by blistered feet. The disease is most prevalent in the large island colonies off west Wales*

## Endoparasites

The main groups of endoparasites in wild birds are the unicellular animals (protozoa), roundworms (nematodes), tapeworms (cestodes), flukes (trematodes) and thorny-headed worms (acanthocephala). The majority have an indirect life-cycle and require an intermediate host for development and birds are infected when that host (a snail, earthworm, fly, etc.) is eaten by the bird. However, many roundworms and some protozoa have a different life-cycle. An intermediate host is not required but a period of development outside the host to an infective stage is needed for the cycle to complete.

*Protozoal infections*. Many species of protozoa occur in wild birds but only a few are regarded as potentially serious pathogens. For example, heavy infections with coccidia may build up during hand feeding of birds and cause high mortality in domestic poultry or gamebirds. This is usually controlled by mixing coccidiostats in the feed. Coccidiosis of the kidney of geese is pathogenic for goslings but even so, severe disease is not a feature of coccidiosis of wild birds.

Various intestinal flagellated protozoa are associated with ill health. Trichomonads (*Trichomonas gallinae*) cause disease of the crop of pigeons, being transmitted by 'pigeons milk' from the adult. Squabs (young pigeons) may die from the infection. Raptors may be infected by predation on infected pigeons leading to disease of the mouth and upper digestive tract. *Histomoniasis* or blackhead is a fatal disease of turkeys (where it is controlled by continuous medication in the feed) but it may also occur in other gallinaceous birds such as quail, grouse, partridges, pheasants and especially peafowl.

Blood protozoa of various species, transmitted by blood-sucking flies, are common but rarely pathogenic. Two forms, *Leucocytozoon* and *Haemoproteus*, may be associated with deaths in ducks and geese, and pigeon squabs and quail respectively.

*Roundworms* (nematodes). Roundworms with direct life-cycles include the ascarids, which are large intestinal forms that can occur in large numbers in birds feeding in groups and which may cause impaction or perforation of the intestinal tract. The gizzard worm of geese (*Amidostomum anseris*), gapeworms (*Syngamus* and *Cyathostoma*) and the small strongyle (*Trichostrongylus tenuis*) of grouse and other birds are particularly important. The gizzard worm causes erosion and ulceration of the horny coat of the gizzard with marked inflammation and frequently ends in death. Heavy mortality may occur in wild geese breeding grounds. Gapeworms cause severe respiratory disease in domestic turkeys and gamebirds but they also occur in Rooks and starlings. Even a few worms may be sufficient to cause gasping (gapes) and death. In ducks and geese, the larger *Cyathostoma* may cause the death of chicks even with one or two worms. Trichostrongylosis, of grouse and many other birds, has been responsible for severe mortality of Red Grouse populations in Scotland. The parasite occurs in the intestine and caeca causing severe inflammation and haemorrhage.

Roundworms with indirect life-cycles frequently require an arthropod as an intermediate host. Important examples are the spirurids which parasitize the gizzard or proventriculus of ducks and geese. One species, *Echinuria*, which uses the waterflea *Daphnia* as an intermediate host, is a particularly serious pathogen of waterfowl.

*Tapeworms* (cestodes) all require intermediate hosts in their life-cycle, these usually being aquatic arthropods or annelids. Many species have been reported from wild birds and their identification is a matter for the specialist. Tapeworms are generally non-pathogenic even though they may be found in large numbers in individual birds. Nevertheless, extremely heavy burdens in waterfowl may be responsible for mortality and ill health.

*Flukes* (trematodes). All these endoparasites require a snail as an intermediate host. Some are small, difficult to see by the naked eye and, though they may occur in large numbers, do not cause severe disease. Others, larger in size, may be more pathogenic and species such as *Echinostomum* may be pathogenic for waterfowl.

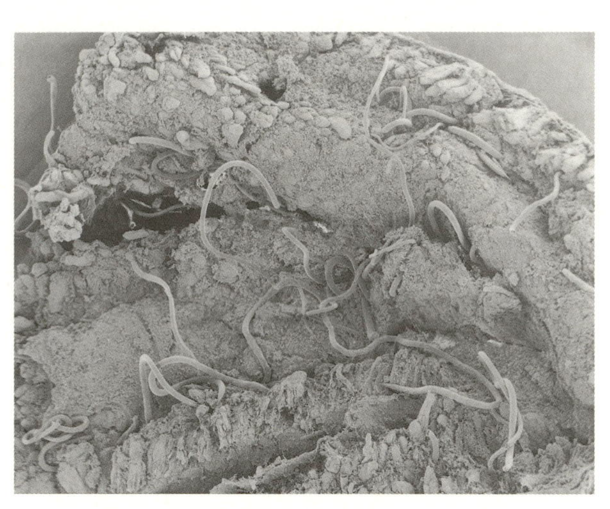

*The gut-dwelling nematode worm Trichostrongylus tenuis is one factor responsible for the ebb and flow of grouse populations. Its biology is therefore of particular concern to those managing shooting moors*

Because flukes use snails as first intermediate hosts and fish, crustacea, annelids and arthropods as secondary intermediate hosts they are particularly common in water birds. The flukes may be found in almost every organ or tissue of the body, including the intestinal tract, gall bladder, blood vessels, kidney, eye, lung, skin, nasal passages and oviduct.

*Thorny-headed worms (Acanthocephala)* are found in the intestine of birds and have a retractable proboscis covered with recurved hooks serving to anchor the parasite in the gut wall. All require intermediate hosts. The most common species in waterfowl is *Polymorphus*, which uses the fresh water amphipod *Gammarus* as an intermediate host and may be found in hundreds in the gut of birds. The parasite is bright orange in colour and, in heavily infected birds, the intestine may be a continuous orange mass.

## Poisoning

There is little evidence that birds die from eating poisonous plants, seeds or berries and most deaths from poisoning are due to human misuse or misapplication of toxic chemicals such as herbicides, fungicides, insecticides or rodenticides. Frequently the toxic chemical intake occurs at the end of a food chain, the organochlorine insecticides being well known examples (e.g. DDT, dieldrin, etc.) causing chronic effects such as delayed breeding, infertile eggs, reduced egg shell thickness and embryo mortality.

Heavy metals may be associated with poisoning. Thus mercury poisoning may result from the ingestion of seeds dressed with fungicides, lead poisoning due to the ingestion by waterfowl of leadshot or lead fishing weights may occur especially in areas where wildfowling and coarse fishing are common. Lead is a cumulative poison and is slowly absorbed from the shot in the gizzard. However, more acute lead poisoning may result from orchard sprays.

Lead poisoning of swans, often by anglers' weights, results in paralysis of the oesophageal sphincter muscle, preventing food from passing into the stomach. The result is a 'kinky neck', and the swan starves to death.

*Botulism* outbreaks may involve thousands of birds. It is a paralytic and usually fatal disease caused by the ingestion of food containing the toxin of the bacterium *Clostridium botulinum*. Several types of the organism occur and type C is the one most commonly associated with disease in waterfowl. The toxin causes muscle paralysis, especially of the neck (limberneck) which leads to an inability to lift the head from water, resulting in drowning. The toxin is formed when the anaerobic bacterium multiplies, especially in warm summer weather, in water containing large amounts of decaying vegetation and dead insect matter. Though ducks are the most commonly affected from feeding on dead invertebrates in the sludge of the water, birds of at least 21 families have been affected by type C botulism. Type E botulism of gulls and other water birds has been reported from the Great Lakes of North America.

Botulism has also been reported in gamebird chicks, especially pheasants, the source being blowfly maggots breeding in dead birds and contaminated by the bacterium. Maggots eaten by the chicks cause death and further maggot-infested carcases. The toxin may also persist through the metamorphosis of the maggot to the adult blowfly and cause poisoning if the adult fly is eaten.

Other rotting debris infested by maggots, such as occurs on rubbish dumps, may lead to botulism in gulls and shorebirds.

EJLS

## Ectoparasites

A variety of insects and mites live on birds and obtain nourishment there. They may live on the bird permanently, such as feather lice, or for a certain part of their life cycle, such as fleas, or they may just visit the bird at times when they need food, such as bed-bugs. Some obtain all their nourishment from the host bird and so are truly parasitic.

Since these insects live on the outside of the bird they are called ectoparasites, as distinct from endoparasites, which live inside the host. However, as biology is never that simple, the term 'ectoparasite' may include animals that may not be feeding directly from the host's tissues, but rather scavenging on organic debris on the host or even predating on the more strictly parasitic animals, and some rather loosely associated species. The term excludes a range of free-living, but blood-feeding insects, such as mosquitoes and black-flies.

Living on or very close to a bird is a risky business and requires special modifications of structure and behaviour to escape the beak and claws and it is perhaps the degree of specialization that decides whether an animal will be called an ectoparasite.

Feather lice (chewing lice or Mallophaga) are clearly highly specialized. They live their entire life-cycle amongst the feathers (with the exception of one genus, which is found in the throat pouch of pelicans and cormorants). Their flattened wingless bodies vary in shape and pigmentation with the colour and structure of the feathers and the position they occupy on the bird. Thus short round-bodied lice tend to be found on the head whereas those on the wings tend to be long and slender.

A single bird species may sustain as many as 12 louse species. Lice lay few, relatively large and well developed eggs, firmly attached to the feathers. The eggs hatch into nymphs, which resemble the adults but have to pass through several moults before they mature. All stages feed on the feathers or on tissue fluids; blood

*Among the common ectoparasites of birds are, from top to bottom, fleas (Siphonaptera), bugs (Cimicidae), flat- or louse flies (Hippoboscidae) and ticks (Acarina)*

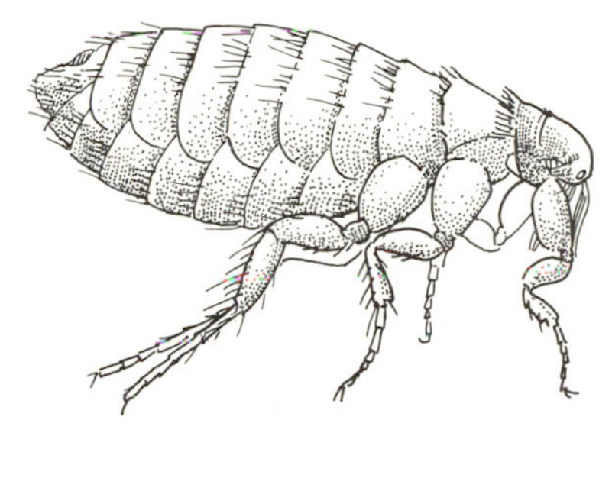

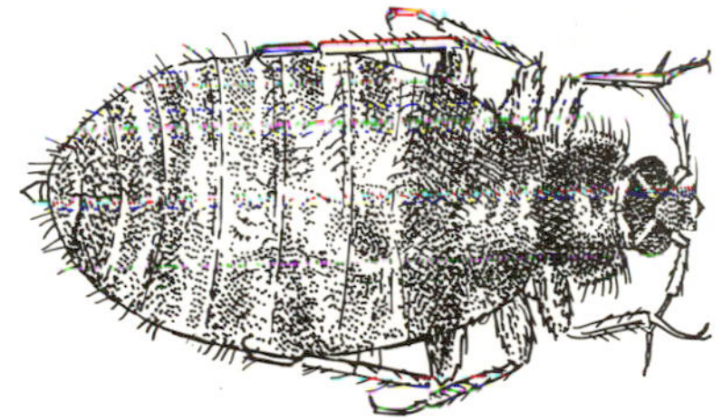

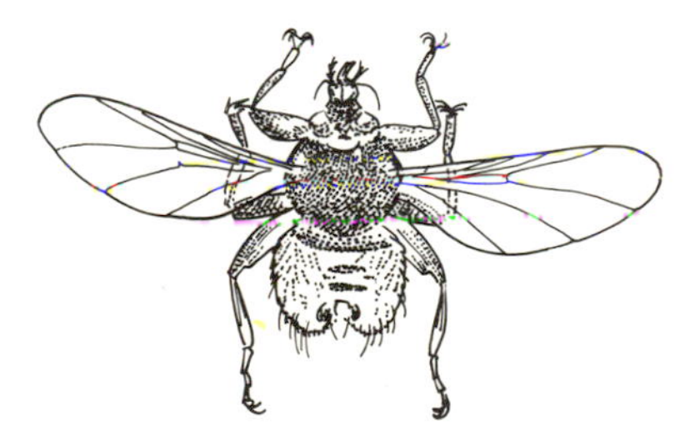

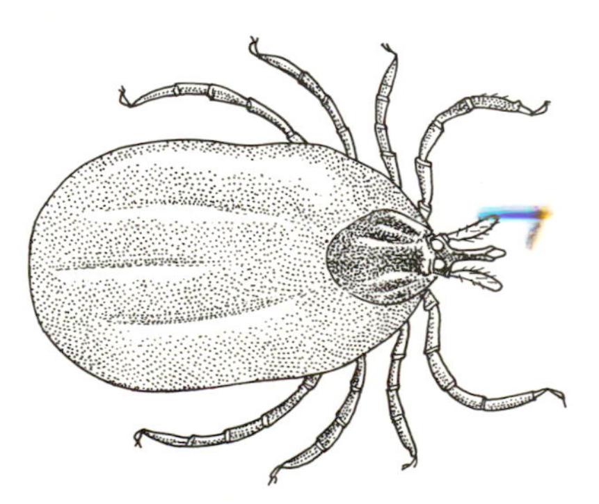

from developing feathers may be taken by some species and a few lice live and feed inside the main shaft of feathers. Lice are frequently very specific, occurring only on one species of host and related hosts tend to harbour related lice. This latter feature has been used to lend weight to arguments about the relationships between the birds themselves, but remains an approach that must be used with caution.

Although the specialization of lice in body form, life-cycle and host specificity enables them to utilize a particular niche, it has severe disadvantages in that, if the host dies, the parasite usually dies too.

Fleas are very different: most flea species are parasites of mammals, but some occur on birds. Generally, bird fleas are not very specific to particular host species, except where the birds are colonial and return to the same nest sites each year, such as some seabirds and hirundines. They are blood-feeders and have one or two additional specializations. They are flattened laterally to allow easy passage through the feathers and they demonstrate one of the widespread features of mammal parasites in having combs of spines, which help protect vulnerable joints (such as between the first and second segments of the thorax) or help prevent the parasite being dislodged. Their ability to jump enables them to reach a new host or to escape predation. It is only as adults that fleas are parasitic. Eggs are usually laid in the host's nest, where the larvae feed on organic debris. Although the time of development from larva to adult can be as little as two to three weeks, the adult is capable of waiting many months to emerge very rapidly at a suitable stimulus, such as the movement of a bird as it returns in the spring to its nest.

The number of fleas found on a bird is generally small (less than ten), despite the fact that there may be up to 2000 in a nest. For bird-fleas, the kind of nest is perhaps more important than is the host species. Thus, the nest of almost any bird on or near the ground will have one species, whereas similar if drier aerial nests of similar (or the same) host species will have another.

Definitely nest-oriented and not well designed to travel with their hosts are the blood-feeding cimicid bugs. This small family of true bugs (Hemiptera) includes our own bed-bug. Most species are parasites of bats, but a number of species occur on birds, notably swifts and hirundines, with one remarkable species (*Caminicimex furnarii*) living in the clay nests of the South American Rufous Hornero or Ovenbird. The eggs are laid around the dwelling place of the host and hatch into small versions of the adult. The adult stage is reached after several moults. Like lice, the bugs themselves are flattened (top to bottom), but they have no other special attributes for travelling with the host. Hence, although they might be quite numerous in the nest, they only visit their host briefly for food and are rarely found on the host away from its nest. They tend to be quite host-specific.

The flat-flies (Diptera: Hippoboscidae) are different again. As tough, leathery, flattened, blood-feeding flies with well developed claws and an ability to move very rapidly in any direction, they are well adapted to living on their hosts. Some are free-flying and hence able to move freely between hosts; these often demonstrate a low level of host specificity. Others, which have reduced wings and are unable to fly, are much more host specific and also show a reduction in development of the eyes – towards the state seen in other ectoparasites. They also have a rather special form of reproduction in that the female develops a larva internally until it is ready to pupate. A single larva develops at a time, but the abdomen becomes extremely distended at the end of each 'pregnancy'.

The larvae of a number of flies are also parasitic on nestling birds, but although many other insects are clearly associated with birds, they are perhaps better termed nest fauna; and discussion continues about the precise relationship between birds and some insects that are not clearly parasitic, nor strictly nest fauna. Thus the tiny wingless flies, *Carna* spp., develop in the nest and spend some time on the bird. They appear to get nourishment from the bird, but exactly what that is remains in doubt. They could also be phoretic, i.e. using birds as a means of transport to get dispersed to new habitats.

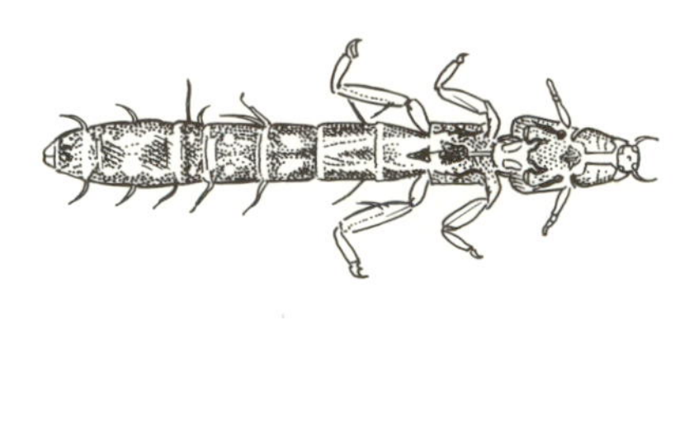

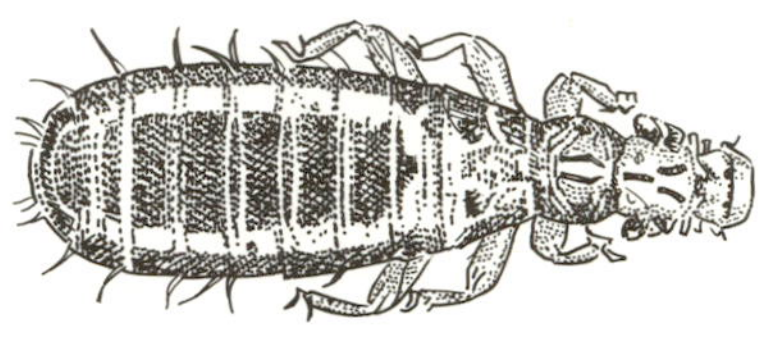

*The Leather-lice (Mallaphaga) that are parasitic on birds belong to two superfamilies, Ischnocesa (above) and Amblycera*

Mites have no wings or jumping ability and so might be expected to be rather more limited in their adaptations and food. Nevertheless, they show an extraordinary range of behaviour, from the ticks, a most distinctive group of mites that are totally obligated to a parasitic way of life, to others that spend only a brief first stage of their life-cycle on birds. There are feather mites, blood-feeding mites, predacious or scavenging mites, mites that are found only in the nares and others only on the feet. Some have abbreviated their basic life-cycle, while some have extended it. It is with this group of ectoparasites that we are least familiar – it will be a long time before we have anything like a complete understanding of even the naming and classifying of many mite species, let alone know of their behaviour and specializations. The relationships between birds and the life that lives on them is already known to be diverse, but there undoubtedly remain many exciting relationships to be discovered.

Generally speaking these parasites live in reasonable balance with their hosts – a healthy host can carry a healthy range of parasites, but will keep them under control by its own preening. The European Swift is host to a relatively large flat-fly (*Crataerina pallida*). About two-thirds of over 4000 swifts examined in one study carried no flies, the others carried up to 33, but there seemed to be no relationship between the heavier infestations and the condition of the bird. Frequently such parasites are at most a seasonal or temporary nuisance and the energy that they cost the host is easily made up. However, if a bird is in poor condition and unable to preen properly, it can accumulate high and damaging ectoparasite populations. High parasite loads may also have deleterious effects in the nest where the young have high energy demands and little self-defence.

Isolated animals with temporary homes give little opportunity for the development of a diversity of ectoparasites or for the build-up of large populations. But coloniality, for all its advantages, does carry the disadvantage of offering greater opportunity for the development of a wider range of ectoparasites and for these to develop higher infestations. A wider range of parasites and higher infestations can in turn promote the transmission of disease organisms. Thus the likelihood of ectoparasites having deleterious effects on nestling birds, either by their direct activity or by disease transmission, is generally greater in colonial nesting birds. In particular, the nesting success of some colonial seabirds and certain hirundines can be affected by the attentions of ectoparasites. For example, Cliff Swallow nestlings infested with swallow bugs (Cimicidae) may be up to 3.4 g lighter (approximately 15% of body weight) than uninfested nestlings. Infested nests produce only half the fledglings of uninfested nests. Also, the level of infestation is higher in larger swallow colonies; around 500 bugs per nest in colonies exceeding 100 pairs compared to only 200 bugs per nest in colonies containing under ten pairs. AMH

# 6 DISTRIBUTION

Although birds have been recorded at virtually every place on earth, their distribution is distinctly uneven. No species is exclusively marine, and birds that spend most of their time at sea account for only 3% of the world's species. Polar regions, deserts, and high mountaintops support relatively few species, whereas hundreds of species may be found within small areas of tropical rainforest. An observer wandering through a rainforest in the Amazon basin of South America, however, is likely to record considerably more species than will be logged in a similar area of western Africa. Moreover, the lists from the two areas will contain hardly any species in common. If one crosses the Andes into rainforests on the western slope of South America, the species list changes again. On the other hand, a traveller recording birds in the northern regions of North America, Europe, and Asia is likely to encounter many of the same species in each area.

Such general patterns of bird distribution reflect three facts. First, the rate of diversification of evolutionary lineages into sets of species and the rate of extinction of species in these groups have not been uniform over the earth. Some areas have been 'hot spots' of evolution, and may therefore contain a great many species. Second, once evolved, species have not expanded their distributions in the same ways, largely because of geographical barriers and differences in their dispersal abilities. Finally, the contemporary distributions of individual species within and among regions vary tremendously, often in response to ecological factors. Understanding how birds are distributed over the earth, then, requires a knowledge of the history of evolutionary radiations of birds in different regions – evolutionary or historical biogeography – and of the factors that determine the distributions of individual bird species – ecological biogeography. Information on the evolutionary component of distribution is often less detailed and direct than that on the more immediate ecological aspects of species' ranges, but is no less interesting or important.

## Evolutionary biogeography

### General faunal patterns

There are over 350 species of tyrant flycatchers (Tyrannidae) in the world, but they are found only in the New World. Woodpeckers (Picidae) occur over much of the world, but are absent from Australia, New Zealand, and Madagascar. The 13 species of vanga shrikes (Vangidae) are restricted to Madagascar, while the helmet shrikes (9 species, Prionopidae) are found only in Africa south of the Sahara. Some 63 species of wrens (Troglodytidae) occur in the New World, but only one species, *Troglodytes troglodytes*, is found in the Old World, where it has a vast distribution. Conversely, there are 75 species of larks (Alaudidae), of which only two occur in the New World. One, the Horned Lark, is widely distributed in North America and occurs in the eastern Andes of Colombia. The other, the Eurasian Skylark, occurs naturally in North America only in the Aleutian and Pribiloff islands during migration.

Naturalists became aware of distributional patterns such as these as soon as they began to explore the world. Following the classificatory urges prevalent at the time, they categorized and mapped regions of the globe according to general similarities in the distributions of species, genera, and families. The system developed for birds in the mid-nineteenth century by P. L. Sclater, modified only slightly by Alfred Russel Wallace to apply to animals in general, is still in use today. In this scheme, the land masses of the earth are divided into six biogeographic realms.

#### *The Palearctic realm*

This includes Europe, Africa north of the Sahara, and all except the south-eastern portions of Asia. Slightly over 1000 species breed in this region; most are migratory, and over half of the species are passerines. Old World warblers (Sylviinae) are especially well represented. Only one family, the Accentors or hedge-sparrows, is restricted (*endemic*) to this realm.

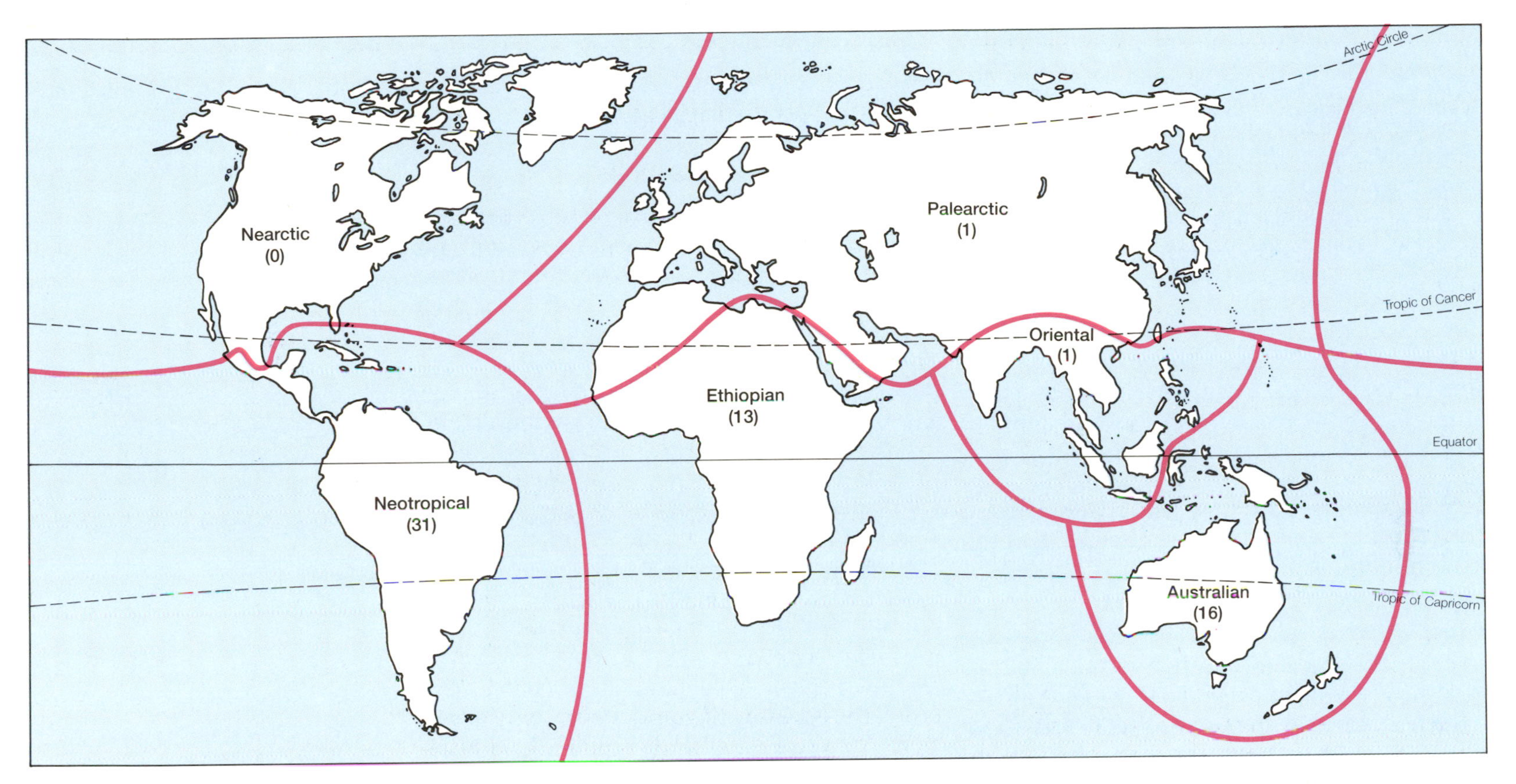


*Major biogeographic realms of the world. In brackets after the name of each region is the number of its endemic bird families*

Perhaps because migratory pathways tend to be predominantly north–south, the closest affinities of this region are with the Ethiopian and Oriental realms, with which the Palearctic shares roughly 15% and 22% of its species, respectively. Ties with the Nearctic are not so great (about 13% of the species are shared), but several families (divers or loons, Gaviidae; grouse, Tetraonidae; auks, Alcidae; and waxwings, Bombycillidae) occur only in the Palearctic and Nearctic. For this reason, these realms are sometimes combined into a single region, the *Holarctic*.

### *The Nearctic realm*

This realm, which includes Greenland and North America north of the tropics, also has a relatively impoverished avifauna, containing perhaps 750 breeding species. Wood warblers (Parulidae), blackbirds (Icteridae), and cardinals and grosbeaks (Cardinalinae of the Emberizidae) are dominant groups, but there are no endemic families. As in the Palearctic, many species are migrants. Such a widespread expression of migratory behaviour in these faunas may reflect the historical influences of Pleistocene glaciations and the fact that the northern land masses are large, supporting many species in the breeding season despite being cold and unproductive during the long winter.

### *The Neotropical realm*

By far the richest realm is the *Neotropical* of South and Central America, southern Mexico, and the West Indies. Over 3000 species, many of them nonpasserines or primitive passerines, occur in this region. This fact, along with the occurrence of some 31 families that are endemic to this realm, suggests that it has an ancient and distinctive evolutionary history. Many of the endemic families, such as the ovenbirds (Furnariidae, 213 species) or antbirds (Formicariidae, 230 species), have undergone extensive evolutionary radiation and contain large numbers of species. Other endemic families, however, are small. There are, for example, two species of rheas (Rheidae), one sunbittern (Eurypygidae), one oilbird, (Steatornithidae), three trumpeters (Psophiidae), two seriemas (Cariamidae), three screamers (Anhimidae), one hoatzin (Opisthocomidae), and four seedsnipe (Thinocoridae), all of which occur only in the Neotropical region.

The Neotropical realm is not totally unique, however. Many species are shared with the Nearctic, including not only Nearctic breeders that migrate to the tropics but several widespread species of New World vultures (Cathartideae), wrens (Troglodytidae), tyrant flycatchers (Tyrannidae), blackbirds, (Icteridae), and hummingbirds (Trochilidae). Several families and genera that occur in the Neotropics (e.g. cuckoos, Cuculidae; parrots, Psittacidae; barbets, Capitonidae; and trogons, Trogonidae) are also found in tropical regions elsewhere in the world, and have a *Pantropical* distribution.

### The Ethiopian realm

Africa south of the Sahara, Madagascar, and southern Arabia comprise the *Ethiopian* realm. Although it occupies similar latitudes to the Neotropical realm, it contains fewer than half as many species. This may reflect the interrelated facts that tropical forests and mountain ranges (the Andes) are much more extensive in the Neotropics and evolutionary diversification has apparently been more rampant there. Excluding Madagascar, only eight families are endemic to this realm, and none of them contains very many species (one ostrich, Struthionidae; one hammerhead, Scopidae; one whale-headed stork, Balaenicipitidae; one secretary-bird, Sagittariidae; twenty-two turacos, Musophagidae; six mousebirds, Coliidae; six woodhoopoes, Phoeniculidae; and nine helmet shrikes, Prionopidae). Of these families, all except the Scopidae, Balaenicipitidae and Prionopidae have been reported as fossils from the Tertiary of Europe. Madagascar, with its distinctive fauna, contributes five additional endemic families, all of which are equally small.

Most of the species of the Ethiopian realm are passerines. Various weaver finches (Ploceidae), starlings (Sturnidae), larks (Alaudidae), shrikes (Laniidae), and sunbirds (Nectariniidae) dominate the fauna. The region has strong biogeographic affinities with the Oriental realm, with which it shares roughly 30% of the genera (but only 2% of the species). Many Palearctic breeders winter in the region south of the Sahara.

### The Oriental realm

The *Oriental* realm of India, South-East Asia, and adjacent islands is separated from the Palearctic by the Himalayan mountains. Nearly 1000 species breed in this region, and there has been a considerable radiation of pheasants (Phasianidae) here. Only one family, the leafbirds (Irenidae) is endemic to the region, although the broadbills (Eurylaimidae) are nearly so. The closest affinities are to tropical Africa, but linkages to the Palearctic and Australian faunas are also relatively strong. The boundary between the Oriental and Australian realms is especially fuzzy. As one travels eastward from Java toward Papua New Guinea, the number of species with Oriental affinities decreases while the number of species characteristic of the Australian region increases. Some Oriental groups, such as the barbets, extend east only as far as the island of Bali and are not found on or beyond Lombok, only 35 km away, while the distribution of some Australian groups, such as honeyeaters and cockatoos, extends west only to Lombok. Noting this abrupt transition, Wallace defined a line of demarcation of the Oriental and Australian realms between Bali and Lombok, although a more complete and quantitative analysis of faunal similarities places this line some 600 km to the east.

### The Australian realm

Finally, the *Australian* realm of Australia, New Zealand, Papua New Guinea, and adjacent islands in the East Indies and Polynesia contains nearly 1600 species, over 1000 of which are endemic to the Australian plate and neighbouring islands. Species of parrots (Psittacidae), pigeons and doves (Columbidae), honeyeaters (Meliphagidae), and kingfishers (Alcedinidae) dominate the avifauna. Overall, this is the most arid of the biogeographic realms. Because making a living in the desert is relatively difficult for birds, however, few of the endemic species are desert dwellers.

Australia has had a long history of isolation from other land masses, and evolutionary lineages there are old. Sixteen families are endemic (or virtually so) to the region. Of the 308 species of passerine birds that occur there, roughly 80% are 'old endemics' that belong to lineages that almost certainly originated and radiated in Australia and New Guinea. Based on recent studies using DNA–DNA hybridization techniques, some workers have suggested that a major lineage of birds (the parvorder Corvi) began an evolutionary radiation some 55–60 million years ago, eventually diversifying into three superfamilies and 10 families. According to this classification, one superfamily, the Menuroidea, contains the Australian treecreepers (Climacteridae), the lyrebirds and scrub-birds (Menuridae), and bowerbirds (Ptilonorhynchidae). The second superfamily, the Meliphagoidea, is composed of fairy wrens (Maluridae), honeyeaters and Australian chats (Meliphagidae), and Australian warblers, pardalotes, and thornbills (Acanthizidae). Finally, the Corvoidea includes Australian robins (Eopsaltriidae), the distinctive logrunners (Orthonychidae), the *Pomatostomus* babblers (Pomatostomatidae), and a diverse set of birds now included in the family Corvidae. This latter family contains not only crows and jays, but other groups still generally classified as separate families, such as birds of paradise (Paradisaeidae), woodswallows (Artamidae), magpie larks (Grallinidae), bell magpies (Cracticidae), and whistlers and shrike-thrushes (Pachycephalidae). This classification leads to the conclusion that Australia is the place of origin of several families like the Corvidae that dispersed to Asia when Australia drifted closer during the late Tertiary and then radiated further there. At the same time, some species of flycatcher (Muscicapidae) from the Oriental region invaded Australia and Papua New Guinea and subsequently differentiated.

Whether or not this scenario of the evolutionary diversification of birds in the Australian region is correct (and there is much debate), it is evident that many of the species and higher taxa in this realm have evolved *in situ* rather than being products of dispersal from other regions.

These categorizations of the distributional patterns of families and species are helpful in establishing a foundation for avian biogeography, but it is important to remember that they are descriptions rather than explanations of patterns. The determination of the boundaries of these regions was made some time ago, using rather subjective criteria. More quantitative and objective methods of biogeographic classification might well yield different patterns. Moreover, the details of these broad biogeographic patterns depend on the taxonomic classifications of birds and their relationships that are in vogue, and as these change, so also will the patterns.

## Biogeographic patterns within continents: areas of endemism

Because birds are mobile creatures, one might expect the distributional boundaries that define biogeographic patterns to be blurred within continents or biogeographic realms. In fact, many species have quite limited distributions. By reviewing the distributional patterns of many such species, regions that harbour unusual numbers of geographically restricted species, or *areas of endemism*, may be detected. Such areas have been interpreted as the centres of evolutionary origin of taxonomic groups, ecological settings that are unusually productive, stable, diverse, or ancient, or locations that have been isolated from nearby areas by ecological or geological barriers.

Over 30 areas of endemism have been recognized for the birds of South America. Many of these involve bird species of lowland tropical forests. The prevailing explanation for these

*Areas of endemism. Probable forest refuges in South America during dry periods of the Pleistocene are shown on the left. During dry periods, the Amazonian forest was divided into fragments (shaded areas) separated by savannas and grasslands. The present distribution of three species of manikin in the genus Pipra is shown on the right. Arrows indicate possible dispersal paths of the species from their presumed Pleistocene retreats*

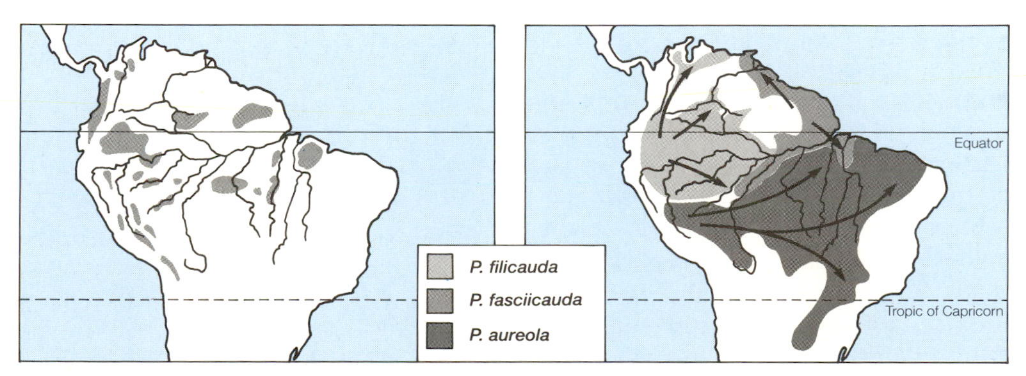

centres of species richness suggests that, during the Pleistocene, dry cycles associated with glacial periods fragmented the previously extensive and continuous lowland forests into forest islands isolated by wide areas of savanna or grassland. Populations in these forest 'refuges' differentiated, some of them becoming new species. With the return of wetter conditions, the forests once again became continuous and the bird species expanded their ranges from these centres of endemism. Some expanded very little, while others extended their ranges dramatically. The number of species would therefore remain greatest at the locations of these former forest refugia – the areas of endemism. In some cases, the differentiated forms enlarged their ranges until they established contact with forms expanding from other refugia. If the forms were not sufficiently differentiated, hybridization might occur along this zone of contact; otherwise, it might persist as an area of geographic replacement of one species by the other. Because many species were expanding from these forest refugia as conditions became wetter, one might expect a rough coincidence in the locations of the contact zones between faunas. The evidence supporting this view, then, is the congruence of the distributions of areas of species richness and of contact zones among many species of birds, as well as of other animals and plants. Similar scenarios have been proposed to explain areas of endemism in tropical Africa and in mediterranean woodland habitats on several continents.

This explanation has been criticized on several counts. It is difficult to document climate changes in South America during the Pleistocene with any precision, and the evidence linking the present distributions of birds to their presumed forest refugia is often indirect. In some cases, geomorphological barriers, like mountain ranges, may explain the patterns of distribution and endemism as well as the postulated Pleistocene refugia, while in other cases the areas of endemism appear to be considerably older than the Pleistocene. Better information about the contemporary and Pleistocene distributions of South American birds and habitats is necessary to resolve the issue.

The debate about the Pleistocene forest refuge hypothesis illustrates a more fundamental difference of viewpoints about the relationship between biogeography and speciation. On the one hand, many evolutionists subscribe to the view that most patterns of species' distributions, endemism, and faunal richness are the results of long-distance *dispersal* of taxa across pre-existing geographic or ecological barriers and subsequent differentiation from their source taxa to form new taxa. The faunal distinctiveness of oceanic islands, areas of endemism, or biogeographic realms is due to barriers to dispersal, which otherwise would blend the faunas together.

On the other hand, a growing group of scientists believes that long-distance dispersal from centres of origin across pre-existing barriers is of minor importance in producing biogeographic patterns of taxonomic diversification, particularly within continental biotas. Instead, they argue that these patterns arise when the distributions of sets of widespread species are subdivided by a series of historical changes in geology, ecology, and/or climate (*vicariance*) that produce barriers to gene flow and permit taxonomic differentiation. Biogeographic distinctiveness is therefore due to the imposition of barriers on species' distributions and the subsequent effects on the evolution of lineages.

If this view is correct, one might expect that speciation would be particularly pronounced in the areas separated by such vicariance processes, creating areas of endemism. Moreover, members of different evolutionary lineages would show the same biogeographic patterns, reflecting the sequential separation of regions by the same events. Both of these expectations appear to be met in the distribution patterns of birds in South America and elsewhere. Explaining such patterns by the dispersal idea needs not only multiple long-distance dispersal into an area of eventual endemism, but also requires that each of the evolutionary lineages should exhibit parallel cycles of dispersal and differentiation.

One way to analyze such patterns is by reconstructing the evolutionary pedigree of several lineages in an area, determining the

*Right above. Distributions and phylogenetic relationships of species and subspecies of wedgebills and whipbirds (Psophodes) in Australia (top). The hypothesized barriers separating areas of endemism are shown bottom left, while the diagram bottom right illustrates the suggested sequential arrangement of these barriers. WD = western desert, ED = eastern desert, SW = south-western corner, EF = eastern rainforest, SE = south-eastern corner, AP = Atherton Plateau*

*Right below. Scales of geographical distribution of the Black-billed Magpie (Pica pica). Diagram (a) shows the overall range of the species, while (b) illustrates range and variations in abundance across a portion of western North America. Densities are average numbers of birds counted per route surveyed during the US Breeding Bird Survey, 1965–79. The hypothetical diagrams (c) and (d) show how the distribution at a regional scale within a small portion of the North American range may be associated in a grassland plains environment with riparian zones along watercourses and how territories of individuals may be distributed within a small section of such zones*

centres of endemism and their historical relationships, and comparing the geography of these patterns with the distribution and history of geological and/or ecological barriers in the region. In Australia, for example, there are six differentiated taxa (species or well-defined subspecies) in the genus *Psophodes*, the wedgebills and whipbirds. One closely related pair, *P. olivaceus* and *P. lateralis*, occurs in the eastern coastal woodlands. This pair is closely related to another, *P. nigrogularis* and *P. leucogaster*, which is discontinuously distributed in southern Australia. The species of the third pair, *P. cristatus* and *P. occidentalis*, have abutting distributions in the arid interior of the continent. Other genera, such as the emu-wrens *Stipiturus* and the quail-thrushes *Cinclosoma*, exhibit similar patterns of phylogeny and distribution. As the map on the right indicates, the evolution and distribution of all of these taxa may have been influenced by the same series of isolating events, beginning with the separation of southern and eastern elements in the mid-Miocene (barrier Y). Separation of the eastern and western desert forms resulted from environmental and geomorphological barriers (barrier X) that occurred from the Oligocene through the Pleistocene, and the separation of the southern forms (barrier Z) may have been produced by climatic and geological changes occurring until the late Pliocene. Finally, the uplifted Atherton Plateau of the north-eastern coast was separated from the humid forests of south-eastern Australia (barrier E) by a wide area of savanna vegetation that probably developed during the Pleistocene.

JAW

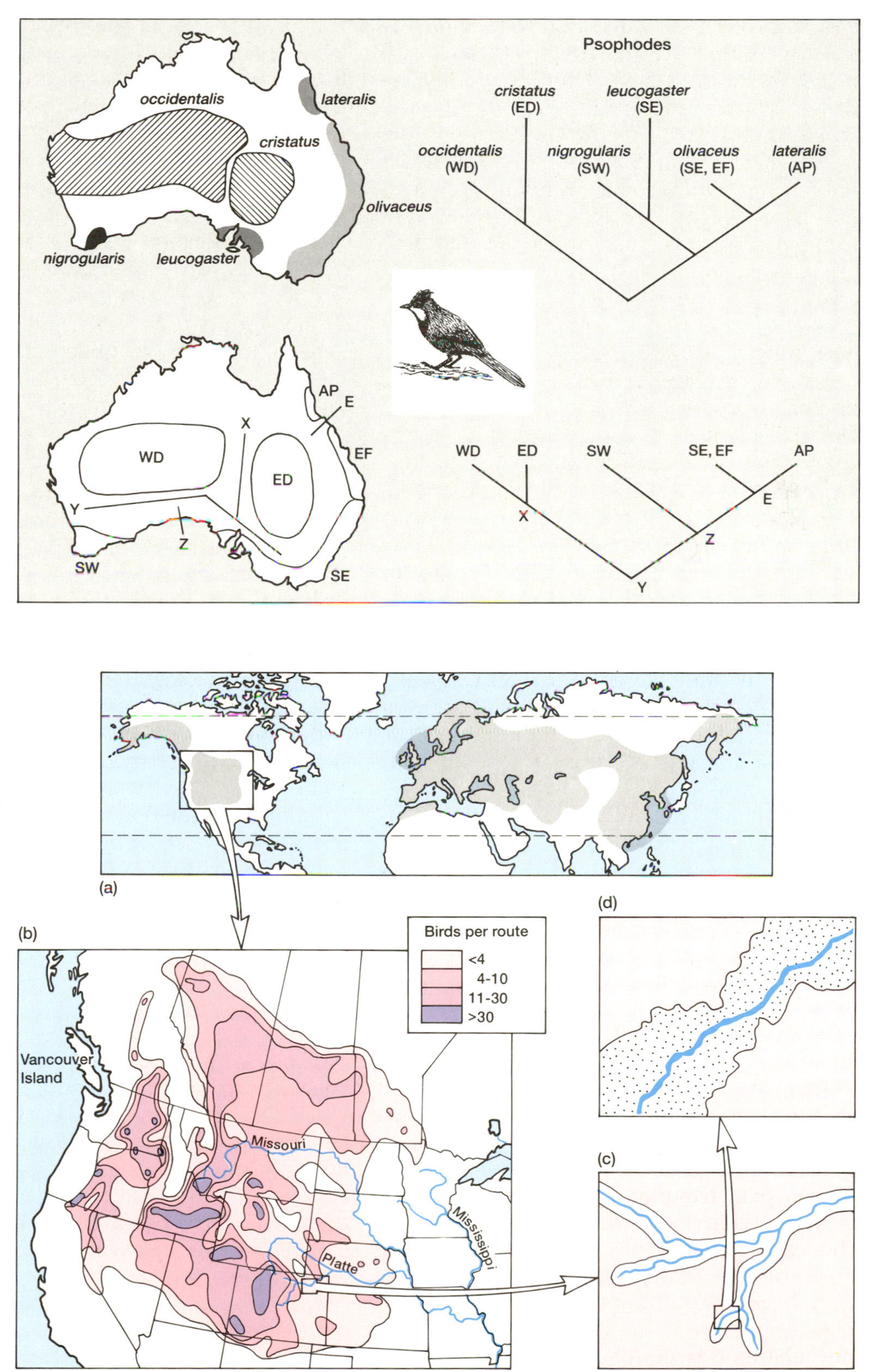


## Ecological biogeography

The contemporary distribution of a species bears the imprint of evolutionary and historical events, but it is also the result of ecological factors. These act at several geographical scales of resolution, and the questions one asks about distribution differ with changes in scale. At the scale of the species' range, we may ask what factors determine the boundaries of distribution, whether the range changes in time, or how the sizes and shapes of ranges vary among species. By shifting to a finer scale, attention is then drawn to gaps in the distribution of a species within its range and variations in its abundance where it does occur. How is abundance related to distribution? How are rare species distributed? At a still finer scale, we may highlight regional and local responses to habitat and other environmental factors and ask what determines how a species is distributed among patches of different habitats in a landscape mosaic or how changes in habitats affect distributions. We could sharpen our focus even further to consider how individuals are distributed in a very local area, but this falls outside the traditional domain of distributional studies.

## Geographic ranges of species

### *Influence of climate*

The distribution of a species is a consequence of its ecological tolerances (which dictate where it can live), the geographical distribution of places satisfying these requirements, and how the matching of species to places is modified by factors such as barriers to dispersal, interactions with other species, or chance. In some cases, the limits to distribution at higher latitudes are set by climate. The northern distribution of the Goldcrest in Norway coincides with the limits of conifer forests, but in northern Sweden, Finland, and the Soviet Union it is absent from the northernmost spruce and pine forests. In these areas, its northward distribution is apparently limited by low mean summer air temperatures. The northerly distributions in Scandinavia of other breeding species, especially those of central European affinities such as Pied Flycatchers, are likewise related to spring or summer temperatures. In North America, the northern limits of many (but by no means all) passerines in winter are associated with mean January minimum temperature, duration of the frost-free period, or habitat, while western and eastern range limits are related to annual rainfall, minimum winter temperatures, habitat, and height above sea level. For a small set of species on which physiological studies have been conducted, the wintering range limits roughly coincide with areas in which the birds must increase their basal metabolic rate by no more than 2.5 times to cope energetically with average winter temperatures.

### *Influence of other species*

Often, however, the range limits of species bear no obvious relationship to climatic factors, and it is widely believed that interactions with other species may determine these boundaries. Charles Darwin observed that 'species in a state of nature are limited in their ranges by the competition of other organic beings quite as much as, or more than, by adaptation to particular climates'. In the upper Amazon basin, three species of curassow apparently replace each other geographically in lowland forests that lack obvious barriers to range expansion, and climatic limitation is unlikely in this region. The Yellow Warbler is widely distributed across temperate North America but occurs in the tropics only in coastal mangrove swamps and islands, to which it may be restricted by competition with a number of other insectivorous bird species. The Boreal Chickadee breeds in coniferous forests across northern North America, but in the coastal mountains of the Pacific Northwest it is replaced by the Chestnut-backed Chickadee. Where the two species meet in southern British Columbia and Washington, the Boreal Chickadee is found inland at higher elevations, the Chestnut-backed at lower elevations in wet coastal forests. The Siberian Tit occurs in interior Alaska and the Yukon; where it overlaps in range with the Boreal Chickadee it is found at forest edges and along rivers rather than in the dense coniferous forests favoured by the Boreal Chickadee. The differences in ranges and in habitat distributions within areas of geographical overlap have all been attributed to competition among the species. In all of these examples, however, other explanations are also possible.

### *Other factors*

Interactions among species, besides competition, may also influence distributions. The absence of Whinchats and Meadow Pipits from the island of Ulversö (Åland archipelago) in the

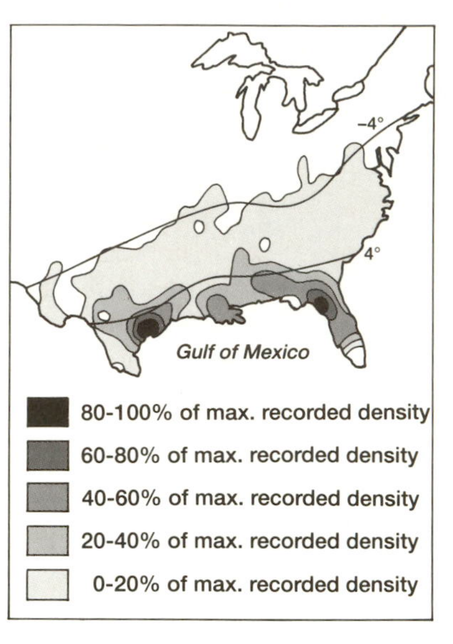


*Winter distribution and abundance of the Eastern Phoebe (Sayornis phoebe) in North America. The northern range limit is associated with the isotherm at which average minimum January temperature is −4 °C, while the northern edge of the area of high abundance roughly follows the 4 °C January isotherm*

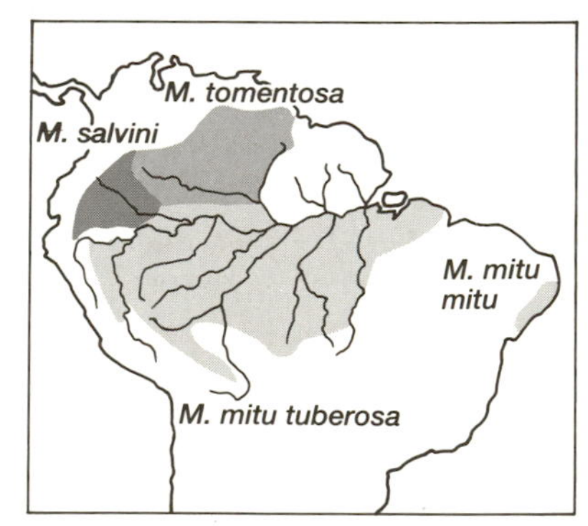


*Distribution of three curassows of the genus Mitu in the Amazon Basin. In some areas, range boundaries coincide with obvious barriers such as the Amazon River, but in others distributions of species apparently abut one another without physical barriers being present. Because knowledge of the distribution of forest birds in South America is fragmentary, these ranges are necessarily approximate in places. Notice also that a disjunct population of Mitu mitu occurs in the extreme east of Brazil*

Baltic Sea has been attributed to nest predation by Hooded Crows, which are abundant there. In Hawaii, the high incidence of avian malaria at lower altitudes may contribute to the restriction of many of the highly susceptible native bird species to higher altitudes. Introduced exotic species, which are resistant to malaria, are most abundant at the lower elevations.

Some species have limited ranges because of narrow habitat or food requirements. The breeding range of Kirtland's Warbler is confined to a small area of jackpine in northern Michigan. In Australia, the Rock Warbler occurs only in a small area of coastal New South Wales, where it is associated with rocky ravines in Hawkesbury sandstone and adjacent limestone areas. The wintering range of Brent Geese (Brant) in coastal areas is closely tied to the distribution of eelgrass, which comprises most of their diet. The widespread disappearance of eelgrass from the Atlantic coast of North America and elsewhere in the world in the early 1930s caused dramatic changes in the wintering distribution of these geese.

Such examples suggest that a small geographic range is associated with habitat specialization. This is generally true, although some habitat specialists (e.g. spruce-forest birds) are widely distributed because their habitat is widespread. Species occupying disturbed or strongly seasonal habitat types may also have large ranges. Other factors, such as body size and abundance, also affect range size. Large species usually occur at lower densities than small species. If their range is very small, so also will be their population size, and the probability of extinction is enhanced. Among North American land birds, there are many large and small species that have large ranges, but relatively few large species have small ranges. In Australia, most terrestrial bird species have relatively small ranges. Not all of these species are small, and it seems more likely that the pattern is a consequence of the geography of Australian environments. Most of Australia is arid or semi-arid, but most of the bird species are forest or woodland dwellers. The few species adapted to dry habitats may therefore be widespread, but the forest species are necessarily restricted to much smaller areas.

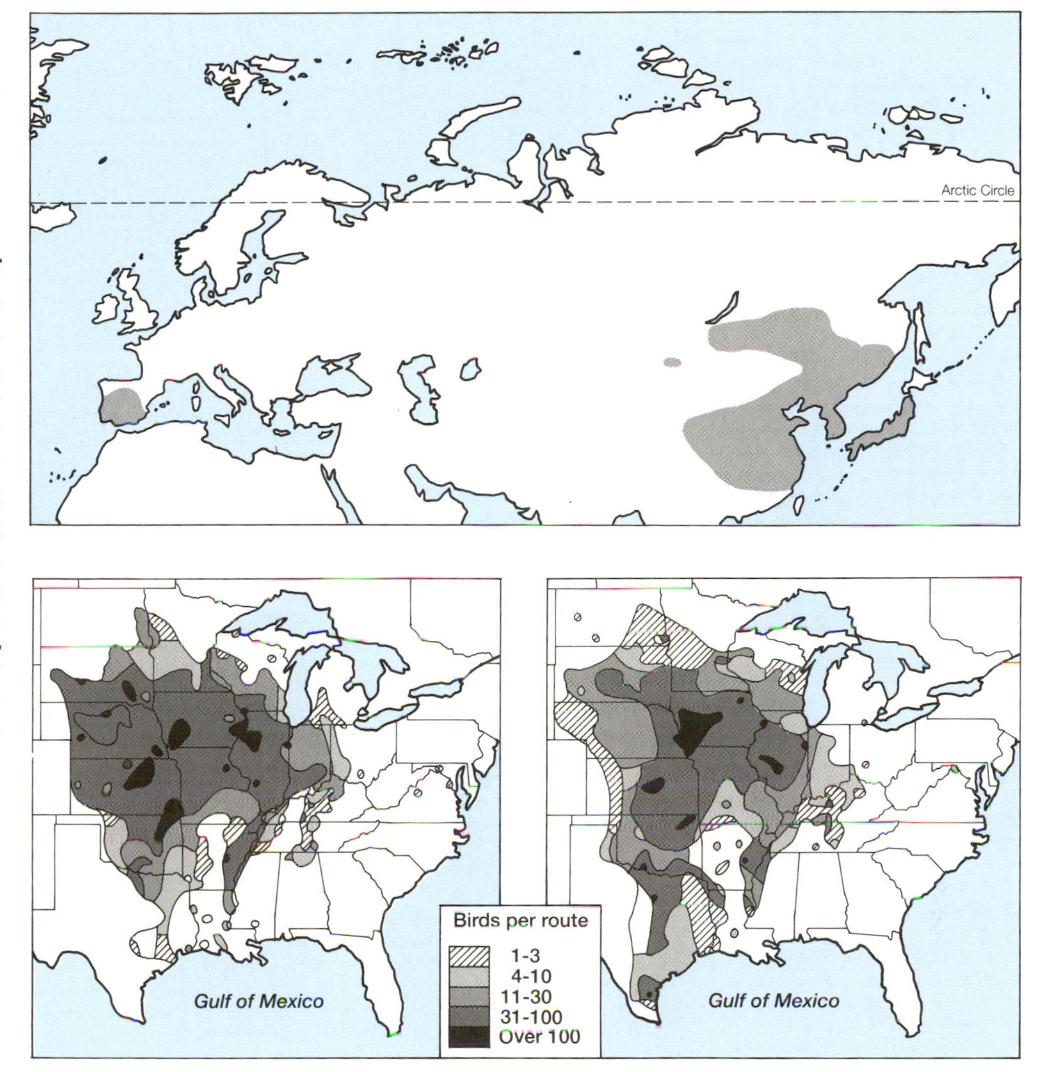


*Top. Distribution of the Azure-winged Magpie (Cyanopica cyana). The main body of the species' distribution is in the eastern Palearctic, but it is also common in portions of Spain and Portugal*

*Below. Breeding distribution and abundance of the Dickcissel (Spiza americana) in North America in 1967 (left) and 1968 (right). Densities are average numbers of birds counted per route surveyed during the US Breeding Bird Survey, 1965–79*

Even at the scale of overall range maps, many species have gaps in their distributions. The Pearl Kite, for instance, inhabits most of non-forested South America but has an isolated population in Nicaragua, while the Scrub Jay is widespread through much of western North America but also occurs in Florida. The range of Black-billed Magpies includes most of the Palearctic, but the species is common in western North America as well. In Australia a diverse group of species including the Little Bittern, Freckled Duck, Shining Bronze-Cuckoo, Laughing Kookaburra, White-naped Honeyeater, and New Holland Honeyeater have ranges in the east and south-east and disjunct populations in the extreme south-west. In this case, the extremely arid Nullarbor Plain, which extends from the interior to the southern coast in west-central Australia, provides no suitable habitat for such species and fragments their distributions. Many migratory species, of course, have temporally disjunct ranges, wintering in locations far removed from their breeding ranges.

*Historical changes in distribution*

The range maps of field guides give a misleading impression of the stability of species' ranges. In fact, ranges are dynamic, on time scales ranging from years to millennia. The nomadic species of the deserts of Australia, Africa, or the Middle East may breed in large numbers in certain areas one year and not be found within hundreds of kilometres the next. In North America, the range boundaries of grassland species such as Dickcissels, Lark Buntings, and Bobolinks may fluctuate considerably from year to year. The entire breeding community of some 10–12 million seabirds disappeared from Christmas Island following the 1982–83 El Niño event, but populations of most species recovered over the next few years.

Other range changes are more persistent and lead to long-term expansions or contractions in distribution. During the past century the Crested Tit, Chaffinch, European Blackbird, Scarlet Rosefinch, Lapwing and several other species have expanded northward in Finland, apparently in response to climate changes and/or human-caused changes in habitats. Similar range expansions have occurred among some North American species. Perhaps the most dramatic range extension (aside from human introductions of species) is that of the Cattle Egret. Originally native to Africa, this heron colonized north-eastern South America in the late 1800s. Populations thrived, and the species spread rapidly throughout much of the New World. It now breeds as far north as southern Canada and west to California.

History also leaves its mark on contemporary bird distributions. During the Pleistocene the West Indies were considerably drier than they are now and savanna-scrub woodlands were much more extensive. As the environment became wetter and sea levels rose, much of this habitat was lost. Species adapted to dry habitats, such as Burrowing Owls and Bahama Mockingbirds, now have a limited and patchy distribution through the Caribbean, the remains of formerly much larger ranges. Similarly, the Iberian population of Azure-winged Magpies is probably a relict of a formerly more extensive distribution that linked it with the main body of the species' range in Asia.

In other situations, contractions in ranges can be linked to eradication of populations by humans before the coming of Europeans. On Henderson Island in the South Pacific, examination of archeological material reveals that more than a third of the landbird species present when the Polynesians first arrived became extinct shortly thereafter, and perhaps half of the native bird species on the Hawaiian Islands suffered a similar fate. Two pigeon species, one small (either *Ducula aurorae* or *D. pacifica*) and one large (*D. galeata*), were eaten to local extinction on Henderson Island by the Polynesians. *Ducula galeata* now occurs only on one island in the Marquesas, where it was previously thought to have evolved as an endemic. *Ducula pacifica* is now restricted to Tahiti and Makatea, whereas *D. aurorae* is widely distributed through the southwestern Pacific. None of these contemporary distributions can be understood without recognizing the impact of the Polynesians.

## Abundance and distribution

When one considers the abundance of species rather than just their presence or absence, our insights about distributions are enriched. Instead of being visualized as a sharp line dividing areas in which a species is generally present from those in which it is not, the range boundary is seen as a rather fuzzy zone in which abundance diminishes to zero. Isolated enclaves occur beyond the main body of the distribution and gaps are also evident within the main range.

Other things being equal, one might expect the abundance of a species to be greatest at the centre of its range, where the combination of factors influencing its population dynamics may be most favourable. If environments vary more or less continuously in space, abundance should decrease smoothly toward the range periphery. These expectations appear to hold at least broadly for many bird species, such as the Dickcissel. In woodland habitats on the Hawaiian Islands, for example, the maximum abundances of the Maui Creeper and the Elepaio are nicely centred in their ranges. However, other species, such as the Common Amakihi and the Apapane, show multiple peaks of abundance or a rapid decline in abundance at the edges of their ranges. These patterns, which typify many other species, may emerge if favourable environmental conditions are patchily distributed within the range or if the factors determining abundance change abruptly in space. In one analysis of the abundance and distribution of wintering species in North America, 85% of the species had multiple peaks of abundance within their ranges.

If a species can tolerate environmental conditions sufficiently well to attain high abundance in some places, one might also expect that they would be able to occur at many locations over a relatively large area. Rare species, on the other hand, may have such narrow ecological requirements that they cannot reach high abundances anywhere, and relatively few locations will meet their needs. Widespread species should therefore be relatively abundant where they occur. Of course, because there are relatively few individuals of rare species by definition, it is likely that they would be absent from many locations within their ranges and would have smaller ranges.

In North America, many widespread species do indeed occur at greater average densities than species with restricted ranges, although there are exceptions. The Tricoloured Blackbird, for example, occurs only in a small area of the Central Valley of California, but it is extremely abundant where it occurs. In British woodlands, some low-density species, such as the Mistle Thrush and the Treecreeper, are relatively widespread, whereas others, such as the Reed Bunting and Collared Dove, occur at relatively few locations within their ranges. However, none of the really abundant species occurs at low frequencies in its range. In

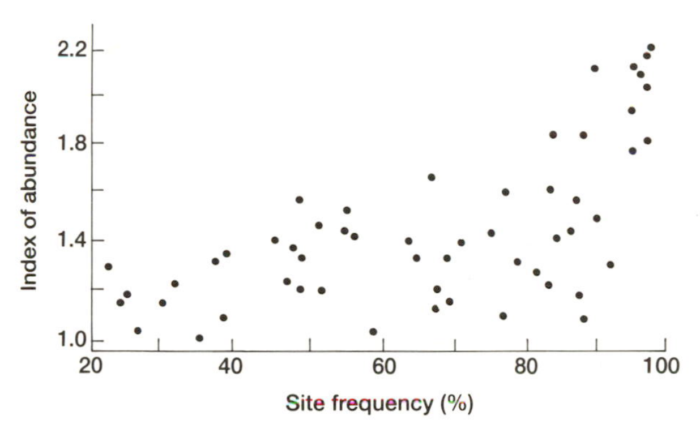


*The abundance of breeding woodland birds and their distribution among British woods. Abundant species are always widespread; less abundant species may or may not be*

Australia, on the other hand, species with large ranges tend to be relatively rare at many places within the range, perhaps because the widespread species mostly occur in the arid regions that support only low densities of birds, or because the species are nomadic and are therefore recorded irregularly at many locations.

Whether a species is rare or not depends on where one looks for it. A species that is locally rare may be abundant elsewhere in its range or rare throughout its range.

Most of the Australian species fit the former pattern. Rarity tends to be more frequent among Australian nonpasserines than among passerines, perhaps because they generally are of larger size and for this reason occupy larger individual home ranges. The result is lower local densities.

## Regional and local distribution: the influence of habitat

In most of the world, habitats occur in a patchwork mosaic rather than in broad expanses covered by a single uniform habitat. This is especially true in areas occupied by humans. Forest birds do not breed in grasslands, and the distribution of such species in a landscape that is a mosaic of forests and grasslands will therefore be quite patchy at regional or local scales, some patches containing high densities, others few individuals, and some none at all.

The patchiness of landscapes is most apparent at *ecotones*, places where one vegetation type grades rather abruptly into another. As one moves up a mountainside, for example, vegetation types give way to one another in a sequence of zones with sharp transitions. How birds appear to respond to such ecotones depends on where one looks. On mountainsides in several regions of North America, the distributional limits of the majority of species coincide with habitat ecotones. In tropical areas such as the Andes of South America or the mountains of Papua New Guinea, on the other hand, the elevational distributions of relatively few species stop at habitat ecotones. Instead, many species reach the boundaries of their local distributions within areas of seemingly uniform vegetation. In a few instances, these distributional boundaries occur where another closely related species suddenly appears, suggesting that the boundaries are set by competition between the species. Competition among species has been offered as the explanation of the distributional patterns of most species in these areas, but the evidence is generally weak and other explanations are equally plausible.

How narrowly or widely a species is distributed among habitats may also be influenced by the density of individuals in different habitat patches. In theory, individuals encountering an unoccupied region (as when returning from migration in the spring or when colonizing a new area) should occupy the most suitable habitat type first. As more and more individuals select this habitat, densities will increase and crowding effects will reduce its suitability. At some point the suitability of this habitat will no longer be greater than that of another initially inferior habitat type, and individuals will now begin to select the second habitat type as well. As a result, breadth of habitat occupancy and local and regional distribution of the species will increase. As densities in the two habitats continue to increase, their suitability will decrease until a third habitat is of equivalent quality. It too will then be occupied. As the abundance of a species in an area increases, its range of habitat occupancy should expand and it should become locally and regionally more widespread, with fewer gaps in its distribution.

There is some evidence to support this idea. In areas in North America, Finland, and Great Britain, locally abundant populations of a species often occur over a greater variety of habitats than sparse populations. The range of habitats occupied at the edge of a species' range, where its abundance is generally low, is often restricted. Perhaps more convincing are changes in habitat distributions accompanying changes in the population sizes of species. Yellowhammer populations were severely reduced in Britain by the extraordinarily severe winter of 1962–63. In subsequent years, densities increased first in farmland (which is seemingly the preferred habitat) but then levelled off. Densities in woodland, however, continued to increase, apparently as a result of spillover of birds from farmlands.

The habitat composition of regional and local landscapes changes through time, especially as a result of human activities. Forests are felled, fields are cleared, croplands and urban areas expand, and, sometimes, disturbed areas are left alone to revert to a natural state. These activities usually lead to a fragmentation of large areas of natural habitat into smaller and more isolated parcels, accompanied by an increase in the area and extent of managed habitats. Some species, such as Red-winged Blackbirds in North America, Eared Doves in Argentina, or House Sparrows in much of the world, benefit from such changes, and their distributions expand. For species using natural habitats such as forests, fragmentation may lead to increased patchiness in their distribution at local and regional scales and, in some cases, to a reduction in their overall range.

The bird species occupying a habitat type such as forest, however, differ in their sensitivity to fragmentation, for two reasons. First, there are differences among species in the amount of continuous area of habitat that is needed to support a local population. In Japan, the Japanese Pygmy Woodpecker is rarely found in forest patches of less than 100 hectares, but the Brown-eared Bulbul occurs in fragments as small as 0.1 hectare. Generally, the larger the body size of a species, the greater its minimal area requirement. Large birds, therefore, are usually more sensitive to fragmentation than are small species. Second, the affinity of a species for the edge *versus* the interior of habitat patches influences its response to fragmentation. As the size of a tract of

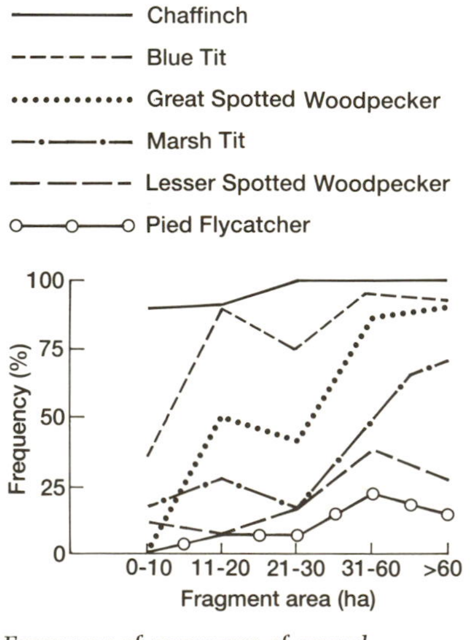


*Frequency of occurrence of several breeding bird species in woodland fragments of different sizes in the Netherlands. The species studied differ considerably in their distribution among woods of different sizes*

habitat is reduced, the proportion of the patch that is interior decreases. Species that characteristically occupy interior locations in forests, such as Hooded Warblers in eastern North America, will be much more sensitive to fragmentation than edge-dwelling species, such as Indigo Buntings. Obviously, the overall loss of natural habitat through fragmentation will alter the distribution of species whether they are interior or edge specialists.

JAW

## The community consequences of species' distributions

The species that occur together in a local or regional bird community are a subset of the species that are present in the regional or biogeographic pools of species from which these communities are drawn. The factors that influence the distributions of species therefore determine community membership at several scales of resolution, and changes in distributions are manifested as changes in the composition of the species pool of a region. Because of range expansions and contractions caused by changes in climate and human impacts on the environment, some 88 species were added to the species pool of Scandinavia during the period 1850–1970, while 22 species were lost.

Perhaps the most basic feature of communities is the number of species present, the *species diversity*. Like distribution, diversity may be considered at several scales, from variations across continents, where diversity patterns may be influenced by speciation and biogeography, to variations from point to point within a small field, where patterns may be consequences of individual habitat selection and territory placement. As one increases the size of area considered, diversity generally increases. Areas of the same size in different habitats or latitudes or on different continents, however, may contain quite different numbers of species.

At a very coarse level of resolution, terrestrial bird diversity increases from the poles towards the equator. A number of explanations of these latitudinal diversity gradients have been suggested: (1) There has been more time without geological disruptions in the tropics, and a richer fauna has evolved; (2) Habitat complexity is greater in the tropics, providing greater opportunity for ecological specialization of species; (3) Competition among species is more intense in the tropics, restricting species to using a narrow range of resources and thereby permitting more species to coexist; (4) Predation and parasitism are more intense in the tropics, reducing population sizes of competitors, though allowing them to overlap more in the use of resources; (5) Climatic stability is greater in tropical than in temperate latitudes, so enabling species to specialize on a narrower range of resources and leading in turn to lower extinction rates; (6) The tropics have greater biological productivity than the temperate zones, and the greater amount of useable energy permits more species to coexist; (7) There are distinctive resources such as fruits or nectar-producing plants in the tropics that support categories of species not found in many temperate locations.

Many of these explanations deal with effects rather than causes of diversity and are thus not altogether convincing. More important, these explanations are all ecological, and they fail to consider explicitly the evolutionary contributions of controls on speciation and extinction rates. Many factors contribute to latitudinal gradients in diversity, but their importance varies among regions or continents.

At a somewhat finer scale of resolution, diversity exhibits complex patterns of geographical variation. In Australia, diversities of insectivorous birds are high at mid-latitudes on the east coast, decreasing both northward and southward and toward the arid interior but rising again on the north-west coast. There is a clear latitudinal change in the diversity of passerines in Argentina, but changes in habitat, terrain, or areas of endemism produce many distortions of this pattern. In North America, diversities are greater in western mountains than in the east, and along the east coast diversity in both deciduous and coniferous woodlands actually decreases from north to south. A similar 'reversed latitudinal gradient' in diversity has been documented in fens and mires in Scandinavia.

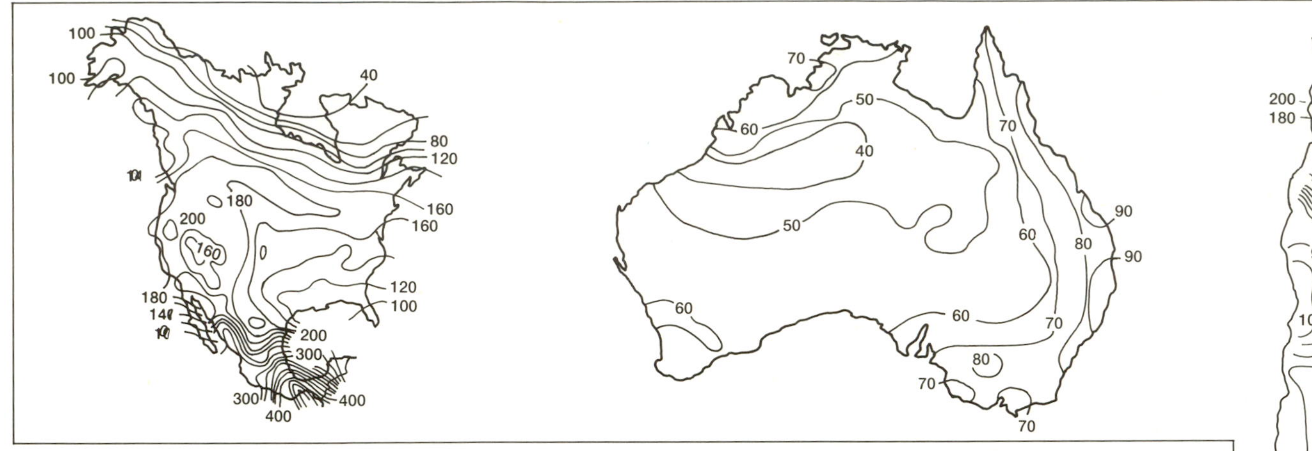


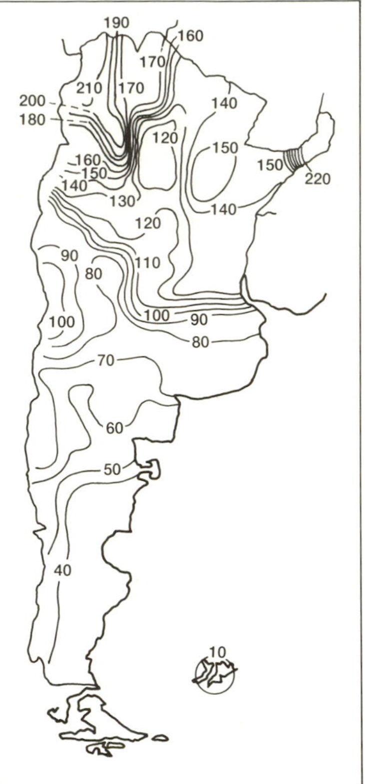


*Geographical variation in bird species diversity. From left, numbers of breeding birds in North America, insectivorous bird species in Australia, and variations in the numbers of passerine species in Argentina*

Patterns of diversity have important conservation implications, especially at the scale on which nature reserves are established. In some instances it is certainly appropriate to focus attention on the preservation of a particular rare species, but conservation efforts are likely to be more productive in the long run if resources are directed toward maintaining populations of large numbers of species (many of which may be common) in their native habitats.

One way of doing this is to locate reserves in centres of endemism or where a maximum number of species with high conservation priority co-occur. Existing reserves often do not meet these needs.

JAW

## The special case of island distributions

Islands are isolated from other land areas by reaches of ocean that are formidable barriers to the movements of many species. Unlike mainland locations, they are essentially closed systems – island populations have nowhere to go, and selective pressures to adapt to local conditions may therefore be more intense. Evolutionary diversification has produced many endemic bird species on islands, and their bird communities are therefore often quite distinctive. On the Hawaiian Islands, which are over 3000 kilometres from the nearest large land mass, 83% of the bird genera and 97% of the species are endemics. Some 39% of the genera and 80% of the species occurring on the Galápagos Islands (1000 km from the mainland) are endemics. On archipelagos that are close to other land masses, such as Fiji (450 km) or the Philippines (350 km), relatively few of the genera are endemic, but many of the species are (44% and 49%, respectively). These percentages are underestimates, of course, because many of the endemic species on such islands suffered extinction at the hands of native or European colonists.

Even where the bird species are not narrowly endemic, island distributions are still remarkable. In the West Indies, most lowland islands contain two (or fewer) hummingbird species, a small species and a large one. High-elevation islands that support humid montane forest

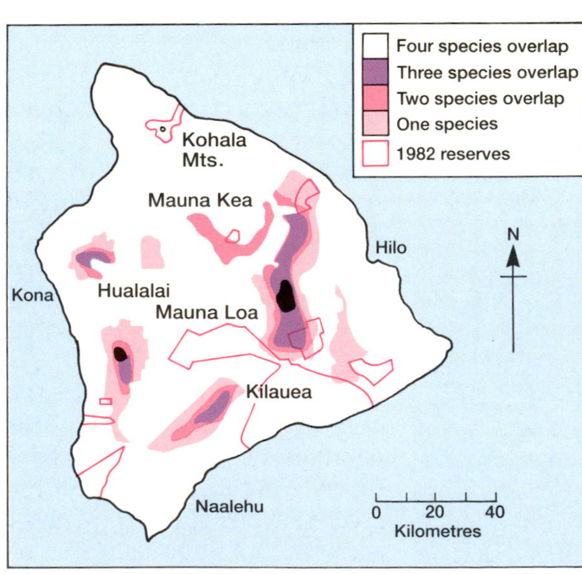


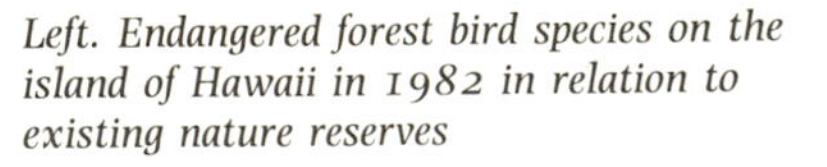
*Left. Endangered forest bird species on the island of Hawaii in 1982 in relation to existing nature reserves*

almost always contain three species: a small species that ranges over the entire elevational gradient and two large species, one restricted to the lowlands and one to the uplands. Overall, there are 15 species of hummingbirds in the West Indies: six small, four large lowland forms, and five large highland species. The two or three species occurring on a particular island are therefore drawn from a considerably larger pool of potential colonists. In other situations, groups of island species belonging to the same genus have 'checkerboard' distributions, in which a given island is occupied by only one of several species, but which species is present varies from island to island. These island distributional patterns have been attributed to a combination of chance and competition: which of several species first colonizes an island may be largely a matter of chance but, once established, the resident is able to exclude competitively other ecologically similar species. It is also possible that the distributions reflect differences among islands in habitats or food resources available.

Island isolation and size (or elevation) and the colonization and extinction rates of bird species figure importantly in these examples. These factors have been synthesized into a general theory of island biogeography that goes like this: colonization rates to an island decrease with increasing distance from the mainland or other sources of immigrants, whereas the likelihood of extinction of populations on islands decreases with increasing island area (larger islands support larger populations that are less likely to be wiped out by chance events such as disease or hurricanes). At some point the rate of addition of species by colonization will be balanced by the rate of loss through extinction, and the number of species reaches an equilibrium. This equilibrium species number will be less for a small island than a large island, and for a distant island than a near island. The species composition of the island community will continue to change after this equilibrium is reached, as new, competitively superior immigrants replace previous residents, but the number of species will remain the same despite this turnover.

Although this theory is intuitively appealing, the evidence supporting it is generally weak. Reports about the degree of equilibrium in species number conflict, and turnover does not occur in some situations and is not balanced in others. In many instances, the role of competition in species turnover is unsubstantiated. Between 1958 and 1978, the endemic Socorro Dove became extinct on Socorro Island off Mexico and the mainland Mourning Dove became established. This would appear to be a classic case of competitive replacement of one species by another. It is more likely, however, that predation by feral cats introduced to the island in the late 1950s led to the demise of the Socorro Dove, while the establishment of the Mourning Dove was due to the availability of free water associated with human settlement of the island.

Perhaps the most troublesome feature of the equilibrium theory is its oversimplification of the factors influencing the distribution of species among islands. Area and distance are surely important, but many other factors may also be significant. In the Åland archipelago in the Baltic Sea, for example, 52 of the 121 species found on the large island of Main Åland are not found on the small island of Ulversö, 10 km away. Perhaps a third of these species are absent because they are rare on Main Åland and would therefore be unlikely to establish populations on Ulversö, while Ulversö lacks suitable habitat for almost half of the missing species. A few species are absent because of the presence of competitors or predators on

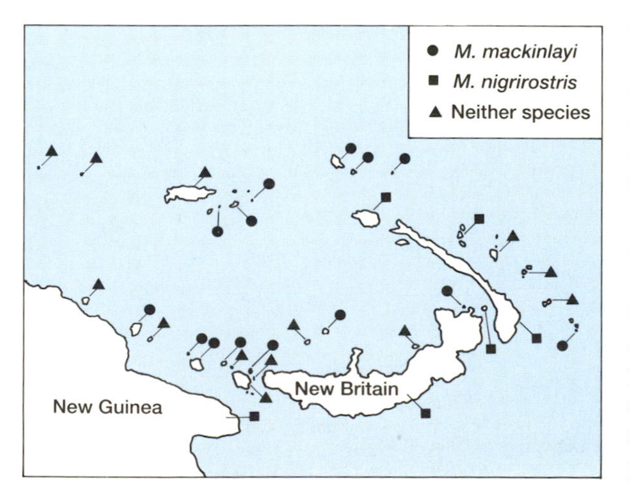


*Left. The checkerboard distribution of Mackinlay's Cuckoo-Dove (Macropqia mackinlayi) and the Black-billed Cuckoo-Dove (Macropyqia nigrirostris) in the Bismarck region of the Pacific. On the islands on which cuckoo-doves have been observed, either one species or the other occurs, never the two together*

Ulversö. Less than 10% of the absences can be attributed to dispersal difficulties. In other situations, extinction rates are lower than would be expected because small populations on an island are continuously being 'rescued', by immigration from elsewhere. This is especially likely to occur among islands within an archipelago. JAW

## FACTORS INFLUENCING BIRD DISTRIBUTIONS

In addition to past history and the effects of barriers to dispersal and of isolation, the boundaries of the geographic range of a bird species may be determined by several other factors.

*Climate* may be the ultimate determinant of the range limits of most birds. A species cannot live in areas where its range of physiological tolerances to temperature or water availability is exceeded.

*Competition* with other species may lead to the exclusion of a species from areas where a competitor is dominant. This is especially likely among closely related species that are similar in their ecology (see pp. 173–4 on niches and ecological segregation).

Other biological interactions, such as *predation* or *parasitism*, may be important in some situations. Species may be absent from areas where mortality from predation or parasitism is great.

Most species have particular *habitat* requirements; if appropriate habitat is unavailable, the species will not occur.

Where key *resources* such as food of a particular type, appropriate nesting sites, or other plant or animal species on which a bird species is dependent are absent, so also will be that species.

*Area* alone may also be important; if the area of a patch of habitat or an island is insufficient to support a breeding population of a species, the species will be unable to persist there. This is increasingly likely the larger the size of the species.

The role of *chance* cannot be overlooked. Sites that are otherwise suitable to a species may not be occupied because, by chance, the species has failed to colonize the area or has been eradicated from it. This is especially likely to be important in small isolated areas of habitat or on islands.

These factors usually act in combination rather than alone, and the contributions of different factors to the combination differ in different areas. The determination of range boundaries is therefore frequently complex.

JAW

## DETERMINING BIRD DISTRIBUTIONS

Documenting the distribution of a species is primarily a matter of determining where it is present and absent. This sounds like a simple task, but it is not. Because abundances are usually low toward the periphery of a species' distribution, establishing whether it is present or absent may depend on how long and intensively one looks for it. Moreover, some individuals may be vagrants and occur well beyond the boundaries of their normal distribution.

When such individuals are sighted thousands of kilometres from the species' range, it is easy to classify them as transients, but what if they are recorded only a few hundred kilometres outside the previous known distribution? Does this represent a *bona fide* range extension, vagrancy, or incomplete knowledge of the species' previous distribution? How many individuals must be seen in an area, over how many years, to justify including the area in the range of the species? There is no hard and fast answer to these questions. For this reason, documenting the distribution of a species' abundance, although more difficult, provides a more accurate image of the species' distribution than a boundary drawn on a map.

Distributions are determined in a variety of ways. In developed areas of the world, most are based on records of *incidental observations* made by amateur ornithologists or field naturalists. In more remote areas, *systematic explorations* help to define where species occur, but our knowledge of the details of range boundaries is still incomplete for many tropical and desert regions of the world.

*Organized surveys* provide more detailed information on both distribution and abundance. In North America, counts have been conducted as part of the *Breeding Bird Survey* over more than 1800 roadside survey routes located randomly within 1° latitude–longitude blocks since the mid-1960s. *Bird Atlas Projects* have been conducted in several countries (e.g. Britain, Ireland, Denmark, France, Natal, Australia) to map the occurrence, frequency of observation, and/or relative abundance of both breeding and wintering birds in latitude–longitude grids. These surveys have enlisted the participation of thousands of amateurs and professionals in cooperative, carefully organized programmes, and provide the best information available about the distribution of many species.

JAW

## BIRD DISTRIBUTIONS AS INDICATORS OF ENVIRONMENTAL CHANGE

The distribution and abundance of birds change through time, and some of these changes have been linked to specific environmental disruptions. The classic example is the disappearance of the Peregrine Falcon from many parts of its range with the widespread increase in the use of agricultural pesticides following the Second World War. These linkages have given rise to the hope that bird populations can be used as sensitive indicators of environmental changes, providing an early warning of environmental degradation that may not yet be apparent to humans.

Birds would seem to be well suited to such a role. They occur widely, the ecology of many species is well known, they often occupy a position high in the food chain and therefore concentrate the effects of pollutants, they are visible and easily monitored, and large groups of volunteer observers are available in many parts of the world to record changes in their distribution and abundance.

Changes in the distribution and abundance of most species, however, are usually due to a variety of factors, some of which act directly, others indirectly through the food chain or the habitat. Different combinations of factors may be responsible for changes in different parts of the range, and changes in one area may frequently be a consequence of environmental events some distance away (e.g. on the wintering grounds). Also, changes in environmental conditions may take some time to be realized as changes in distributions. Acid rain, for example, may take several years or decades to affect vegetation, and bird populations may not respond to the vegetational changes for several additional years (or generations).

Because of these difficulties in linking distributional changes with their environmental causes, it is not likely that changes in bird distributions will normally be of much value in providing an early warning of environmental changes. Features of the breeding biology of birds, such as reproductive rates or survival rates, offer greater promise. In order to use any aspect of the biology of birds as an indicator of environmental change, however, widespread, systematic, and long-term monitoring programmes are necessary.

JAW

## ISLAND BIOGEOGRAPHY THEORY AND THE DESIGN OF NATURE RESERVES

Ecologists have used island biogeography theory to derive several predictions about how nature reserves should be designed to maximize their effectiveness:

1. Reserves should be as large as possible (to reduce extinctions).
2. A single large reserve is better than several small ones of the same total area.
3. Reserves should be as close together as possible (to increase colonization rates). If the reserves can be linked by 'corridors' of natural habitat, so much the better.
4. Reserves should be as circular as possible (to maximize the amount of interior habitat).

There is relatively little argument about the first and third predictions, which agree with common sense. Evidence supporting the fourth prediction is inconclusive. The second prediction, however, has been the most controversial. Proponents argue that a single large reserve provides a larger area for species with large individual home ranges and more interior habitat, as well as more habitat variety. Opponents believe that it is less likely that species distributed over several small reserves will suffer local extinction in all of them – a variation on the adage that you should not put all your eggs in one basket. Several small reserves will therefore contain more species, overall. Both the theory and the evidence behind these opposing positions are inconclusive.

It is unlikely that a single general theory, especially one as simple as island biogeography theory, will be useful in all situations. Different sets of species in different regions respond to reserve area, isolation, shape, and habitat in different ways, and management must be attuned to the species and habitats involved in particular situations. Different conservation goals, such as maintenance of an entire bird community, preservation of habitat diversity, or sustaining viable populations of a particular species, may also dictate different reserve designs.

JAW

*Ecological segregation is well illustrated by the tits (Paridae) of British woodland. Blue Tits (P. caeruleus) forage along twigs towards the treetops, Marsh Tits (P. palustris) at an intermediate level, and Great Tits (P. major), in the winter, on the ground. The bill of each species is appropriately adapted to its feeding needs*

## NICHES AND ECOLOGICAL SEGREGATION

The concept of the *ecological niche* is central to our thinking about a variety of ecological topics – how species use resources, the coexistence of similar species, the composition of communities, the role of species in food webs, or patterns of distribution within and among habitats or geographically. The niche of a species is a description of the range of environmental conditions in which individuals can survive and reproduce or, said another way, a depiction of how the species is distributed on gradients of various environmental factors. Although the niche can be thought of as a statement of the sorts of environmental conditions that enable populations of a species to persist independently of any other species, it is most often considered in relation to interactions among species that share limiting resources and therefore may compete with one another. Species that have similar niche requirements or a high degree of niche overlap are likely to use resources in very similar ways; if the supply of those resources is insufficient to meet the demands of the species, competition will ensue. This may lead either to exclusion of one species by another so that their distributions on local or even geographic scales do not overlap, or to a shift in the niches of either or both species so that the niche overlap between them is reduced. In this case, the species do not occupy the full range of environmental conditions that they might inhabit in the absence of the other species.

Ecological theory suggests that species that exhibit a high amount of overlap on many niche dimensions may nonetheless coexist if they differ on a single important dimension, such as food type or size or nesting habitat. Theory also indicates that the limits to the niche similarity of coexisting species are determined by the degree of difference between their average positions on such important dimensions and the breadth of their occupancy of the dimension (for example, the range of prey sizes eaten and differences in the average prey size).

There are many examples of such *ecological segregation* among species that coexist in local habitats. In British lowland woodlands, one may commonly find four or five species of tits (Paridae): Blue Tits, Great Tits, and Marsh Tits occur regularly, Coal Tits somewhat less so, and Willow Tits are sparse and irregular. The smallest species, the Coal Tit, feeds chiefly on the branches of oak and ash trees, and most of the insects it captures are tiny. The slightly larger Willow Tit feeds primarily on birch, somewhat less on elder, and avoids oak trees. During winter, it uses branches even more extensively than the Coal Tit but, like all of the species, it shifts its attention to leaves in summer. By contrast, the Blue Tit feeds primarily on oak throughout the year, foraging toward the tops of trees on twigs or buds or, in summer, almost entirely on leaves. It is an agile species, often hanging from twigs or foliage to feed. Most of the insects it captures are small. The Marsh Tit is the same size but has a longer bill. It feeds extensively on the shrub layer or on twigs and branches in the lower canopy of oaks or in the understorey herb layer. It captures somewhat larger insects than the other species and also eats various seeds and fruits. Finally, the large Great Tit feeds mainly on the ground in winter, although it shifts to the leaf canopy in summer when gathering caterpillars for its young. Most of the insects it eats are relatively large, and it eats more seeds (especially hard ones) than the other species. Collectively, these differences in niche characteristics are presumed to segregate the species sufficiently to reduce competition among them to levels that permit them to coexist in the same woods. The situation is actually more complex, however, as there are seasonal and yearly variations in the abundance of resources and in the foraging behaviour, diet, and habitat use of the birds that alter the niche patterns of the species and their potential competitive relationships. Moreover, these characteristics of the species vary geographically, so that their relationships in Spain, Belgium, Denmark, or Finland may be different from those one sees in Britain. Other species, such as Goldcrests, Treecreepers and Pied Flycatchers may also exhibit high niche overlap with one or more of the tit species in different areas and also enter into the competitive milieu.

Although the niche concept is usually applied to species, it is by no means unusual for the sexes within a species to occupy different niches (or to occupy different ranges outside the breeding season). For example, the larger female sparrowhawk brings larger prey to the nest than her smaller mate. During their breeding season in the Southern Ocean, male Giant Petrels are predators and scavengers at penguin and seal colonies, while females forage more at sea for fish and squid. A remarkable example of a sexual difference in niche accompanied by a morphological difference is provided by the extinct Huia of New Zealand. Females had a slender curved bill for probing crevices while the stout chiselling bill of males broke into rotten bark.

Despite its intuitive appeal, the niche concept and the predictions of niche theory are not without problems. Usually, niche dimensions that relate to resources and their use are emphasized, but the resources themselves are rarely measured. If resources are so abundant that they are not limiting to the species, high niche overlap between the species will not lead to competition. One cannot predict whether or not competition is likely to occur from measurements of niche overlap alone – the resources must also be considered. By the same token, the observation that species differ in aspects of their behaviour or ecology and therefore appear to be segregated cannot be used to conclude that the species are competitors or that the differences between them represent shifts in niches to avoid competition. The differences could easily be due to other factors in the evolutionary history of the species that are unrelated to competition. Finally, there may be many aspects of environments other than resources that are important determinants of the niches of species – climatic factors such as temperature, wind, or precipitation, salinity or water acidity for aquatic species – and a preoccupation with resource aspects of niches and competition may yield only an incomplete view of a species' ecology. The value of the niche concept is in drawing attention to the multiplicity of factors that may influence the ecology of individuals and, ultimately, the distribution and abundance of species.

JAW

# 7 MIGRATION AND NAVIGATION

## Migration

### Reasons for migration

Migration is the term traditionally applied in ornithology to two-way seasonal movements of birds between their breeding and non-breeding areas. The functions of such movements are twofold: (1) to enhance the chances of survival of those birds that leave the breeding areas before seasons of unfavourable weather arrive there; and (2) to permit birds to utilize temporarily available food resources, such as seasonal abundances of seeds, fruits, fish or insects, to enhance the number of young they can raise successfully. These benefits of migration must be viewed against the costs of such extended flights, during which birds may die for a variety of reasons – shortage of fuel, adverse weather, failure of navigation systems, and so on. In some species, both migrant and non-migrant individuals are found in the same breeding population (see 'partial migration', p.182). This emphasizes that migration confers benefits and costs on individuals and is not merely a population phenomenon.

Another use of the term 'migration' is gaining recognition in ornithology. It refers to the movement of a young bird from its birthplace to find a suitable breeding place. (This has often been termed 'natal dispersal', an incorrect use of the word 'dispersal' which means to move away from other animals of the same sort and not to move away from a particular place.) This behaviour occurs even in species traditionally described as resident, since it is most unlikely that a young bird will be able to nest for the first time in exactly the site in which it was born (because, for example, it may be excluded by surviving parents or other older birds). This form of migration may also have more positive functions – to promote outbreeding or to ensure that individuals raised in temporarily favourable habitats look elsewhere for breeding sites, since the habitat of their birthplace may no longer be suitable a year later. This behaviour will not be considered further here, however.

Two-way migration has probably evolved many times and it now appears in a bewildering variety of forms. At one extreme are the migrations of those species like geese and swans, in which young birds travel with their parents for at least one and perhaps several years, thereby learning in detail the routes to be followed and the staging posts at which feeding is possible. At the other extreme are the movements of species in which the young are abandoned by their parents just after they have fledged (or even before they have fledged in the case of parasitic species like the European Cuckoo) so that the young have to migrate without any assistance from older birds of the same species, let alone from their parents.

Most birds require a high-protein diet whilst breeding. This is normally provided by insects or vertebrates, both terrestrial and aquatic; seeds and fruits are not normally sufficiently nutritious, though they provide excellent sources of energy for survival outside the breeding season. Both terrestrial and aquatic insects show seasonal peaks of abundance in most parts of the world, and thus strictly insectivorous bird species, such as many small songbirds, migrate each year when or just before insects become scarce, i.e. in autumn from mid or high latitudes. In contrast, those birds that change to a diet of fruit and seeds in autumn may not need to migrate each year, for many trees fruit heavily at intervals of two to three years. Those birds of prey that feed upon small mammals whose populations reach peak levels at intervals of several years also migrate irregularly; in contrast, those that feed primarily on insects or insect-eating birds, migrate each year. Many seabirds migrate, either in response to the migrations of the fish upon which they feed or to changes in the availability of fish related to spawning. Most wildfowl are herbivorous and migrate away from high northern latitudes in autumn because the aquatic plants they eat become locked in ice during the winter, and the grasses and sedges they graze are covered by snow. In tropical regions, migrations between savannah and forest occur in many species, in response to seasonal peaks in availability of termites and of seeds of grasses. Wildfowl migrate to large safe feeding areas to moult, for they are particularly vulnerable to predators when they are flightless, as they grow new wing feathers.

## Historical ideas and early observations of bird migration

Man must have been aware of the seasonal migrations of birds for as long as he has lived in the parts of the world through which they passed. Indeed, in those bygone eras when he was a hunter-gatherer, his survival may well have depended on his abilities to utilize seasonal abundances of migratory birds as well as of fish, mammals, seeds and fruit. Hunting of migratory wildfowl is depicted on some ancient Egyptian paintings and there are numerous incidental references to migrants in the Old Testament – particularly to conspicuous species like storks and turtle doves.

The first systematic treatment of migration is usually credited to Aristotle, in one of the volumes of *Historia Animalium*. Not only did he classify a large number of birds as migrant species, but he also recorded some information on times of arrival and departure in the Greece of more than 2000 years ago. Although there was obvious proof that some species – notably those that migrate by day – left Greece in the spring or autumn, migration was not the only phenomenon he invoked to explain the changes between the summering and wintering bird fauna. Aristotle supposed that some pairs of similar species (that we now know to migrate primarily at night) 'transmuted' one into the other at about the times of the equinoxes. Thus a summer visitor to Britain, the Common Redstart, was thought to change in autumn into the European Robin, a winter visitor to Greece!

Aristotle's other proposal was that some birds hibernated instead of migrating. The disappearance of Barn Swallows and Sand Martins into large reedbeds to roost on an autumn evening must have lent credence to such an idea. Furthermore, the late autumn migrants of these two species are sometimes overtaken by cold weather during migration through mountain passes in southern Europe and may then be found in a torpid condition (from which they do not normally recover). It was not until the beginning of the nineteenth century that scientific evidence was assembled to refute the idea that hibernation accounted for the disappearance of many European species in autumn and their reappearance in spring – though it is now known that a few species of American nightjars are able to enter and emerge from a torpid state.

By the second half of the nineteenth century and in the first few decades of the present century a considerable quantity of observations was collected on the timings and routes of migration in both western Europe and North America. More scientific study of other features of migration – its timing in relation to weather conditions, the altitude of flight, weights and measurements of migrants shot or killed at lighthouses – followed the establishment of bird observatories on Heligoland and Fair Isle and the studies of sportsman-naturalists like Abel Chapman in north-east England. Some of the questions posed at those earlier times were solved by the use of marking techniques, particularly numbered aluminium rings (bands) attached to the legs of migrant birds that were later found on the breeding or wintering grounds. Other questions, such as how migrants react to changing wind conditions once airborne, have still to be answered unequivocally.

## Different forms of migration and their possible evolution

On an evolutionary time-scale, the ultimate determinants of migration in a particular species are those environmental conditions that at least some individuals in every generation of that species should avoid or utilize. Many hypotheses about the ultimate causes of migration are not amenable to direct testing, but almost all assume that the tendency to migrate is heritable. For example, the benefits and costs of residency or migration in partial migrant populations probably vary from species to species, and with geographical location. For species like the Goldcrest in Finland, the benefits of migration are clear in those years in which a majority of the resident population is killed by the severity of the winter; but the costs are also high. Of all species, this small bird (mass five to six grams) appears one of the least well designed aerodynamically for long-distance migration. In certain autumns, large numbers are deflected westwards across the North Sea by strong easterly winds and some undoubtedly perish; though others reach Britain. In contrast, residency in southern French populations of Blackcaps, some of which migrate, is less likely to involve heavy mortality through severe weather, since some migratory Blackcaps from northern Europe come south in autumn to winter in the same parts of France. Possibly food resources there in winter are patchily distributed and resident birds have access to the best areas. Certainly in summer the resident French Blackcaps appear to outnumber the migrant Blackcaps in evergreen forests but *vice versa* in broad-leaved forests. If indeed there is some spatial separation in the breeding habitats occupied by resident and migrant fractions of a partially migrant population then it seems likely that competition for the better habitats has, at least in part, interacted with genetic variation in the disposition to migrate to produce the situation seen today.

## Age and sex differences in the extent of migration

More than two centuries ago, the Swede Linnaeus noted that partial migration of a special kind occurred in the Chaffinch, namely that all females and juveniles left Scandinavia in winter but that some males remained on their breeding grounds. He named this finch species *coelebs*, meaning bachelor, in recognition of this behaviour. A suggestion that male sex hormones, acting to inhibit migration, may be responsible for these sexual differences in behaviour has been examined in a partially migratory population of European Blackbirds, but no consistent differences in hormone levels between migrants and residents have been detected. The advantages of residency to male blackbirds, and perhaps other songbirds, appear to be greater success in establishing territories, earlier breeding and higher nesting success. In blackbirds, however, resident females do not have higher breeding success than migratory females when paired to resident males; this implies that resident males obtain better quality territories than migrant males. It is not clear whether these benefits are counterbalanced by higher mortality and, if so, from what causes.

Often, birds of different ages and sexes within a migratory species differ in the distance travelled from the breeding grounds; and in some species also in the timing of migration. In several seed-eating species, for example some finches, whose food resources on the breeding grounds can be covered by winter snowfalls, juveniles winter further south than adults and females further south than males. The proximate factors that lead to these differences are not known, but various evolutionary arguments can be offered to account for the behaviour. If each individual bird is seeking to maximize the number of young that it produces during its lifetime, then the crucial feature for juveniles is to survive until they can breed for the first time. This means that they should travel sufficiently far from the breeding area to be no longer at risk from severe winters. Adults that have bred already can 'afford' to take

*For the male European Blackbird (Turdus merula) in the western part of its range, residency means greater success in maintaining a territory and breeding*

greater risks and winter nearer their breeding sites, if by doing so they improve their chances of breeding successfully and reduce the risks associated with migration itself. (This implies that the hazards of migratory flights increase with the total distance travelled.) But why do males incur greater risks than females by wintering further north? As will be shown below, this is probably attributable to some increasing advantages of wintering closer to the breeding area.

Segregation between adults and juveniles in winter is not universal among seed-eating species; it is not obvious in the Dark-eyed Junco which breeds in southern Canada and the northern states of the USA and winters further south in the USA. Social interactions between individuals, which affect their chance of obtaining access to limited food resources, may alter the balance point between the costs of migrating further and the benefits of an assured food supply. Intensive studies of the juncos have shown that juveniles are not always subordinate to adults and so do not necessarily have to migrate further.

## Differences in the extent of migration between individuals

Unlike most songbirds, which tend to migrate in a generally southerly direction within a continental land-mass, some Palearctic shorebird species or populations migrate on a predominantly east to west axis, to shores of the southern and western North Sea and around the Irish Sea, as well as southwards to Africa. In one species, the Grey (Black-bellied) Plover, there is competition for wintering sites in north-western Europe, as close to the western Siberian breeding areas as winter weather conditions normally allow survival. Juveniles reach these potential wintering areas before adults and there is competition, first amongst the juveniles but later between juveniles and adults, for feeding territories; as a result of this, many juveniles are displaced further from the breeding grounds (often to the south). It is not known whether the juveniles that manage to winter furthest north and east are predominantly males, but it has been established that once a Grey Plover has survived one winter at a particular place, it will return there in future years. This faithfulness by individuals to a wintering site occurs in many shorebird species. Perhaps the fact that predictably rich intertidal feeding sites are much more restricted geographically than most terrestrial habitats places an extra premium on site fidelity in shorebirds, a premium which overrides the increased risk-taking with increasing age that is seen in many songbirds.

The advantages to birds of wintering closer to the breeding area could be: (1) to allow them to return earlier than longer-distance migrants and so obtain limited resources on the breeding grounds, such as nest-cavities or territories; (2) if they remain in the same climatic zone as their breeding grounds, to enable them to respond to year-to-year variations in weather so as to return and commence breeding as soon as this is possible; and, finally, (3) to minimize the distance moved if migration is hazardous. This may be particularly true for juveniles.

There is considerable evidence for the existence of competition for resources on the breeding grounds in some bird species, but also evidence in a few species that older males may be able to re-establish the territories they held the previous breeding season, no matter when they arrive. The necessity to arrive early is not absolute, therefore, but it should apply more severely to males. This could lead to males wintering nearer to the breeding areas, on average, than females. However, another alternative is that birds could winter further away but set out on their return journey at an earlier date!

For species in which there is no marked segregation in wintering areas between the different ages, and particularly for those in which suitable winter foraging habitats are scarce, the advantages of wintering as close to the breeding areas as possible are probably to be found in reducing the chances of death during migration, particularly for juveniles. (The mortality of adult shorebirds during migration is probably low. Losses on flights both ways between western Europe and the Arctic, together with those on the breeding grounds, amount in total to no more than seven to eight per cent per annum.) On their first migrations, juvenile songbirds have been shown to fly on a pre-programmed compass course (or sequence of courses) from their birthplace; information from marked juvenile shorebirds suggests that they do likewise. This means that if they were to fly past a suitable over-wintering site, they have no means of knowing on their first journey how far it will be to the next suitable site. Although adjacent good intertidal feeding areas, notably estuaries, in western Europe are relatively close, in the Mediterranean and north African regions they may be separated by many hundreds of kilometres. Thus if a juvenile migrant overshoots one suitable site it will require considerable fuel reserves to reach another. Examples of birds reaching sites in an exhausted condition are well-known, and mortality of juvenile migrants during migration and before they find a suitable wintering site is likely to be considerable – though difficult to measure in practice.

## Leap-frog migration

Several examples are known in which the most northerly breeding populations of a particular species (in the northern hemisphere) winter furthest south and the most southerly breeding populations migrate little if at all. This phenomenon is also seen amongst ecologically related pairs of species; for example, the Whimbrel breeds chiefly to the north of the larger Curlew in Europe, but winters further south (in Africa). Several explanations have been advanced for this behaviour.

One of the first ideas arose from a study of the Ringed Plover in which there appeared to be strong separation between the wintering areas of different breeding populations, such that the largest birds wintered furthest north, whereas on the breeding grounds the largest birds nested furthest south. It was suggested that selection for size was operating on the wintering areas and that large birds were better able, physiologically, to withstand cold and so could winter further north. This argument is not now thought important, since many arctic-breeding birds face colder conditions during the early part of the nesting season than at any other time of year.

A second hypothesis suggests that birds which breed in areas with marked between-year variation in onset of spring (and therefore in optimal nesting times) benefit from staying within the same climatic zone in winter, so that they can respond quickly to favourable conditions. Temperate zone birds can do this, but Arctic birds cannot since conditions are too severe. Therefore, it is argued that Arctic birds lose nothing by migrating south of the temperate zone in winter, whereas temperate zone birds that did so would be at a disadvantage compared to those that stayed. Hence Arctic populations or species, it is suggested, should overfly temperate zone equivalents. This idea ignores the costs of extending the migratory journey.

A further suggestion uses the costs of migration as one of its starting points. Noting that it is normally the largest of a set of populations or species-pair that travels the shortest distance

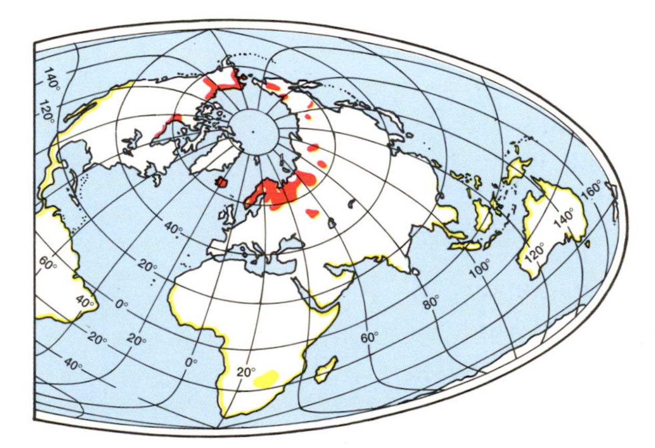

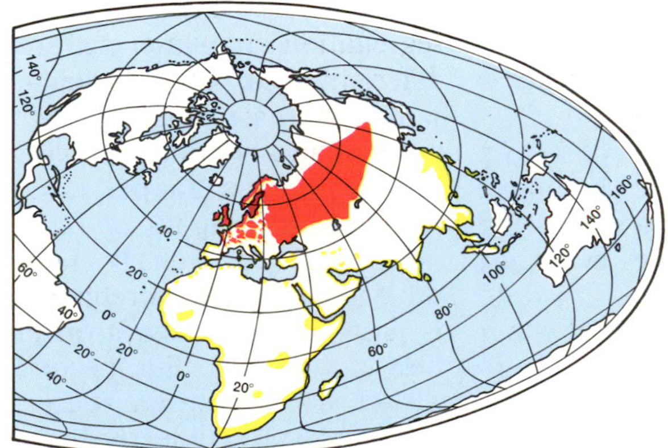

*Leap-frog migration. The Whimbrel (Numenius phaeopus) breeds further north than the Curlew (N. arquata), but tends to winter further south, so leap-frogging the Curlew. The breeding areas of both species are shown in red, wintering areas in yellow*

from its breeding grounds, the hypothesis involves competition between individuals as the basis for their winter distributions. As mentioned earlier, some juvenile Grey Plovers have been shown to displace others of the same species from a potential wintering site; the displaced birds tend to be smaller. If extended to populations or species-pairs, this could explain why the largest birds winter furthest north.

Avoidance of competition on spring staging-grounds could be a further reason for smaller birds wintering further away and passing through to higher latitude breeding grounds after larger birds have left to go to their nesting habitats. This might apply only to species in which breeding and wintering habitats differed.

PRE

## The stimuli determining the timing of migration

*The need for precise timing*

The evolution of migration systems has enabled birds to cover virtually all parts of our planet. Migrants have even settled into the remotest corners of the Arctic and Antarctic, overcoming geographical barriers such as deserts, mountains or oceans. Some species, like the Arctic Tern, can travel half of the earth's circumference during a single seasonal migratory movement.

As we have seen, there are various kinds of migration, including seasonal 'there and back' migration between summer breeding grounds and winter quarters, moult migration where birds fly to appropriate moulting grounds, and intratropical migration to suitable feeding areas. All require an extraordinarily high degree of temporal organization. Long-distance migrants, for instance those breeding in the Arctic, have to arrive on their breeding-grounds within a few days every year even though they start their journey many thousand kilometres away. If they arrive too early the birds risk adverse weather conditions, but if they arrive late they would shorten the already brief breeding period and thus reduce their chances of successful breeding.

*'Internal calendars' in 'calendar birds'*

Only a small percentage of migratory birds, the so-called winter migrants (e.g. a number of waterfowl species), perform their movements in direct response to winter conditions. These show considerable variation in their departure from year to year and may even stay in their breeding grounds if winter is abnormally mild. The great majority of migrants, however, leave the breeding grounds well ahead of seasonally deteriorating conditions and demonstrate a high degree of consistency in both departure and arrival times. Arctic breeders, like northern populations of the European Cuckoo, regularly arrive in a given area on the same date, give or take three to four days, year after year. The seasonal precision of these species is so high that they were called 'calendar birds' in former times.

In 1702 Ferdinand von Pernau assumed that early departing long-distance migrants could hardly be triggered directly by unfavourable environmental conditions because some, like the Marsh Warbler, leave Europe in mid-July. He proposed that it was much more likely that some 'endogenous factors' in the bird itself might be the stimulus triggering migration.

European warblers were the first birds in which the existence of real 'internal calendars' could be demonstrated. Nestlings were hand-raised in constant experimental conditions and were subsequently kept under these constant conditions, with no seasonal changes in photoperiod, temperature, air-humidity or food supply. The results were astounding: despite the absence of any environmental seasonality the birds produced seasonal rhythm, including migratory behaviour. After their juvenile development (which ends with the juvenile moult) they spontaneously deposited fat as a reserve for the migratory period. Both the amount and the pattern of this fat deposition were similar to that of their wild counterparts. Moreover, at the beginning of the migratory period these caged birds also demonstrated what we now call migratory restlessness (see Chapter 10). This activity consists mainly of caged birds wing whirring (beating their wings while in a sitting position). Among those individuals kept for long periods under constant conditions, the migratory processes occur in regular intervals at approximately the appropriate time. This internal rhythmicity is, as genetic and physiological experiments have since shown, inherited and due to rhythmic physiological processes in which the hypothalamo–hypophysical system of the brain and a series of other hormonal feedback mechanisms play an important role. The existence of endogenous calendars sheds light on the question of how migration is temporally organized. The necessary time mark is set, as both field observations and experiments have shown, by the internal physiological 'clock'. Deviations of this endogenous calendar from the calendar year and thus from the biological seasons are corrected by environmental synchronizers, mainly photoperiod: the seasonal change of day length is one of the most reliable fluctuations.

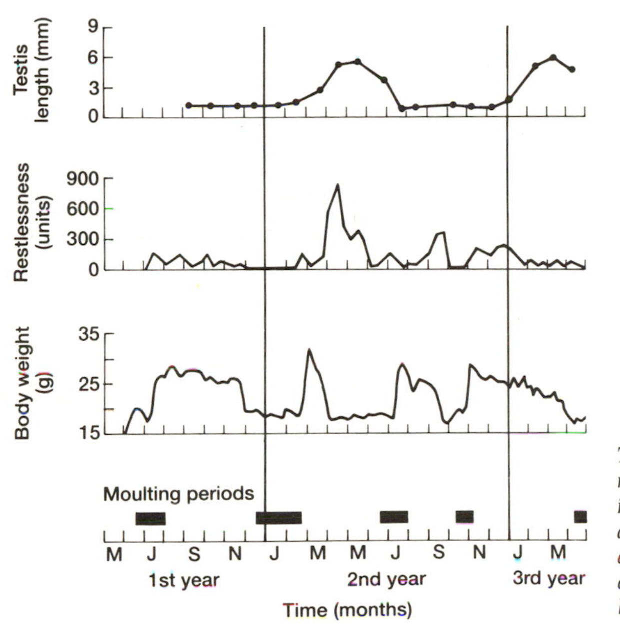


*The timing of migration. Endogenous rhythms of four annual events in an individual Garden Warbler (Sylvia borin) are shown left. In a long-term experiment, the bird was kept in a constant daily light-dark ratio of 10:14 hours from juvenile development onwards*

## *Inherited time-programs for migration*

The discovery of internal calendars shows how migrants can be triggered to start migration. But how long should they migrate and when should they end the journey? The idea that migrants have an inborn knowledge of environmental cues of their winter quarters, such as characteristic star patterns, has not been supported by experiments. However, a detailed investigation of migratory restlessness yielded results which shed more light on the question. Comparative studies of migratory restlessness have shown that different species, and even different populations, with migratory journeys of different lengths produce corresponding amounts of restlessness. A detailed analysis demonstrated a close positive correlation of both features: the longer the migratory journey the lcnger lasting the restlessness. In other words, a migrant, at least on its first migration, probably halts when migratory restlessness fades out.

In the Garden Warbler, migratory restlessness has been analyzed by video-recording under infrared illumination. It proved to consist entirely of wing whirring. When the total average wing whirring time during an autumn migratory period of an experimental group of central European birds was multiplied by the average flight speed (about 30 km/h), it was clear that the birds would have just reached the centre of their winter quarters in central Africa with that amount of flight. Thus migratory restlessness could be closely related to the distance the birds have to cover.

In a related species, the Blackcap, cross-breeding experiments have shown that the amount and pattern of migratory activity are genetically determined. These results suggest that migrants are equipped from birth with inherited time-programs for migration in which the duration of the journey is essentially preprogrammed.

*Below. Migratory restlessness over time in groups of Blackcaps (Sylvia atricapilla) from three European and one African population (left) and from hybrids of two populations (right). The migratory stimulus appears to be genetically determined*

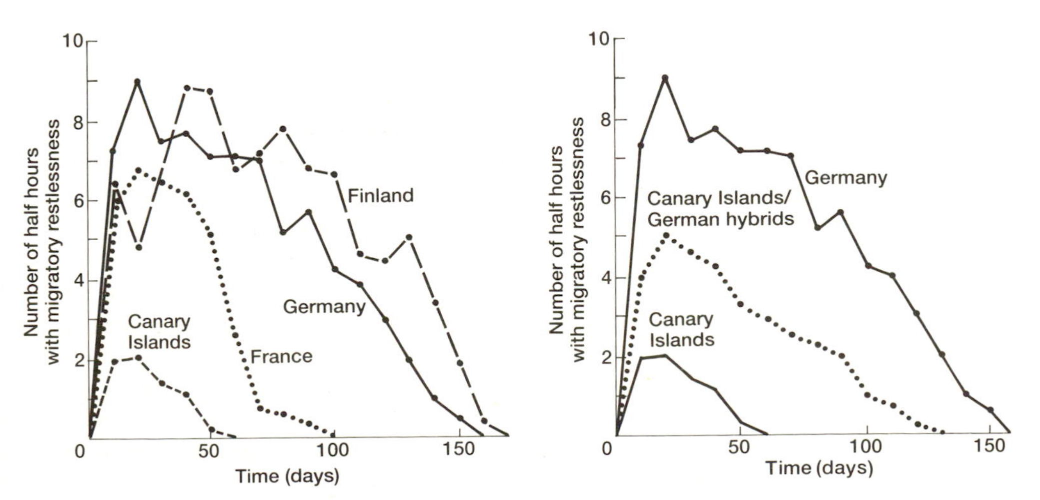

### *The combination of temporal and spatial programs: vector navigation*

A detailed study, again on Garden Warblers, has demonstrated that the endogenous time-program providing appropriate temporal orientation during the course of migration is closely linked to another endogenous program that can provide proper spatial orientation. Garden Warblers from central Europe leave their breeding grounds in a south-westerly direction heading towards the Iberian peninsula. Then, during the second half of their migratory journey, they have to shift their migratory direction from south-west to south to reach their central African quarters. Otherwise they would fly into the Atlantic Ocean, a lethal decision. When Garden Warblers were tested in orientation cages for migratory restlessness and preferred migratory direction during their first migratory period, they adopted a south-westerly direction during the first half of the migratory period. In the second half, they changed 'automatically' to a southern direction even though they still were sitting in central Europe. The most parsimonious interpretation of this finding is that the migratory direction can be preprogrammed and the birds have something like an 'internal turntable' for an appropriate shift in the preferred migratory direction.

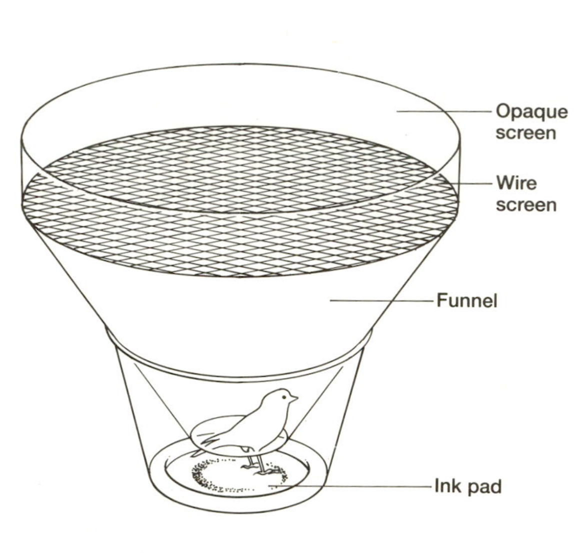

An Emlen funnel is used to study the orientation of migratory birds during migratory restlessness. The scratch marks made by the bird as it attempts to jump up the side of the funnel (above left) can be quantified and translated into vector diagrams (above right)

These findings can be summarized in the so-called 'vector-navigation hypothesis'. It postulates that in inexperienced first-year migrants, the onset of migration is normally set by the physiological 'internal calendar'. Migration then is performed according to inherited programs for the temporal and spatial course which lead the migrants 'automatically' to the hitherto unknown winter quarters. The internal program produces endogenous vectors, each composed of a time and a direction-schedule. Subsequently, it is open to modification by environmental factors. For example, there is evidence that the amount of fat reserves can influence the length of resting periods and unsuitable conditions can cause additional migratory activity.

Left. Migratory direction. The arrows indicate directional preferences shown by experimental Garden Warblers in southern Germany, during periods at which free-living warblers were passing through particular banding stations. Average passage dates are indicated and winter quarters shown

### *Decision making in partial migrants: solving the problem at the appropriate time*

Many migratory species are so-called partial migrants. Part of the population leaves the breeding ground and flies some distance to winter quarters and part stays within the home range and winters there. Two questions arise with respect to this strategy. How do individuals decide whether to leave or stay respectively and when do the migrants decide to leave?

After decades of speculation the problem has now been elucidated for a few European species, notably the Blackcap. Selective breeding experiments in which migrating individuals in captivity were bred together and non-migrants bred together showed that both behavioural traits rapidly increased in the respective offspring. These results indicate that the decisions are largely inherited. In other words, the decision whether to migrate or not and the time of migration are genetically preprogrammed.

The selection experiments also showed that behaviour can adapt rapidly to environmental changes. Imagine a partially migratory species during a series of severe winters. The residents and the early arriving migrants would be decimated. The late arrivals, on the other hand, will survive. As a result, their offspring will compose a higher proportion of the next generation. Selection experiments have further shown that partially migratory bird populations can change their behaviour from almost exclusive residency to almost complete migratoriness with early departure within just five to ten generations! Thus birds are able to adjust their migratory behaviour and the required temporal precision amazingly rapidly and effectively to varying environmental conditions. A good example for this is the European Blackbird, which a few centuries ago was exclusively migratory north of the Mediterranean. Benefiting from orchards, feeders, etc. most populations reduced their migratory activity on a genetic basis and some, as in Britain and central Germany, have become resident.

PB

## Preparation for migration and the flight itself

### *Fat deposition*

The fuel used by migrant birds is fat. This can be stored in an almost dry form and provides roughly twice as much energy per unit mass as any other biochemical fuel available. Fat is laid down chiefly in well-defined deposits under the skin, but also between the wishbone and around the liver and gut. Many small migrant species that undertake long flights may almost double their body mass before departure. Most of this increase is comprised of fat, but in several species flight muscles also increase in mass before migration. Part of this may be necessary to provide extra power to carry the large mass of fuel at the start of the flight, but in some Arctic-nesting species the additional muscle also serves as a protein store to be used in case of poor feeding conditions on arrival at the breeding grounds.

Although a general relationship exists between the mass of fat (expressed as a percentage of total body mass) that a bird lays down before migration and the distance it will travel to a refuelling site or its destination, it has not been clear until recently whether migrants carry any safety margin of fuel. Most of this uncertainty has arisen because of difficulties in predicting, from existing aerodynamic models of bird flight, the energetic costs of travel over particular distances. It has now proved possible to catch certain shorebirds, notably Red Knots, immediately before departure from their final migration staging posts and immediately after arrival on their Arctic breeding grounds. These birds had used during flight only about half the fat stored before departure. The remainder was available as an energy reserve for use on the Arctic breeding grounds to aid survival in bad weather or possibly to accelerate egg production. Indeed, certain geese species that nest in the Canadian Arctic lay down enough fat before migration to provide the constituents for egg formation and a reserve of energy to last them through incubation.

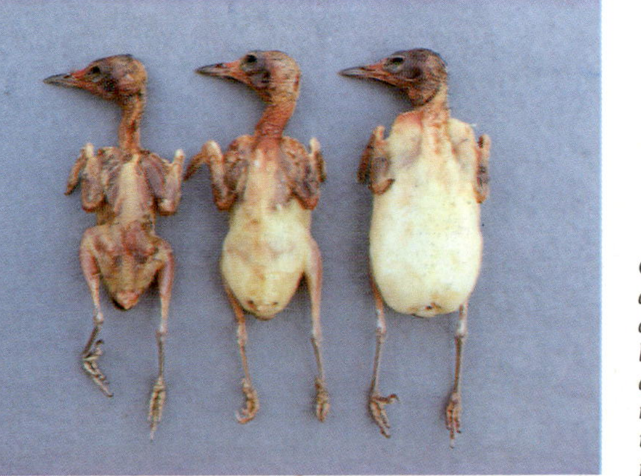

*Garden Warblers like many migrants, lay down large amounts of fat to serve as fuel during their journey. These dead, plucked birds were found in the warblers' winter quarters in Uganda, just prior to migration north. Notice the contrast in the amount of fat among the three individuals*

Fat reserves are laid down quite quickly, often in no more than two to three weeks before departure. Absolute rates of storage depend on the size of bird concerned. During a two-week stay at a refuelling site in north Norway, Red Knots gained about 70 g in mass (i.e. 5 g/day), most of it fat, from a starting mass of about 150 g. Migrant Sedge Warblers, less than a tenth of the size of Red Knots, have been shown capable of increasing in mass at 0.7 g/day. It seems likely that the rate of fat deposition is closely regulated and does not depend critically on food availability, for neither the rate nor the quantity of fat deposited by the Red Knots in Norway differed when recorded in successive springs which were amongst the coldest on record and much milder, respectively.

Because it is normally very difficult to capture migrants just before they depart, it is uncertain whether, in the wild, individuals of a particular species have to accumulate a minimum level of fat reserves before they depart. Recent studies of songbird migrants that have interrupted their flights, however, have shown that fat birds resume flight as soon as weather conditions permit, but that lean birds stay to refuel, provided that suitable foods are available. If not, they too move on. In species that migrate in flocks, for example Chaffinches, social behaviour may also cause lean birds to resume migration.

*Weather conditions on setting out*

Radar studies indicate that, early in the migration season, birds normally wait for very favourable conditions before they set out. Such conditions generally include good visibility and a following wind; these are often associated with anticyclonic weather or, in spring, the warm sector of an approaching depression. In autumn, small songbirds that fly over the Atlantic from Nova Scotia to the West Indies, or over the Gulf of Mexico, usually depart in the strong following winds behind cold fronts, even though these rarely persist until the birds make landfall. Later in the migration season, when birds are at risk in spring of failing to reach their breeding grounds by the most appropriate time for nesting, or, in autumn, of failing to reach migration staging posts or non-breeding areas before the best feeding sites have already been depleted, migrants are prepared to set out in less favourable weather. In those species in which juveniles and adults migrate at different times, young birds often depart under inappropriate conditions and, if migrating over the sea, may encounter changing wind conditions which, if they become adverse, blow them many hundreds of kilometres off-course. In this way, large numbers of birds on their first migrations from Scandinavia to Africa *via* Iberia, for example Pied Flycatchers, reach the eastern coast of Britain in some autumns.

*Altitude of flight*

Once aloft, migrants concentrate at those altitudes at which the following component of the wind is strongest. This indicates that they must explore a range of altitudes before making their choice. Such behaviour is most important when birds depart in changing weather conditions, for example in advance of a depression, in which wind direction and strength alter significantly with altitude. It is a commonplace observation that day-migrants are scarcely ever seen (except with powerful binoculars) if winds are following, because they fly high, but if they are flying into headwinds they fly low, because wind speeds near the ground are reduced by friction and so less unfavourable to progress. It is only in adverse winds that migrants follow topographical features such as mountain ridges, or, more particularly, sea coasts or lake shorelines, to any great extent. (Such 'leading-lines' may cause birds to deviate considerably from their normal directions of flight.)

Information on altitudes of migratory flights was scarce and probably severely biased before the advent of radar studies. Even now, quantitative information is limited. What there is suggests that most songbirds generally fly at altitudes of between 1000–2000 m over land at the start of their journeys. After several days of non-stop overwater flights, however, migrants travelling south from eastern Canada have been recorded at altitudes of up to 6000 m off the east coast of the southern USA. The species concerned are not known for certain but were probably shorebirds. They had probably flown a 'cruise–climb' strategy, flying higher as their fuel loads steadily lightened.

Although the height of flight above ground is influenced by wind conditions under some circumstances, another determining factor may be physiological. As fat is burned, water and carbon dioxide are produced and heat is released by the rapid actions of the wing muscles. To dissipate this heat, birds could have recourse to evaporation of water, but the amount of water produced by oxidation of each gram of fat is thought to be insufficient to compensate for the heat produced, so that migrants would dehydrate during flight. To circumvent this problem they rely on direct cooling in flight – the equivalent of an air-cooled engine – and to achieve this must fly sufficiently high to encounter low air temperatures. It appears they should choose altitudes at which the temperature is no higher than 10°C, to avoid dehydration in flight.

*Speed of flight*

The other aspect of flight behaviour on which radar has provided quantitative information is speed of flight over the ground. This is the outcome of the interaction between the migrant's airspeed (which determines the energy cost of the flight along its heading) and the speed and direction of the wind. Unfortunately, precise measurements of wind velocity at the height at which the migrant is travelling have rarely been made alongside records of the migrant's ground speed, so airspeeds have usually been recorded only during calm weather. Nevertheless, there are a few studies which indicate that a given migrant species reduces its airspeed somewhat when helped by a following wind and increases its airspeed to combat an opposing wind, altering the cost of flight accordingly.

Most radar equipment is unsuitable for determination of precise species of birds during migration. Consequently, only broad generalizations about speed of flight in calm air are possible. Small songbirds, for example European warblers, have airspeeds of about 20–25 mph (32–40 kph); larger songbirds about 30 mph (50 kph); and most shorebirds and ducks between 40–50 mph (64–80 kph). Relationships between airspeed and body mass have been predicted, such that a migrant should fly progressively more slowly as it uses up its fuel load, but it has not yet been possible to examine this in the field.

The mechanical power required to fly at different speeds can be calculated – several aerodynamic mathematical models of bird flight have been devised in the last twenty years. However, these models do not allow precise prediction of the rate at which chemical energy (derived from oxidation of fat) must be supplied to provide the requisite mechanical power and so cannot be used to estimate the rate at which birds use fat during migratory flights. Direct field measurements, however, are another possibility. Decreases in mass of fat during a flight of known length and duration can be determined, with difficulty, and provide an estimate of the costs of flight only at a certain *average* speed. For example, Red Knots travelling from the southern North Sea to northern Norway (1800 km) in two days, used about 60 g of fat during the flight. A more promising approach for the future is provided by measurements of oxygen consumption and carbon dioxide production during flight.

In spite of these difficulties, all models and measurements of the costs of flight agree that costs are minimal at a certain speed for a bird of particular body mass. Above the 'minimum power' speed there exists another speed, characteristic of the species and the individual's body mass, at which the distance that can be flown for a given fuel load is maximized. It is tempting to assume that migrants fly at this 'maximum range' speed, but this cannot be tested at the moment because of disagreements between different theoretical models and difficulties of making all the relevant measurements on the same individual migrant whilst it is in flight. For this reason, estimates of the potential flight ranges of migrants departing with different lipid indices (the proportion of total body mass made up by fat reserves) are not precise at present, although the *relative* flight ranges of different individuals can be deduced. In practice, field studies of distances flown between refuelling sites by several shorebird species suggest that many theoretical estimates of flight ranges are underestimates. For a Turnstone that travelled 3635 km from the Pribilof Islands to Hawaii in 3 days, the flight ranges (estimated from its 40% lipid index at departure) varied from 2750 to 4450 km.

Many species, particularly those undertaking long overwater flights, may have to migrate non-stop for several days and nights. It has been assumed that this applied also to small songbird migrants crossing large areas of inhospitable ground, for example the Sahara desert, but recent observations indicate that at least some migrants descend to spend the day sheltering in oases and fly only by night. This accords with radar observations in northern Europe, which indicate that most small insectivorous songbirds depart just after dusk but descend shortly before dawn, thus covering no more than 200–250 miles (320–400 km) in a single flight. Flight by night also allows the day to be used for feeding, avoids aerial predation, reduces potential problems of overheating by using the cold night air and may provide a better range of cues for accurate navigation. Migration by day is the norm for species that feed on flying insects, like swallows and martins, and for many seed-eating species such as finches and buntings that tend to move relatively short distances. The latter often fly for only a few hours after dawn, before stopping to feed if they encounter suitable habitats en route. Another important group of migrants that fly by day are soaring birds – birds of prey, cranes, storks and herons – that utilize thermals, up-currents of warm air that assist the birds to climb to great heights before they glide across-country for many miles to the next up-draught. This is an energetically inexpensive form of migration because most of the power needed to keep airborne is provided by the air currents themselves. However, because these are relatively few and far between, they are difficult for a single bird to locate. Many soaring birds therefore travel in flocks, which allow the search for thermals to be shared.

## Migration routes

Migration patterns are constantly being shaped by natural selection. This means that migration routes, staging posts and destinations are constantly being re-evaluated. The flocks of geese that head for the Canadian Arctic in spring use the prairie provinces on either side of the US/Canadian border as staging posts. The wheatfields that now provide their feeding grounds were scarcely there a century ago. On a longer time-scale, many European estuaries that now provide staging posts for migrant shorebirds were almost totally covered by ice until 7000 years ago and have been altered dramatically by natural processes ever since. Other habitats have been altered by human activities, particularly in the last few centuries, so that total dependence on a fixed migration route by a particular species could have been disastrous. Although the migration patterns we see today must have been influenced dramatically by the final retreat of the glaciers, this is not to say that all migrations originated then and that they had not been a phenomenon of earlier epochs in the existence of birds on the Earth.

Seasonality must have been present south of the limits of the ice-sheets even before the glaciers retreated, but migration routes for many species must now be much longer than they were before deserts existed in tropical latitudes. The extensions of some routes must have been subject to certain constraints which still exist today. For example, soaring birds concentrate their flights over the two ends of the Mediterranean – the Straits of Gibralter and the Bosphorus – because these are the shortest sea-crossings and thermals, upon which these birds depend, are normally absent over the sea.

Although many songbirds are seen by radar to travel on a broad front over whole landmasses, particularly at night, the routes followed by many shorebirds and wildfowl are much narrower, at least during certain parts of their journeys. Inevitably, those species that are widely scattered while breeding over large geographical areas (e.g. many ducks) do not follow tight flyways immediately upon departure, whereas those that nest in large concentrations, like geese, cranes or herons may well do so. Three distinct wildfowl flyways are recognized in North America, along the east and west coasts and following the

Mississippi/Missouri river, and one in north-west Europe, from Russia to the British Isles; but not all species follow them rigidly. Similarly, although many waders from western Siberia and European Russia pass through the Baltic and along the North Sea coasts on their way to north African wintering grounds, some species fly overland on a broad front, eventually crossing the Sahara. Many northern European and Russian populations of insectivorous songbirds pass westwards into Iberia before refuelling and then heading southwards into west Africa; but a few pass south through Italy or the eastern Mediterranean countries. Thus it is difficult to make reliable generalizations about routes, just as it is about the destinations of different populations of a single species. The winter quarters of swallows breeding in different European countries tend to be fairly well segregated, whereas Yellow Wagtails from breeding populations throughout Europe, from northern Scandinavia to southern Spain, 'winter' alongside each other in Zaire.

The total distances flown by individual birds during the journey between their breeding and non-breeding areas can be spectacular. An oft-quoted example is the Arctic Tern which covers more than 10 000 km, between nesting areas in high northern latitudes and the Antarctic pack ice, twice each year. Since individual terns may well live to be ten years of age or more, each must cover more than 200 000 km in its lifetime. Many seabirds cover very long distances each year, for example various species of shearwaters, but these can rest on the water if necessary. It is claimed that Sooty Terns, however, cannot do so because their plumage becomes waterlogged if they alight. Some of the longest non-stop flights by land birds are undertaken by migrant shorebirds, for example by Hudsonian Godwits from Canada to the southern coasts of South America. During this flight, they are believed to travel over the western Atlantic Ocean, a distance of some 4500 km and a probable journey time of about three days.

Many of the longest non-stop flights that involve land birds follow a broadly north–south path, but a few wildfowl and wader species travel in a series of flights, chiefly east–west (up to 5000 km between breeding grounds in Siberia and wintering sites in western Europe) or south-west–north-east (between Siberia and South Africa, a total of some 13 000 km) and some songbirds travel south-east–north-west (up to 4000 km, between central Europe and Asia). For these, there is considerable evidence that their migrations may approximate to Great Circle routes – the shortest distance between two points on the earth's surface. To fly such a course accurately requires a continuous change of compass direction, but it is not known whether this is the mechanism used in practice.

As explained earlier, experimental studies indicate that the main means by which young birds find suitable non-breeding areas for the first time is to follow a pre-programmed compass direction, characteristic of the species and population involved. There is, however, considerable variation between individual juveniles in directions followed. In later years of life, adults show much greater unanimity, perhaps indicative of heavy selection against the most way-out juveniles. In some species adults and juveniles may travel by different routes to the same non-breeding area, for example, Dunlin travel from Arctic Russia to eastern Scotland via Norway (juveniles) or the southern coast of the Baltic (adults). This occurs particularly if adults from a wide variety of breeding areas congregate to moult in a few sites before moving on to their winter quarters, as in Dunlin, and several other shorebird species, which seem to require moulting sites that are large, open, productive intertidal flats like the Wadden Sea, which provide ample food yet give safety from predation.

Both adults and birds returning from their winter quarters for the first time may use routes in spring that differ from those followed in the previous autumn. The reasons for this are various. Migrants returning northwards from west Africa in spring would encounter headwinds if they retraced their autumn route through Iberia; but a more easterly route through Tunisia allows species such as Curlew Sandpipers heading for western Siberia or Pied Flycatchers heading for northern Europe to find a following wind in a narrow altitude band.

Species dependent upon inland wetlands have a much greater choice in spring than autumn in the Mediterranean and north African countries. And food resources may be present in only one migration season at a particular place, as, for example, the eggs of horseshoe crabs that congregate in Delaware Bay on the east coast of the USA in May at the time that shorebirds are passing northwards.

## Duration of migration

Autumn migration appears to be much more leisurely than spring passage, but this is in part an illusion. The migration season in western Europe extends from mid-July almost until the year's end with the earliest birds to return from high latitudes being those that failed to breed successfully. The last are often those juveniles that were reared late and perhaps had difficulty in finding enough food to lay down fat for the migratory flight before severe weather arrived on the breeding grounds. Another feature that extends the autumn migration season is that some species moult into non-breeding plumage, including renewing their flight feathers which takes a minimum of five weeks, before they leave; others pause to moult at an intermediate staging post; and yet others delay moult until they reach their final destinations. Such variation may also occur amongst populations of the same species breeding in different geographical areas, for example the Dunlin. Once migration is underway, the time-course of movement of an individual bird is probably no less rapid in autumn than in spring (allowing for pauses related to moult). Amongst seabirds, Manx Shearwaters, for example, reach their winter quarters off the eastern coast of South America less than two weeks after leaving their Welsh breeding colonies. For land birds, there may be advantages in reaching staging posts or non-breeding areas early if feeding territories must be established there. Juvenile birds often occur in a variety of atypical habitats during autumn migration, and are consequently much more widespread than adults; but the lengths of each stage of their migratory flights may not be significantly less than those of adults.

The return passage in spring occurs during a shorter time-period than in autumn, particularly for individual species, since the period during which successful breeding can occur is short, particularly at higher latitudes, and migrants must ensure that they arrive at or before the optimal time. Particularly in the high Arctic, there are penalties for arriving too early – shortage of food and severe weather – so migration schedules must be tightly controlled. Further south, in temperate latitudes, migrants may move back to their breeding areas by a series of progressively shortening flight stages, each timed by the arrival of improvements in the weather, particularly a rise in temperature. Thus the arrival schedules of these species at nesting areas may vary more from year to year than those of higher latitude species.

## Navigation

The capabilities of migrants to undertake non-stop flights of several thousand kilometres, mentioned earlier, are remarkable enough. What is even more impressive is the ability of individual long-distance migrants to return to the same nesting areas, and in some cases also 'wintering' sites and migration staging posts, year after year. Mutton Birds, Australian shearwaters, return to their nesting islands in the Bass Strait after overwater migrations that take them through the north Pacific, almost to Alaska and back, probably with no sight of land. European Swallows often return not only to the same nesting sites in successive years but also to the same parts of southern Africa in the northern winter. The capability of accurate homing by individual migrants has now been shown in a wide variety of songbird, shorebird and seabird species as a result of ringing (banding) studies. For example, Little Stints from Siberian breeding grounds have returned to the same beach in Cape Province, South Africa year after year after journeys totalling more than 10 000 km each way. More exacting, perhaps, are the navigational feats of land birds travelling long distances over featureless seas, for example the Shining Cuckoo of New Zealand that migrates over 3000 km northwards to the Solomon Islands. PRE

## BIRD OBSERVATORIES AND RINGING

The first bird observatory was established on the German island of Heligoland in the mid-nineteenth century, by Heinrich Gätke. The work he carried out formed a model for subsequent studies of migration in Europe; that is, the systematic daily recording of grounded migrants in a defined area and the observation of visible movements. Following the introduction of large-scale bird-ringing (banding) in Denmark by Mortensen (1899), this technique was incorporated into bird observatory work at Rossitten (then in Germany) and also at Heligoland, where large cage-traps were used to catch migrants for marking and release. Such 'Heligoland traps' were built at other bird observatories around the Baltic coasts and on islands off the British Isles during the 1930s to 1960s. Many British bird observatories were established at places first visited by William Eagle Clarke and others at the turn of the century, for example Fair Isle, the Isle of May and Skokholm. Subsequently, observatories have been established in many other parts of the world. Ringing and recording of visible migration of songbirds and seabirds at such sites are often organized now by a resident warden and still provide valuable data for the study of migratory movements of different species. However, with the introduction of more mobile forms of bird-catching equipment, such as mist-nets, the contribution of observatories to the annual totals of birds ringed has declined in percentage terms, though not in absolute numbers.

*Ornithologists ring large numbers of birds like this European Cuckoo (Cuculus canorus) each year. The subsequent recovery of ringed birds, whether alive or dead, provides invaluable information on migration and survival*

*Far left. The bright beams of a lighthouse have a powerful attraction for migrating birds, especially during misty conditions around the new moon. Unfortunately, many birds die as a result of flying against the glass of the lantern*

*Left. Observatories like Southeast Farallon Island Field Station, part of the Point Reyes Bird Observatory, California, play a crucial role in the study of migration by ringing large numbers of birds*

Another important role of bird observatories, particularly as practised under successive wardens at Fair Isle, has been to stimulate interest in detailed descriptions of bird plumages, measurements and morphology, as seen in live birds rather than museum specimens. Additionally, by weighing birds caught for ringing and release, information on rates of gain of body mass (and on minimum fat levels before departure) has been collected systematically. A question mark remained over the representativeness of such data, however, until radar studies indicated which bird observatories lay in the path of normal broad-front migrations and which sites probably attracted chiefly lost or atypical migrants. Undoubtedly, many observatory sites were selected originally because of their reputations as 'magnets' for rare birds, rather than for their locations in relation to migration flyways, but most British ones have proved to be sited appropriately on routes between Iceland or Scandinavia and south-west Europe.

PRE

## IRRUPTIONS

At irregular intervals of several years, a variety of bird species that nest in sub-arctic and northern boreal regions irrupt from their breeding areas and invade the northern temperate zones. Amongst the best-known of these irruptive species are the waxwings, in both Europe and North America, as their tameness and striking plumage brings them to the attention of a wide cross-section of the general public. Irruptive species rely on a few types of food, usually tree seeds and fruits or small rodents, that fluctuate in abundance markedly from year to year and/or place to place. The birds involved, chiefly finches and some owls and raptors, probably begin migration every autumn but move variable distances from year to year, stopping as soon as they encounter abundant food. This leads them to winter in different areas in different years. In spring, most begin a return movement but again may not move far if they encounter good food resources, with the result that they often breed in different sites in different years.

*This Nutcracker (Nucifraga caryocactes) from Siberia has turned up in a garden in Suffolk, eastern England. When their main food source, pine seeds, fails in the northern USSR, Nutcrackers irrupt southwards, often in large numbers*

The directions of movement involved in irruptions of a particular species, especially amongst the finches, are much more diverse than those found in a regular migrant species; and the navigational mechanisms involved (if any) are unknown. Young birds often predominate in irruptions of seed-eaters, and females usually outnumber males. These features point to successful breeding as a forerunner to large-scale emigration, perhaps with a greater proportion of young than adults leaving any one area after the breeding season. However, food shortage alone may stimulate emigration, as happens in the invasions of western Europe by Rough-legged Buzzards in years when their chief prey, voles and lemmings, become very scarce.

An exception to the general rule of autumn irruptions are the crossbills, which feed on conifer seeds. These finches breed very early in the year, since the cones of most of their food-trees open in mid-winter. As a result, if the population of birds is high but food supplies poor, emigration may start as early as midsummer. Birds encountering good cone crops in autumn may stay to breed in areas well beyond their normal breeding range. Interestingly, at least some crossbills lay down fat stores for their flights, which may take them several thousand kilometres during an irruption (see Chapter 10).

PRE

## RADAR

Radar, as applied to bird migration studies, is an immensely useful technique which permits migrants to be followed by night as well as day, and at altitudes above the range of normal eyesight, even when aided by binoculars. The equipment available for Radio Detection And Ranging (RADAR) enables single birds to be followed over short distances, say up to 10 km, and flocks to distances of over 100 km, by which distances the focussing of the radio beam is not fine enough to resolve individual birds flying relatively close together in a flock. Birds reflect a radio beam chiefly because they contain water. The strength of the 'echo' depends on the cross-sectional area presented by the bird, and so varies with the angle between the direction of the beam and the bird's flight direction. Often a radar beam will be set either to scan a range of altitudes in one particular direction or to scan progressively around all points of the compass (without distinguishing echoes from different altitudes). The latter method produces an image of migration over a large area of land, as though viewed from a satellite, and enables the track directions flown by individual migrants or flocks to be measured. From these, if wind speed and direction are known at the height at which the migrants are flying, the birds' headings can be calculated.

Radar has been used to investigate several distinct problems related to migration. These include:

1. The weather conditions under which most migration occurs in spring and autumn.
2. The orientation behaviour of migrants in flight – their reactions to changes in wind velocity, topographical features and cloud.
3. The main routes taken by different groups of migrants, such as thrushes, waders or small songbirds (distinguished chiefly by flight speed).
4. The altitude of migration in different parts of the world – over the oceans, near mountain ranges such as the Swiss Alps and over lowland areas of Europe and North America.

PRE

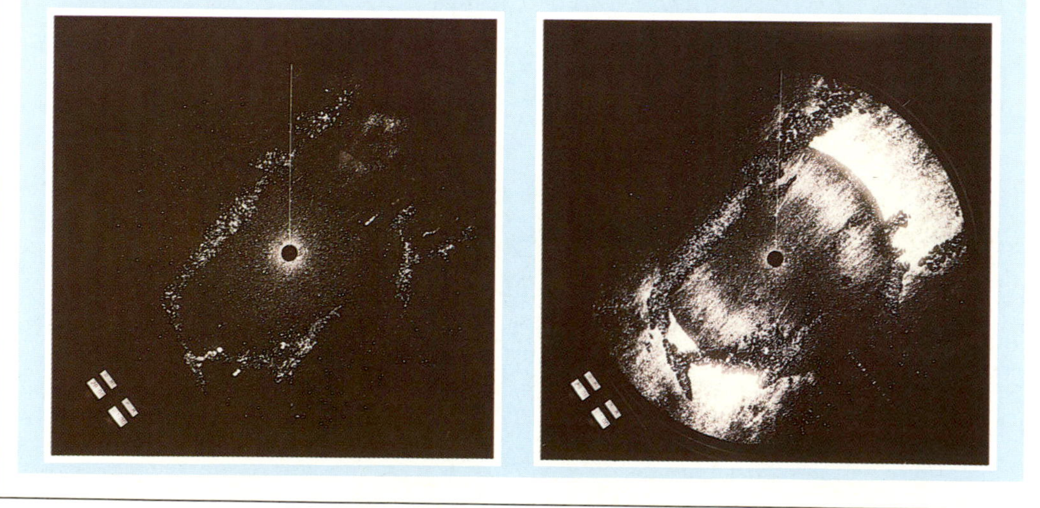

*Radar is invaluable in the study of bird migration. These two pictures are of the radar screen at Geneva airport at 2300 h on two consecutive nights in late August. The first screen shows little migratory activity (the white points are mountains, screen coverage 60 km); the second shows, as short white lines, massive numbers of migrating birds flying south-west*

## Navigation

The French Revolution of 1789 had many consequences. One of the most unlikely was evidence that a bird could return to its nest site after spending six or more months away. A nobleman, in hiding from the mob, placed a copper ring on the leg of one of a pair of Barn Swallows nesting in his château in Lorraine. This simple experiment revealed that the same bird returned to the same nest site in three consecutive years.

That birds could and would return to some fixed home base after being elsewhere had been known long before the French Revolution.

Centuries earlier, the Romans had used swallows to carry the winning colours of chariot races. On the other side of the world, Pacific Islanders used tamed frigatebirds for communication between islands. No other bird, however, has proved more suited to the carrying of messages than the pigeon. Descended from the Rock Dove, the species has been used as a 'pigeon-post' since the days of the ancient Egyptians. Later, it graduated from carrier to racer and, in 1818, the first pigeon race over 100 miles was held in Belgium. Nowadays, in continental North America, pigeons often race over distances of 1000 kilometres or more. Yet even these impressive feats of homing are dwarfed by those of long-distance migrants.

Slowly, during the nineteenth and early twentieth centuries, it dawned on ornithologists that migratory birds must possess some sophisticated ability for finding their way. However, the awe-inspiring scale of their navigational prowess only really became apparent when ringing (banding) programmes showed that some birds, such as swallows, may travel thousands of kilometres during the course of their annual migrations. Having finally accepted the magnitude of the phenomenon, ornithologists, in the 1940s and 1950s, began experiments to study how navigation (i.e. the act of finding one's way to a familiar destination over unfamiliar terrain) was achieved.

Early experiments were largely preoccupied with the discovery of some discrete, specialized and rather mysterious 'navigational sense'; a search that actually dragged on until the late 1970s. Yet, as long ago as 1955, the path to the modern view had been initiated by the German ornithologist, Gustav Kramer. He argued that bird navigation involved just two steps: first the bird referred to a *map* to determine the direction of home, then used a *compass* to set off in that direction. The following sections discuss the senses and mechanisms by which birds come to possess such maps and compasses, and how these may then be used. Of the two, much more is known about avian compasses.

*Far left. The remarkable navigation and homing abilities of pigeons have meant that from ancient times they have been used by man to carry messages. Pigeons like this were used extensively during the Second World War*

*Although pigeons are rarely used to carry messages today, pigeon racing remains extremely popular internationally, as the sheer number of birds being released in this race shows*

## Compasses

In everyday language, a compass is an instrument that indicates north, south, east and west, etc. The most familiar is the magnetic compass; a magnetized needle free to align itself with the lines of force of the Earth's magnetic field. However, there are others, and biologists often refer to a sun, star or even a wind compass. All of these sources are capable, given sufficient experience and knowledge on the part of the user, of showing the direction of north.

*Investigating the navigational abilities of birds requires sophisticated equipment. Here an orientation cage is surrounded by a Helmholtz coil which has the effect of shifting true magnetic north to the southeast. The ingeniously simple Emlen funnel provides information on the direction and amount of migratory restlessness. An ink pad at the bottom means that as a bird responds to the star patterns above, it leaves visible foot marks on the sides of the funnel*

### *Magnetic compass*

The fact that birds have a magnetic compass was discovered as a result of an incredibly fortunate quirk of bird behaviour first noticed by Gustav Kramer in the early 1950s. As already described, caged individuals of migratory birds show a 'migratory restlessness' at the time of year they would normally begin to migrate.

Special cages, known as orientation cages, have been designed to take advantage of this propensity of caged birds to show preferred directions of orientation during migratory restlessness. The primary function of such cages is to record automatically how often the bird visits the different parts of the cage, thus allowing the experimenter to calculate the bird's preferred direction. It is possible, therefore, to study a bird's orientation mechanism while the bird is confined in a small space on the ground rather than flying over long distances hundreds of metres up in the air. Now, the experimenter can alter the information available to the bird to find just how it is managing to recognize direction.

The first indication that migrant birds might be able to 'read' the Earth's magnetic field came with an observation in the late 1950s by two Germans, working at Frankfurt. They noticed that caged European Robins in autumn could continue to orient to the south-west during nocturnal migratory restlessness, even when the cage was in a dark room with no view of the night sky. Merkel, and his student, Wolfgang Wiltschko, then took the crucial step of surrounding the orientation cage with electromagnetic coils, thus causing the magnetic field passing through the cage to be changed. With magnetic north now altered to be in some direction other than true geomagnetic north, the birds changed their direction of orientation accordingly. This first experiment has since been repeated many times on many species by various workers in Europe and the United States. By the early 1980s, all but a handful of sceptics were convinced that birds possessed some specialized sense organ, a 'magnetoreceptor', that allowed them to read compass directions from the Earth's magnetic field.

Small magnetic coils on the heads of pigeons can influence their sense of direction. In consequence, it is generally assumed that the magnetoreceptor lies somewhere in the bird's head. Neurophysiologists, again at Frankfurt, have found that when the magnetic field around the head of a pigeon is changed, nerves leading from the back of the eye fire in a way that could mean they are connected to the magnetoreceptor. This, along with the observation that pigeons seem to have difficulty reading the Earth's magnetic field in total darkness, has led some people to think that the magnetoreceptor is sited at the back of the eye and needs energy from light in order to function.

A more popular suggestion, but with even less supporting evidence, is the magnetite hypothesis. Magnetite is a magnetic iron oxide, also known as lodestone. The hypothesis suggests that, somewhere in the head (perhaps in the sinus region), lies specialized tissue containing large numbers of tiny particles of magnetite between which ramify nerves. Whenever the bird turns its head, the array of tiny magnets moves through the Earth's magnetic field. All manner of electrical, magnetic and even pressure changes take place around the particles; changes that trigger the ramifying nerves in a pattern that depends on the direction the bird is facing. Nerve impulses then travel from the magnetoreceptor to the brain where the pattern is decoded into compass directions.

It may be a few years yet before we know which, if either, of these two suggestions is correct. For the moment, however, we have to be content with the simple knowledge that birds can, and do, judge compass direction by sensing the Earth's magnetic field.

It has not yet been proved beyond doubt that birds have a magnetic compass sense the moment they crawl out of the egg. However, when only two days old, chicks of the Ring-billed Gull in the north-eastern United States show what seems to be magnetic compass orientation to the south-east, the direction in which eventually they will migrate. Certainly, as soon as Homing Pigeons are old enough to be taken out on homing experiments, it is clear that they have a magnetic compass. It seems that a bird's sense of magnetism provides its basic compass sense. The bird then uses this basic compass to learn other, more sophisticated, ways of judging direction.

*Sun compass*

As a source of compass information, the sun has two tremendous advantages: it is conspicuous and predictable. Unfortunately, it is only conspicuous if not covered by cloud and is only predictable if the observer not only has a good sense of time but also has used this sense to learn those features that impart predictability.

All over the earth (excluding the Arctic and Antarctic), the sun rises in the morning over the eastern horizon and sets below the western horizon. At mid-day, north of the Tropic of Cancer, the sun is always due south; south of the Tropic of Capricorn, it is always due north. At mid-day between the tropics, the sun is sometimes due north, sometimes due south, and on two days a year is exactly overhead. Nothing is known about how, or whether, birds in the tropics learn this rather complex pattern. It is known, however, that birds can learn the simpler pattern found in temperate regions.

The fact that the compass bearing (or 'azimuth') of the sun changes during the day allows a neat test of whether temperate birds can use a sun compass. The experimental method is known as 'clock-shifting' and involves, in effect, changing by a few hours a bird's natural, internal, clock. To do this, an experimenter keeps a bird for a few weeks in a windowless aviary. First, the bird's existing clock is confused by giving it continuous light or dark, or short, variable-length, dark/light sequences. Next, the light in the room is switched on and off on a 24 hour cycle, but out of phase with the real day/night cycle outside. A common regime is, each day, to switch the light on and off six hours before the real sun rises and sets. 'Bird time', the time of day registered by the clock inside the bird, thus runs six hours ahead of 'real' time outside.

If clock-shifting has no effect on a bird's sense of direction, the bird is not using a time-compensated sun compass. On the other hand, if advancing a bird's clock by six hours causes it to misjudge compass direction by 90° (anti-clockwise), then that is good evidence the bird is using a time-compensated sun compass. Similarly, retarding the clock by six hours should cause the bird to misjudge compass direction by 90° (clockwise). Advancing or retarding the bird's clock by 12 hours should cause the bird to misjudge direction by 180°.

Training and clock-shifting experiments on caged birds show that European starlings, meadowlarks, Homing Pigeons and others all behave as if they are using a sun compass. The important question, however, is not 'Can caged birds learn to use a sun compass?' but 'Do they use a sun compass during navigation?' This idea has been investigated in navigation experiments on Homing Pigeons.

The experiment was straightforward. The pigeons to be tested were divided into two groups. For a few weeks before the experiment, half the birds (the controls) were kept in a room in which the light was switched on and off as

*A family of Bewick's Swans (Cygnus columbianus) arrive at Slimbridge, western England, having migrated together from their breeding grounds in the northern USSR. Relatively few birds migrate in family groups like this*

the real sun rose and set; the other half (the experimentals) were clock-shifted by six hours. On a sunny day, the two groups of birds were then driven away and released. The control birds flew off towards home; the experimentals flew off in a direction that was in error by 90°. Evidently, when the sun shines, Homing Pigeons use a sun compass to help them find their way home.

In the laboratory, Homing Pigeons can judge direction with an accuracy of 3–5°; or, at least, they can if held steady on a turntable and tested against an artificial sun. Moreover, they seem to achieve this level of accuracy, not by observing the sun itself, but by watching the shadows cast by the sun. Whether free-flying birds use the same method and are as accurate in their use of the real sun is not known.

Laboratory tests also suggest that birds should be able to see the sun through thicker clouds than humans. The sun gives off a wide spectrum of wavelengths of light, some of which humans can see, some of which they cannot. In particular, humans cannot see ultra-violet light (wavelength about 350 nanometres) but some birds, for example hummingbirds and Homing Pigeons, can. Such ultra-violet light penetrates cloud more than longer wavelengths. Hence, birds should be able to use their sun compass on more days than if they could not see ultra-violet light.

Although birds are able to judge compass direction from the sun, they are not born with this ability; it has to be learned. To do this, Homing Pigeons need to observe the sun move across the sky and to relate this movement to some other fixed reference system. Moreover, they need to observe the sun throughout the day, it is not enough just to see the sun in the morning or afternoon; they cannot extrapolate from one time of day to another.

Evidence from Homing Pigeons shows that the fixed reference system used to calibrate the sun's movements is the Earth's magnetic field. In other words, pigeons use their magnetic compass to set up their sun compass. Yet, once they have learned how the sun and magnetic field relate to one another, experienced pigeons take more notice of the sun than the magnetic field. However, even on sunny days, pigeons do not abandon their magnetic compass and, on overcast days, the Earth's magnetic field continues to be their main source of compass information. In addition, pigeons do, from time to time, recalibrate their sun compass against their magnetic compass, perhaps to keep track of the way the sun's position changes with the seasons.

*A polarized light compass?*

As long as cloud is broken with patches of blue sky, birds may not need to see the sun's disc in order to work out compass direction. Due to atmospheric scattering, the light from blue sky is linearly polarized. Moreover, the plane of polarization is closely related to the position of the sun and conveys compass direction in much the same way. Birds can detect these polarization patterns. Whether they use the patterns as a compass in their own right, or simply to work out the location of the sun's disk when hidden behind thick clouds in some other part of the sky, is not known. As long as there is some break in the cloud, however, birds should be able to judge compass direction from the sky.

Polarization patterns are particularly strong and useful during twilight, when the sun is below the horizon but the sky is still light. Adult birds of species that migrate at night seem to make particular use of these patterns in checking their night-time compasses.

*Night-time compasses: stars and moon*

The discovery that birds could orient by the stars was made by the German ornithologists, E. G. and E. M. Sauer in the 1950s. However, it was not until the more rigorous work of Stephen Emlen in the United States in the late 1960s and Wolfgang and Roswitha Wiltschko at Frankfurt University in the 1970s that the possibility gained general acceptance and the way in which stars were used began to be understood.

Basic experiments showed that migrant birds in orientation cages could orient at night in the direction appropriate to the season, both under natural starlit skies and in a planetarium. Moreover, when the planetarium sky was rotated through 180°, the birds changed their orientation accordingly.

Stephen Emlen hand-reared Indigo Buntings in a planetarium under a facsimile of the natural starlit sky. A month or so after fledging, such birds were able to orient appropriately during migratory restlessness, but only if, during their first few weeks of life, the planetarium sky had been made to revolve like the natural sky. Buntings raised under a stationary sky never developed a star compass.

Most people know little about stars, and few probably realize that the night sky appears to rotate. Yet the stars actually behave like a billion suns, all apparently fixed in their positions relative to each other, like luminous dots stuck on the dome (known as the celestial sphere) of the night sky. Due to the Earth's rotation about its axis once every 24 hours, this sphere, like some huge planetarium dome, appears also to rotate about an axis once every 24 hours. Fortuitously, the axis of rotation of the northern hemisphere of the celestial sphere is marked by a relatively conspicuous star. This is the Pole Star, Polaris. Unfortunately, the axis of rotation of the southern hemisphere of the celestial sphere is not so marked, its position being located in a barren area of dark sky.

Evidently, rotation of the celestial sphere about its axis was important to Emlen's Indigo Buntings. Without such rotation, they were unable to judge direction by the stars; with it, they were. In a series of elegant experiments, Emlen eventually showed that, during their first few weeks of life, his birds were observing the rotation of the sky and noting which part of the sky, the axis of rotation, did not move. They also learned which star marked the axis of rotation and how to locate and recognize that star from the pattern of surrounding stars. Different individuals used different sections of the surrounding stars to locate the Pole Star. Having learned how to locate the axis from the pattern of surrounding stars, rather than from the movement of the sky, Emlen's buntings no longer needed the sky to rotate. From that stage onwards, a stationary sky was as meaningful to the buntings as a sky that rotated.

As the axis of orientation of the celestial sphere is fixed, day and night, a sense of time is unnecessary to use stars for compass orientation. At night, therefore, clock-shifting has no effect on birds' judgement of direction; as long, that is, as the moon is not shining. If the moon is shining, clock-shifting experiments on Mallard are as effective as those experiments on birds judging direction from the sun. It seems that, if the moon is shining at night, birds may use it in much the way that they use the sun. However, if the moon is beneath the horizon at night, birds use the axis of rotation of the celestial sphere.

In the same way that Homing Pigeons use their basic magnetic compass to calibrate and recalibrate their sun compass, so some European passerines, such as Garden Warblers and Robins, seem to use their magnetic compass to locate the axis of rotation of the night sky. Moreover, every night (in the case of Garden Warblers), or every three or so nights (in the case of robins) the birds recalibrate their star compass against their magnetic compass. This recalibration is necessary during migration for star patterns change as birds migrate south or north.

In contrast, Emlen's work on Indigo Buntings suggests that this species, even though it is known to have a magnetic compass, may be able to develop a star compass without any reference to the Earth's magnetic field, solely through observation of the rotation of the night sky during the first few weeks of life.

Unlike the Indigo Bunting of America, the compatriot White-throated Sparrow is more like European passerines in that it calibrates its star compass against its magnetic compass when young. As it ages, however, it seems to rely more and more for this calibration on twilight polarization patterns as the sun sets, rather than magnetism.

The final picture, therefore, is that birds are born with a magnetic compass. This they then use during their first few weeks and months of life to learn how to use other compass reference systems based on the sun, twilight polarization patterns, stars and moon. The magnetic field appears indispensable to most young birds but, as they age, they become more reliant on celestial sources of compass information. Even then, the magnetic compass is used when celestial compasses cannot be seen and, from time to time, for their recalibration.

Compasses alone, however, even as efficient and integrated an array as this, are not enough for birds to recognize their location and find their way around. For this, they need a map.

## Mental maps

One of the early suggestions to emerge from the work of ornithologists in the 1940s and 1950s was that perhaps birds have some form of inbuilt 'universal' map that allows them to recognize their position from anywhere on the Earth's surface. The favourite version pictured a bicoordinate grid, much like a system of latitude and longitude. In support of this view, experiments are often cited in which birds have homed after being displaced to locations that it was most unlikely they had ever visited previously. A classic example is the Manx Shearwater that returned to Skokholm Island off the Welsh coast after being taken to Venice.

An embellishment of this idea of an inbuilt universal map was that migrant birds could be born with the coordinates of breeding and wintering grounds, etc. encoded into their genetic material. Their migratory lives thus become a succession of navigational episodes as they attempt to find their way from one encoded location to another.

A few ornithologists are still seeking evidence for the existence of universal maps. However, as emerges later, there are simpler explanations for the navigational feat of the Manx Shearwater from Skokholm and, so far, no critical experiment has yet provided the necessary supporting evidence. Moreover, young naive migrants, when displaced from their migration route or wintering ground, do not navigate back to these traditional areas but instead readily adopt the new routes and areas to which they are displaced. Most people, therefore, reject the idea of an inbuilt map in favour of one based on the bird's own experience, details of which are stored in the head.

The maps carried in a bird's head, stored probably in the hippocampus of the brain, are usually known as 'mental' or 'cognitive' maps to contrast them with the printed maps used by people. One simple experiment with clock-shifted Homing Pigeons tells us a great deal

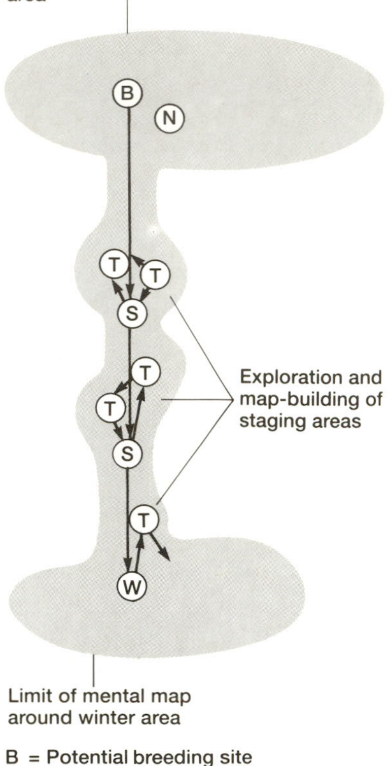


*A young bird of a migratory species builds up its mental map by exploration and navigation in the manner shown above. The mechanisms it uses are the same as those available to a Homing Pigeon*

about the form of a bird's mental map and the way it is used. Perhaps surprisingly, this experiment shows an important role for the bird's array of compasses.

When a clock-shifted Homing Pigeon is released on a sunny day just a kilometre from its loft its behaviour depends on whether or not the loft is visible. If, on the bird's release, it can see the loft, the pigeon behaves as expected and flies directly home. On the other hand, if the loft cannot be seen, then even though the bird has flown home from the release site many times before, when clock-shifted by six hours the bird flies off on a bearing in error by 90°. The only reasonable interpretation is as follows. First, the bird checks its surroundings against its mental map and correctly recognizes its location. Then, instead of saying something like 'to fly home from here I must fly towards that clump of trees' the bird says 'to fly home from here I must fly south'. However, as, in this experiment, the pigeon's sun compass has been confused through clock-shifting, it misjudges the direction of south and flies off in the wrong direction.

This experiment tells us a great deal about the way the pigeon's mental map must be organized. First, it must consist of a mosaic of familiar landmarks that allow the bird to recognize its location. It is a mosaic because, instead of linking these major landmarks together with yet more detailed landmarks, the bird uses compass directions. In effect, reference to the mental map must generate a series of instructions such as: 'to go from the wood shaped like an egg to the pointed mountain, fly south; to go from the pointed mountain back to the loft, fly east'; and so on. Such a map is very adaptable, very economic, and most importantly would not require the bird to waste time and energy in 'steeplechasing' from one major landmark to another. Use of simple surveying principles would allow the bird to work out novel compass bearings, such as 'to fly back to the loft from a point half-way between the wood and the pointed mountain, fly east–south-east'. This interplay between map and compass would allow the bird to find its way around large areas of countryside with economy of both movement and memory. Each area of, say, 100 km² would require the bird only to memorize two or so major landmarks and the compass bearings that link them.

We have already seen the compass information a bird may use. The next question is: what makes a landmark and how might it be sensed?

### *Visual landmarks*

Humans have a mental map based largely on visual landmarks. Hills, mountains, woods, individual trees, rivers, cities, buildings, rock formations, each with their own individual characteristics, provide the features by which we identify where we are. It seems reasonable to assume that the mental map of birds will be similar. Firm evidence, however, is not easily obtained.

*To navigate, migrating birds use olfactory, auditory and visual landmarks – like those around this estuary in Wales – in conjunction with their mental map. The mudflats around estuaries are important feeding stations between the Arctic and southerly wintering stations*

Returning Homing Pigeons clearly adjust their flight direction as familiar buildings around their loft come into view. There is also the anecdote concerning an American pigeon fancier with a loft in the middle of a large forest. His birds seemed reluctant to explore far from the loft, performed badly in races, and his stock suffered heavy losses through birds never returning. The problem seemed to be that his pigeons were unable to recognize when to break away from the racing flocks flying along the nearby coast. Eventually, the fancier erected a tower above the trees to support a huge golden ball. There was an immediate improvement. Young birds explored further and racing birds homed faster and more often, breaking away from returning flocks at the correct time.

By day, migrating birds fly above and along leading lines such as rivers, coasts, hills and mountain ranges. At night, radar observation reveals less clear-cut a pattern, but still with signs that the birds are aware of the inky landscape beneath them.

Thus, although there has been little by way of formal experiment, it seems fairly clear that, by day and night, birds do observe and use visual landmarks. In which case, it seems likely that these form at least a part of their mental map. Most interest, however, has centred on other types of landmark.

### *Smell landmarks*

In a classic experiment in the early 1970s, Floriano Papi and his Italian colleagues raised two groups of Homing Pigeons in a specially constructed loft. Above each loft, A and B, was a tunnel aligned north–south and into which the pigeons could climb. Whenever the wind was from the north the smell of olive oil was added to the air being blown through the tunnel of loft A; when the wind was from the south, turpentine was added. The pigeons in loft B received the same two smells but reversed; olive oil with south winds, turpentine with north. Both sets of pigeons were then taken together to a release site to the east of the loft. Before release, half had turpentine rubbed on their bill, half had olive oil. Birds with olive oil on their bill that had been exposed to olive oil on north winds departed to the south; those that had been exposed to olive oil on south winds departed to the north. Birds with turpentine on their bills responded comparably.

Papi interpreted these results as follows. While in their loft, each pigeon learned the smells associated with winds from different directions. When released, the bird noted that one of these smells (because it was actually on the bill) was stronger than ever experienced at the home site. The corollary was that the pigeon was now nearer the source of that smell than when it was at the loft. Thus, if the smell had always arrived on winds from the north, it followed that the bird was north of its home loft. To return home, therefore, it should fly south. The conclusion is that the pigeon's mental map has smell, as well as visual, landmarks.

Italian workers have produced a whole range of further evidence to support this conclusion. Following various experimental manipulations, pigeons that were unable to smell were less able to navigate than pigeons with their sense of smell intact. Indeed, Papi goes so far as to suggest that, without a functional sense of smell, pigeons are unable to navigate. Other researchers, on the basis of their own experiments in Germany and the United States, do not necessarily agree and the role of smell landmarks in the pigeon mental map is still contentious. The answer seems likely to be a compromise.

There seems little doubt from Papi's results that pigeons are able to detect and recognize different smells and to link particular smells with winds from particular compass directions. It also seems clear that not all pigeons at all lofts find this ability useful. Probably pigeons from lofts surrounded by countryside with clear, consistent and different smells include this information on their mental map. However, pigeons from lofts in which the surrounding countryside has weak, uniform or confusing smells do not.

### *Sound landmarks?*

In the same way that the mental maps of birds and people may both involve visual landmarks and smells, so too may they both involve sounds as landmarks. People living near airports, or near the sea, may often be able to orient themselves by noting the direction of the sound of planes taking off and landing or the sound of foghorns. Birds may do likewise but, just conceivably, they may be able to do so over much longer distances than people.

Everything rests on the use birds make of their ability to hear those sounds, called infrasounds, with very long wavelengths and therefore low frequencies (below about 10 Hertz). Humans can hear infrasounds, but only if they are very loud. Birds can hear infrasounds 50 decibels quieter than humans, even those sounds with frequencies as low as 0.04 Hertz. This ability could potentially be very useful in the economic formation of a mental map covering huge areas. The reason is that major landmarks, such as mountain ranges and coastlines, all give off infrasounds as air passes over. Most usefully, however, these sounds show little attenuation with distance and may be heard over very long distances, perhaps 1000 km or so. Here are potential landmarks to be placed at the limits of a bird's mental map; landmarks that the bird never needs to visit. To be able to incorporate them on the map, however, the bird needs to be able to detect their direction, and here lies the first snag.

The problem with detecting the direction of a source of infrasound is that the wavelength is so long, ears would need to approach being a kilometre apart in order to judge direction by the conventional method of phase, intensity and/or time differences between the two ears. However, the problem may not be insuperable if the bird can make use of the well-known Doppler effect. Everyone has noticed that when a train or ambulance travels towards them, the sound it makes is different from that when it is travelling away. If birds could fly fast enough, and their ability to detect differences in pitch were acute enough, then by flying in a circle they would be able to detect when they were flying toward a particular infrasound and when they were flying away from it. The necessary experiments have been done and the calculations made: it seems that Homing Pigeons at least have the speed and sensitivity necessary to detect the direction of infrasounds by this means. However, detecting direction is not the only problem to be overcome before infrasound can find a place on a bird's mental map.

Not all infrasounds are useful. In addition to those given off by geographical features, which a bird could find useful for navigation, and those given off by weather fronts and storms, which it could find useful for weather forecasting during migration, there are many useless and distracting infrasounds in the environment. These are generated not only by the air moving over trees and buildings but worst of all by the air moving over the birds' own wings. To place infrasound landmarks on their mental map, birds would need to be able to filter out the useful from the useless. Whether birds can do this is still unknown. On the whole, most ornithologists seem to assume that the bird's mental map does not contain infrasounds; but they could always be wrong!

## Maps and compasses in action

Birds are now known to possess, not 'a' navigational sense, but a navigational armoury based on all the usual senses: sight; smell; hearing; plus the previously unknown sense, magnetoreception. Together, these senses provide an array of map and compass information which the bird then uses to find its way around. The challenge now facing ornithologists is not the discovery of new senses and abilities but the unravelling of how maps and compasses are used.

A major clue came with the discovery that Homing Pigeons find their way home by following their outward journey while being driven to the release site. In effect their technique seems to be: 'first I went east, then north, so, to return home, I need to fly roughly south-west until I encounter familiar landmarks'. In addition, pigeons may judge their direction from home each time they pass through a familiar smell. The result is that, even before being taken out of the car in which they have been transported, pigeons have often determined the compass direction in which they should fly to return home. Maybe, as it circles round before heading home, a pigeon also registers the landmarks around the release site to check whether these are familiar and, if so, if they indicate the same home direction. If the landmarks are unfamiliar, the site, the landmarks by which it may be identified, and the compass bearing of the direct route to home, may then all be added to the bird's mental map which is thus expanded.

If pigeons can follow a tortuous outward journey when driven by car, it seems likely that they and other birds can and do so during their natural explorations. Here, then, is the key to the way birds form a mental map which, for some species, allows them to navigate their way over thousands, if not tens of thousands, of kilometres.

The modern picture, then, is of a bird travelling across the countryside, comparing the landmarks that it sees, smells or hears beneath it against the reference mosaic of its mental map. At intervals, it recalls from the map the appropriate compass bearing to take it to its next destination. It then refers to the Earth's magnetic field, the sun, stars or moon in order to orient in that compass direction. Exactly similar systems allow a sparrow to explore and exploit the fields, woods and buildings surrounding a house; a Homing Pigeon to return to its loft after being transported and released in the surrounding countryside; and a swallow to migrate from a wintering area in South Africa back to a nesting site in a Lorraine château during the French Revolution.

RRB

*Migrant birds often use water courses and bodies of fresh water as navigational reference points. This Barn Swallow (Hirundo rustica) is taking the opportunity to skim the surface to drink*

# 8 BIRD POPULATIONS

The world holds about 8800 bird species. These species vary in abundance from extremely rare, consisting of a few individuals on the verge of extinction, to extremely widespread, numbering many millions of individuals. Some species, restricted to small oceanic islands, have naturally small populations, consisting in some cases of less than 100 breeders. But most rare species owe their present status to past persecution or habitat destruction by man. Extreme examples include the Japanese Crested Ibis, which in 1984 numbered only 22 individuals, and the California Condor, which in 1984 numbered only 17 individuals, and is now extinct in the wild. If rare species are protected, however, they usually increase again, providing that suitable habitat remains available. Thus, the Whooping Crane in North America increased from 33 individuals in 1959 to more than 100 in 1984.

Generally speaking, accurate counts exist only for those species which are rare and restricted in distribution, or are concentrated at certain seasons in a limited number of sites. Many species of seabirds range widely over the oceans, but while nesting gather in vast colonies, at traditional locations. Some rookeries of Adélie Penguins in Antarctica contain more than a million pairs, while colonies of Sooty Terns on tropical oceanic islands may contain up to a staggering ten million pairs. Perhaps the most numerous seabird in the world is the Antarctic nesting Wilson's Petrel which is thought to number more than 50 000 000 pairs, distributed in a range of colonies.

*The rarest bird in the world? Extinct in the wild, in 1984 there were just 17 captive California Condors (Gymnogyps californianus) surviving*

Among land birds, contenders for the most numerous species would surely include the House Sparrow and European Starling, both of which are widespread in the Old World and, through introductions, in the New World too. But the most likely candidate is the Red-billed Quelea, a seed-eater which breeds in much of Africa south of the Sahara. In this widespread species, roosts, breeding colonies and even feeding flocks, may contain more than a million individuals. In general, however, colonial land birds breed in smaller concentrations than seabirds, but some species form spectacular roosts. Nightly gatherings of more than a million individuals are regular among starlings in some cities, and in the reedbed roosts of Sand Martins in Europe and Red-winged Blackbirds in North America.

Through the combined efforts of hundreds of bird-watchers, attempts were made in 1968–72 to assess the distribution and numbers of all the bird species which breed in the British Isles. If the estimates for 176 land bird species are added together, they give a grand total of 90 million pairs, a mean density of 286 pairs per square kilometre. This total implied only three birds to every person in the British Isles. About half the total was contributed by only seven species, and three quarters in all by some 14 species. The most numerous bird was the Wren, with an estimated ten million pairs, followed by the European Blackbird and Chaffinch, with seven million pairs each, and the European Starling and House Sparrow, with more than five million pairs. The total seabird population of 24 species was estimated at around 3.3 million pairs, plus an unknown number of non-breeders. About half this total was made up by only four species, namely the guillemot (577 000 pairs), puffin (490 000 pairs), Black-legged kittiwake (470 000 pairs), and fulmar (305 000 pairs). The contributions made to the totals by different types of birds are shown in the table opposite.

## Some general patterns

Whilst such counts are interesting, and useful in defining conservation priorities, they provide little understanding of the factors that influence bird numbers. From general studies of bird communities around the world, several patterns have, however, emerged:

*Among the most abundant birds in the world, the Red-billed Quelea (Quelea quelea) of Africa breeds, feeds and roosts in vast numbers, making it a serious pest of cereal crops*

1. In any one area, abundance is related to body size, and in general small species are more numerous than are large ones. This is presumably because small birds require less food and other resources than large ones, so small birds can live at greater densities.
2. Each bird species has special habitat requirements, so that any particular habitat, such as conifer forest, has a distinct bird community. Many of these species would also occur in other types of forest, but few (if any) would occur in a markedly different habitat, such as open grassland. Moreover, in any one habitat, the composition of the bird community remains fairly stable through time, some species being consistently among the commonest and others among the rarest.
3. The numbers of species, and individuals, tend to be greater per unit area in complex habitats, such as woodland, than in simple habitats, such as grassland. Thus in Britain, open habitats usually hold less

*Breeding birds in Britain, based on numerical estimates in the Atlas Survey of the British Trust for Ornithology, 1968–72*

| Group | Numbers of pairs (thousands) | Commonest species (pairs in thousands) |
|---|---|---|
| Divers and grebes | 19 | Little Grebe (13.5) |
| Raptors | 129 | Kestrel (100) |
| Gamebirds | 1478 | Partridge (500) |
| Crakes | 409 | Moorhen (300) |
| Waders | 569 | Lapwing (200) |
| Gulls and terns | 1245 | Black-legged kittiwake (470) |
| Other seabirds | 2115 | Guillemot (577) |
| Pigeons | 4360 | Woodpigeon (4000) |
| Owls | 79 | Tawny Owl (55) |
| Swallows, martins and swifts | 1550 | Swallow (750) |
| Woodpeckers and others | 104 | Great-spotted Woodpecker (35) |
| Crows | 3356 | Rook (1500) |
| Songbirds | 77 547 | Wren (10 000) |

The above figures give totals of about 90 million land bird pairs, and over three million seabird pairs. Of 208 regular breeders, 18 had breeding populations exceeding one million pairs, 40 were in the range 100 000 to one million pairs, 52 were in the range 10 000–100 000 pairs, 33 were in the range 1000–10 000 pairs, 28 were in the range 100–1000 pairs, and 37 had less than 100 pairs.

than 15 breeding species, at total densities of less than 100 pairs per km², while woods may contain more than 50 species, at densities of 250–500 (occasionally up to 2000) pairs per km². Moderately complex habitats, such as marshes and heaths, tend to be intermediate in both respects.

4. In any type of habitat, the number of species rises with increasing size of patch. Thus more species are found in large woods than in small ones, or in large reedbeds than in small ones. In contrast, densities of birds tend to be larger in the smaller patches. This curious finding is due partly to the so-called 'edge-effect', in which birds are more numerous at the edges of a habitat patch than in the middle, and smaller patches have more edge per unit area than large ones. In addition, some birds may breed within one habitat patch but forage outside.

5. Within comparable habitats, the number of species, and overall densities, decline from tropics to poles. A tropical forest may hold more than 200 bird species at more than 5000 pairs per km², but a northern forest may hold less than 20 species at 200 individuals per km². This is part of a general pattern, for productivity and diversity of life to decline with rise in latitude. A similar trend occurs in any one region with rise in altitude, and affects all plant and animal life, as may be seen by climbing a high mountain.

IN

*Populations of some species such as the Fulmar (Fulmarus glacialis) have increased dramatically during the present century. Fulmars are thought to have benefited from man's fishing activities, as here where they scavenge offal from a fishing boat off Shetland*

## Counting birds

For individual species, the most interesting question is what determines the population level. In practice this entails understanding what causes the variations in density which occur from place to place, and from year to year.

Except for very localized species, it is not normally possible to study the whole population. Instead, the biologist has to restrict himself to a defined area, whose avian occupants may form a tiny part of a much wider population. Individuals may move freely in and out of his study plot, and those birds that breed there may occupy a wider area, or even a different part of the world, outside the breeding season. Hence, most 'population studies' are concerned with the numbers found in a defined area at a specific time.

In any population, numbers change during the course of the year, reaching a peak at the end of the breeding season, and declining again to the start of the next breeding season. The extent of these seasonal fluctuations depends mainly on the reproductive rate of the species concerned, the most prolific species showing the most pronounced seasonal peaks. Most birds are easiest to count when they are breeding, because they are most conspicuous then, and tied for long periods to specific locations where they nest. Hence, knowledge of population trends from year to year, or from place to place, is usually based on counts of breeding numbers, or on indices of breeding numbers, such as displaying males or nests. Some bird populations contain large numbers of non-breeding individuals too; these are often difficult or impossible to count and tend to get ignored.

IN

## Population trends

The breeding populations of most birds normally remain fairly stable through time. Extreme examples are provided by some birds of prey, such as Golden Eagles, in which the density of territorial pairs may vary by no more than 10% on either side of the mean level over long periods of years. The majority of populations fluctuate somewhat more, perhaps halving or doubling in size from one year to the next, or declining after hard winters and increasing again in subsequent years.

Small finches and other birds, which depend on tree seeds, can fluctuate in density by twenty-fold or more from year to year, depending on the size of the seed crop. But even these fluctuations are small compared with those in other creatures, such as insects, in which annual changes of a hundred-fold or more are not uncommon.

In most bird species the year-to-year fluctuations in numbers are irregular, but some northern gamebirds, such as grouse and ptarmigan, undergo regular cycles, with peaks of abundance about every four or ten years, depending on region. These cycles in gamebirds are matched by their predators, which go up and down in step with the prey. Yet within suitable habitats the mean level of abundance of these various species does not change much over the years.

Although a general stability in bird populations is probably the norm, it would be expected only in stable environments. Over much of the world, human activities are leading to continual changes in bird habitats and food supplies, and, where long-term counts are available, they nearly always reveal some marked change in populations. During the last

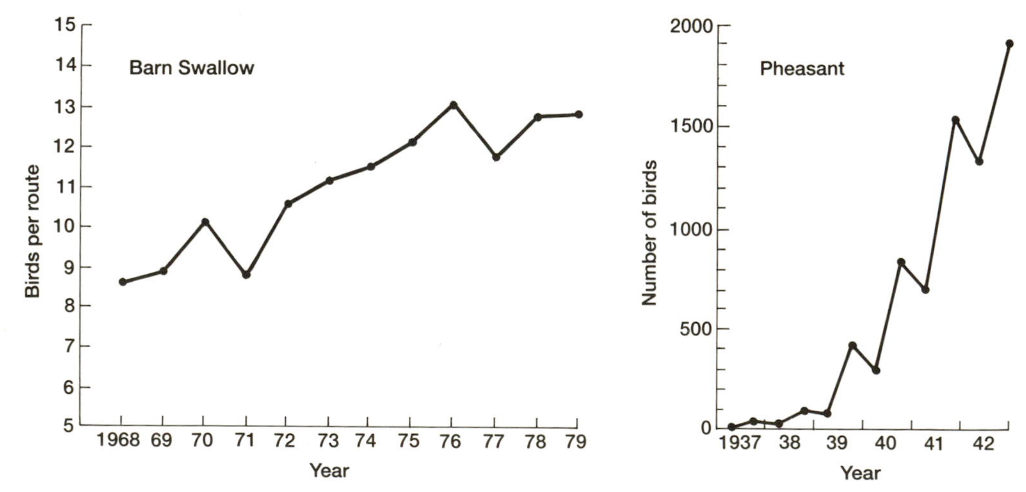


*Increasing bird populations. The Barn Swallow (Hirundo rustica) of North America is steadily increasing in numbers. The Pheasant (Phasianus colchicus), in contrast, exhibits very rapid population growth following the introduction of two males and six females to an island off Washington State*

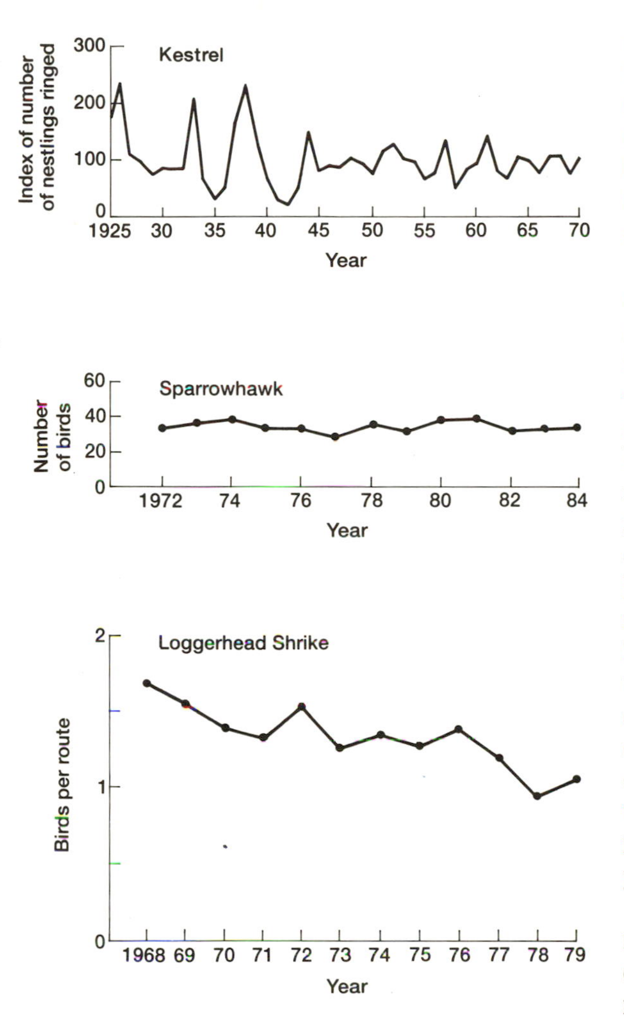


50 years in Britain, agricultural developments have caused the decline of first the Corncrake, then the partridge and Linnet and latterly the Lapwing. Meanwhile, most seabird species have increased, some quite dramatically, as human fishing activities have made their foods more available.

Herring Gulls have also benefited from foraging on the growing number of rubbish dumps, resulting from the increasing affluence of an expanding human population.

On a wider scale, during the present century the White Stork in Europe has greatly contracted its range and declined in numbers. This has been linked with falling food supplies, resulting from the draining of meadows in Europe and the control of locusts in African winter quarters. Several other species which winter in Africa have also declined, notably the Red-backed Shrike, Wryneck and European Nightjar. The most recent include the Whitethroat and Sand Martin, whose declines have been linked with droughts, superimposed on human-induced desertification in the Sahel zone of Africa.

Other European birds have greatly expanded their ranges in recent decades, becoming common breeders in areas from which they were formerly absent. Examples include the Serin and the Collared Dove. Until 1930 this last species was restricted to the Balkans in Europe, but then spread rapidly across the continent, reaching Britain about 1955. For many years its numbers increased exponentially (by a constant percentage each year), until habitats became filled and numbers levelled off. Most species which have spread in this way clearly benefit from human activities, which may therefore have facilitated the spread. Range expansions of other species may well have occurred in the past, long before there were ornithologists to record them. Indeed some of the most familiar birds of today in 'developed' regions may have been rare or absent there 500 years ago.

In other species, expansions appear entirely natural, in that the birds concerned do not depend entirely on man-made habitats in their new range. Quite early in the present century,

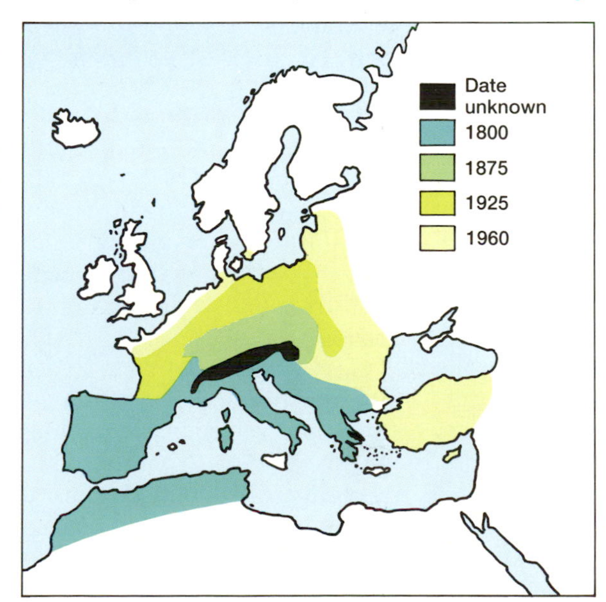


*Above left. Stable and declining populations. The Common Kestrel (Falco tinnunculus) graph uses numbers of nestlings ringed in Britain, 1925–70, expressed as a percentage of an 11 year sliding average. The population fluctuations have been attributed to cyclic changes in the numbers of voles, which form the Kestrel's main prey. In the case of the European Sparrowhawk (Accipiter nisus), the number of nests identified in Eskdale, southern Scotland, 1972–84, shows the stability of breeding population that is possible in a period of habitat stability. The Loggerhead Shrike (Lanius ludovicianus), regretably, is in straightforward decline*

*Left. The northerly spread of the Serin (Serinus serinus) across Europe*

some Cattle Egrets crossed the Atlantic to Venezuela, from which they spread over much of North and South America. In a similar period, New Zealand acquired at least ten new breeding birds, which moved in naturally from Australia. They include such diverse species as Silvereye, Spur-winged Plover and White-faced Heron.

Within a human lifetime, such spectacular expansions in range and numbers are exceptional, however, as are precipitous declines to extinction. Yet every species when unchecked has an intrinsic rate of natural increase (called '$r$'), until it reaches a level ($K$) determined by the carrying capacity of its habitat. (The value of $K$ is determined partly, or entirely, by the food and other resources available in the habitat.) Thereafter, in the absence of marked environmental changes, the species tends to remain fairly constant in numbers over the years.

## Density-dependent regulating factors

General stability in bird numbers can be brought about in various ways. If the resources on which a species depends remain constant through time, they can prevent numbers exceeding a certain level. For example, Jackdaws nest in large tree-holes, and in many areas the numbers of tree-holes are limited. Every one is occupied by a single Jackdaw pair, and without increasing the numbers of holes, the numbers of Jackdaws can never exceed the observed level. But resources seldom impose such a definite and consistent ceiling on numbers, and stability in population is brought about by the action of 'density-dependent' regulating factors. Any regulating factor, which acts more severely to reduce numbers as the population rises, is termed density-dependent. Three mortality agents which can act in this way include predation, disease and food shortage, while emigration might be a secondary factor, mitigating the effect of food shortage. Such factors contrast with 'density-independent' factors, such as severe weather and other natural disasters which can kill large numbers of birds, regardless of their density. Such factors act to de-stabilize numbers, causing large and random fluctuations. Both density-dependent and density-independent factors can act on breeding and mortality or on immigration and emigration.

Some years ago, arguments were frequent on whether animal numbers were limited primarily by density-dependent or by density-independent factors, some people being impressed by the relative stability of populations, and others by their instability. It now seems more sensible to ask, for any one species, what is the relative importance of the two types of factors in causing the changes observed. In general, the more dominant the influence of density-dependent factors, the more stable the population, but even in populations which fluctuate widely, some aspects of breeding or mortality may be density-dependent.

There are two main problems in detecting density-dependent regulation in populations, apart from the fact that in some species its influence may be small. First, density-dependence may not occur over the full range of densities found; as it often involves competition, it may operate only at the highest population levels observed. Second, any resource over which competition is occurring may itself change in abundance from year to year, leading to variations in the density of population at which competition sets in. For example, in a species which eats tree-seeds, competition may be apparent at low bird densities in years with light crops, but only at much higher densities in years with heavy crops. One way round this problem is to express bird numbers, not in absolute terms, but as numbers per unit of food. It then becomes easier to detect any density-dependent influence there might be. If mortality rises (or reproduction declines) with increase in numbers per unit of food, then density-dependence can be taken as proven.

Having detected density-dependence, a second question is whether the response is strong enough to regulate the population at the levels observed. In some populations, the increase in mortality (or reduction in breeding rate), which occurs as numbers rise, is so slight as to have little or no detectable impact on numbers. For example, Great Tits lay slightly larger clutches

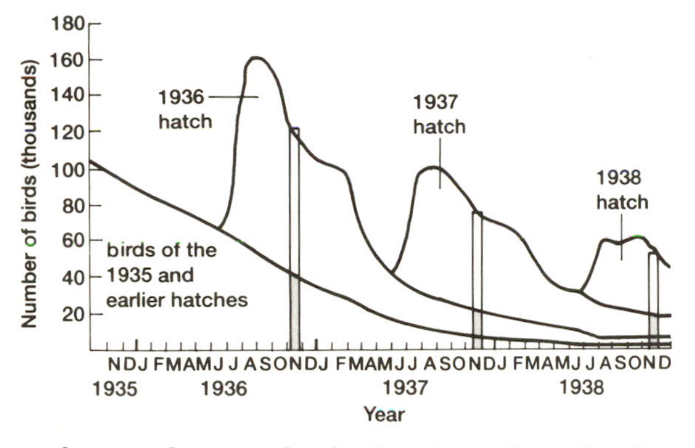


*Left. A classic study of population renewal in the California Quail (Lophortyx californica). The vertical bars represent the age-structure of the population in November: adults shaded, immature birds unshaded. Eggs and chicks are not included*

at lower than at high densities, but this has little influence on subsequent population levels.

There are other possible relationships between density and mortality factors. 'Inverse density dependence' occurs when a mortality factor acts more strongly at low densities than at high ones.

This can happen, for example, if a predator takes a constant number of a prey species, irrespective of its density. It thereby removes a greater proportion of individuals at low densities than at high ones. Such factors tend to destabilize a population, causing it to fall at low densities and rise at high densities, unless other factors intervene.

'Delayed density dependence' is also observed in wild populations. This occurs when there is a lag in the operation of a regulating factor, so that it relates to previous densities not to present ones. For example, the numbers of predators rise and fall, in parallel with, but somewhat later than the numbers of their prey. This occurs because a rise in prey numbers leads to enhanced breeding, and survival or immigration of predators, while a decline in prey numbers has the opposite effects. Thus predation pressure always relates to pray numbers sometime previously. Northern Goshawk predation on Hazel Grouse in Sweden shows such delayed density dependence, and the numbers of the two species have been observed to rise and fall in regular cycles.

Biologists use the term 'limiting' for any factor which can set a ceiling on density, whether or not it acts in a density-dependent manner. They usually restrict the term 'regulating' to factors which operate only in a density-dependent manner acting to prevent both overpopulation and extinction. In many species densities remain more stable (and hence more regulated) in main habitats, and are more fluctuating (and less regulated) in subsidiary habitats.

## Social behaviour as a factor in population regulation

Whenever individual birds come into contact with one another, they usually reveal some form of dominance hierarchy, or 'peck-order'. This can lead to an uneven sharing of resources, and provide a basis for the regulation of density. Territorial behaviour is an obvious example, as it serves to space out breeding pairs through the available habitat, and to exclude surplus individuals once the habitat is saturated. Such behaviour can thus be regarded as a form of contest, in which each individual attempts to corner the resources it needs. Behaviour can also be adjusted to correspond with resources, such as food supplies, as pairs defend smaller areas in food-rich habitats than in poor ones.

In such cases food supply is the environmental factor determining density, while territorialism is the behavioural mechanism through which this adjustment is made. The experimental removal of territory holders often reveals a surplus of birds waiting to occupy the vacant sites. This does not mean that territorial behaviour evolved in order to regulate population density. Ownership of territory confers great advantage on the individual, including the ability to breed, and natural selection acts on individuals not on populations. Hence, the regulation of density is an incidental consequence of behaviour evolved to benefit the individual.

Not all territorial behaviour is concerned with protection of a food supply, as some species defend other resources, such as nest sites. This may still serve to limit densities, and, in colonial species, even the sizes of breeding colonies, where space is limited. Birds which feed in flocks also show dominance hierarchies, and individuals which win fights take precedence over others at food sources. This leads to the emigration or starvation of the weaker individuals, and a readjustment of numbers to correspond with food supplies, as shown for example in Woodpigeons. So even in species which are not markedly territorial, social behaviour can still act as a mediator of the effects of density-dependent competition.

IN

## Factors limiting population levels

### Resources

Resources, especially food supply, could limit the size of any bird population. The question of interest is which species are normally limited in this way, and which are held at a lower level by natural enemies, such as predators and pathogens. To be limited by food, a species need not be up against the food limit all the time. Shortages may occur only at certain times of year, under specific weather conditions, or irregularly every few years. Some European birds are cut back so severely in hard winters that they may then take several years of increase before they again reach the food limit. Examples include the Grey Heron and Common Kingfisher.

The evidence that certain bird species are limited by food is mostly circumstantial: (1) densities are higher in areas where food is abundant than where food is scarce; (2) densities are higher in years when food is abundant than when food is scarce; and (3) sudden or long-term changes in densities often follow sudden or long-term changes in food supply.

*Differences between areas*

Correlations between bird numbers and food supplies in different areas have been noted in various groups, including seabirds, waders and raptors. Thus the densities of European Sparrowhawks in woodland in various parts of Britain are correlated with the densities of songbirds which form their prey. Within the woods the hawk pairs nest only 0.5 km apart in the richest prey areas, and more than 2 km apart in the poorest areas. Likewise, Great Tits breed at higher densities in mixed broad-leaved woods than in conifer woods, where the insects they eat are scarcer.

Such relationships occur even at the level of the individual and its territory. In the North American Ovenbird, for example, the sizes of 13 territories in one wood varied inversely according to the density of invertebrate prey in the local leaf litter. Similarly, the Golden-winged Sunbird in Kenya feeds on nectar, and in one study the size of territories varied more than tenfold; at any one time each territory contained enough nectar to supply the owner's needs, and over some days territories were expanded as flowers died and their nectar contents fell. In cases such as these, with each bird (or pair) holding a territory of a size dependent on its food content, the density of birds can become related to food supplies over wide areas.

In other studies, variations in bird densities have been related, not to food directly, but to some more easily measured index of food, such as soil fertility. In terrestrial habitats, birds reach higher densities on productive than on less productive soils. Not only is plant growth better on the fertile soils, but populations of insects and other organisms on which birds feed tend to be higher there. This is apparent in woodland birds and in open country ones. A well-researched example is the Red Grouse, which in Scotland reaches higher densities on moors overlying basic rather than acidic rocks, linked with differences in the nutrient content of its food plant, heather. Similarly, for any given size of lake, waterbirds generally occur at higher numbers on nutrient-rich (eutrophic) waters than on poor acidic (oligotrophic) ones. This can be related to the more abundant plant-life, invertebrates and fish in eutrophic lakes. Finally, seabirds are also most numerous near upwellings and other sites of nutrient-rich waters, which support an abundance of plankton and fish.

Such correlations between bird numbers and food are consistent with the idea of food limiting numbers, but they do not conclusively prove it. Alternative explanations might be that birds are limited in each area by some unknown factor which varies in parallel with food, or that birds already limited by some other factor then distribute themselves according to food. Only further study will clarify the matter.

*Differences between years*

In some birds, breeding density varies in the same area from year to year in parallel with fluctuations in food. Striking examples are provided by predatory birds which depend on cyclic prey. Two main cycles are recognized: (i) an approximately four-year cycle of small rodents on the northern tundras and temperate grasslands; and (ii) an approximately ten-year cycle of Snow Shoe Hares in the boreal forests of North America. Certain grouse species are also cyclic, but whereas in some regions they parallel the four-year rodent cycle, with peaks usually in the same years, in others they parallel the ten-year hare cycle. The role that food plays in the cycles of these prey species is uncertain, but there is little doubt about the predators, all of which tend to breed most densely and prolifically in years of peak prey numbers.

In years of shortage, some predator species move out, appearing in large numbers well south of the usual range. Such migratory 'irruptions' are well known in Rough-legged Buzzards and Snowy Owls, and also in goshawks in North America. In contrast, raptors that have fairly stable food supplies have stable breeding densities. Even the same species (such as Northern Goshawk) may fluctuate in one region, but not in another, depending on whether its prey are cyclic or stable.

Other birds whose numbers fluctuate greatly are the northern finches that depend on the seeds of conifers and other trees. Such species increase greatly in years of good crops, and migrate southward on large irruptions in years of widespread crop failure. Crossbills, siskins and redpolls are well known examples. In these seed-eaters, as in the vole-eaters, increases in local populations between years are often too great to be accounted for by good survival from the previous year, and must be due partly to immigration. In all these species, the local correlations between numbers and food are sometimes impressive, with changes in numbers keeping strictly parallel with changes in food. They are still correlations, however, and so open to the possibility that some other unknown factor fluctuates in parallel with food, so causing the changes.

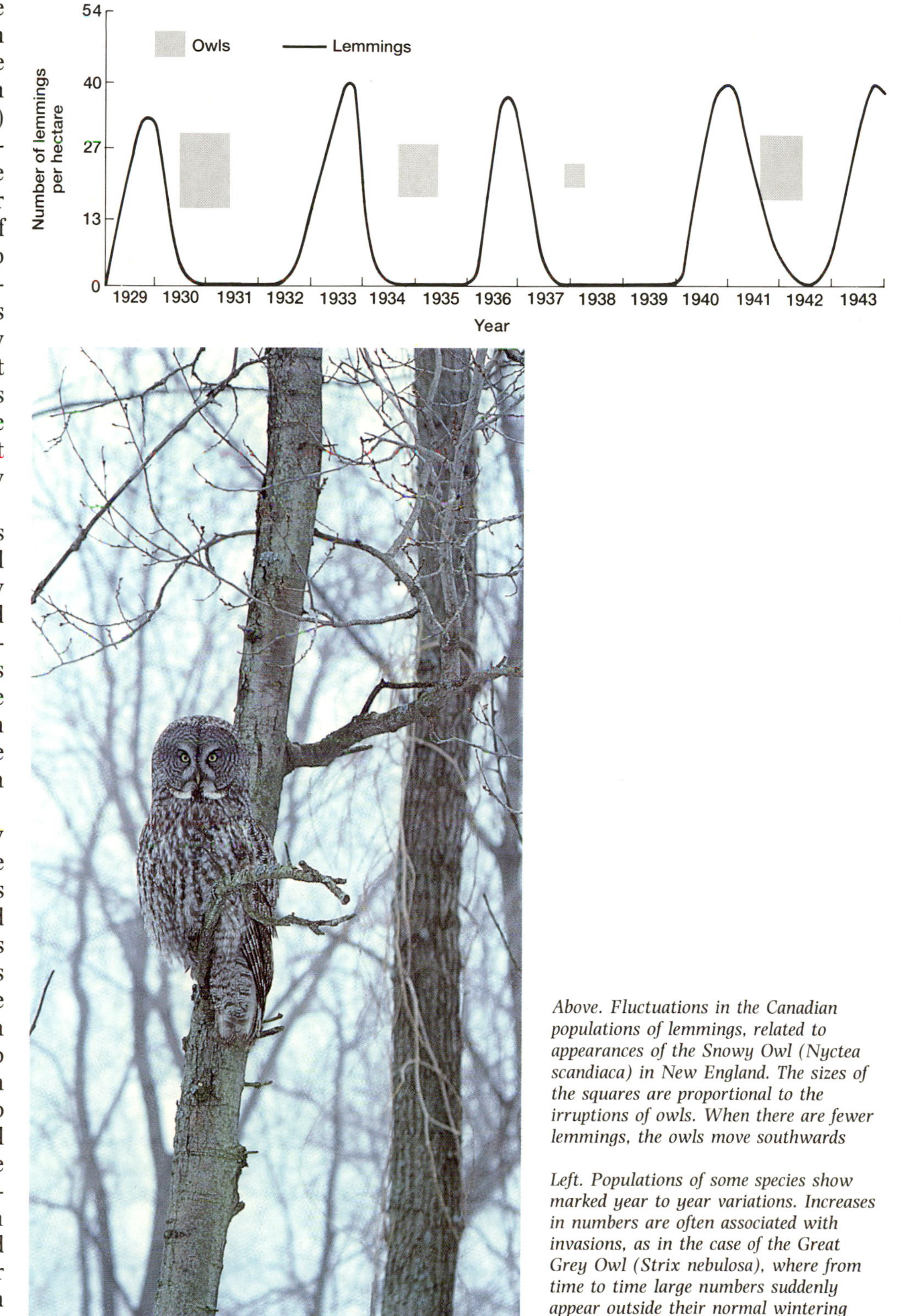


*Above. Fluctuations in the Canadian populations of lemmings, related to appearances of the Snowy Owl (Nyctea scandiaca) in New England. The sizes of the squares are proportional to the irruptions of owls. When there are fewer lemmings, the owls move southwards*

*Left. Populations of some species show marked year to year variations. Increases in numbers are often associated with invasions, as in the case of the Great Grey Owl (Strix nebulosa), where from time to time large numbers suddenly appear outside their normal wintering range*

### *Sudden and long-term changes*

Periodic crashes in the numbers of some South American seabirds have been documented since the last century, because the birds and the fish they eat are commercial assets. Three seabird species breed in huge numbers off Peru in the rich cold waters of the Humbolt current. In the poor warm waters beyond, other seabirds occur in the low numbers which are typical of most tropical seas. In certain years, the warm waters come further south than usual, and envelop the Peruvian Guano Islands. Then the abundant plankton disappears, as does the anchovy which supports the seabirds, and millions of birds die from starvation. Declines from 27 million to 6 million birds occurred between 1957 and 1958; after which numbers increased steadily to 17 million in 1965, when the current again shifted, causing another decline in fish and another crash in bird numbers to 4.3 million.

Many other examples of long-term or sudden changes in bird populations and their food can be found in ornithological journals. They suggest that food limited numbers at different levels in different periods, and that sudden declines in food supply caused sudden declines in numbers; in some cases (as in the seabirds) other mortality factors, such as disease and predation, were clearly not involved. Such events form natural experiments, demonstrating the effects of food supply, but the possibility remains that some unknown mortality agent changed at the same time.

### *Experimental Evidence*

Designed experiments have also been used to test the effect of food supply on bird densities, but have given variable results. In Red Grouse, burning or fertilizing areas of heather promoted an increase in breeding numbers one year later over the previous level, as well as over that in other areas nearby. This occurred when the experiment was started in years with low or moderate densities, but in a later trial, fertilizing an area failed to halt a big decline.

In another experiment, sunflower seeds were provided to tits in winter to find whether their local breeding numbers then increased. By comparison with other areas, the numbers of Blue Tits increased after this treatment, but the numbers of Great Tits did not. Perhaps Blue Tits were limited in breeding density by the winter food supply, while Great Tits were limited by some other factor. But in similar experiments with Great Tits in other areas, breeding density did increase, as did individual survival. That food can act at the level of the individual territory was shown in an experiment in which sunflower seeds were provided in winter to nuthatches, which then reduced the size of their territories.

Because of the cost and practical problems, such experiments can be difficult. In further attempts to assess the effects of food supply on bird numbers, some biologists have measured the rate of decline in food stocks over several weeks, as birds feed. This is possible only for certain species, and results may be hard to interpret. Thus in one species the removal of 90% of food might cause no obvious mortality, while in another the removal of only 10% might cause a lethal reduction in feeding rate. Such measurements are useful chiefly in defining periods of likely shortage.

### *Other resources*

While food supply is potentially limiting to all birds, in some species breeding density is held at a lower level by shortage of some other resource. That nest sites can act in this way is shown by two types of evidence: (1) breeding pairs are scarce or absent in areas where nest sites are scarce or absent, but which seem suitable in other respects (non-breeders may live there); and (2) the provision of artificial nest sites is sometimes followed by an increase in the number of breeding pairs. The species concerned mostly use special sites (such as tree holes), and include kestrel, Wood Duck, Eastern Bluebird and Pied Flycatcher. In woods with nest boxes, this last species has reached densities equivalent to 2000 pairs per $km^2$, greater than all other species together in those woods.

Some species not only increased in density, but also spread over thousands of square kilometres following the provision of nest sites. Examples include the Mississippi Kite and other raptors which spread in Western American grasslands following tree planting. In all these cases, the food was present beforehand, but lack of nest sites precluded breeding. Similarly, European Swallows and House Martins, which now rely mainly on buildings for nest sites, must presumably have been much more localized when they had only caves and cliff faces respectively. The same holds for the North American Purple Martin which now uses special purpose-built multi-celled nest boxes. Other aspects of habitat structure, involving cover or roosting sites, can also be considered as resources, occasionally limiting bird density. Thus Scaled Quail increased in numbers after piles of brushwood were added to their desert habitat in North America.

In arid lands water-shortage can limit bird numbers, partly through an interaction with food supplies. The need to drink forces birds and other animals to remain within regular travelling distance from water. This in turn may lead to a depletion of local food supplies, with effects on bird numbers, while in more distant areas food remains abundant but out of reach. There is a parallel here with seabirds constrained to concentrate near breeding colonies for part of the year.

## Predation

Breeding densities of some birds may be reduced by predators below the level that food would permit. The most dramatic cases are from islands, where new predators, such as cats or rats, were introduced by man, and subsequently devastated the local bird fauna. This is in fact the most important cause of extinctions among island birds. Cases such as these, where predators eat out their prey, are likely to occur only as a result of some human action, or as a result of some natural spread of a predator. Once the predator has eaten the prey, it is likely to become rare itself, or even extinct if alternative food is lacking. Indeed the only predator–prey systems which persist in the long term are those in which the predator is incapable of eating its prey to extinction. This happens, for example, where predation is density-dependent, so that below a certain level the prey become almost immune to further attack. In some areas the mortality inflicted by goshawks on pheasants is in this category. Predators have two responses which lead them to act in a density-dependent manner. The first is the so-called functional response, in which individual predators include more of a particular prey in their diet as its numbers rise. The second is the numerical response, in which predators concentrate temporarily in areas where prey are plentiful.

There are three other types of evidence that predators do sometimes hold their prey below its food limit:

1. *Cyclic systems.* Those Canadian predators which eat cyclic hare species can in certain years reduce the numbers of gamebirds. As the numbers of hares rise, so do the numbers of goshawks and Great Horned Owls which eat them. Then, when hare numbers crash, the abundant predators switch to gamebirds, reducing their numbers, and eventually declining themselves through starvation and emigration. In these circumstances, gamebird numbers can be said to be limited by predators, at least in the decline phase of the hare cycle. The same may happen in northern Europe, in association with the four-year vole cycle, as foxes and other predators turn to gamebirds when vole numbers fall.
2. *Stable systems.* Predators which eat a wide range of food can depress the numbers of particularly vulnerable prey species. Partly because of their wide diet, foxes can achieve high densities in the English countryside. They can also prey efficiently on incubating partridges whose nests are easy to find. It is not unusual for foxes to destroy all, or almost all, the partridge nests in large tracts of countryside, taking both female and eggs. This is possible only because foxes are maintained at a high density by alternative foods; if they depended on partridges alone, they would soon eat themselves to extinction. Their efficiency may be enhanced by the present structure of the English countryside, where nesting cover for partridges in some areas is extremely limited and easy to

search. Foxes and other generalist predators may have similar effects on other ground-nesting birds, again depressing their numbers below the food limit.

3. *Experimental evidence.* The most convincing evidence for predators limiting prey numbers is obtained when the experimental removal of predators is followed by an increase in prey numbers. In several such experiments, egg and chick survival was better than in other areas, leading to an increase in the post-breeding population, but this was not maintained in the breeding population of the next spring. In at least one experiment, however, breeding populations of Capercaillie and Black Grouse increased over those in other areas, following the removal of foxes and other predators.

In many bird species, predators concentrate on young individuals, and merely alter the seasonal pattern of mortality, without greatly affecting breeding numbers. Each year, a large proportion of individuals in any bird population is destined to die between the end of one breeding season and the start of the next. In the absence of predators, the surplus individuals would emigrate, or die from other causes, such as starvation.

In Britain and some other European countries, it is customary for gamekeepers to destroy predators, in order to increase the numbers of game-species available for hunters to shoot. This is done even though the predators do not necessarily reduce their prey in the long term. However, the hunter is concerned with the post-breeding population of adults and young, so anything that reduces breeding success, through eating eggs or chicks, can potentially affect his bag.

The conservationist, in contrast, is not concerned with post-breeding populations, which are destined to decline anyway, but with the maintenance of breeding numbers from year to year. Hence, it is not contradictory that predators might substantially lower game numbers at the time of the seasonal peak, but have no measurable effect on breeding numbers at the seasonal low.

Furthermore, predators need not have a wholly negative effect on a population, but could in some cases promote an increase in breeding numbers. Imagine a population, such as that of a seed-eater, which has a fixed stock of food to last the winter. High numbers in autumn may cause the food supply to be depleted quickly, and few birds to survive the winter; but lower numbers in autumn (resulting from predation) could ensure that the food supply lasted for longer, so that more birds survived the winter than might otherwise have done so. Similarly, if predators selectively removed diseased individuals as they appeared, they might suppress epidemics, and thus help to maintain numbers. Although little or no information is available on these aspects, the influence of predators is clearly complex.

## Diseases and parasites

Diseases may affect wild populations either as widespread epidemics, as local outbreaks, or as sporadic individual mortality, and cause a range of effects from negligible to fatal. They may be caused by a variety of organisms or their toxins, including viruses, bacteria, other micro-organisms and parasitic worms. Diseases may be further considered as 'population-dependent' or 'population-independent'. The first type depends for its spread on contact between individuals of a species, and can thus act in a density-dependent manner. An example would be the infectious viral disease, psittacosis, which affects parrots and other birds, and occasionally humans. In contrast, population-independent diseases are contracted from the environment, food or water, and so are likely to act in a density-independent manner. An example is botulism, a form of food poisoning caused by a toxin from the anaerobic bacterium *Clostridium botulinum*. This organism thrives in stagnant water and mud, and periodically kills large numbers of waterfowl, chiefly in North America. In one severe outbreak, on the Great Salt Lake in 1929, botulism killed between one and three million birds.

In general, though, disease is not very evident in bird populations. Most known instances of disease causing lasting declines refer to island species exposed to a new disease, usually through some human action. The best documented case concerns endemic Hawaiian birds, which were exposed to avian pox and malaria, following the introduction by man of alien birds, mostly from Asia. The alien birds were resistant to such diseases, but the native birds were not. The two diseases evidently led to the rapid extinction of several native species, and much restricted the distribution and numbers of others. Their effects provide a parallel with the first exposures of vulnerable island prey species to predators, as described above.

One species whose numbers may be regularly reduced by parasites and disease is the Red Grouse. On some moors heavy infections with the strongyle worm *Trichostrongylus tenuis*, are associated with periodic crashes in the grouse population, and on other moors, a tick-borne flavivirus disease called 'louping ill' may result in much reduced grouse densities compared with virus-free moors. Some other bird species may be subject to epidemics of various kinds, which serve to reduce the numbers at the time, but may then experience many years before the disease strikes again, so that for most of the time their populations may be limited by some other factor.

The view is now generally accepted that most disease organisms and parasites evolve in such a way as to become less harmful to their host with time, since the organisms have a better chance of surviving if they do not destroy their host. Likewise, the host species tends to evolve resistance to its parasites and to the toxins they produce, but it may be vulnerable when weakened by other factors, or when exposed to a new disease. In general, therefore, diseases are like the predator–prey systems mentioned above, in that the only ones which survive in the long term are those which do not lead to extinction of both. Almost certainly, disease plays an insignificant role in the control of most bird populations, and accounts for only a small part of the total mortality, but more information is needed.

In the foregoing, disease was described as an interaction between the bird and some pathogenic organism. But in human medicine the term embraces other conditions, such as heart disease and cancer, which are usually associated with particular diets and life-styles rather than with pathogens. Among wild birds long-lived individuals may occasionally suffer from similar degenerative disorders. Examples include some seabirds nesting on St Kilda, which show kidney damage resulting from high natural cadmium levels in their diets. There is no reason to suppose that such disorders are common, however, or sufficient to affect population levels.

## Problems of assessing the relative importance of different limiting factors

One further point needs stressing, that just because the bulk of mortality in a species is due to predators, parasites or diseases, this does not mean that such factors are limiting numbers. If the breeding density of the species is limited by the carrying capacity of its habitat (i.e. resources), so that surplus individuals are bound to leave or die, they may then be removed by any of these mortality agents, depending on species or circumstances. Removal of one mortality agent, such as predators, would not necessarily increase prey numbers, as the surplus birds would then die from some other cause. Such inverse relationships between mortality factors are termed 'compensatory'. It is in fact extremely difficult, without carefully controlled experiments, to assess the role of particular factors in influencing the numbers of any bird species.

Another problem is that starvation, predation and disease may often interact. For example, starving birds may be more vulnerable to these other factors, so that while food shortage may be the predisposing cause of death, predators, parasites or disease may deal the final blow. Conversely, continual harassment by predators might prevent a prey species from feeding sufficiently, leading to death by starvation in the midst of plenty.

Any effects of food shortage on numbers are greatly influenced by movements, both on a large scale, such as seasonal migrations, and on a small scale, by local dispersal. The frequency of such movements makes the importance of food shortage on numbers much harder to assess. Another problem is that, except in rare circumstances, starving birds are seldom found. But this may be because few are starving at any one time, making it easier for predators and scavengers to quickly remove the evidence.

## Pesticides and pollutants

Increasingly in recent decades, bird populations have encountered a new threat, namely chemical poisoning from pollution and pesticide use. Such chemicals act in a density-independent manner, and in extreme cases have reduced, or even eliminated, populations over wide areas. The effects of organochlorine pesticides on birds of prey provide a well known example. These chemicals, which include DDT, aldrin and dieldrin, were widely used in agriculture in the 1950s and 1960s over much of the northern hemisphere. They caused widespread population declines among birds of prey, especially in bird-eating and fish-eating species. The Peregrine became extinct as a breeder in much of Europe and North America, including the whole of the eastern half of the United States. Other species, including finches, doves and waterfowl, were also affected, wherever these chemicals were used as seed-dressings, as unsown seeds were available as food on the soil surface.

Population declines were caused mainly by increased mortality, but some birds of prey also suffered a decline in breeding rate. This was brought on by the chemical DDE (a breakdown product of DDT), which caused eggshell thinning and egg breakage, so that insufficient young were produced to maintain the population. Indeed this mildly toxic chemical may have caused extinctions of some regional populations entirely through its sublethal effect on reproduction.

Once the use of organochlorine pesticides was reduced, the affected species began to recover in numbers. Meanwhile, the use of such chemicals had shifted southward, to tropical and subtropical areas. Here they are still having similar effects on birds as they did in the north 30 years earlier. Some other types of pesticides have periodically killed large numbers of birds, but have not produced the widespread population declines associated with the organochlorines.

*Numbers of Peregrine Falcons (Falco peregrinus) decreased dramatically in Europe and North America following the widespread use of toxic chemicals as insecticides and herbicides in the 1950s. Numbers are now recovering*

Sometimes quite unexpected activities can lead to poisoning and population decline in birds. The shooting of waterfowl is affecting numbers, not only through birds that are hit, but also through the poisoning by lead shot of those that are left. At some waters, shooting has been so frequent over the years that large amounts of spent shot have accumulated in the bottom sediment. Waterfowl consume this shot, along with grit (which they need to digest their food), and over a period of weeks become poisoned and die. A similar problem has reduced the numbers of Mute Swans on some English rivers, but here the lead is derived from the shot and weights discarded by fishermen. In North America some of the waterfowl that are shot but not recovered by their human hunters are subsequently eaten by Bald Eagles and other scavengers, leading to secondary poisoning. All these problems could be solved if lead were replaced by some other less toxic material.

With the continuing increase in various kinds of pollution and in pesticide use, toxic chemicals are likely to become increasingly important in affecting the numbers and distribution of birds, banning them from some otherwise suitable areas. Indeed, as the effects of toxic chemicals are so difficult to detect, they could be much more important now than we realize.

IN

## Population dynamics

There are two ways of explaining a change of population size. The first is in terms of external factors – basically changing environmental conditions – such as food supply or predation pressure, which permit an increase or force a decline. Then there are the internal demographic factors – the changes in births, deaths or movements – which are the immediate cause of the population change.

In a stable population, births and immigration are balanced by deaths and emigration. A change in numbers can be brought about by change in any or all of these various parameters. Many biologists have studied only the internal demography and, while this has shown what happens within a population during a change, it has revealed nothing of the underlying cause of the change.

Nor can demographic studies alone explain why densities remain much higher in certain habitats than in others. Populations in two areas may remain at markedly different levels, yet if they remain stable, they could show identical rates of births, deaths, immigration and emigration. Again, such differences in mean densities can be explained only in terms of external factors, such as food supply.

### Dispersal

Movements form an integral part of population regulation, because they enable rapid changes in local densities in response to changes in conditions. They enable birds to escape areas with high numbers and low food supplies and find other areas where conditions are better. Indeed species which occupy temporary habitats, such as an early stage in forest growth, rely on their dispersive powers to survive long-term.

Whatever the species, most individuals leave the place where they were raised and move some distance to breed. They then tend to return to the same breeding locality in subsequent years. As a rule, fidelity to an area (or 'philopatry') occurs in both resident and migrant, solitary and colonial species. Typically, the distance between natal site and breeding site is greater than the distance between successive breeding sites. In the Great Tit, for example, young males may shift some four to seven territory widths between natal site and breeding site, but then breed in the same or a neighbouring territory for the rest of their lives.

In most birds that have been studied, females generally disperse further than males, but in some, the males move further. The first pattern is found mainly in species in which the male defends the territory, and the second mainly in species such as waterfowl. The reason for this difference is that in these species the male defends the female and it is the female that takes the lead in deciding on the breeding area, the nesting site and where to take the brood after hatching. Clearly, under these circumstances it is females that will benefit from being on familiar ground, and therefore from not dispersing. One result of a sex difference in dispersal is that the chance of inbreeding is reduced, but this is probably an incidental consequence of movement patterns serving some other purpose.

In some migrant species individuals return to the same wintering site. There are many examples among birds of prey, including Steppe Buzzards which migrate between Siberia and South Africa. The main advantage of strong site fidelity presumably comes from familiarity with the area.

Site fidelity is most apparent in species whose food supplies are fairly predictable, giving some advantage in returning to the same area. Other species, which depend on sporadic food sources, are much less faithful to particular sites. Finches which eat tree-seeds, and raptors which eat rodents, often encounter abundant food supplies in different areas in different years, and may shift several hundred kilometres between successive breeding or wintering sites, as ring recoveries have shown. Such exceptions only serve to emphasize the role of food supplies in influencing the movements, and hence the local densities, of birds.

## Life history features

Flying birds range in weight from about two grams (some of the hummingbirds) to 17 kilograms (Great Bustard), and flightless birds range up to 150 kilograms (Ostrich). Among birds as a whole, body size largely determines life history features. The larger the species: (1) the longer it tends to live; (2) the later the age at which it begins breeding; (3) the longer the breeding cycle; and (4) the fewer the young produced at each attempt. Near one extreme, the small Blue Tit lives up to about eight years and begins breeding in its first year. It lays 10–12 eggs in a clutch, with one-day intervals between each egg, and has incubation, nestling and post-fledging periods lasting about 14, 16 and 10 days respectively, bringing the total breeding period from the first egg to about 50 days. Moreover, two broods may be reared in a single year. The maximum increase in population possible in a year is thus 10–12 times the breeding population, assuming that both the parents survive. Such fast-breeding species are often said to be '*r*-selected', from the technical notation for the intrinsic rate of increase, *r*. At the other extreme, the large Wandering Albatross can probably live more than 50 years, and does not begin breeding until it is about 10 years old. It lays only one egg at a time, and has incubation and nestling periods lasting about two months and up to twelve months respectively, so that annual breeding is impossible. Ignoring the non-breeding immatures, the maximum possible increase in a population of albatrosses is 50% in two years, or 25% in one year. Such slow-breeding species are often said to be '*K*-selected' from the technical notation for carrying capacity, *K*. These two contrasting species lie at opposite ends of a spectrum, between which other species show a continuum of variation in life history strategy.

Although in general, life history features relate to body size, exceptions occur. Thus swifts are fairly small, but are long-lived with small clutches, while grouse-like birds are short-lived with large clutches. Also, the smallest of all birds (the hummingbirds) lay only two eggs, while the largest (Ostriches) lay up to 10. Still, however, the main features are correlated, so that long-life, delayed maturity and small broods go together, as do short-life, early maturity and large broods. Similar patterns occur in mammals, where they also relate to body size.

Within this general trend is another, for tropical species to be more '*K*' in character than their high latitude equivalents. Thus tropical birds are generally longer-lived, with smaller clutches and longer breeding cycles than similar temperate zone species. Such trends are apparent even within species, and in many, mean clutch sizes decline with increasing distance from the equator.

The particular life history strategies shown by different species greatly affect their population dynamics. In big long-lived species, population turnover is generally slow, with more overlap between generations and a more stable age structure, all of which tend to dampen short-term fluctuations in numbers. There also tends to be a large non-breeding sector, consisting mainly of immatures. Less than half the total population may breed in any one year, producing only a small number of young. In small short-lived species, by contrast, population turnover is rapid, with less overlap between generations, a less stable age structure, and a high production of young, all of which facilitate short-term fluctuations in numbers. Most individuals that survive a winter will have the opportunity to breed in spring, so that the non-breeding sector remains small. And because of their fast breeding rates, small species can recover from a population low more quickly than can larger species. They are therefore better able to withstand heavy predation and persecution by man.

Information on longevity, annual mortality and age of first breeding in birds comes mainly from ringing studies, in which individuals are marked as nestlings and reported at later stages in life. Details are given for a range of species in the table, together with their clutch sizes. Several patterns have emerged. Comparing species, maximum life-span in the wild is roughly proportional to the fifth root of body mass: $L = 17.6M^{0.20}$, where $L$ is longevity and

*Life history features and body weight of various non-tropical bird species. Longevity records are likely to increase as more ringed birds are recovered*

| Species | Maximum recorded longevity (years) | Annual adult mortality (%) | Age of first breeding (years) | Clutch size | Body weight (g) |
|---|---|---|---|---|---|
| Blue Tit | 10 | 70 | 1 | 12–14 | 11 |
| European Robin | 13 | 52 | 1 | 4–6 | 18 |
| Song Sparrow | 8 | 44 | 1 | 4–6 | 30 |
| House Sparrow | 12 | 50 | 1 | 3–6 | 30 |
| European Starling | 20 | 50 | 1–2 | 4–6 | 80 |
| American Robin | 10 | 48 | 1 | 4–6 | 100 |
| European Blackbird | 20 | 42 | 1 | 3–5 | 80–110 |
| Barn Swallow | 16 | 63 | 1 | 4–6 | 20 |
| Common Swift | 21 | 15 | 2 | 2–3 | 36–50 |
| Tawny Owl | 18 | 26 | 2 | 2–4 | 680–750 |
| Mourning Dove | 17 | 55 | 1 | 2 | 140 |
| Woodpigeon | 16 | 36 | 1 | 2 | 450–550 |
| Atlantic Puffin | 22 | 5 | 4 | 1 | 350–550 |
| Black-legged Kittiwake | 21 | 14 | 4–5 | 2–3 | 300–500 |
| Herring Gull | 36 | 6 | 3–5 | 3 | 750–1250 |
| Curlew | 32 | 26 | 2 | 4 | 575–800 |
| Redshank | 17 | 31 | 1–2 | 4 | 110–155 |
| Lapwing | 23 | 32 | 1–2 | 4 | 200–300 |
| Avocet | 25 | 22 | 2–3 | 4 | 250–400 |
| Pheasant | 8 | 58 | 1–2 | 8–15 | 900–1400 |
| Kestrel | 17 | 34 | 1–2 | 4–6 | 190–240 |
| Buzzard | 26 | 19 | 2–3 | 2–4 | 550–1200 |
| Osprey | 32 | 18 | 2–3 | 2–3 | 1200–2000 |
| Mallard | 29 | 48 | 1–2 | 9–13 | 850–1400 |
| Tufted Duck | 15 | 46 | 1–2 | 8–11 | 550–900 |
| Eider | 18 | 20 | 2–3 | 4–6 | 1200–2800 |
| Barnacle Goose | 23 | 9 | 3 | 3–5 | 1400–1600 |
| Mute Swan | 22 | 10 | 3–4 | 5–8 | 10000–12000 |
| Grey Heron | 25 | 30 | 2 | 4–5 | 1600–2000 |
| White Stork | 26 | 21 | 3–5 | 3–5 | 3000–3500 |
| Shag | 21 | 16 | 3–4 | 3–4 | 1750–2250 |
| Short-tailed Shearwater | 31 | 5 | 5–8 | 1 | 530 |
| Royal Albatross | 36 | 3 | 8–10 | 1 | 8300 |
| Yellow-eyed Penguin | 18 | 10 | 2–4 | 2 | 5200 |

*The life cycles of small and large birds differ significantly, as a contrast between the European Robin (Erithacus rubecula) and the Common Buzzard (Buteo buteo) shows. Small birds generally start breeding at an earlier age, have shorter lives and enjoy a higher reproductive output than larger birds*

*M* is body mass in kilograms. Thus, on average, a doubling of body mass would lead to a 15% greater longevity, and body mass has to increase 32-fold for a doubling in longevity. In captivity many birds have lived longer than 50 years, including various parrots, large owls, large raptors, pelicans and penguins.

Although potential lifespan is of interest, in the wild extremely few individuals reach old age, as losses are high, especially among the young. Typically, in small songbirds with open nests, about one half, and in hole nesters about two thirds, of the eggs laid give rise to flying young, some of which then die before becoming independent of their parents. Similarly, in precocial species, whose young leave the nest soon after hatching, such as waders and ducks, only about one quarter of the eggs laid produce fledged young. Moreover, mortality is normally much higher in the naive juveniles than in the experienced adults. As a result, only between 5 and 20% of the eggs laid give rise to adult birds (although this is, of course, a much higher proportion than in insects or fishes). The annual death rate is also high among adult birds of some species, ranging from 40–60% in various songbirds, ducks and gamebirds, and 30–40% in various wading birds and small gulls, to 20% in swifts, and 5–10% in some large seabirds and raptors.

Most small birds breed in their first year of life, unless they are unable to get a mate or territory. In large species, however, all individuals may defer breeding for one or more years, even when territories and mates are freely available. Many such species have distinct immature plumages, which they wear during this non-breeding phase. Typically, geese start breeding at two to three years, large gulls at three to four, swans at three to four, eagles at four to six, and large seabirds at five to ten. Some species may visit breeding areas for one or more years before they actually nest, and individuals may breed at a younger age than usual if the population is depleted, or if resources are otherwise abundant. The main advantage of long deferred maturity is the greater experience, particularly of foraging, that the bird can gain. The demands of breeding are high, and within such long-lived birds, it may in the long run be advantageous to wait, rather than to breed when young, and jeopardize future survival.

Some birds, like the albatross mentioned above, lay only a single egg at each attempt, while others lay up to 20. In most species clutch size is variable within limits, and may change with latitude, habitat, date in season and female age. In altricial species, which feed their young in the nest, clutch sizes are thought to correspond roughly with the maximum number of young that the parents can feed at one time. In precocial species, clutch sizes are thought to correspond with the maximum number of eggs which the female can produce and incubate effectively. However, clutch size is obviously related to other factors, such as the size of the eggs and the growth rate of young. Its evolution is therefore best not considered in isolation, but as one component in a particular life-history strategy.

In some bird species attempts have been made to find whether reproduction has costs to the individual which are manifest in reduced survival or breeding success in later years. Studies on some species, such as House Martin, have indicated that such costs do indeed occur, whereas studies on other species have indicated that they do not. Most such work has been on small short-lived birds, however, and more studies of this question are needed in long-lived species, with deferred maturity.

IN

## Management of bird populations

The study of bird populations has practical relevance in conservation, pest-control and game management. The preservation of rare birds usually presents no problems in principle, as the species concerned have nearly always become rare because they have been killed or because their habitats have been destroyed by man (usually for agriculture). Hence, conservation is achieved by stopping the killing, so that the species can increase again, or by setting aside and maintaining areas of suitable habitat. Conservation is most difficult for large long-lived species, because these have slow rates of increase, and in addition, the big raptors require large areas of land. One complicating factor for migrant species is that habitat must often be provided in both breeding and wintering areas, and at staging posts en route, as is commonly practised for waterfowl in North America. With species that have become rare for some other reason, such as exposure to pollutants or new disease, research may be needed to ascertain the cause, and special measures to remedy it.

To reduce the numbers of birds which have become crop pests, the same principles can be applied in reverse, namely the removal of habitat or the enactment of control measures. Habitat removal provides a long-term solution, but it may be impractical or harm desirable species too, while killing entails a sustained commitment. Since many pest species, by their nature, can reproduce rapidly, control operations need to be continued indefinitely, without respite. Such programmes can easily become more costly than the damage done. Scaring pest species off vulnerable crops can sometimes work, so long as there are alternative feeding areas nearby.

In game management, the objective is the same as in the harvesting of any other animal, to obtain the maximum possible yield each year without depleting the stock long-term. It is like living off the interest of a bank account, while maintaining the capital. Without effective regulations, the over-harvesting of birds (like fish and whales) is all too common, and population decline sets in. Some of the most successful game management is practised on private land, where hunting is regulated, so as to maintain the stock long-term. Management of habitat is necessary to ensure high carrying capacity, and control of selected predators will often help to improve the production of young. To obtain the largest possible bag, hunting should then be concentrated in the period immediately post-breeding, before dispersal and other mortality have taken much toll.

Some biologists have tried to bring the findings from birds and other animals to bear on the problem of human population. The same principles of resource limitation hold, but the issues and solutions are complicated by sociological, political and religious aspects. As the Reverend Malthus put it nearly 200 years ago: 'the ultimate check to population appears to be want of food . . . but the immediate checks . . . are all resolvable into moral restraint, vice and misery'. Clearly, birds are simpler.

IN

## Timing of the breeding season

The time at which birds have their eggs or young in the nest is termed the breeding season. The timing of breeding is set by two types of factors. The first directly determine the success or otherwise of the breeding attempts made. For example, birds that lay and rear young in a seasonal environment at the time when most food is available are more successful and so leave more offspring. Since food availability directly sets the selective value of this timing it is called an *ultimate factor*. Several studies have shown that breeding seasons are often timed to the seasonal peak in food level in this way. Crossbills in northern Europe may lay in February to take advantage of the spruce cone crop. Elsewhere in Europe the mammal-eating Common Kestrel starts breeding in late April when small mammals are relatively easy to catch in short grass, but European Sparrowhawks, which feed on small songbirds, breed slightly later, in early May, and their young are in the nest just as many woodland songbirds fledge, creating a large pool of easily caught prey.

In the same areas the insect-eating European Hobby breeds a week or ten days later again, deferring its nestling period to the time when large insects are abundant on the wing. Yet another raptor in southern Europe, Eleanora's Falcon, does not lay until mid-summer but preys on the many young and tired migrants passing through the area in early autumn.

A similar gradient in time of breeding is apparent among the aerial insectivores. The Barn Swallow, for instance, takes many small early-emergent insects and breeds earliest. The House Martin takes rather larger insects that emerge only later in the spring and early summer and breeds correspondingly later, whilst the European Swift, which depends on large numbers of high-flying insects, does not lay until late May and produces its young some three weeks later when aeroplankton is abundant.

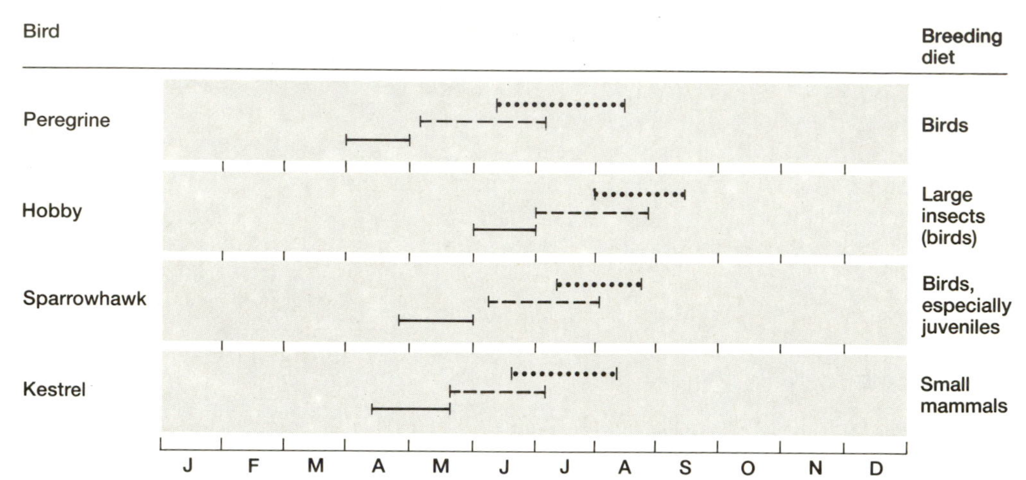


*Breeding seasons of raptors in Britain. Species are arranged approximately in order of body-size. For each species, the lower continuous line shows the period when clutches are started, the middle broken line the period when young are in nests, and the upper dotted line the period when flying young are fed by parents. The main trend is for large species to begin breeding earlier and have longer breeding seasons than smaller ones, but among species of similar size, timing is also related to diet, shown on the right*

Despite the general ability of species to breed when food for rearing young is most abundant, there are anomalies. For example, the young of Great Tits and Manx Shearwaters that fledge earliest are most likely to survive to breed, because they are heavier at fledging. It would be to the benefit of the late-laying tits and shearwaters to lay earlier when the prospects for raising a heavy youngster were better. Why do they not do this? One idea is that, early in the season, food conditions are too poor to allow the female to form eggs. Once conditions improve, the female lays but, by that time, the opportunity to rear nestlings at the peak of food abundance has been lost.

The ultimate factors explain why birds breed when they do, but not how birds time their breeding seasons. Before young birds hatch, their parents will have had to take a territory, build a nest, lay the eggs and incubate them. Before copulation and egg laying, the gonads of both parents will have had to grow. These events can take many weeks or even months. So for young to hatch when food becomes abundant, it is essential that birds predict the onset of the breeding season well beforehand by responding to environmental stimuli called *proximate factors*. Food supply itself cannot be used as the main proximate factor.

The environmental variable used as the proximate factor by birds outside the tropics is daylength, since this changes absolutely predictably during the year, and between years. Increasing daylength during spring causes the gonads of both sexes to grow at a rate that ensures that they will be fully mature at the time when the eggs should be laid. Near the time of egg laying, other environmental variables may act as modifying proximate factors. Bad weather, low temperatures or lack of food may delay (or even prevent) egg laying. However, there is no evidence that any environmental stimuli can initiate gonad growth in the absence of appropriate daylength.

Species within the tropics experience no marked seasonal changes in daylength. Many such birds show approximately yearly cycles of gonadal maturation and regression largely in the absence of external cues. The amplitude of changes in gonad size is much less than in temperate-zone birds. The gonads remain large for a longer period of the year, so that birds are ready to breed as soon as conditions become favourable, such as immediately after rainfall. It is interesting that, in experimental conditions, such species do respond to long days even though they never normally experience them; if they are exposed to long days outside the breeding season, the gonads grow rapidly.

Tropical seabirds often experience even less environmental change than terrestrial birds; frequently their food supply remains constant. Consequently, there is little selective pressure to breed solely at one time of the year. Some such birds breed at non-annual intervals. For example in the Sooty Tern on Ascension Island, the whole population breeds together every nine months. In other species (e.g. tropicbirds) breeding is asynchronous so that, at any one time, some individuals will be breeding.

Although the increase in gonadal size during spring in temperate-zone birds is largely stimulated by increasing daylength, gonadal size is not simply proportional to daylength, otherwise breeding seasons would always be symmetrical about the summer solstice. This is in practice rarely the case. In the majority of species, breeding ends well before the return of short days during the autumn. If it did not, young raised late in the summer would have little time to enable them to attain an adequate body condition to survive the winter. Thus ultimate factors impose selection pressure on the timing of the end of breeding, just as they do the beginning. And again, the food supply itself cannot be used to time the end of breeding, otherwise eggs would be laid at the end of the period of good food abundance, leaving young to be raised when food supply was poor. Instead, change in daylength acts as the proximate factor to end seasonal breeding, just as it does to start it.

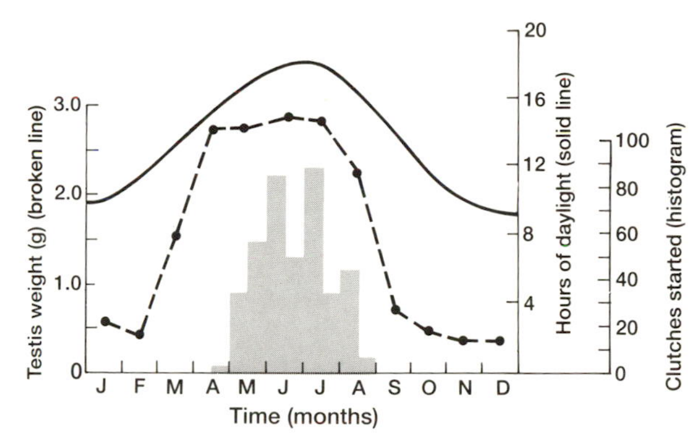


*Seasonal changes in day length, testis weight and egg-laying in House Sparrows (Passer domesticus) in Cambridgeshire, eastern England*

Birds can measure daylength accurately. Exactly how they do this remains unclear, but it is thought to involve photoreceptors within the brain. These receptors receive light directly through the skull rather than via the eyes. Also within the brain there are cells which synthesize and release a hormone called gonadotrophin-releasing hormone, GnRH. The size and maturity of the gonads is ultimately determined by the amount of GnRH being released from the brain.

The way that seasonal changes in daylength control the annual cycle of gonadal size is thought to be as follows. During the autumn, as daylength shortens, synthesis of GnRH begins. But because daylength is short, the release rate of GnRH is low, so gonadal maturation occurs only slowly during winter. However, this is often enough to initiate sexual behaviour. Many birds begin to sing during autumn and winter. In species in which the bill changes colour with sexual maturity, this colour change begins during autumn or winter. Later, as daylength increases during spring, the release rate of GnRH increases, the gonads grow more rapidly and eventually copulation and egg laying occurs. As daylength increases further, and critical daylength is exceeded, synthesis of GnRH ceases, so the gonads regress and breeding ends. Although GnRH synthesis is switched off by long days, it can be some weeks after critical daylength is passed before synthesis finally stops. The breeding season of many species continues beyond the summer solstice.

The rate at which release of GnRH increases with increasing daylength, and the critical daylength which causes GnRH synthesis to stop, have evolved in each species, so that each begins breeding at a particular time, and its breeding season lasts a particular time. These timings will exactly suit the food supply and other ultimate factors affecting that species.

RJO and AD

## Nests

Having embarked on breeding activities, the great majority of birds build a nest of some sort. A nest is a special construction in which eggs and young develop. Nests, designed by natural selection to help parents meet the needs of their young, are guides to really significant ecological factors in the life of birds. In the different nests of related species one may discern the directions nest evolution has taken, whereas convergent evolution in the nest building of unrelated species gives clues to the nature of the selection pressures involved.

### The principal types of nests

#### *Mound builders: eggs generally buried in the ground*

In the early stages of avian evolution from reptiles, some birds probably continued or reverted to burying their eggs, as many reptiles and as megapodes (Galliformes) among modern birds do today. The eggs are incubated by heat from decomposing vegetable matter, or by heat from the sun, volcanic activity, or hot springs. Nests of the nine species of megapodes vary from small pits dug in the sand just large enough for one egg, to gigantic mounds of sand and decaying vegetation as much as 18 metres long and 5 metres high, perhaps the largest structures made by birds. The Mallee Fowl, a megapode, lives in semi-arid parts of Australia where the temperature ranges from below freezing to above 38°C during its breeding season. It buries its eggs in a sandy sun-lit mound into which the male scratches moist leaf and twig litter, the decomposition of which generates heat in the fashion of a compost heap. By scratching sand on or off the mound the male closely regulates the temperature about the eggs, keeping it between 32 and 35°C, aerates the soil and reduces high carbon dioxide concentrations. The great variation in hatching times (40–90 days) and the great labour involved suggest that mound incubation is not as efficient as is incubation by sitting on the eggs.

*The male Australian Malee Fowl (Leipoa ocellata) builds a mound of earth and vegetation in which females lay their eggs. The decomposing vegetation generates sufficient heat to incubate the eggs, and the young birds hatch and fledge with no other parental care*

#### *Hole nests: eggs incubated in an enclosed cavity or burrow*

Hole nesters include whole orders of birds such as kiwis, parrots, trogons, coraciiforms (kingfishers, bee-eaters, rollers, hornbills and allies), and piciforms (woodpeckers, toucans, puffbirds and allies), as well as many passerine species. Nesting in holes provides shelter from the weather and the eyes of predators and conserves energy. In the wet tropics of Costa Rica, Alexander Skutch, over more than 30 years, found 61% nest success for birds nesting in holes (145 nests, 16 species) and only 31% for species with various nest types in the open (885 nests, 49 species).

The European Bee-eater breeds in southern Europe during the hot dry season in colonies in earthen banks, laying its eggs in an unlined chamber at the end of a tunnel one to two metres long that the birds dig by loosening the sand with the beak and kicking it back with the feet. Despite intense heat, up to 50°C soil temperature at the surface of the ground, the temperature inside the tunnel remains close to a comfortable 25°C. Moisture from the breath of nestlings and adults keeps humidity in the nest chamber at or near saturation. Gusts of wind across the mouth of the tunnel rapidly ventilate the nest chamber and keep carbon dioxide levels generally tolerable.

The male European Starling places bits of fresh green vegetation into a treehole or nest box a month or more before the bulk of the nest of dried grasses is built by both sexes. The birds continue to add bits of fresh green plants into the matrix of dried grasses until the eggs hatch. Recent evidence indicates that the green plant material is selected from plants that release chemicals with a fumigant action effective against such arthropod ectoparasites as bird lice. Many passerine birds, especially those nesting in holes, incorporate fresh green vegetation into the nest.

In Africa, the female hornbill walls herself up in a nest cavity in a tree, using pellets of mud and droppings to reduce the entrance to a narrow slit through which she can insert her bill to be fed by the male. She and her brood peck their way out later. Temperatures taken at the nest cavity of a Grey Hornbill in South Africa showed that the low thermal conductance and the thermal lag through the thick walls of the tree trunk helped keep daily changes inside the chamber between 21°C and 31°C whereas the tree surface varied from 15°C to 44°C. Warm air from the bird's body rises and flows out of the top part of the entrance slit; cooler air enters at the bottom of the slit.

Kingfishers nest in holes, and about 25% of the world's kingfisher species nest in termite nests. The termites seal the exposed portion of the walls of the hole dug by the birds, so there is no actual contact between the birds and the insects.

*Open nests: a scrape, platform, cup, or bowl*

In the Atacama desert in Chile where the Grey Gull nests, there is no rain and no vegetation, and its nest is a mere scrape in the ground. Similarly, in grassy salt marshes of the eastern United States, subject to floods or tides, Laughing Gulls add plant materials to their ground nests throughout incubation, and heavy rains are followed by increased nest building. Nests of diving ducks on platforms over water are generally safer from predators and more successful than nests of dabbling ducks on dry land.

The problem of placing a nest in a tree varies with body size. Large birds can use twigs and branches not easily blown out of the tree. The large platform nests of the American Bald Eagle and of the European White Stork are added to year after year and may last for decades. Some birds use mud; Australian Magpie Larks plaster their mud bowl nests on horizontal branches and on various man-made structures. Many very small birds use spider or insect silk to attach the nest and bind the nest materials together, for example, hummingbirds in the Americas and many honeyeaters in Australia.

The Goldcrest, the smallest European bird, breeds as far north as northern Scandinavia. Its deep cup nest, suspended in a spruce tree, is made of mosses, lichens and cobwebs, and is heavily lined with hair and feathers which have very high insulating value. One possible reason

*Grey Gulls (Larus modestus) breed in Chile's Atacama Desert. Their nest is nothing more than a scrape on the ground*

*Far left. A number of birds build their nests in or close to the nests of social insects and benefit from the protection they provide. Here a male White-tailed Kingfisher (Tanysiptera sylvia) sits outside a nest built in a termite mound*

*Left. Many hummingbirds, like this Blue-tailed Emerald (Chlorostilbon mellisugus) use silk threads from spiders' webs to construct or attach their nests*

for the prevalence of open nests among small birds, especially in temperate and cool climates, may be a need for the warming rays of the morning sun.

*Domed nests; roof constructed by the bird*
Small birds especially benefit from the protection that a roof gives from sun, rain and the eyes of predators. Domed nests are most frequent among passerine birds in the tropics. The ancestors of passerine birds were probably coraciiform (roller-like) birds, which are hole nesters. It is uncertain whether the earliest passerines had hole nests, cup nests, domed nests, or a combination of these as do present-day tapaculos (Rhinocryptidae), a small family of relatively primitive passerines from Central and South America, which often place their nests in holes.

Domed nests may be of very different materials in different birds: mud in some swallows and in the Rufous Ovenbird or Hornero of South America, a mass of short, heterogeneous plant materials bound with spider web in sunbirds, matted plant fibres in some New World flycatchers, or woven of long leaf strips in some weaverbirds and orioles. Convergent evolution in these diverse instances emphasizes importance of a roof in the life of many small birds.

The roof of domed nests helps shed rain. The Village Weaver of sub-Saharan Africa breeds during the rainy season in colonies in isolated trees, often in villages. Beneath the thin woven roof, the male thatches a thick ceiling of small tree leaves or of broad, short segments of grass leaf, so arranged that rain water is diverted off to the sides of the nest.

Snakes are more numerous and varied in the tropics than in cooler climates, and domed nests probably give some protection from snakes as well as other predators. There is an interesting convergence in the presence of two entrances (or exits) in the domed sleeping nests of a weaverbird in Africa and of an ovenbird in South America. Weaving is a more or less regular pattern of interlocking loops of flexible materials and enables nests to be fastened at or near the drooping ends of long branches in relatively safe places. Nests of some weaverbirds placed in trees have a bottom entrance at the end of a long vertical entrance tube. At the nest of a Baya Weaver observed in India, when a snake attempted to crawl down the outside of the long entrance tube, the tube buckled inward and the serpent fell off. The entrance tube may be over 50 cm long in the nest of Cassin's Malimbe, a red and black weaver of tropical Africa, which constructs one of the most skilfully made nests of any birds.

*Multi-chambered or compound nests with a communal roof*
Sociable Weavers of south-western African deserts are sparrow-like birds which reside the year round in nest chambers, up to 100 of which open separately on the underside of a huge nest mass which is thatched, rather than woven, of straws, dry grassheads and fine twigs.

The thick domed roof is built communally and helps protect the birds from strong sun, rain, wind, and many predators. In winter when air temperatures may sink below freezing in the Kalahari desert, the nest chambers are kept as much as 18 to 23°C above external air temperatures. The birds usually sleep in groups of four or five to a chamber in winter when not all chambers are occupied, and one to three to a chamber in summer when almost all chambers are occupied. The more birds that sleep together the more the heat produced. The larger the nest mass the more heat is retained in the winter and the greater the protection from strong sun in the summer, producing a selection pressure in evolution for increased nest size. The entire top of the supporting tree may eventually be covered by the nest mass as the birds keep adding materials over successive generations.

*This gigantic structure is the communal nest of the aptly named Sociable Weaver (Philetairus socius). Underneath a weatherproof thatch, several nests, each with a separate entrance, are constructed*

## How nests are built

*Description and development of nest building behaviour*
Both inherited actions and experience are of importance in nest building. The location of suitable nest materials must be learnt, and the performance of the normal activities of nest

building are probably self-reinforced, with the result that such actions are more likely to be repeated.

Most birds that build nests on the surface of the ground begin by making a circular scrape with the feet. The bird often lines this hollow with plant materials and may build up a nest rim around itself by passing materials back along one side of the body with the beak while sitting in the nest, turning around periodically. Many species add an inner lining of fine grasses which give good insulation, are light and easy to carry, and have high tensile strength.

The cup nest of many passerine birds is built by the integration of similar stereotyped movements, as by the female domestic Canary. An initial platform is built up into a cup, and loose strands projecting from the rim are pulled in and tucked down into this cup.

The female then presses down into the cup with her breast, pushes back hard with each leg alternately, and turns around periodically while sitting in the nest shaping the cup. As the diameter of the nest cup decreases she brings a higher proportion of lining materials such as feathers.

Domed nests are built in various ways: the walls of an initial cup may be gradually built up to form the roof as in the Cliff Swallow's mud nest and in the domed nest of the Long-tailed Tit which is built of mosses and lichens bound with spider silk; the entrance and nest cavity may be hollowed out from an initial, solid mass of various plant materials (as in some New World flycatchers and Old World sunbirds); or the roof, walls, and bottom of the nest may be built out from an initial loop or ring (as in most true weavers).

A male Village Weaver weaves his roofed nest of long, green, flexible strips that he tears from the leaves of grasses or palms. In weaving, he pushes and pulls ends of strips with his beak through the meshwork of the developing nest. The rounded form and normal size of the egg chamber result when the male stands on the bottom of the ring and pushes out the chamber with his bill as far as he can reach.

Each stage in building of the nest automatically provides the external stimulus for its own termination and for starting the next stage, and there is some apparent awareness of what a finished nest should look like, since experiments have shown that the adult male Village Weaver will replace parts cut away from almost any part of the nest. He generally adds a short entrance tube after a female has accepted his nest. Nicholas and Elsie Collias, of the University of California, inadvertently gathered some evidence for the theory of W. H. Thorpe that actions performed while building which make the existing structure approximate closely to a finished nest are reinforced and thus repeated. They repeatedly threaded one end of long strips of nest material into the rim of the entrance leaving the other end dangling. Each time the male would weave in these loose ends in an attempt to finish the rim, until the entrance tube was some 30 cm long, instead of the normal 5 to 10 cm. Many nests were built afterwards by this male without the Collias's intervention and, surprisingly, the entrance tube in some of them was as long or longer than in the abnormal one he was first induced to build.

The crude appearance of nests built by yearling male Village Weavers is caused by insufficient practice in selecting and preparing flexible nest materials, and in weaving. Young males manipulate all sorts of materials, but a preference for green materials is self-reinforcing and quickly increases with the manipulation of green as opposed to materials of other colours. Longer strips are used in later nests.

Internal factors in nest building have been most studied in domestic African Collared Doves. During courtship, both sexes have a reciprocal effect on each other stimulating a rise in secretion of sex hormones. Testosterone secretion from the testes stimulates the male to gather nest materials and bring them to the female. Estradiol and progesterone from the ovaries are both required to stimulate nest building (working material into the nest) by the female.

During incubation, secretion of the hormone prolactin from the pituitary gland gradually increases and inhibits secretion of the sex hormones, so sexual behaviour stops, while nest building (which is sensitive to lower hormonal levels) declines more gradually.

*The commonest form of bird's nest, the conventional open-cup, here belonging to a Red-winged Blackbird (Agelaius phoeniceus)*

*A male Black-headed Weaver (Ploceus melanocephalus) weaves fresh, green grass stems to construct a nest. At an early stage in construction (above), the male stands on the ring prior to starting the nest chamber itself. On completion (below) this is provided with an entrance tunnel*

*Who builds the nest?*

Cooperation between male and female in nest building is the rule among birds, occurring in more than 80% of families. Building of the nest by the female alone is fairly common (28%), especially among passerine birds (48%). In only 6% of bird families are there species in which only the male builds the nest, as among jacanas and phalaropes, polyandrous waterbirds in which the male alone incubates.

In many species the male brings nest materials to the female, who builds the nest. The male may do most of the building if he uses the nest in his courtship. The male Village Weaver displays his nest to unmated females. If a female accepts the nest, she lines the nest with soft grass tops and mates with the male. If his nest is repeatedly rejected the male tears it down and puts up a fresh model. In many tropical and sub-tropical species, more than two birds, particularly a pair and their young from previous broods, may help build a large communal nest.

*Work involved in nest building*

The amount of work needed to build a nest can be measured, theoretically at least, by measuring and calculating the metabolic cost (in terms of calories, joules and watts) of different building activities. Many species of passerine birds make over a thousand trips to gather the materials used in making a nest, and the distance required to travel for materials is an important factor in deciding nest sites because of the relatively high energy demands of flying. Some birds, such as the Cliff Swallow, economize by using the same nest for a second brood, but then run the risk of increased infestation by nest parasites.

## Nest sites

Competition for nest sites has resulted in the evolution of a great diversity of specialized nesting sites for different species. The need for safety from predators has often led to the location of nests in harsh climates or in relatively inaccessible places. In turn, the nature of the nest site profoundly influences the whole behaviour and life history of a species.

*Factors affecting choice of nest sites*

Birds breeding in cold climates or on high mountains seek shelter from cold winds and storms and from the cold night sky by nesting under or in sheltering vegetation, in crevices, holes or caves.

Nests of many desert birds are placed on the ground next to a sheltering bush or rock and receive the warming rays of the early morning sun, but are shaded in the hot afternoon. On dry sloping ground, some wheatears, rock wrens, finch-larks, and desert larks use small stones to build up and level the substrate, presumably making the nest more stable and possibly aiding temperature control in the nest.

The Horned Coot in Andean lakes builds a small islet of stones on which it places its nest. The Egyptian Plover, or Black-backed Courser, which breeds on hot sand-bar islands along rivers in tropical Africa, buries its eggs and even small young under the sand, and then cools them by water carried to the nest site in its belly feathers.

Many birds derive protection from wind and rain by placing their nest in trees on the side opposite to the direction of prevailing winds and storms. In the rain forests of South-East Asia the spider hunters, members of the sunbird family (Nectariniidae), fasten their nests to the underside of a broad leaf giving shelter from the frequent heavy rains.

Sites which are safe from predators, such as cliff ledges, often pose problems in fastening the nest. Kittiwakes fasten their nests on cliff ledges with algae, mud and their droppings. The Rock Warbler of Australia uses spider silk which has great tensile strength to hang its domed nest from the roof of a cave. Nests made of saliva by the Edible Nest Swiftlet are the commercial source of bird's nest soup in the Orient (p. 308). The saliva contains viscoelastic mucoproteins that hold the bracket-shaped nest to the wall of the cave.

Some tropical birds, particularly species with roofed nests, build their nests on or near human habitations, or close to nests of ants, bees, wasps, termites, or large birds of prey. The habit is widespread and in some instances has been shown to be more than a matter of chance and to reduce nest losses from predation.

Nest sites may vary among different subspecies, in different parts of the range of the same subspecies, locally in different habitats, or among individuals. The normally tree-nesting Ospreys and American Robins nest on the ground on Gardiner's Island, New York, in the absence of mammalian predators. Different individual birds vary greatly in how well they hide their nest and consequently in vulnerability to predators.

A Hawaiian honeycreeper, the Common Amakihi, builds open nests of twigs and grass stems, lined with lichens, rootlets, or wool in cold, dry savannah forest on the volcano, Mauna Kea. With an increase in altitude and cold, nest walls become denser (but not thicker) and have diminished thermal conductivity. But nests in the warmer rain forest on Kohala Mountain with four to five times as much rainfall were scarcely lined and had more porous walls permitting the nest to dry more rapidly after a shower.

*Nest-site selection*

In most birds nest-site selection is by the female, although the male may show her various prospective sites in his courtship. She bases her selection on such things as the local availability of food sources, nest materials and shelter, and the safety of the nest site.

The complexity and inter-relations of the many factors involved have been well shown by a recent six-year study in the lowland rain forest in Peru of colour-banded Yellow-rumped Caciques, large orioles in which the females weave their long, hanging nests in colonies, often located along rivers.

The most successful and most preferred sites were in trees on river islands, within one metre of occupied wasp nests, and within a cluster of interwoven nests. Such nests were safest from monkeys, hawks, and snakes, and their female owners were chosen as mates more often by the dominant males. More scattered or isolated nests within the colony tree were less likely to be successful. Failed nests were deserted and the female would then leave the area or try to select a site in a safe place for her next nest, provided she was not excluded by more dominant females.

Early experience has a strong and lasting influence on such factors in nest-site selection as return of migratory birds to the locality and habitat in which they were reared. Zebra Finches even prefer nest materials of the same colour as those in the nest in which they were raised.

*Competition for nest sites*

Replacement of one species from part of its range by another competing for more or less similar nest sites has been shown especially for hawks, gulls, and various hole-nesting birds. Population density of birds that nest in tree holes can often be dramatically increased by putting up suitable nest boxes, suggesting that nest sites are limiting. Predation rates are frequently higher in natural holes than in nest boxes, suggesting that hole-nesting birds must often do with what is available rather than with what is optimal. Competition for nest sites is reduced by different species preferences.

A recent study in western Canada by Peterson and Gauthier of the use of natural tree cavities by four species of cavity-nesting birds, including the introduced European Starling, showed that despite some overlap between species, cavity volume and to a lesser extent entrance area, were the most important variables characterizing nest sites. Starlings prefer to nest in nest boxes of a certain size and avoid boxes with a smaller volume in which their young may become overcrowded and die of overheating.

A nine-year study of hole-nesters in south Sweden showed that the rates of nest failure and predation were greater in nest holes or in nest boxes situated low in trees than in higher ones for the Starling, Blue Tit and Marsh Tit. The starling reduced the breeding success of the European Nuthatch by taking over holes occupied by the latter, especially in higher nests. The defence of the nuthatch is to reduce the entrance hole of its nest by plastering it with mud until starlings are excluded, while the smaller nuthatch can still slip through. All four species preferred higher nest holes when there was a choice, and on the average occupied decreasing nest heights according to the aggressive dominance status of each species.

*Tradition and evolution of nest-site changes*
Originally, the Grey Gull in Chile probably flew inland to nest around lake borders, and this traditional habit persisted as the lakes gradually dried up and gave place to the driest desert of the world. The gulls make daily feeding flights to the coast and back, taking advantage of the scarcity of predators in the desert nesting grounds.

Traditional nest sites may change as new more suitable sites become available. Chimney Swifts of eastern North America now generally nest in chimneys, only very rarely in hollow trees as they commonly did in Audubon's time. Vaux's Swift of western North America, which looks very similar, still usually nests in hollow trees. The Mauritius Kestrel, one of the rarest birds in the world, used to nest in tree cavities where it was vulnerable to introduced monkeys. Eventually, one pair nested on a sheer cliff face, safe from monkeys, and the young kestrels apparently became imprinted on cliff nesting sites resulting in a tradition shift and a dramatic increase in population growth.

NEC

# Eggs

## Laying behaviour

Nests are built to receive eggs. The physiology of egg formation is described in Chapter 2. Forming the perfect egg within the body requires some change in the behaviour of the female bird. The first requirement is to gather large amounts of food material to build up the eggs. A Blue Tit lays a clutch of eggs weighing considerably more than herself over the course of about two weeks and she needs 40% extra food at this time. Female birds therefore spend more time feeding in the days before laying. Many species are assisted by their mates who bring them food. This so-called 'courtship feeding' can significantly contribute to the nutrition of the female at a time when her needs are greatest.

Some birds have difficulty finding enough calcium to form the eggshell. The domestic chicken, for example, requires about two grams of calcium for each egg, but only 25 milligrams are circulating in her blood. The rest of the calcium is provided from a reserve laid down previously in her bones. Small birds, however, are unable to store sufficient calcium in their bones to supply a large clutch of eggs and they turn to special food sources. Sandpipers nesting on the Arctic tundra swallow the bones of long-dead lemmings; they can then lay a clutch of eggs containing more calcium than the adult's skeleton. Female Blue Tits and House Sparrows characteristically collect snail shells before retiring to roost and digest them to manufacture the shell of the next egg.

Once the first egg of the clutch has been laid, subsequent ones appear at regular intervals, the length of which is probably governed by the time taken to lay down various parts of the egg. The interval is about one day for most passerines, many ducks and the domestic chicken, and rises to a week in the Blue-faced Booby. If food is in short supply, some birds lay smaller clutches; others lay smaller eggs, or lay at longer intervals. The European Swift changes from laying three eggs at two-day intervals to laying two eggs at a three- to four-day interval in bad weather.

If, as in oystercatchers, the interval is not 24 hours, the time of day when an egg is laid changes through the clutch. However, many birds lay at a set time. Most passerines such as tits, wrens and finches lay in the early morning, while most non-passerines lay at other times; pheasants for example lay in the evening. Small birds probably lay in the morning for two reasons: the shell can be laid down overnight while the bird is sleeping, and the bird avoids having to carry extra weight during the day.

RWB

## Eggs: strength, shape and colour

*Strength*
The shell provides the mechanical strength of the egg and therefore varies with the risks of breakage prevalent in different species. For instance, guillemot eggshells are reinforced at their narrow end where they are in constant contact with the rocky ledges on which they are laid. In contrast, the eggs of megapodes

*The size, shape and colour of eggs vary enormously. From top to bottom: a clutch of African Jacana (Actophilornis africana) eggs, with their distinctive pencil-like scribbles, presumably for camouflage; the Common Guillemot's (Uria aalge) single, very pointed egg with more varied background colour and markings than any other species, aiding individual recognition on crowded breeding ledges; the large dark eggs of the Emu (Dromaius novaehollandiae); the unmarked, elongated eggs of the Shag (Phalacrocorax aristotelis); the almost spherical, white eggs of the Tawny Owl (Strix aluco); and the eggs of the little Ringed Plover (Charadrius dubius) beautifully camouflaged for a nest scrape on a stony shore*

are 'incubated' in mounds of rotting vegetation or in sun-warmed soil, are hardly touched by the adults, and are accordingly thin-shelled. Thick shells provide protection against breakage but are correspondingly difficult for the chick to cut through. Shells are weakened during incubation, though, for the embryo draws on the shell for calcium for incorporation into its developing skeleton.

*Shape and colour*

The shape of birds' eggs is very varied. Nevertheless, there is uniformity within a species and among related species. For example, owls and kingfishers lay almost spherical eggs, while aerial species like hummingbirds and swallows lay long elliptical eggs. It is possible that shape is related to the geometry of the pelvic bones – the deeper the pelvis, the rounder the egg, but there are also some correlations with breeding ecology.

Plovers have pyriform (pear-shaped) eggs that fit neatly together in clutches of four, points inward, so they are easily covered. Guillemots nesting on narrow rock ledges have tapered eggs that may allow the parents to incubate in an upright posture. The eggs roll in tight circles and are less likely to fall off, but other cliff-nesters have conventional eggs.

The colour of the egg is laid down in the testa of the shell, while the patterning is in the cuticle; in falcon and plover eggs it may be so superficial that it can be rubbed off. Colours are created essentially from two groups of pigments: reds and browns from porphyrin and blues and green from cyanin.

However, the ancestral egg was probably white, and pigmentless eggs are still common among birds nesting in holes (e.g. kingfishers, woodpeckers) where they may show up better and so be less liable to trampling. For example, Jackdaw eggs, normally bluish-white, were more likely to be broken by the parents when experimentally painted black. Birds that cover their eggs on leaving the nest (ducks, grebes) or incubate on laying the first egg (owls, penguins, pigeons) also have white or whitish eggs.

Camouflage is obtained by the ground colour which in some species closely matches the background. In species that lay on bare ground the outline of the egg is broken by the pattern of spots and scrawls. Some birds, such as guillemots and Royal Terns, recognize their own eggs by individual patterning.

Ecological correlates of colour patterns are widespread. Ground-nesters on dead vegetation often produce cryptic brown eggs but those on beaches or deserts lay pale sandy eggs. Thrushes nesting in tree-forks lay eggs heavily blotched to resemble shadow marks but those in crevices have speckled white or pale blue eggs and, as mentioned above, those in hole nests have pure white eggs.

## Egg recognition

Egg colouration is under genetic control and eggs of individual females may be readily recognizable, especially where confusion of egg ownership is possible, as with guillemots, where the eggs are laid on bare cliff ledges crowded with other incubating guillemots. Extreme variability in egg colour and pattern is present in this species: the eggs are heavily marked with blotches, spots or line patterns of a variety of colours and the ground colour may be blue, green, red, brown or white. The individually variable egg pattern and colour assists each female to locate her own egg on her return to the nest.

Experimental manipulation of the eggs has shown that the birds will accept only eggs similar in pattern and colouration to their own. Royal Terns nesting in dense colonies also recognize their own eggs.

Egg recognition (or, at least, discrimination) is most pronounced amongst those species regularly exposed to brood parasites. However, not all species identify foreign eggs by recognizing them individually. Garden Warblers, for example, reject foreign eggs appearing in their nests during laying, apparently on the basis of their difference from eggs already laid. When all eggs of an incomplete clutch were experimentally removed from a nest and replaced with Lesser Whitethroat eggs, the female Warbler ejected her own egg the following day, clearly regarding it as the 'odd man out'.

## Egg size

Egg size varies across five orders of magnitude, from the 0.3 g egg of a hummingbird to the 1600 g egg of the Ostrich. Over this entire range, egg weight is an allometric function of female body weight within each order or family, with an exponent of 0.67. This means that for each doubling of female weight, egg weight increases on average by a factor of $(2.0)^{0.67}$ or 1.59. Egg weight also varies substantially between taxonomic groups. A 100 g female cuckoo lays an egg of only 4.5 g, a 100 g dove lays a 6 g egg, a 100 g falcon a 15 g egg, and a 100 g petrel a 21 g egg. These differences are partly due to different modes of development within these groups. An altricial species is reared in the nest and hatches at a less developed state from a smaller egg whilst a precocial species of the same weight requires a larger egg to allow it to attain greater maturity within the egg and still have reserves left until it finds its own food. In addition, the extent to which the chicks are covered with down on hatching is correlated with the size of the egg.

Two groups lay unusually small eggs. First, compared to non-parasitic cuckoos, parasitic cuckoos lay relatively small eggs, a trait adapted to their parasitizing species smaller than themselves. Second, swallows lay the lightest eggs for their size among passerines, presumably because a heavy egg could interfere with their agility in flying.

Within an individual species egg weights also vary considerably, both between females and between clutches laid by a single female. Individual females tend to lay eggs of a consistent size. Even so, females breeding for the first time lay smaller eggs than do older females, though in very long-lived species the very oldest females may again lay smaller eggs. Such variations in size are, however, slight since the survival of a nestling or chick is often correlated with the weight of the egg from which it hatched. This means that when a female has difficulty in acquiring enough food for egg formation, it is more productive to lay fewer eggs of relatively standard weight than to reduce the weight of each egg. Thus when European Swifts experience bad weather and food shortage after starting a clutch, the female takes three or even four days (rather than the usual two) to form the remaining egg and does not proceed to a third egg if the second is thus affected.

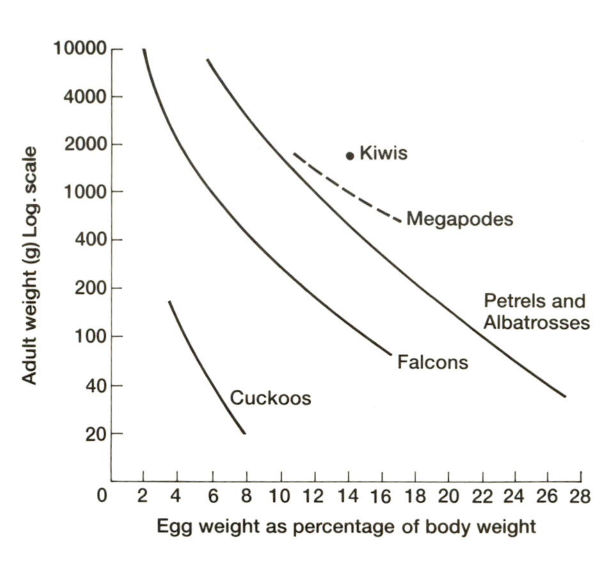


*Variation in the size of eggs relative to body weight across several families of birds*

Variations in egg weight within a clutch may also be physiological or morphological in origin. In some species the first egg of a clutch is often slightly smaller, perhaps because the oviduct stretches slightly with its laying. Small final eggs may occasionally be due to reserve depletion, in extreme form manifested as pathologically small 'runt' eggs.

Egg formation is a difficult task for the females of many species, particularly the smaller ones. The kiwi is a family with proportionately one of the largest eggs among birds, weighing about 25% of the female's weight. Some hummingbirds and storm petrels also produce an egg of this relative weight. Larger species have relatively smaller eggs because of the allometric scaling of egg weight to female size, and an Ostrich's, for example, constitutes just 1% of her weight.

For many species, though, more than one egg is laid in a clutch and the total energetic input required is correspondingly larger. For small birds the metabolic performance involved is impressive. A 17 g Great Tit, for example, lays more than its own weight of eggs; each egg weighs 1.7 g and the average clutch is 11 eggs. During egg formation females need 40% more food each day.

*The Brown Kiwi (Apteryx australis) produces the largest egg, relative to body size, of any bird, as clearly shown in this X-ray of a female shortly before laying*

## Number of eggs

All bird species lay a characteristic number of eggs in a clutch. The average number varies from just one egg in many seabirds to as many as fifteen (and even occasionally higher) in some gamebirds. Clutch size is strikingly related to taxonomic group, with virtually all plovers laying clutches of four eggs and all petrels and albatrosses laying just one egg. In a few groups, however, clutch sizes vary very markedly, often in response to food levels.

Clutch size decreases allometrically with body weight, so that within ducks and pheasants, for example, clutches are about 29% smaller in a 2000 g species than in a 200 g species. In addition, species laying large eggs have somewhat smaller clutches than similarly-sized relatives laying small eggs.

Hole-nesting species lay larger clutches than do species in more open sites. Within the duck subfamily Anatinae, for example, clutch size averages 10.4 across hole-nesting species but only 5.0 across all other species. In Central Europe the hole-nesting passerines typically average 6.9 eggs per clutch against just 5.1 for the open-nesters.

Several theories as to the size of avian clutches have been advanced. These include the ideas that birds lay as many eggs as they can cover and incubate properly; that they lay the maximum number they are physiologically capable of laying; that they lay the number needed to balance the mortality experienced by their population; and that they produce the number that maximizes the number of young reaching maturity. The first is unlikely to be true since hatching success varies little with the number of eggs laid. The second is also unlikely since the physiological capacities of the adults of various species have no relationship to their clutch sizes.

If eggs are experimentally removed daily, females often continue to lay further eggs and greatly exceed their normal clutch size. For example, a Northern Flicker is on record as laying 71 eggs in 73 days, despite normally laying only six to eight eggs. Similar experimental removals with other species have yielded comparable results, though not as extreme as with the Flicker.

The third idea – that clutch size is set to balance an independently controlled mortality – implies that populations are in general at some ideal size and that breeding adults rear fewer young than they are capable of rearing to prevent the population exceeding that size. A problem here is that any adult with a genetic make-up that led it to 'cheat' by rearing more than the required young would increase the representation of its own genes in the population. Hence the gene concerned would spread throughout the population, there being no mechanism selectively to penalize the cheaters (see Chapter 1).

The fourth theory, proposed originally by the Oxford ornithologist David Lack, has been substantially supported by field evidence and by experiment. Lack's hypothesis was that clutch sizes evolve to maximize the numbers of offspring in future generations. Too small a clutch size limits the number of offspring available, even were all of them to survive to maturity, whilst too big a clutch size decreases the probability of the young concerned reaching maturity.

Lack thought that the ability of the parents to get food to their young was the limiting factor. If so, too big a brood would mean that the parents would be unable to feed them all properly, so that some would die either in the nest or after fledging but before breeding. Lack's ideas were initially supported by studies of European Starlings, European Swifts, and Great

*Reproductive success of the European Starling with respect to brood size. The figures demonstrate that the maximum productivity corresponds to the modal brood size*

| | | | Recovered more than three months after fledging | | |
|---|---|---|---|---|---|
| Brood size | Number of broods | Number of young | Number | Per cent of individuals | Per cent per brood |
| 1 | 65 | 65 | 0 | — | — |
| 2 | 164 | 328 | 6 | 1.8 | 3.7 |
| 3 | 426 | 1278 | 26 | 2.0 | 6.1 |
| 4 | 989 | 3956 | 82 | 2.1 | 8.3 |
| 5 | 1235 | 6175 | 128 | 2.1 | 10.4 |
| 6 | 526 | 3156 | 53 | 1.7 | 10.1 |
| 7 | 93 | 651 | 10 | 1.5 | 10.2 |
| 8 | 15 | 120 | 1 | 0.8 | 6.4 |
| 9 | 2 | 18 | 0 | — | — |
| 10 | 1 | 10 | 0 | — | — |

Tits, with in each case the greatest number of surviving young coming from an intermediate clutch size.

Several problems with Lack's hypothesis emerged as further research was undertaken. One significant difficulty was that the most productive clutch size was not always the one most frequently observed in the population. In general, a clutch size slightly smaller than the most productive was most frequent. One possible explanation for this discrepancy is that individual females differ in their ability to rear young and therefore adjust their clutch size to reflect this. The larger clutches laid by the best females would then be more successful than average clutches laid by average-quality females.

This idea has been tested by experimentally manipulating natural broods of titmice so that they were larger or smaller than their original size. The reduced broods proved more successful, and the enlarged broods less successful, than were natural broods of the same size. This can be explained if each pair expected to be able to rear optimally a brood of the original size. Birds given an experimentally reduced workload then did better whilst those with increased workloads were too stretched to cope properly. A similar effect occurs with Magpies in territories of different quality, with birds given enlarged broods faring less well, irrespective of their initial clutch size.

Another problem with Lack's hypothesis is that it focuses on success within a single season. However, the clutch that yields the higher lifetime success rather than single season success is likely to be favoured. As a simple example, a bird that put enormous effort into rearing a brood of twelve young and then died through exhaustion would leave fewer offspring than a bird that reared only eight at a time but did so in each of two seasons. Hence if rearing a larger brood is accompanied by increased chances of adult mortality, selection will favour a slightly smaller clutch than the most productive.

An alternative but related explanation for an average clutch size smaller than the most

*Average clutch-size of nidicolous land birds in tropical Africa and mid-Europe. The bracketed figures give the number of species in each subfamily*

| Subfamily | Equatorial Africa | Mid-Europe |
|---|---|---|
| Accipitrinae (hawks) | 2.4 (5) | 4.3 (2) |
| Buteoninae (buzzards) | 1.8 (8) | 2.2 (3) |
| Milvinae (kites, sea-eagles) | 2.0 (2) | 2.5 (3) |
| Columbinae (pigeons) | 1.9 (14) | 2.0 (6) |
| Striginae (owls) | 2.5 (6) | 4.6 (8) |
| Caprimulginae (nightjars) | 1.9 (9) | 2.0 (1) |
| Apodinae (swifts) | 2.1 (5) | 2.6 (2) |
| Alcedininae (kingfishers) | 4.5 (3) | 6.5 (1) |
| Meropidae (bee-eaters) | 3.2 (6) | 7.0 (1) |
| Coraciinae (rollers) | 3.1 (3) | 4.0 (1) |
| Upupidae (hoopoes) | 5.0 (2) | 6.5 (1) |
| Picinae (woodpeckers) | 2.3 (8) | 5.9 (6) |

productive size lies in nest predation. Nest predation is particularly high for most birds and may fall especially heavily on larger broods. First, such broods are often hungrier and noisier than smaller broods and therefore more at risk of attracting predators. This also puts the parents concerned at greater risk of being captured on the nest by the predator. Second, large broods require more feeding visits than do smaller ones, allowing sit-and-wait predators more frequent opportunities of locating the nest. Third, since most species lay one egg per day larger clutches take that much longer to lay and eggs are in the nest for a longer time than with smaller clutches. With high predation rates this can significantly increase the risk to the brood.

Precocial species are not fed in the nest but travel about in search of food. Species differ in the extent to which they assist their young with feeding but most guard their young. With a large brood the young must cover a greater area in search of food and so are at greater risk of encountering a predator by chance. In addition, with a larger, more dispersed brood the adults are less likely to be able to guard each chick as carefully as within a smaller brood, again increasing the chances of predation. In an experimental test of this idea with Semi-palmated Sandpipers the success of artificially enlarged broods of five was only 57% that of broods of four (the natural brood size).

That predation may have influenced clutch size is demonstrated by the fact that species nesting in protected sites, such as hole-nesters, produce larger clutches than open-nesters. As the table above indicates, clutch sizes vary markedly within species, particularly with latitude. Among European Robins breeding in North Africa the average clutch is 4.2 eggs; in central Europe it is 5.8 eggs and in northern Europe 6.3 eggs. In principle the large clutches observed at high latitudes might be due to the longer daylengths there. The adults would therefore have more time in which to collect food for the additional young. However, the increases in clutch sizes are not proportional to the increase in daylength involved. What is more, even nocturnal species show clutch size increases, despite the shorter nights they experience. A more likely explanation lies in relative seasonality, an idea now known as Ashmole's hypothesis. At high latitudes seasonality is extremely marked, with winters being especially severe. Few species are capable of overwintering successfully in such conditions but those that do have access in spring to very large seasonal increases in productivity. Their *per capita* share of these resources is very high and allows each pair a large clutch size. Conversely, at lower latitudes more birds overwinter successfully and the summer resource peak is less pronounced. Hence a smaller summer bloom is shared amongst more birds, yielding a lower *per capita* share and smaller clutch size. In practice the differences in clutch sizes in areas as diverse as Costa Rica, Kansas, and Alaska have proved related to seasonality in this way. Similar seasonality may also account for another pattern in clutch size, of larger clutches in savanna habitats and smaller clutches in less seasonal rain forest.

The relative ease with which clutches can be laid and reared in areas of seasonally abundant resources also contributes to an abundance of migrants in such areas. These come in to exploit the summer resources but leave again in autumn rather than compete with residents through the winter bottleneck.

A major determinant of clutch size is population density. At high densities territories are relatively small and each pair may be unable to find as much food as when they possess a larger territory. For example, in Great Tits breeding in the Netherlands, clutches average ten eggs at low population densities (1 pair/10 ha) and about seven eggs at the highest densities (19 pairs/10 ha).

This explanation has also been advanced to explain the tendency for birds on islands to lay fewer eggs than those breeding on adjacent mainlands. Because islands have fewer species, the species that do occur sometimes exist at relatively high densities, creating relatively intense intraspecific competition, which may reduce clutch size.

In many species clutch size changes in the course of the season, usually with smaller

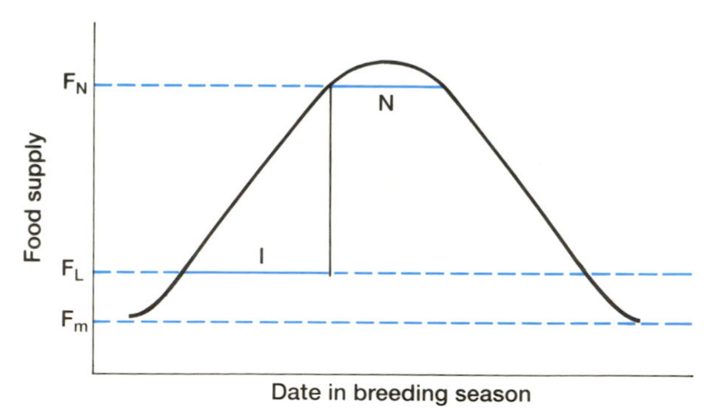


*The effect of changes in seasonal food on the timing of breeding. Normal female food requirements are met as long as the supply is above the level $F_m$. However, because egg formation costs energy, laying cannot start until the food supply has increased seasonally to level $F_L$. With a time (I) required to lay the clutch and incubate it to hatching, a period N is available with food level $F_N$ on which to rear the nestlings. After this, food supply has fallen below the minimum level needed for feeding the brood optimally. If food requirements for egg formation rise, say for a larger female laying larger eggs, relatively more of the rearing period will take place in inadequate feeding conditions*

clutches among the later breeders. Among European Starlings breeding in southern England, the earliest first clutches average about 5.5 eggs, the latest about 4 eggs. In single-brooded species this implies that late pairs breed at less suitable times. In several multi-brooded species such as the European Blackbird, clutch sizes first increase with date, then decline. Here the very early pairs produce their first brood in conditions that are less than ideal but are able to fit in a further and larger brood later in the season when conditions have improved. Timing of breeding is generally less critical for multi-brooded than for single-brooded species. By breeding in less than optimal conditions they can fit in two (or more) smaller broods which together can yield more offspring than a single optimally timed brood.

In many species young females lay smaller clutches than older birds and are often less successful in rearing even this smaller brood. Young birds usually lay later than older birds and so should have smaller clutches for this reason alone but the age difference in fact persists even among old and young females nesting at the same date.

## Number of broods

Most non-passerines and the larger passerines produce just one brood in a year, probably because their long nest cycles preclude a second attempt within a season. By the same token, species that provide extensive parental care may be unable to attempt a second brood. The Tawny Owl, for example, cares for its fledglings for nearly three months. In the same way,

*Effect of age on clutch size*

| Species | Clutch size: First breeding attempt | Clutch size: Later breeding attempts |
|---|---|---|
| Pied Flycatcher | 5.6 | 6.3 |
| Great Tit | 8.4 | 9.1 |
| Sparrowhawk | 4.8 | 5.4 |
| Black-billed Magpie | 4.0 | 6.0 |
| Black-legged kittiwake | 1.8 | 1.94–2.39 |
| Red-billed Gull | 1.33 | 1.69–2.09 |

the Wandering Albatross has a nest cycle of a year and successful birds can breed only in alternate years. Small species often attempt a second or even third brood and the Stock Dove in Britain may attempt up to five broods in a single year, largely because its dependence on grain and on arable weed seeds allow it a very long breeding season (from mid-February to mid-November). Extended food availability also allows other species additional breeding attempts in some years, with Barn Owls and Short-eared Owls rearing second broods when the rodents that form their main prey are especially numerous. The European Blackbird likewise attempts more broods in wet years when the soil is moist and earthworms are more easily extracted from the soil.

The cost to the parents of attempting a second brood may be quite significant. For example, House Martins that attempt a second brood in a year are significantly less likely to survive to return the following year than are birds that attempt only one brood. Hence second broods may be most profitable in such species only in years where conditions are particularly favourable for doing so. The Red-legged Partridge times both of its breeding attempts to the seasonal optimum by means of the female laying two clutches in quick succession, the cock then incubating one and the hen the other.

Further broods are clearly impossible once the food supply falls below the female's egg formation threshold. However, other factors may intervene before this. Many birds moult before migration or before the winter sets in. Moult is energetically demanding since a new set of feathers must be grown and for many species breeding must stop early to allow time with good food levels to get through moult.

Moult and breeding frequency interact in an interesting way in over 100 species in tropical Africa and America. These species often have small clutch sizes relative to temperate zone species and they often lack a sharply defined moult period. Instead, a small number of feathers are being replaced at any one time. One possibility is that these species are trading clutch size for breeding attempts. By spreading the entire moult process over the combined breeding and moult periods, more nesting attempts can be packed into each year. Each clutch has to be smaller than with separate nesting and moult periods, since energy and nutrients are being diverted from eggs to feathers. Against this, nests have a much higher risk of being predated in the tropics than at higher latitudes, so more frequent attempts with smaller investment per attempt is the best strategy to adopt.

## Inheritance of breeding characteristics

If natural selection is to work on the aspects of breeding biology just discussed it is essential that the relevant characteristics be inherited.

Only recently has evidence been obtained to show that egg size, adult body size, laying date and clutch size are inherited. Given the varied influence of the environmental factors discussed above, these various traits clearly cannot be inherited precisely. What is inherited is a tendency to lay earlier or to lay more eggs (and so on) than other members of the population under the prevailing circumstances.

Another point of interest is how quickly breeding biology can respond to changes in environment. Some such changes can take place very rapidly. Thus in Britain the breeding season of the Woodpigeon has advanced by nearly two months over a 20-year period, doing so in response to changed methods of cereal production that result in most grain being available earlier. In the Netherlands the breeding season of several meadow-nesting species has advanced by about two weeks over a 50-year period as mechanization of mowing allows early cutting, destroying proportionately more of the late broods. In such conditions any bird with a gene for early laying will benefit. Eventually, of course, the benefits of early laying will be countered by the adverse effects of some factor penalizing very early nests and the new breeding season will stabilize. Longer term changes in the traits of bird populations have also been documented. Thus House Sparrows introduced to Costa Rica only 120–150 years ago already have the small (two egg) clutches typical of tropical species in the area instead of their usual one, of four to five eggs.

The diagram on p. 236 shows the implication of these ideas for the timing of breeding. Each spring, food levels start to increase again. Below a certain level the female can meet her own maintenance needs each day but does not have enough left over to form an egg each day. Once food abundance reaches this level, however, egg formation can proceed. If the incubation period is long relative to the width of the seasonal food pulse, low egg formation requirements allow early laying (and therefore hatching), so that the chicks do not miss the food peak. A high threshold for egg formation may mean that the food peak is over before the eggs hatch. Hence a female's energy needs may preclude the breeding season being optimally timed for the young.

What evidence is there in support of these ideas? Since energy requirements increase with body size, one can expect from this model that species that share the same food resources should differ in their timing of breeding. A small species crosses its daily egg formation threshold sooner than does a larger species and so should lay correspondingly earlier. This is exactly the case among the tit species breeding in Wytham Woods, Oxford, all of them species-dependent on each spring's population of defoliating caterpillars.

Other evidence comes from studies in which extra food has been made available to laying females in the period prior to egg laying. In seven studies laying was advanced by periods mostly of two to eight days but by up to 26 days in the case of the Red-winged Blackbird and by a month in the case of the Common Kestrel. In only one case was clutch size also affected by the provisioning. Thus these species appear to have been limited as to when they could breed by the absence of enough food to allow the females to start laying.

One alternative to being so dependent on the seasonal increase of food in spring is to lay up a reserve, either as fat reserves or as food caches. Fat reserves are probably a limited option for most species: since clutch weights can exceed body weight, fat reserves would need to be higher still, necessitating impossibly high body weights. Moreover, residents would need to lay down these reserves in late winter, a time when food is often extremely hard to come by, and migrants would need to do so in parallel with depositing reserves for their journey back to the breeding grounds.

One group of birds that is capable of laying down reserves on the scale needed are the migratory Arctic-breeding geese. Thus Lesser Snow Geese accumulate reserves on their wintering grounds in the southern United States and then migrate to the Arctic where they arrive with substantial reserves – as much as a fifth of their body weight – remaining. Individual females lay different numbers of eggs according to how much fat they arrive with, deriving the necessary nutrients and energy from their reserves, then continuing to draw on the remaining reserves through incubation, until snow-melt exposes fresh vegetation. Again, among Brent Geese (Brant) trapped and weighed on spring migration northwards through Holland, those birds that were heaviest were more likely to return with young than were lighter birds. Presumably these latter had inadequate reserves to complete incubation and rearing successfully. The effects of size and scale on power requirements mean that only large birds are able to carry and retain fat reserves over and above their base body weight. Most small birds are unable to use this particular strategy.

One partial exception to this is the Red-billed Quelea which stores protein in its flight muscles for subsequent use in egg formation. Queleas are itinerant breeders, following the rain belts of central Africa and breeding in the fresh grasslands that spring up after the rains arrive. With just a short time before the grasses and their seeds disappear after the rains move on, speed is of the essence and the head start obtained by bringing in reserves gathered on a previous breeding ground has a clear selective advantage.

An alternative tactic is to cache foods early in the non-breeding season, when food sources are generally plentiful, and to draw these down at the start of the breeding period. Clark's Nutcracker depends almost entirely on caches of pine seeds for egg formation and for feeding its young. The European Jay similarly stores acorns in autumn for use the following spring.

Another tactic used by some species is for the male to feed the female, thus augmenting the amount of food she receives each day and allowing her to cross the egg formation threshold sooner. This behaviour was originally termed 'courtship behaviour' because it was thought to be merely part of the mating ritual. Subsequent studies have shown that the behaviour is in fact intense shortly before laying and that the female may receive from 40 per cent to 100 per cent of her daily requirements in this way. Courtship feeding is frequent in passerine birds, in seabirds, and in birds of prey, all of them groups in which the female might be expected to have difficulty obtaining enough food through her own efforts. In Common Terns females that receive more food from their mates also lay larger clutches and bigger eggs and do so earlier than do females with smaller male contributions. RJO

## Incubation

Embryos do not produce enough metabolic heat during their development to compensate for heat losses from the eggs which must therefore be incubated to provide additional external heat. This is normally provided by a parent (or foster parent in the case of brood parasites) keeping the egg in close contact with its own body, either at one or more brood patches or at its feet. The megapodes are unusual in relying entirely on natural sources of heat, such as solar radiation, volcanic warmth, or the fermentation heat of decaying vegetation. Grebes also make use of vegetational decay but only as a supplement to normal incubation.

Most species have a brood patch, an area of ventral skin that loses most or all of its feathers and becomes engorged with blood vessels during the breeding season. By keeping the egg or eggs in contact with such areas their temperature is raised to near body temperature to allow embryonic development. Thus brood patches allow House Wrens to make their eggs some 5.6°C warmer than would be without the feather loss. Brood patches normally develop only in the sex that incubates, for example in male phalaropes. In pigeons and doves the brood patch areas are bare throughout the year, possibly because these species often have very prolonged breeding seasons (eight or more months). Some waterfowl do not lose these feathers in response to hormonal changes but the female plucks the area herself. Pelicans, boobies, and some penguins lack brood patches and incubate with the egg either under or on their feet. In the latter situation the eggs are surrounded from above by the sitter's ventral surface which in some penguins develops into a specialized fold of skin for the purpose. The relative extent to which heat reaches these eggs through the adult's feet and through its ventral surface is uncertain.

Birds vary greatly in the manner in which the adults incubate the eggs. Both sexes may share incubation (about 54% of families), the female alone may incubate (25%), the male alone may do so (6%), or all three patterns may occur within the same family (15%). The primitive pattern is probably shared incubation, either with the parents alternating incubating spells or with one parent sitting whilst the other brings food to it. Hornbills provide an extreme form of the latter pattern. In this group the female is walled into a tree cavity for the 6 to 17 weeks (depending on species) of incubation by plastering most of the entrance with mud, regurgitated food, and faeces. The male feeds her through a residual opening small enough for the female to guard with her powerful bill.

The most unusual incubation method, recently discovered, occurs in the White-rumped Swiftlet in northern Australia and Fiji. In this species the young of the first brood incubate the eggs of subsequent broods: the second clutch is laid at such a time that the first chicks fledge before those of the second clutch hatch. The parents then complete the incubation. This strategy saves at least three weeks of incubation time. This is important because the species has a long incubation and chick-rearing period but only a limited time when there is sufficient food for feeding young.

In species where one parent incubates alone, it must either go largely or entirely without food for the duration of incubation, as is the case with the Eider, or resort to intermittent incubation, leaving the nest untended whilst the sitter forages. Such a strategy is usually associated with a relatively complex nest structure that reduces the rate of heat loss from the untended eggs. Mallard nests, for example, are heavily lined with down and this is pulled over the eggs when the female leaves during her brief foraging trips. In small passerines in which the female alone incubates, recorded nest attentiveness ranges from 64% in the Carolina Wren to 95% in the Pied Flycatcher. Even where both parents share incubation, however, the duration of spells may vary hugely, from just a few minutes in hummingbirds to two to five days in shearwaters and petrels and to as long as 30 days in some albatrosses. Large species are, of course, better able to sustain such long spells without food.

*Unusual among birds, female hornbills, like the Red-billed Hornbill (Tockus erythrorhynchus) shown here, seal themselves into the nest chamber at the start of incubation, leaving only a narrow slit through which the male feeds them*

Egg temperatures are regulated by changing the amount of time spent in contact with the eggs: at low ambient temperatures spells of egg contact lengthen and spells out of contact are reduced, and conversely at high temperatures. Above about 25°C, however, brooding decreases sharply and egg cooling behaviours may be needed. These include (depending on species) shading the eggs, defaecating on the nest and its contents to increase evaporative cooling, and flying to nearby water to wet the plumage and return thus to the eggs.

Embryonic development essentially ceases below about 34°C and for most birds incubation temperatures, although varying across species, range from this level to about 39°C. Egg temperatures rise slightly through the incubation period as embryo metabolism increases. In the Herring Gull the increase is from 37.6°C in the first week to 39°C at the end of incubation but in the Great Skua the recorded range is from 29°C initially to 39°C at the end of incubation.

Unincubated eggs are quite resistant to chilling. In many species early eggs in the clutch are deliberately left without incubation until the last (or the last but one) has been laid. Such eggs then commence their embryonic development in synchrony, resulting in a near-synchronous hatching. Conversely, in species in which brood reduction is common, eggs are incubated from their laying, resulting in the asynchronous hatching needed to yield the requisite brood hierarchy.

In most species the embryos become increasingly sensitive to chilling as development proceeds but in a few particularly prone to environmental disruption of food supplies – swifts and petrels for example – the embryos are adaptively resistant to chilling, even over spells as long as seven days. This permits the parents to suspend their incubation duties to forage far afield. Such species may be able to continue development at temperatures as low as 30°C but metabolic rates fall off very steeply with further cooling. The incubation period is correspondingly prolonged but the energy cost of the interruption is quite low, as little as 4% over the uninterrupted budget.

The incubation period is defined as the average interval between the laying of an egg and the emergence of the young bird. Incubation periods are correlated with egg weight, increasing by about 16% for each doubling of egg weight and ranging from 11 days in the smaller passerines to about 80 days in the Royal Albatross. In general, therefore, small species have shorter incubation periods and large species longer ones. This may also be in part because many of the smaller species are passerines with higher body temperatures than non-passerine species.

As the table below indicates, species that nest in secure sites such as holes or crevices or in domed nests in inaccessible locations tend to have longer incubation periods than species nesting in more exposed sites.

The shortening of incubation with increased insecurity of site is presumably an anti-predator adaptation. Such shortening is well correlated with corresponding shortening of the nestling periods of these species. Incubation periods also decrease seasonally, at least in the Temperate Zone. Wrens in the Netherlands, for example, decrease in incubation period from 17.5 days in April to 14.5 days in July. Brood parasites constitute an additional group of species that develop quickly in the egg. In the European Cuckoo this may be because the egg is partially incubated in the oviduct of the female before being laid into the host nest. As a result the larger parasite hatches sooner than the eggs of its host and the young cuckoo then ejects these from the nest.

The immediate pathways for the adjustment of incubation period may lie with the thickness and porosity of the egg shell, both of which affect the water metabolism and gas exchange of the developing embryo. Water losses over the incubation period are relatively constant, at about 15% of the initial fresh weight, in a wide

*Incubation and fledging periods in open-nesting and hole-nesting birds*

| Nest site | Bird Family | Incubation Period | Fledging period (days) |
|---|---|---|---|
| Hole nesters | Kingfishers | 22 | 29 |
| | Owls | 30 | 30 |
| | Rollers | 18 | 28 |
| | Hornbills | 35 | 46 |
| | Bee-eaters | 20 | 23 |
| | Swifts | 20 | 44 |
| Open nesters | Anis | 13 | 11 |
| | Pigeons | 15 | 17 |
| | Cuckoos | 12 | 22 |
| | Turacos | 17 | 28 |
| | Passerines | 13 | 13 |

range of species differing markedly in shell porosity. This suggests either that nest humidity is regulated or that the shell porosity is adaptively variable across species. In fact, species nesting at high altitudes have lower egg shell porosities than their conspecifics at sea level, thus compensating for the change in embryonic gas exchange that would otherwise result from the lower atmospheric pressure.

The diagram on the right summarizes the pattern of heat losses and energy cost through incubation of a domestic duck egg. Initially, heat is lost primarily through evaporative water loss but this decreases to around only 10% in late incubation. Radiation (the emission of energy as electromagnetic radiation) and convection (the movement of warm air from close to a warm object and its replacement by colder air) instead take over as important channels of heat loss. By the end of the incubation period the embryo is providing about 75% of its heat requirements. The balance is provided by the adults. In the case of the Herring Gull this amounts to 8.5 to 14.5 kilocalories per day for the clutch of three eggs.

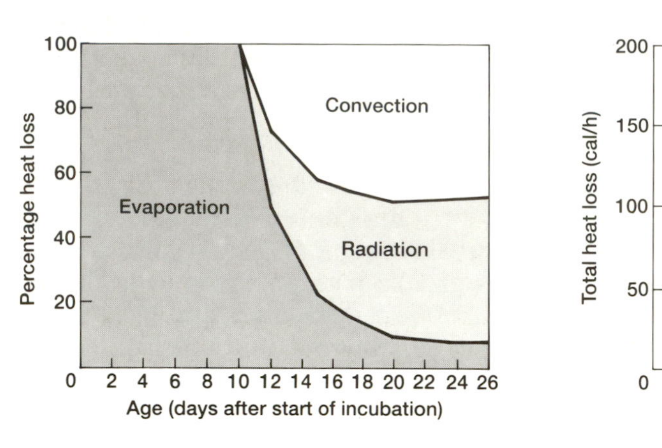




*Above left. Principal channels of heat loss in domestic duck eggs in relation to stage of incubation. The graph immediately above shows total heat losses in calories per hour*

## Hatching

Hatching is a major transition for the young embryo and several behavioural and morphological adaptations have evolved to assist it in making the transition. Many species have a small hatching tooth present on the upper mandible at hatching, greatly enhancing the embryo's ability to cut through the membrane and shell.

A special muscle – the hatching muscle – is often markedly developed at the back of the head and neck of the embryo, greatly increasing the force with which the embryo can work the egg tooth. Shortly after hatching, most species lose their egg tooth and the hatching muscle reduces in relative size.

The time taken between the first splintering of the egg shell and the actual emergence of the chick – the hatching climax – ranges from about 15 hours in the House Wren to 41 hours in the Bobwhite Quail. In grebes a rather short interval is the rule, presumably because the nests of this order are extremely wet.

Hatching efficiency is also enhanced by various specialized behaviours. First, the embryo changes its position in the egg shortly before hatching begins, to enable it to attack the shell more effectively. In the last stages of embryonic development, the embryo moves about until it has taken up a position with its head under its right wing. In this position bill movements over the next few hours gradually erode the membrane between the embryo and air space, eventually penetrating the membrane and starting lung respiration. These movements continue until the shell eventually pips, creating the first tiny starring of the shell. Eventually complete emergence is achieved. During hatching this stereotyped behaviour results in a steady rotation of the chick about the egg, so that the chipping is directed systematically around the circumference of the shell, eventually cutting a small cap of the eggshell away. In a few species, such as the puffin and the Common Rhea, a cap is not cut off but the shell shatters when stressed by the internal flexing of the embryo's body. In megapodes, which hatch at the bottom of a fermentation mound of vegetation or sand, the chick must work its way unaided to the surface. In this species the head is held between the embryo's legs throughout both hatching and climbing to the surface, thus avoiding ingesting the mound material during the climb.

### *Hatching synchrony and asynchrony*

In many precocial species the eggs hatch together, despite each egg being laid one or more days after its predecessor. This synchrony is brought about by vocal communication among the developing embryos. When Bobwhite quail eggs are experimentally incubated in contact with each other, they hatch synchronously, within an hour or so of each other. However, if the eggs are isolated but otherwise held in identical conditions, they have a much greater hatching spread, though with the same mean incubation time. This means that eggs that were laid late must accelerate in development when in contact with other eggs and eggs that were early must slow down their development when in contact with other eggs. Audible sounds (clicking) and low frequency vibrations have been recorded from the eggs in the last few days of hatching after the chicks have penetrated into the air space of the egg and started respiration. Recordings of these sounds and vibrations have been shown to synchronize the hatching of the eggs.

The synchrony brought about in this way is adaptive in precocial species where it is essential for all the young to leave the nest together. This is often necessary if the nest site is remote from natal feeding areas; the brood must hatch together if they are to get to the feeding areas before the earliest hatched young exhaust their fat reserves.

In altricial species, on the other hand, no evidence of egg clicking has been found, and hatching spreads within a clutch are much greater. Indeed, in some such species the spread of hatching induced by starting incubation with the laying of the first egg is combined with other characteristics to facilitate *brood reduction*. Species that display this pattern include herons, raptors, gulls, and other species with unpredictable food supplies. Frequently the final egg of these clutches is smaller than the others.

Since the chick's hatching weight and the egg weight are correlated, the smaller final egg produces a smaller and weaker nestling that hatches late into the brood because of the hatching asynchrony. If the food available for the entire brood is inadequate such chicks die quite quickly. In this way the size of the brood can be decreased without first starving and weakening all the young. If food is plentiful, however, the chick from the last egg survives. In yet other species, of which owls and coots are examples, the later eggs of the clutch are progressively smaller with laying order, creating a brood hierarchy in weight and thus extending the process to the penultimate and earlier chicks should food be so short that additional young must die. In a number of penguins, particularly those of the genus *Eudyptes*, a small egg is laid first and the chick from it survives only if the much larger (by about 72%) second egg fails to hatch.

RJO

## The young bird

### Precocial and altricial development

Two extremes of development are found in birds. *Precocial* species hatch from the egg in an advanced state, largely capable of leading an independent existence. Since they usually leave the nest soon after hatching they are termed *nidifugous*. *Altricial* species, on the other hand, hatch as very dependent young, born blind, with little or no down, and without the ability to walk or regulate their own body temperature. They therefore need prolonged parental care before they become capable of independent existence outside the nest. They are therefore termed *nidicolous*.

Many species show some but not all of these characteristics and six intermediate categories are recognized (overleaf). Each category is largely restricted to certain taxonomic groups, except that the Gruiformes (cranes and their allies) and Charadriiformes (waders, gulls and auks) are quite diverse in development mode across species.

Altricial and semi-altricial species spend a variable period of time in the nest after hatching and are dependent on their parents for food, brooding, and defence. The length of this nestling period varies substantially between species but is fairly well correlated with the duration of the incubation period. Many of the ecological factors discussed in relation to the duration of incubation, therefore, are also relevant here. Nestling growth, however, is subject to different constraints, particularly reflected in growth patterns and rates.

| Mode | Type | Down | Eyes | Mobility | Parental nourishment | Parental attendance | Examples |
|---|---|---|---|---|---|---|---|
| Precocial | 1 | ○ | ○ | ○ | ○ | ○ | Megapodes |
| | 2 | ○ | ○ | ○ | ○ | ● | Ducks, Shorebirds |
| | 3 | ○ | ○ | ○ | ○ | ● | Quail, Grouse |
| | 4 | ○ | ○ | ○ | ◒ | ● | Grebes, Rails |
| Semi-precocial | | ○ | ○ | ◒ | ● | ● | Gulls, Terns |
| Semi-altricial | 1 | ○ | ○ | ● | ● | ● | Herons, Hawks |
| | 2 | ○ | ● | ● | ● | ● | Owls |
| Altricial | | ● | ● | ● | ● | ● | Passerines |

## Growth rates

Most young birds initially increase in weight rather slowly, then accelerate rapidly before finally levelling off. This pattern is referred to as sigmoidal growth. The diagram below right shows some typical examples. For a number of species the final weight achieved before fledging is below adult weight, but in others the young may fledge above adult weight.

One of the main factors influencing the growth rate of young birds is adult body weight: the young of small birds grow more rapidly than those of large species. It seems likely that the ability of young birds to gain weight is proportional to the metabolic rate of adults of their particular species. This means that it is possible to estimate the growth rate for any species simply from its adult body weight. One can then look to see how much the observed growth rate for particular species deviates from the value predicted by its body weight. In other words, the relationship between body weight and growth rate can be used to determine whether particular species have relatively high or low growth rates and then to see what ecological factors are associated with them.

Of particular interest here is the difference in growth rates between altricial and precocial species. Altricial nestlings are pretty helpless at birth, but precocial chicks can provide a great deal of self-care. The price of this greater maturity, however, is a much slower growth rate. Size for size, precocial chicks gain weight more slowly than do altricial species, growing at about 30–35% of the latter's rate. This higher growth rate of altricial chicks is obtained by delaying the onset of temperature regulation and locomotion (i.e. walking and flight). Among semi-altricial species, such as falcons and owls, some ability to regulate their body temperature develops relatively early, but both forms of locomotion are deferred.

Among the semi-precocial gulls (Laridae), the ability to walk develops early, thermoregulation develops in intermediate fashion, and the development of flight is late. Almost all precocial species can walk from an early age; some, like the megapodes, can also fly, and many can thermoregulate at that stage.

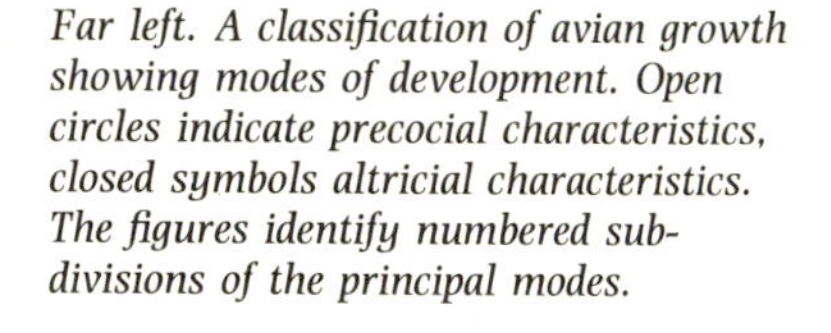

*Far left. A classification of avian growth showing modes of development. Open circles indicate precocial characteristics, closed symbols altricial characteristics. The figures identify numbered sub-divisions of the principal modes.*

*Above. The range of developmental stages of newly-hatched chicks is remarkable, ranging from the precocial young of megapodes and shorebirds, like the Black-winged Stilt (Himantopus himantopus) shown centre, to naked and helpless altricial passerines like the Jay (Garrulus glandarius), immediately above*

*Early growth. Postnatal growth curves for a variety of species are illustrated below, with weight at each age expressed as a percentage of ultimate adult weight*

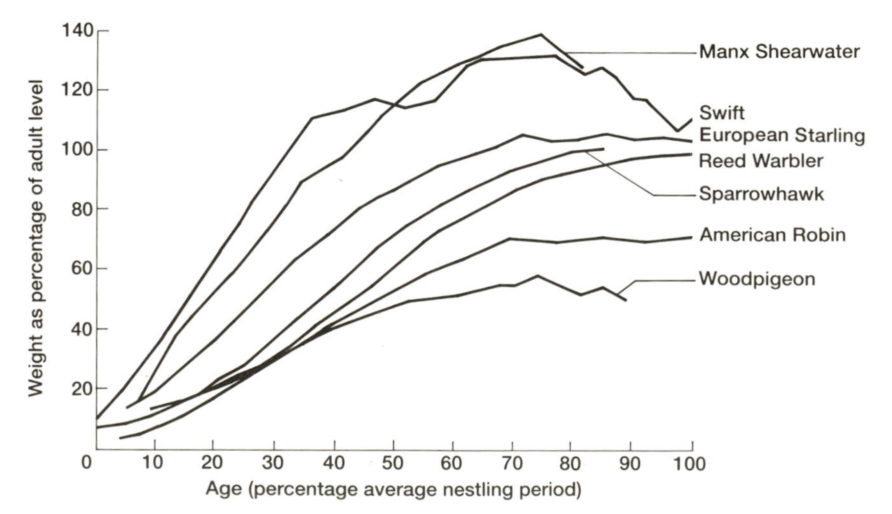

The ability to thermoregulate, walk and fly involves the development of fairly large muscle masses, used for heat production in temperature balance and for locomotion. In most adult birds the flight muscles account for 15–39% of adult body weight. The need for the early development of these muscle masses diverts growth potential away from general size increase, and so the early development of flight tends to depress growth rate among birds.

The internal organs of young birds also develop differentially in such a manner as to promote overall growth, with the gizzard, alimentary tract, and liver all growing faster than the body as a whole during the early period of development. This increase in size helps the young bird to digest the foods brought by its parents, thus sustaining a high rate of conversion of food to tissue and heat. Before the chick can maintain a stable body temperature, however, heat losses need to be in balance with heat production. Another suite of developments assist this. In general, the growth of the body covering occurs at a constant rate relative to overall body size and does not appear to develop differentially. In a few passerines, such as House Martins, the early development of thermoregulation is accompanied by the growth of a special cover of down that greatly reduces heat losses before the integument is fully developed.

In most species, therefore, the resources received by the nestlings are devoted to those aspects of the chick's development that have most immediate benefit to it. The early development of food processing organs enhance the rate of subsequent growth, the development of increased heat production then requires better heat conservation tissues and processes, and finally the development of the flight feathers and flight muscles speeds up as nest departure approaches. Progress through this sequence also results in more and more growth potential being diverted to maintenance costs.

Fast growth rates (and therefore short nestling periods) are adaptive where the young are exposed to high mortality (principally predation). Since rapid growth increases the amount of energy needed by a nestling, rapidly growing broods may need to be smaller than slow growing broods. Indeed, among hole-nesting species and seabirds growth rate and brood size are inversely correlated.

Slightly more than half of all species (about two thirds of all passerines) leave the nest with some additional growth in weight to complete as juveniles. About one-fifth of all species grow to maximum weights that exceed adult weights by 10% or more. Among swallows, swifts, and other aerial insectivores, the excess nestling weight has been due to the high water content of growing tissues, much of which is lost as the tissues mature. Similarly, nestlings of oceanic seabirds often exceed adult weights. In some of these species such as shearwaters, the excess is due to stored fat. These reserves serve as an insurance against periods of bad weather which prevents the chicks' parents from foraging and feeding them. Many of the species involved are those that nest either in holes or in extremely secure sites, thus permitting them the long nestling periods that are needed if the flight muscles are to mature fully before fledging. Species that feed mainly on the ground, such as finches and thrushes, rarely have nestling weights which exceed the adult weight. This is consistent with their reduced dependence upon flight; the young fledglings of these species are able to walk into cover or to run away from a predator.

## Energetics of growth

There are several stages involved in converting the food consumed by the young bird into new tissue. The efficiency with which the energy in the diet is utilized (assimilation efficiency) has been measured in a number of birds and ranges from 57% in the Dunlin to 82% in the House Sparrow. These are average rates, and individual efficiencies as high as 88% have been recorded for the Double-crested Cormorant and 91% in the Eurasian Tree Sparrow. Even within a species assimilation efficiency may vary substantially with diet, meal size, season, and so on. The differences in assimilation efficiency between species determine the energy intake needed in the diet, for low efficiencies mean more food must be consumed for the same growth. The energy requirements for growth

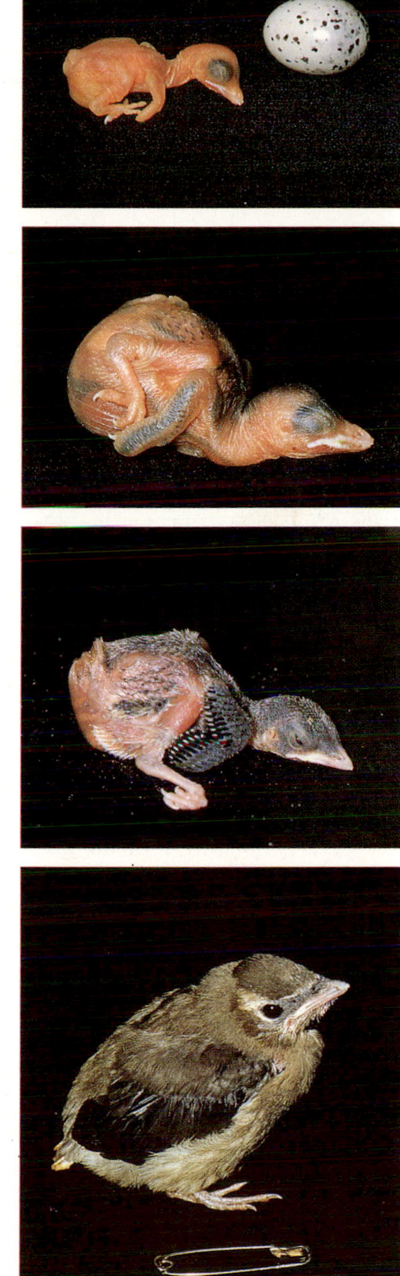

*The transition from egg to fledgling is extraordinarily rapid, as illustrated here by the Cedar Waxwing (Bombycilla cedrorum). The eyes open seven or eight days after hatching, the first feathers burst through their sheaths at about 12 days, and fledging occurs when the young are just 16 to 18 days old*

*An adult Sunbittern (Eurypyga helias) in Costa Rica feeds its two chicks in the nest. The meal here is a frog*

have been measured as daily energy budgets (DEB) and are closely linked with body weight.

Variation between species in this relationship is relatively small and can be ignored. Daily energy budgets rise more steeply with weight increase in the growth of young than they do with weight increase between species: an increase in nestling weight requires a bigger increase in daily energy budget than the difference in DEB between two adults of those weights. The extra costs are probably associated with greater heat loss through the vascularized feather sheaths. For example, fledgling Red-backed Shrikes expend energy at a rate 20–30% higher if their feathers are still growing.

In the course of development the energy intake of young birds is devoted to supporting growth of new tissue, respiration and maintenance, and activity. Not surprisingly, very young nestlings devote proportionately more of their metabolized energy to growth, this accounting for 73% in the American Robin and 62% in the Brown Thrasher, for nestlings aged between one and three days old. This proportion falls off with age, to 13% in European Starlings aged 10–12 days old. Over the entire period from hatching to independence, House Sparrows devote 11% of their daily energy budget to growth, while in the Black-bellied Tree Duck the equivalent figure is 25%.

Altricial and precocial species differ considerably in their pattern of energy intake. Thus, for the Black-bellied Tree Duck DEB increases more or less steadily with age, but in the altricial House Sparrow it levels off about half way through the nestling period.

RJO

## Parental care

### Brooding

Newly hatched young are unable to produce or retain sufficient heat and need to be warmed (or cooled) by their parents. However, a brooding adult is obviously unable to search for food for its nestlings. Several morphological and physiological developments in the growing nestlings improve their powers of thermoregulation and reduce their need for parental brooding, freeing the parent to search for food.

Parental brooding increases in cold weather in all species. In aerial plankton feeders such as swifts and swallows, and in petrels and shearwaters which take marine food, adverse weather may disrupt feeding and force the adults to forage farther afield: the nestlings therefore possess several adaptations which permit them to survive without parental brooding. Such nestlings often have a dense coat of down from an early age, have large fat reserves, and have the ability to enter a torpor-like state with reduced body temperatures during food shortages. These adaptations exist at the cost of reduced growth rates.

## Feeding

Altricial and semi-altricial young depend on their parents for food. Most young have characteristic begging behaviours which are often triggered by the arrival of their parents at the nest. Stimuli such as the jarring of the nest as the adult lands, or the darkening of the entrance in hole-nesting species, elicit begging behaviour.

In the Chimney Swift, where the adults have to flap their way down a vertical shaft, the young respond even to air currents. As the young develop, however, they become more selective in the stimuli to which they respond. In many species the adults give a feeding call as they arrive, further reinforcing this selectivity.

The actual rate at which parents bring food to the nest depends, where environmental conditions permit, on the intensity of begging from the young. This was demonstrated by the Finnish ornithologist Lars von Haartman in an elegant experiment with a nest box population of Pied Flycatchers. He replaced the young in one of these nests with a succession of hungry young. Von Haartman was able to show that the parents increased their feeding rate response to the begging calls of the young: over a three hour period the feeding rates increased from 12 visits per hour to 30 visits per hour. In another experiment with a two-compartment nest box, von Haartman monitored the visiting rates to a single nestling while varying the number of young, placed in a hidden compartment, whose hungry cries the adults could hear. With hungry young in this second compartment the feeding rate to the accessible chick greatly increased, the parents attempting to feed it even after it had satiated. The crucial role of the begging calls of nestlings is also indicated by experiments with deafened Turtle Doves: unable to hear their chicks begging, deafened adults fed their young to only 56% of normal weight after two weeks.

Precocial young actively solicit feeding from their parents, being particularly sensitive to the parent's beak, which they often peck as a food soliciting behaviour. In a number of experiments in which naive chicks are offered model beaks of various colours, the chicks most frequently responded to the bill with colour and pattern resembling that of their parents.

Feeding is usually unevenly distributed among the brood, the chick begging most intensively being fed first. With asynchronous hatching this is usually the first hatched chick which has had longer to grow and which is usually bigger as a result. This system has the advantage that when food is short, the smallest and later hatched young do not receive any food since the oldest chick has resumed begging before the late chicks get their turn. Food is thus concentrated into a small number of young best able to survive rather than redistributed evenly over the whole brood, weakening them all and diminishing the chances of some young fledging successfully.

Nestling birds are reared on protein-rich diets even where the adults themselves may take lower quality diets. Although the diet of adult House Sparrows consists of 97% vegetable matter, the nestlings receive foods that are 68% animal matter and only 31% vegetable. Young Mallards fed on low protein diets grow more slowly than do ducklings on high protein diets.

A small number of species, however, are adapted to the use of low protein diets. Thus the young of fruit-eating manakins hatch from a disproportionately large egg (2.6 g instead of the 1.9 g expected for their size) and have a very long incubation period (18–20 days instead of the 15 days expected for their egg size). The nestling is correspondingly more advanced at hatching, and is presumably therefore capable of processing fruit. Feeding on fruit can be especially advantageous for tropical birds since fruit is often readily available for long periods of time. On the other hand, the protein in a fruit diet is low and the chicks need correspondingly longer development periods. In the tropics the great abundance of predators makes this option an extremely risky one, and practically all species rearing their young on fruit have highly secure or very cryptic nest sites.

Pigeons, flamingos and seabirds address such problems in a different way: the adults consume their normal foods but concentrate its products into 'pigeon milk', oesophageal fluids or crop fluids, or into stomach oils.

Some evidence suggests that young birds require a fairly diverse diet to meet all their needs during growth. In addition to the obvious needs for protein and calories for tissue synthesis and metabolism respectively, calcium is needed for skeletal development, and a variety of trace elements are needed as well. Several studies have shown birds apparently deliberately diversifying the diet they rear their young on. In one Dutch study a single leatherjacket (cranefly larva) species accounted for up to 79% of the foods brought to starling nestlings, particularly in the largest broods. Even though these leatherjackets could be readily obtained in pastures close to the nest sites, the parents nevertheless periodically travelled to a remote pasture to collect caterpillars; these apparently provided some dietary element not available from the leatherjackets. Similarly, titmice in Britain have been recorded ceasing to visit oak trees to collect defoliating caterpillars, instead travelling elsewhere to find a few spiders to feed to the young before resuming feeding on caterpillars. Many species engage in this behaviour in the early days of the nestling period when tissue synthesis is intense; diet breadth narrows as the nestlings grow older and energy needs become dominant. Several studies have shown that nestlings deprived of these 'special' items develop pathological symptoms.

Practically all species with altricial young bring various particles of grit and other hard materials to their young, and fledglings show a predilection for pecking at grit and other hard objects. Ingestion of grit may be important in increasing the grinding power of the gizzard, particularly for species with hard foods. In a species with calcium-deficient diets parents may bring articles of bone or snail shells to the young to compensate for the deficiency. Young Pied Kingfishers on Lake Victoria digest virtually all of the bones in the fish they consume, even though their parents typically regurgitate two or three pellets of bones per day.

As the chicks grow their food requirements increase and feeding frequency and food items brought to them increase. In addition, as the chicks develop the ability to thermoregulate better, both parents are freed to assist in finding food for the brood. Feeding rates are highest in the morning, when the chicks are especially hungry after a night without food, and again toward evening when the young need to stock up ahead of the night, but feeding rates generally are low in the middle of the day. With the warmer temperatures that prevail then, the adults may be better able to leave the chicks unattended while they forage for themselves.

Young birds must metabolize the food supplied by their parents and the quality of these foods may limit the resulting growth rate. One example of this comes from a study of late broods of Great Tits. These broods are fed primarily on caterpillars and these in turn feed on oak leaves. By the latter part of the breeding season, however, oak leaves have accumulated a high phenolic content of condensed tannin. These are part of the oak's defences against being eaten and act by hindering growth and reducing survival in those caterpillars that do

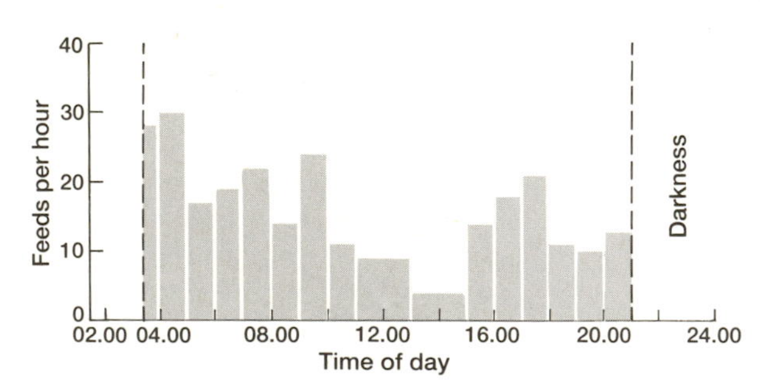


*Diurnal variation in feeding frequency of a European Robin (Erithacus rubecula), bringing food to a brood of six 12-day-old young*

eat them. Since the caterpillars are, in turn, eaten by the Great Tit nestlings, it is possible these chemical compounds could be passed up the food chain and impair the development of the Great Tits. When broods of Blue Tit nestlings were experimentally fed mealworms either with or without extracts of oak-leaf tannin added, growth rates were sharply reduced in birds reared on the diet that contained tannin: by the third day of the experiment weight increments were 21% higher in the tannin-free birds. Interestingly, before feeding natural caterpillars to their broods, the parent Great Tits typically behead the caterpillar and pull out the gut and its contents, those parts where tannin may be concentrated.

Another group of species where nutrient take-up may be a problem is in the albatrosses and petrels, the procellariiforms. In these species the adults partly digest the food intended for the nestlings, subsequently regurgitating an oil-like substance to their young. Presumably this processing reduces the weight of material that must be carried back to the nest from the remote feeding grounds, up to 1000 km distant in the larger albatrosses. Leach's Petrel chicks receive an oil that consists of 33% water, 63% liquid, and only 4% non-lipid dry matter. This means that for every gram of protein present in the diet, the chicks have to process about 130 kcal of metabolizable energy. Such fat-rich diets appear to be very difficult to metabolize, and it is possible that the very large fat deposits in chicks of these species actually serve the purpose of 'sinking' the excess lipid calories.

## Parental defence

Parental defence of young is a prominent feature of bird behaviour. Chick survival can be enhanced by parental care but the provision of that care has a cost. That is, taking a risk on behalf of the current offspring endangers the adult's chances of surviving to rear future offspring. In terms of natural selection the parent should risk its own life on behalf of its young only if the payoff by doing so is likely to be greater than it could achieve by abandoning that young to its fate and reinvesting in other breeding attempts. This concept predicts rather different patterns of parental aid to precocial than to altricial chicks. As a nesting attempt progresses through its cycle less and less time is left to the parent in which to attempt a further brood should the current one fail. Once the young hatch, on the other hand, their prospects of surviving on their own, should the parent be lost while defending them, begin to increase. However, the change occurs more rapidly in precocial than altricial young. Even from an early age the young of precocial species can fend for themselves to some degree and older ones may have quite high chances of surviving to breed, given parental assistance in escaping a current encounter with a predator. Young of altricial species, on the contrary, have little chance of independent survival until quite late in the nest cycle, so that parental risk-taking on their behalf could, in an evolutionary sense, be wasted as compared to starting over with a fresh brood. Several studies of anti-predator behaviour patterns in altricial and precocial species confirm the reality of this theoretically deduced difference between the two groups. Precocial species intensify their anti-predator behaviour around hatching whilst altricial species increase their efforts more slowly.

The form of anti-predator behaviour varies substantially between species. Grebes and other species that cover their eggs usually slip off the nest quietly if they can. Birds startled by predators, however, leave the nest or brood with conspicuous distraction displays, the best known of which is the 'broken wing' display of various plovers. These displays attract the

*An Avocet (Recurvirostra avosetta) defends its nest with elaborate postures and calls against a potential predator, a domestic cow, which might tread on the well-camouflaged eggs or young*

attention of the predator which may follow the apparently injured bird in an attempt to catch it and thereby be led from the area of the eggs or young. With precocial chicks these displays permit the young an opportunity to seek cover before the predator can search the area carefully. Other forms of defence behaviour by the adults may include swooping at or even striking the predator. Thus an entire colony of Fieldfares may join in mobbing and defecating on an intruding predator. In gull or skua colonies, adults with nests in the immediate vicinity of an intruder will fly at or strike it, particularly so among those birds with large young. In one study of Common and Arctic Terns faced with models of intruding Great Black-backed Gulls the attack rates increased from 24 attacks per minute in early incubation to 41 per minute in the fledging period, with the rate of actual strikes on the model more than doubling.

A particularly interesting form of parental defence is where the adults carry their young around on their backs, thus removing the young from the reach of small predators unable to tackle the full-grown adult. The behaviour has been recorded in more than a dozen species of waterfowl, particularly in temperate areas with several hours of darkness when predators may be able to approach the young unseen. For the adults to be able to defend the young in this way key prerequisites are that the adults must be large and the brood size small and the adults must not need to fly frequently. For waterfowl, an additional benefit of carriage on the parental back may be removal from the cold water. A pattern of interest here is that relatively few Arctic species nesting in continuous daylight carry their young in this way, despite the low temperatures; the anti-predator function may therefore be met by the constant illumination rather than the carrying.

An area rich in anecdotal observations is the carrying of young other than on the adult's back. Over two dozen species, including woodcock, have been recorded moving their young from a disturbance by predators or potential predators, chiefly by carrying the young in the bill though occasionally between the legs or (for more terrestrial species) under the wings. At least a proportion of these cases have been sufficiently documented to be credible. No passerines seem to undertake such behaviour.

In certain precocial and semi-precocial species the dependent young often come together in creches, groups of several (or even many) young attended by one or more adults or 'nurses'. This behaviour is common amongst penguins, pelicans and flamingos, and is also regular in some ducks such as the Eider and in

*The small young of many water birds are protected from predators by being carried on their parents' backs, as with this Eared Grebe (Colymbus nigricollis)*

*A Canada Goose (Branta canadensis) cares for the young from several broods. The young of some semi-precocial or precocial birds, which are able to feed themselves, form creches, where one or a few adults are able to provide sufficient protection from predators*

some terns such as the Sandwich Tern. In wildfowl the creche young are self-feeding and the nurse adults serve the role of guards but among penguins parents continue to feed their own young whilst the nurses provide collective guarding of the creche. This guarding usually provides warning of predator approach, evoking a tight bunching of the young which then makes it difficult for the predator to pick off individual chicks. In wildfowl such creching is commonest among diving species in which the adults and young have different food requirements. In Eiders nesting along Scottish estuaries, for example, young take small invertebrates such as amphipods, gastropods and winkles, foods found on estuarine flats, but the adults specialize on mussels found lower on the shore. In this species the creche nurses are a sequence of successful adults, each accompanying their own young to the flats but remaining there for just a few days. These females have spent the previous weeks incubating their nests largely without food and are near starvation. They are thus in desperate need of food and must finally abandon their young to be guarded collectively by later arriving females. Other populations of Eiders nest along rocky coasts, however, and here the young feed predominantly on gammarids and other small crustaceans from the seaweed whilst their mothers, themselves able to feed alongside their young by taking mussels from the holdfasts and rocky substrata of the seaweed beds, continue to guard their own broods.

## Nest sanitation

Most passerine nestlings produce their faeces encased in a gelatinous sac. These are normally consumed by the adults when the young are small, perhaps because the digestive efficiency of very small young is so low that significant nutrients remain in the faecal sacs at a time when the adults are limited in foraging time by the need to brood the young. With older nestlings the sacs are removed and dropped away from the nest. Nestlings often signal the impending production of a faecal sac, usually after a feed, by special postures, though in other species the sacs are excreted even in the absence of an adult, but are then deposited on the rim of the nest to be removed by the adult on its return.

Such sanitation behaviour is most pronounced in species with open nests vulnerable to predation. Species with more secure nest sites often simply excrete watery faeces either through or over the edge of the nest structure, though the young of hole-nest species usually direct their excretions in the direction of the nest entrance, identified as the source of the brightest lights. In species in hot climates subsequent evaporation of nest whitewash may help cool the nest and its contents. In the Cactus Wren of the American Sonora Desert, faecal pellets are removed from nests in the cooler parts of the year but are left in the nest during the hottest part of the season, when evaporative cooling is most beneficial.

RJO

*Nest sanitation is important. A Wood Warbler (Phylloscopus sibilatrix) removes a nestling's faecal sac from the nest*

## Fledging period

A young bird after hatching is referred to as a 'pullus' until it is full-grown and can fly. The fledging period refers to the interval between hatching and being able to fly (excluding flightless species), and varies markedly between species. Young megapodes receive no parental care and can fly within hours of hatching. At the other extreme, the Wandering Albatross fledges after 280 days, and the King Penguin after 360 days.

The variation in fledging period between species has probably evolved in response to two sets of factors: food and predation. Most seabirds which feed off-shore and experience relatively poor food supplies, have markedly longer fledging periods than inshore species. This effect can be seen by making comparisons within a single family, the gannets and boobies. The young of the inshore feeding Blue-footed Booby fledge after about 100 days, compared with 130 days for the off-shore feeding Red-footed Booby. Even greater differences occur between the Wideawake Tern, an offshore feeder whose chicks fledge after 60 days, compared with 30 days for inshore Common and Arctic Terns.

Hole nesters have longer fledging periods than open nesters, and this may be due to predation. Open nesters are more vulnerable to predation so selection will favour more rapid development.

Fledging periods are only weakly related to body size, and for a given body weight fledging periods can vary markedly. For example, the fledging periods of species weighing about 400 g range from three to fifteen weeks. As the table above shows, there is however a close correlation between the duration of incubation and fledging periods. This presumably occurs because both embryonic and post-hatching growth rates are similar.

*Incubation and fledging periods in seabirds*

| | Incubation (days) | Fledging (days) |
|---|---|---|
| King Penguin | 53 | 360 |
| Adélie Penguin | 33 | 51 |
| Wandering Albatross | 78 | 280 |
| Giant Petrel | 59 | 108 |
| Fulmar | 49 | 49 |
| Storm Petrel | 41 | 63 |
| Gannet | 44 | 90 |
| Shag | 30 | 53 |
| Common Tern | 23 | 30 |
| Sandwich Tern | 23 | 35 |

## Transition to independence

Nest departure at the end of the fledging period is a major event in the life of the young bird but parental care often continues well beyond that point. Fledging in fact poses a problem for the parents, for their broods do not always fledge synchronously and parents may have to divide their attentions between continued feeding of unfledged nestlings and care of the new fledglings. In precocial species, where synchrony of nest departure is vital, special behaviours have evolved to ensure coordinated departure but these are uncommon among altricial species. Nevertheless, adult passerines have been seen apparently luring tardy young from the nest by offering food from outside the nest rather than from within and have even been recorded physically ejecting such young.

Post-fledging care typically lasts two to four weeks in temperate zone passerines but longer in larger non-passerines, in species with complex foraging skills, and in conditions of food scarcity. In Sarawak, for example, many passerines still feed fledged young after ten weeks. Here invertebrate prey are both sparsely distributed and protected by a variety of protective adaptations such as camouflage, mimicry, and so on. In such circumstances experience may be important and learning through association with parents may be especially beneficial. Parental assistance is also at a premium where foraging skills must be learned, permitting the young birds to survive a lengthy period of poor success at self-feeding. Species in this category include frigatebirds, terns, owls, kingfishers, herons, and raptors. In these species brood sizes are typically low and second broods non-existent, permitting the adults to give individual attention to their young. These species are also characterized by much play activity related to foraging-oriented actions. Parental care here can persist well into the following winter season and onto wintering grounds thousands of miles away.

In a few species the parents actually attack their young and drive them from their territory but in most species the break is more gradual. Theoretical biologists believe the parent birds benefit sooner from stopping further investment in an existing brood and starting a new brood than do the members of the current brood. Hence the young currently cared for should attempt to trick their parents in continuing to care for them rather than for future siblings, at least for a while. The Cambridge zoologist Nicholas Davies tested this by hand-rearing Great Tits until they could find at least some food for themselves, albeit inefficiently. He then behaved as either a 'mean' or a 'generous' parent in response to their continued begging, by adjusting the amount of begging he required of the young before he would feed them, but leaving them with access to food from which they could feed themselves. He found that young exposed to parental meanness at an early age switched to self-feeding sooner than did those not so exposed until later and that self-feeding in each of three test groups became the norm as soon as feeding rates in self-feeding became more profitable than continued begging. Since the three groups differed substantially in their development stages at the time each switched behaviour, it is clear that self-feeding was not controlled by maturation but was an active switch in behaviour by the young on the basis of the relative profitabilities of the two strategies. In the field young Spotted Flycatchers behave in just this way. This species takes aerial insects by flying out after them from a perch. The young leave the nest at about 12 days of age and are fed for a further 9 days by the parents. The parents then switch first to feeding the young only if they beg and then to feeding them only if they fly to the parents, with an increasing proportion even of those flights going unrewarded. Once self-feeding becomes more profitable than soliciting such increasingly reluctant parents, the young feed themselves completely. Similar changes in parental willingness to feed a begging young have been documented in species as diverse as chickadees, Oystercatcher, and Bewick's Swan.

RJO

## Imprinting

Very young birds respond socially to a wide range of stimuli, even those very different in appearance to their own species, but over time they become selective in their responses through a process known as 'imprinting' described by the pioneer ethologist, Konrad Lorenz. Especially interesting points about the phenomenon are that it occurs only during a restricted time period – a sensitive period – during the birds' development and that preferences thus formed are particularly resistant to subsequent eradication.

*The young of precocial species imprint onto the first moving object they see after hatching. Under natural conditions this is usually a parent, but here a Red-breasted Goose (Branta ruficollis) has imprinted on the shoes of the person that hand-reared it*

Newly hatched ducklings or chicks exposed to moving objects shortly after they hatch tend to treat those objects as their parent. Such imprinting was most effective experimentally with ducklings 13–16 hours old and was absent in very young birds and in those more than 28 hours old. Similar results have been obtained with other species. The most surprising finding has been that realistic parental models are no more effective than other stimuli in eliciting imprinting. Movement enhances the process, as does conspicuousness and the presence of auditory as well as visual cues. Each of these features matches the natural context of imprinting, in which the newly-hatched bird normally encounters its mother as a large moving object uttering maternal calls.

Although the original concept of imprinting was of instant (or at least very rapid) learning of every detail of the imprinting object, the modern view likens imprinting to sketching: an initial outline is quickly prepared and details are later filled in. The delimitation of a sensitive period is achieved through two processes. First, a very young bird may be limited by incomplete morphological or physiological development and thus be unable to respond fully to any stimulus. Second, the older birds come to learn the characteristics of their environment and to avoid newly-met stimuli that have very different characteristics. Nonetheless, young birds retain a preference for novel objects that are slightly different from the imprinting stimulus with which they have become familiar and will, under experimental conditions, work to obtain views of such objects. Thus the two processes between them limit the period of initial learning. Nonetheless, the continuing ability and willingness of a chick to investigate minor differences from the initial object allow it to develop appropriate modifications. It can, for instance, accommodate to the different views of its mother close to and from afar.

Imprinting has proved to be of major importance in species recognition in birds. At the heart of the concept is the idea that a young bird learns what its mother (or father) looks like and retains that memory until it must recognize a conspecific mate when it in turn comes to breed. Such a process would be of particular value in a species undergoing rapid evolutionary changes: a purely genetic recognition process might be slow to track environmentally induced changes in the appearance of newly successful members of the population. An imprinting process, on the other hand, allows the new members of the population to track these changes, the fact of their own birth ensuring that these variants are successful. Imprinting has in fact proved to be widespread in taxonomic groups independently known to be undergoing extensive radiations, examples being ducks and geese, gallinaceous birds, pigeons and doves, and the estrildid finches.

Several cross-fostering experiments have provided insight into the processes of sexual imprinting. By fostering Zebra Finches, for example, under Bengalese Finches, with the fostering terminating after various periods spent with the foster parents, and testing the sexual preferences of the young when they come to maturity many months later one can determine the limits to the sensitive period. In such a set of experiments Zebra Finches remained permanently imprinted on Bengalese Finches if fostered with them until 40 days of age; the sensitive period had closed by that age. However, Zebra Finches initially reared under their own species but then transferred to Bengalese Finch foster parents imprinted on the foster species only if transferred prior to day 20 i.e. the sensitive period for imprinting on their own species appearance was much shorter. Some genetic factor thus predisposes the Zebras to imprint on their own species.

A variety of experiments with these estrildid finches has shown that it is the (foster) parent–offspring relationship that sets the subsequent sexual preference of the young finches. In other species, however, sibling associations set these preferences. In Snow Goose colonies on Hudson Bay in the Canadian Arctic two colour morphs are present, one a 'blue' phase, the other a 'white' phase. Mated pairs are disproportionately of like colour because young birds form sexual preferences based in part on parental colour and in part on the colour of their own siblings.

Birds from families where siblings and parent were all alike rarely mated with birds of a different colour but birds from mixed families in which the adults and siblings were of different colour frequently chose mates of either of the sibling colours. Moreover, the rates of mixed matings in the wild were closely in line with what would be expected if young birds were choosing their mates on the basis of their familial colours.

Imprinting processes allow a means of avoiding inbreeding. Were all young to choose their mates on the basis of exact resemblance to their own kin, species-specific characteristics would be conserved but genetic variability, a reserve of diversity against environmental change, would be lost.

Yet variability maintained by outbreeding implies the conservation of currently suboptimal phenotypes. Imprinting allows a young bird recognize close kin and adjust its preference for mates of slightly different characteristics from those of its own family, thereby maintaining diversity at low cost. As in filial imprinting, sexual preferences in birds favour minor differences from the original imprinting object. It has been shown that Japanese Quail avoid breeding with close siblings and totally unfamiliar individuals. Instead they prefer to mate with birds that resemble somewhat those with which they were reared. The timing of the sensitive periods for sexual imprinting also varies between species, coinciding with the emergence of sufficiently adult-like characteristics in young females to allow males recognize adult females later. Mate recognition processes must also allow for sexually dimorphic species in which only one sex looks after the brood. For example, drake Mallards imprint on their mothers but female preferences are largely independent of early experiences. In monomorphic species, on the other hand, both sexes imprint on the parent's appearance.

RJO

## Brood parasites

Some 80 species of birds are obligate brood parasites, wholly dependent on birds of other species to raise their young. (This total does not include the increasing number of birds known to lay or 'dump' their eggs in the nests of fellow members of their own species.) As the table below illustrates, brood parasitism has evolved in seven different families or sub-families; it is a phenomenon that raises fascinating questions about why the hosts tolerate the parasites and why only certain groups of birds have evolved the parasitic habit.

### What adaptations do parasites show?

Nests are usually parasitized during the hosts' laying period, which suggests that the parasite female has 'earmarked' her victim during its nest building. With the help of a protrusible cloaca the parasitic female lays her own relatively thick-shelled egg directly into the host's nest. Laying can take place very rapidly; it lasts about 10 seconds in the European Cuckoo. It is doubtful if any brood parasites deposit their eggs on the ground and then transfer them to the nest by bill. European Cuckoos were thought to follow this procedure until the observations of Edgar Chance immediately after the First World War confirmed that the parasite laid directly into the host nest.

During her laying the parasite commonly removes one or more of the host eggs. Experiments with European Cuckoos have shown that removal is not done because hosts can count and then reject an extra unexpected egg. Instead the behaviour may ensure that the number of eggs in the nest does not exceed the total the host can incubate efficiently. Alternatively, when the parasite is reared alongside the host young, egg removal may ensure that the clutch does not yield a brood too big for the hosts to feed.

*The brood parasites*

| *Scientific name of family or sub-family* | Anatidae | Neomorphinae | Cuculinae | Indicatoridae | Viduinae | Ploceidae | Icteridae |
|---|---|---|---|---|---|---|---|
| *Common name* | Black-headed Duck | Ground-Cuckoos | Cuckoos | Honeyguides | Widow-birds | Cuckoo-weaver | Cowbirds |
| *Number of parasitic species* | 1 | 3 | 47 | 11 | 10 | 1 | 5 |
| *Breeding range* | S America | S America | Old World | Africa and Asia | Africa | Africa | N & S America |
| *Principal hosts* | Mostly coots | Furnariidae (ovenbirds) & Tyrannidae (flycatchers) | Small passerines; crow family | Various, especially hole nesters | Estrildine finches | Warblers | Other icterids & small passerines |
| *Parasite young reared alongside host young?* | No | No | Usually not | No | Yes | Yes | Yes |

*Left. The brood parasitic European Cuckoo (Cuculus canorus) swoops down onto the nest of its host, a Reed Warbler (Acrocephalus scirpaceus), swallows one of the host's eggs, and within a few seconds lays its own and departs*

The parasite's egg is often small for the size of bird, rendering it more likely to hatch ahead of the host young, an advantage to those cuckoos which eject the host eggs or young. More remarkably the parasite egg colour often mimics the host egg colour extraordinarily closely. Using model eggs it has been shown that many hosts of the European Cuckoo throw out badly-matching eggs from their nests, and incubate well-matching eggs – making it advantageous for cuckoos to lay mimetic eggs. On the other hand most hosts of the Brown-headed Cowbird in North America happily incubate badly-matching (model) eggs. There is little selection on Cowbirds to lay mimetic eggs, and they don't.

Some host parasites are reared alone in the nest. Hatchling honeyguides kill host nestlings with special hooks on the upper and lower mandible. Similarly, about a day after hatching, a young European Cuckoo manoeuvres the host eggs or young into a sensitive hollow on its back, climbs to the rim of the nest and tips out its cargo.

Other parasites are reared alongside the host young, and may mimic them closely. For example, the pattern of tongue spots inside the mouth of young widow-birds matches that of the young of their finch hosts. The Black-headed Duck young is so precocious that within one and a half days of hatching it leaves its foster parents to fend for itself and grow up alone.

Since parasitized hosts suffer the cost of rearing fewer of their own young, why do they tolerate the parasite? The best available answer is that they are in fact fighting back, for example by rejecting non-mimetic eggs of European Cuckoos. But as the host becomes more discriminating, so the parasite will evolve more refined means of circumventing host defences; it is like a biological arms race.

A remarkable exception to this state of conflicting affairs occurs in Panama where the Chestnut-headed Oropendola and the Yellow-rumped Cacique are parasitized by the Giant Cowbird. In some circumstances the young cowbird may protect its fellow nest occupants, host young, from harmful botflies *Philornis* sp., so the host then benefits from being parasitized.

It is not clear why only certain bird groups have adopted the parasitic route. Possibly the diet of cuckoos and honeyguides, noxious hairy caterpillars and wax respectively, is unsuitable for small nestlings. That predisposing factor could have meant that, in the past, individuals which foisted some of their eggs onto other species brought off more young than those individuals that reared all their own young.

MdeLB

# 10 BEHAVIOUR

## Mates and mating

Like people, most birds are monogamous, with male and female working together to rear a family. Over 90% of all bird species are categorized as monogamous; the rest are polygamous, and frequently have more than one mate.

Ther term polygamy has been subdivided into a number of categories: (1) *polygyny* is when one male mates with several females (e.g. Village Indigo-bird); (2) *polyandry* is the reverse, one female mating with several males (e.g. Galápagos Hawk); (3) *polygynandry* is when several males mate with several females (e.g. Dunnock); and (4) *promiscuity* is where there is a mixture of polygyny and polyandry, and usually means there is no lasting bond between any pair (e.g. some hummingbirds).

In monogamous species the bond between partners may last for a single nesting attempt, or for several breeding attempts within a season, or may persist as long as both birds are alive – over 20 years in some seabirds and raptors. Even though pair members may spend their winter apart they may reunite the following breeding season. Most seabirds re-use the same tiny territory from year to year, and one purpose of doing so might be to enable partners to meet again each year.

Why should pair bonds persist over several breeding seasons? The probable answer is that by doing so both partners rear more offspring. Several studies have shown that members of long-established pairs get used to each other and work well together. Typically, if one partner dies and the remaining one remates, the new pair's breeding success is generally quite low for a year or so until the birds get used to each other's routine. For example, kittiwake pairs that remain together fledge 17% more chicks each year, on average, than those that change mate.

Pairs that fail to establish a working routine tend to produce relatively few offspring. Natural selection would hardly favour such individuals remaining together, and the most productive option is divorce. Studies of long-lived birds show that divorce is most likely to occur among pairs that fail in their breeding attempt. In Manx Shearwaters for example, only 5% of successful pairs divorced between years compared with 27% of unsuccessful pairs. The corresponding figures for kittiwakes are 17% and 52%.

Several bird species, such as the Mute Swan, are said to pair for life, and it is often believed that if one of the pair dies the other will either pine away or spend the rest of its life in celibacy. In fact there is little evidence that such behaviour occurs out of choice; some birds may remain unpaired after loss of a partner simply because no unattached individuals are available at the time. However, natural selection will always favour single birds attempting to re-pair. Having said that, the bond between partners of long-lived species can be remarkably strong, as illustrated by a pair of Brent Geese (Brant) observed on Cornwallis Island in the Canadian Arctic. The pair returned from their wintering grounds and settled on a small frozen pond not far from the local Inuit settlement, and waited for conditions to improve. A few days later someone had shot one of the pair, which lay dead on the ice while the other bird stood beside it. A week later the live bird was still standing beside its dead partner.

## Mate acquisition

How do birds find a partner? In general what happens is that the male obtains a territory or nest site and then starts to advertise his availability. A small passerine often does this by singing, a seabird by exaggerated postures, movements and calls. Male sparrowhawks and goshawks advertise themselves by soaring over their territory.

Often the same signals, songs or postures, serve both to attract females and to deter other

*Ritualized displays play a central role in maintaining the pair bond. This Blue-footed Booby (Sula nebouxii) is parading in the Pelican posture and displaying its brilliant blue feet, a form of ritualized walking representing an appeasement display normally performed for the mate*

males. Once a female appears, the male may then perform the same or other displays, generally referred to as courtship. The types of display used in courtship often reflect the internal conflict which both partners are undergoing, so the displays may simultaneously contain hints of fear and aggression as well as of the urge to mate.

Not all species follow the same pattern of mate acquisition. In the Northern Gannet the process of pair formation is relatively obvious. An unpaired male first obtains a nest site. Then he advertises from it to every female that flies past, until eventually a female alights beside him. If she considers him acceptable she will continue to return there so that the two birds spend more and more time together, thereby forming a pair bond. In other birds pair formation is more subtle. Among species such as the European magpie, in which young birds spend their first winter as part of a mixed-sex flock, some birds may simply gravitate towards each other and gradually, over days or weeks, spend increasing amounts of time together. When the breeding season starts the pair is more or less established and together the birds set up a territory in which to breed.

Courtship displays serve a number of purposes; initially they help to ensure that members of the same species breed together. Often the displays of morphologically similar or closely related species are quite distinct and this minimizes the risk of individuals pairing with the wrong species. A clear-cut example involves the Willow Warbler and Chiffchaff; these two species look almost identical but their songs are quite distinct. However, mistakes do occasionally occur, for example Common Guillemots have been seen copulating with Brünnich's Guillemots. Hybrids between grouse species, such as the Black Grouse and Capercaillie are also reported from time to time. In general though, such mistakes are extremely rare and most individuals mate with their own species.

The second purpose of courtship displays is to help synchronize the reproductive cycles of the male and female. As pointed out in Chapters 2 and 9 the reproductive systems of most birds undergo an enormous increase in size at the start of the breeding season. Clearly, it is vital that the cycles of pair members are closely linked so that the male has sperm available when his partner's eggs are ready to be fertilized. It has been demonstrated in Turtle Doves that the presence of a bird of one sex can affect hormone levels in the other, thus bringing the birds into reproductive condition together.

A third function of courtship displays is to help partners to assess each other. Each individual aims to ensure that the chosen partner is of high quality either as a parent (in monogamous species) or in terms of his genes (in species where the male contributes nothing more than his sperm to the female and plays no part in rearing the young). The spectacular courtship displays of raptors like the Peregrine Falcon, in which the male performs fabulous aerial acrobatics, may enable the female to assess how good he is at hunting. A male unable to stoop at high speed in a courtship display may be a poor provider!

Another courtship 'ritual' that allows females to assess male quality, is courtship feeding. It occurs in a wide range of birds, from grebes to bee-eaters, and from skuas (jaegers) to eagles. Courtship feeding prior to mating and egg laying may enable females of some species to assess a male's quality and whether he is worth pairing with. In the Common Tern, the more food the male brings the female prior to

*Courtship feeding can play an important role in pair formation. The rate at which a male Arctic Tern (Sterna paradisaea) provides a female with food prior to egg-laying should indicate his potential as a food-provisioning father*

egg laying, the larger clutch size and eggs she produces, and the more food he brings once the chicks hatch. In other words, courtship feeding is not, as was once thought, a mere ritual, but instead has important consequences for both partners.

Obviously the amount of food a female receives through courtship feeding varies from species to species. On average, female Blue Tits receive 40% of their total food intake prior to laying from the male. In raptors the figure may be much higher. For example, in the Osprey the female barely leaves the nest or hunts for herself throughout the entire four month breeding season, but is fed entirely by her partner. An experimental study, in which female Ospreys were provided with extra food prior to egg laying, showed that well-fed females were more willing to mate with their partners and were less unfaithful than underfed ones. The primary function of courtship feeding in Ospreys appears to be to ensure mate fidelity. Male Marsh Tits feed their mates during incubation. The more often the male delivers food, the shorter is the incubation period, because the female can spend more time keeping the eggs warm. Here the courtship feeding by the male directly benefits the eggs he has fathered.

## Mating systems

We will now consider each of the main types of mating system in more detail. We will also examine the sets of adaptations associated with each of them and pinpoint the ecological circumstances that determine which system(s) a species opts for.

### *Monogamy*

This term is really something of a dustbin for all the mating systems that are not polygamy or promiscuity. In other words 'monogamy' covers a range of mating relationships. Basically it means that a male and female remain together for at least one breeding cycle, have a more or less exclusive mating relationship, and work together to rear the young. However, the amount of work the male does varies enormously; in the Osprey just mentioned, the male performs most of the hunting needed to rear the young. In the case of the Willow Grouse (Ptarmigan), however, the male does virtually nothing but defend the territory.

Because monogamy is so widespread among birds and encompasses such a range of male–female partnerships, it is unlikely that it has a single explanation. In fact it must have evolved several times.

1. Monogamy occurs in cases where male assistance is essential for the female to rear any offspring at all, either because eggs or chicks need to be protected, or because both parents are needed to find sufficient food for the young. For example, in the colonial Herring Gull one partner has to stay with the eggs or chicks at all times to prevent them being eaten by other colony members or predators. Male assistance is also essential in many seabirds which have prolonged incubation shifts. In several albatrosses and petrels, food is situated so far from the breeding areas that each parent incubates for two weeks at a time. Clearly, under these circumstances there is no alternative but for a male and female to work together to rear young.

2. Monogamy is also the best option for a male when defending a female rather than a territory. This is nicely illustrated by several ducks, including the familiar Mallard. At first sight it seems odd that Mallards should be monogamous since the male plays no part whatsoever in helping to rear the young; he does not incubate, nor feed or defend his offspring. Because females undertake all the incubation and rear the offspring alone they take all the risks. As a consequence females suffer a higher mortality than males and in most populations the sex ratio is strongly male-biased. Females are therefore in short supply. Competition for females is intense and males pair with a female as early as possible, that is, as soon as they have completed their moult in late summer. Once paired the male then defends his prize from other males, throughout the winter and spring. The male is concerned only with monopolizing his female and making sure that he alone fertilizes her eggs. This is no easy task because there are always more males around than females, and those males whose females have started to incubate are still keen to fertilize

any other females they can. A male Mallard sticks extremely close to his female during the week or so before, and during laying, when fertilization of the eggs occurs, and vigorously attempts to deter all other males. The female benefits from this system too, since her partner is basically a 'hired gun' who allows her to feed in comparative peace. Without him she would be subject to numerous forced mating attempts (see below). Her feeding would be continually interrupted and she would also be in danger of physical injury or even drowning.

3. The third reason why birds are monogamous is related to the quality of their territories. As we will see below, if adjacent territories differ a great deal in quality, some males are able to attract several females, and breed polygynously. If however, territories vary rather little, this is unlikely to occur. Lack of variation in territory quality is thought to be the single most important factor determining monogamy in birds. It is thus likely to be a major reason why the familiar passerines of garden, hedgerow and woodland are mostly monogamous.

*Polygyny*

Here a male mates with several females in one of two types of situation. He either defends a large territory and attempts to attract as many females as possible into it, or he competes with other males for status at a communal display ground called a lek, and mates with as many females as possible as they visit the lek.

The best-studied species in the first category is the Red-winged Blackbird. This species breeds in marshy areas in much of North America. In spring the males return a week or so ahead of the females and establish their territories. When the females arrive they settle in the male's territories, and some males end up with up to 10 females while others get none. For a long time ornithologists wondered why females should choose to settle on the territory of a male that already had one or more females, while there were still unpaired males available. After all by joining a male who has say, two females already, a female can expect to receive only a third of his help in rearing the young. If she paired with a bachelor male she would receive his undivided attention. The answer to this apparent paradox lies in the quality of the male's territory. If the difference in territory quality is great enough, she will do better (in terms of resources) by mating with the already-paired male. This is because even though she has to share his resources with other females, the territory is so good that sharing is the most profitable option.

If this idea is correct, we would expect most females to occur on the best territories. In several North American grassland passerines, such as the Bobolink and Dickcissel, territory quality is determined by the amount of vegetation cover that exists. This is probably because vegetation provides shade and protection from predators. Unpaired males were found to occupy territories with relatively little vegetation, whereas those with the most females (two or three) had territories with notably dense vegetation.

In another similar species, the Lark Bunting, it was possible to go one stage further and to conduct some field experiments to examine the role of territory quality. As in the other species vegetation cover was important: the shade it provided prevented chicks dying of heatstroke. When plastic rosettes of leaves were added to territories before females arrived in the spring the males in those territories subsequently attracted more females than control males. The rosettes were perceived by the females as nesting cover and improved the apparent quality of the territory.

A rather different type of polygyny occurs in the Pied Flycatcher. Here the male pairs with one female and then as soon as she has started to lay her clutch he flies off to a different area and starts advertising for another female. If he is successful he actually abandons this second female after she has laid, leaving her to rear the young alone. He then returns to his original female and helps her. This system is referred to as male deception polygyny because the male deceives the second female into thinking he is going to pair with her monogamously and help her rear the chicks. Inevitably, secondary females are less productive than primary ones, but this does not matter to the male since he does better out of this system than being monogamous. From the female's point of view this sounds like a poor deal. However, secondary females are usually those that have returned late from migration. Because breeding success declines through the season, time is at a premium. These females therefore pair with the first male they encounter and simply cannot afford to spend time choosing between different males.

*Leks.* This form of polygyny involves several males congregating to defend small territories which contain no resources and together comprise a 'lek'. Lekking occurs in a wide range of species, including gamebirds, waders, several manakins and cotingas, birds of paradise, and some species of weaverbird.

The one feature that all these birds have in common is that the male plays no part in incubating the eggs or rearing the young. Leks therefore occur only in those species in which the female can rear the young unaided, either because (1) the young are precocial (as in the Sage Grouse, Black Grouse and Ruff); (2) a very small brood is produced; (3) food for the young is easily obtained (as in the White-bearded Manakin, which breeds in the tropics: here the female lays a clutch of just two eggs and, by rearing the altricial young on fruit – which is superabundant) – is able to raise them by herself); or (4) the species is a brood parasite (as in the Village Indigo-bird, a widow-bird).

Lekking may have evolved for several reasons. Males displaying in groups may be safer from predators; it may also be safer for females to visit males that are in groups. Several males displaying together may be more conspicuous to females. Alternatively, males sometimes congregate at locations ('hotspots') which are especially attractive to females.

Females visit leks only to mate, and their visits characteristically produce a frenzy of display from the males as they try to outcompete each other in their efforts to lure the female into their territory to mate.

Males may spend part or most of the day on the lek, for weeks or months. In contrast, the females visit leks for just a few hours. This difference in male and female behaviour occurs because females of lekking species tend to come into reproductive condition over a long period of time. As a result, males have the opportunity to mate with several females in the course of the breeding season.

One of the consistent findings from all studies of lekking birds is that some males are very successful at attracting females to their territories, whereas others may spend the entire season on the lek and not obtain a single mating. In the North American Sage Grouse, a typical lek species, 10% of the males in fact obtain 75% of all copulations.

This variation in male success is similar to that in the first category of non-lekking polygynous birds mentioned above. A further similarity is that the females of both categories can usually rear their offspring with little or no help from the male.

The majority of lekking birds, like the ones just described, defend territories which are tightly clustered and territory holders may be just a few metres, or even centimetres apart.

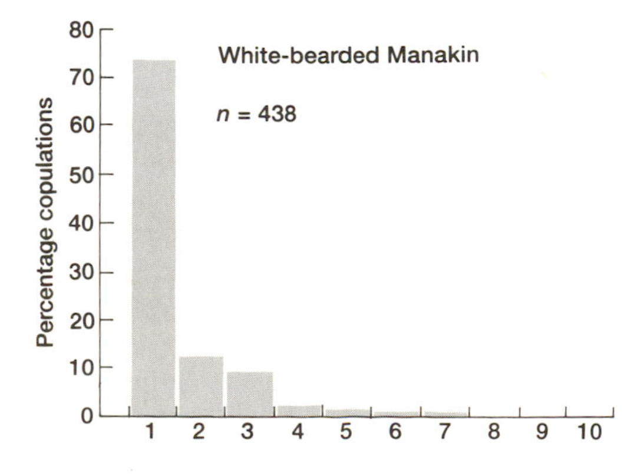


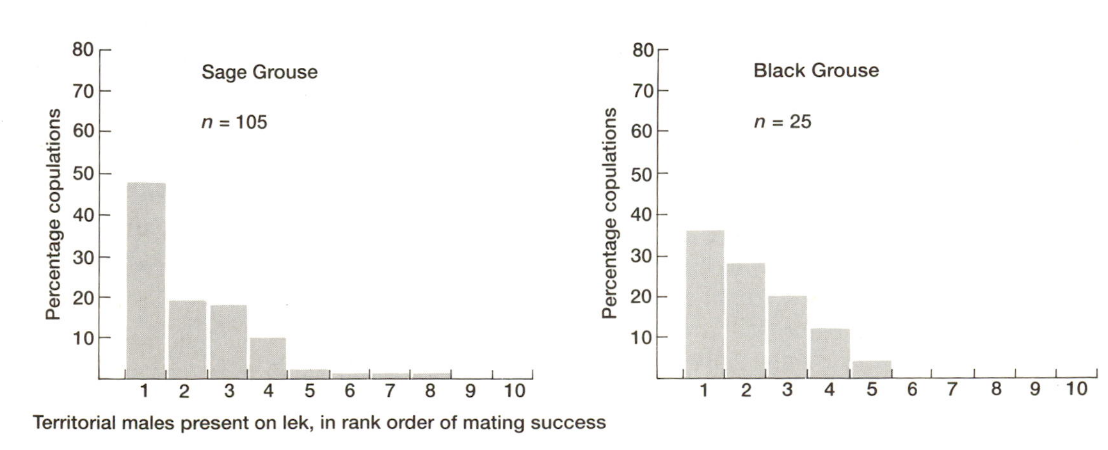


*In each lek system males are ranked in descending order according to the proportion of total copulations each performs in a breeding season. Almost all of the matings are in practice performed by a few males, who defend the central territories on the lek*

There are however a few species which utilize dispersed leks, in which territories are separated by hundreds of metres. Examples include the Capercaillie, the peacock and that most spectacular of passerines, the lyrebird. Similarly, Village Indigo-birds, which breed in Africa, defend call sites in trees several hundred metres apart during the six-month-long breeding season. In these, as in other lek species, females may visit several males before finally mating with one.

In a study of 14 individually marked Indigo-birds, one male obtained over half the matings, while eight were never seen to mate at all. In this species, as in other lekking birds, the male plays no part whatsoever in rearing the young, and in fact neither does the female. This is because the Village Indigo-bird is a brood parasite; after mating the female goes off to find the nest of a Red-billed Firefinch in which to lay an egg; the young Indigo-bird is reared along with the young Firefinches.

In any situation where one or a few males monopolize most of the females, competition among the males for mates will be intense (exactly the same is true for females competing for males in polyandrous species). The evolution of brightly coloured or elaborate plumages in male birds has occurred through a process known as sexual selection. Brightly coloured males may be more attractive to females, or they may be better able to defeat other males when competing for females. As this implies, sexual selection has two components: male–male competition and female choice (see Chapter 1). Both processes occur but at present it is not known which is more important.

If sexual dimorphism occurs as a result of competition for mates we would expect it to be most pronounced in those species in which competition is most intense. This is exactly what occurs: compared with monogamous birds the males of polygynous species are relatively much larger than females. Similarly, in polyandrous species, where females compete for males, the females are larger and more brightly coloured than the males.

The idea that females might prefer males with the most elaborate or brightly coloured

*Black Grouse (Tetrao tetrix) congregate at traditional sites, called leks, where they display and call, to attract females*

*In many polygynous birds, like these wild Turkeys (Meleagris gallopavo), sexual selection has resulted in the males being more brightly coloured and larger than the females*

plumage, or the most elaborate song, has been tested in just a few bird species. The use of elaborate songs by male Sedge Warblers to lure females is described later.

Interesting experiments on elaborate plumage have been performed on two species, one polygynous, and one monogamous. The polygynous Long-tailed Widow-bird, which is just 7 cm long in the body, supports an enormous (50 cm) tail. This is used in display as the male floats across its territory on the African savanna. Males which had their tails artificially lengthened attracted more females than unmanipulated birds, which in turn attracted more females than males with experimentally shortened tails.

A very similar experiment was conducted on the monogamous Barn Swallow. In this species sexual dimorphism is slight, but the outermost tail feathers of males are longer than those of females. In this experiment the two outermost tail feathers were either lengthened or shortened with effectively the same result as in the Widow-bird; males with elongated tails paired up more rapidly and were therefore more successful in breeding than controls which were more successful than males with shortened tails. This experiment is particularly revealing because it demonstrates that female choice operates among monogamous species as well as among polygamous ones.

In bowerbirds, females choose males not on the basis of their plumage, but on the quality of their bower. Males build bowers out of twigs which they then paint and decorate. In the Satin Bowerbird the bower is a short avenue of twigs (aligned north–south) painted black (with a mixture of saliva and charcoal dust; the latter occurs naturally from forest fires). At the north end of the bower is a display area covered by yellow leaves or straw, and decorated with flowers, feathers, berries etc., which are mostly blue. Females visit the bower only to mate; they build the nest, incubate and rear the young unaided. When a female visits a bower the male displays vigorously. If mating takes place, it does so in the bower. Given that males with the best decorated bowers achieve most matings, it is not surprising that males frequently steal from and attempt to destroy each other's

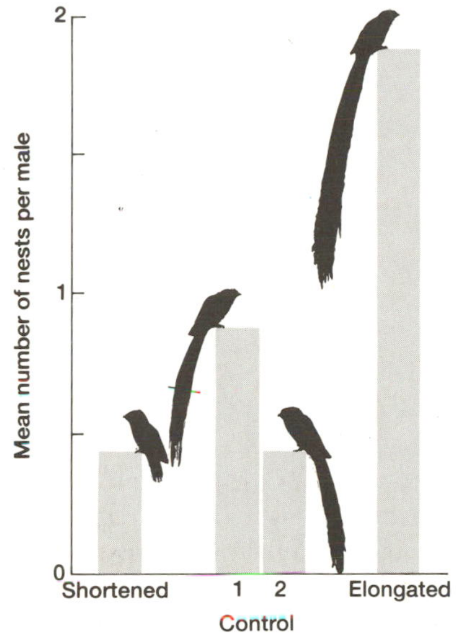


*In the Long-tailed Widow-bird (Euplectes progne) the male's long tail is a sexually selected ornament. Females prefer to mate with the males with the longest tails. As the diagram above illustrates, birds whose tails were experimentally shortened were much less successful in obtaining mates than those whose tails were significantly lengthened*

*The male Satin Bowerbird (Ptilonorhynchus violaceus) builds an elaborate bower to attract females. This avenue of twigs is decorated with a variety of blue objects as a lure: if the female is impressed, copulation takes place within the bower. The female then goes off to build a nest and rear her young alone*

bowers. The importance of the decorations has been demonstrated experimentally. Males from whose bowers decorations had been removed subsequently obtained relatively few matings. Females assess males not only by the number of decorations but by their scarcity, preferentially mating with males that have proved themselves by finding rare objects.

*Polyandry*

In this mating system the female forms a bond with and mates with several males. This is therefore the female equivalent of polygyny. Polyandry is rare, much rarer than polygyny (because it is usually males that compete for females, not vice versa: see Chapter 1) and it has been recorded in less than 50 species. Polyandry can take several forms, as follows: (1) Simultaneous polyandry is where a female mates with several males during a single breeding attempt, as occurs in several jacana species. (2) Sequential polyandry is where a female mates with one male, lays eggs for him to tend and then courts another male, as in the Spotted Sandpiper. (3) Cooperative polyandry is where all males mate with the female and all help to rear a single set of chicks, as in the Galápagos Hawk. (4) Polygynandry is where females are sequentially polyandrous but males are simultaneously polygynous.

There are two important features of polyandry: one is the ability of females to lay a lot of eggs, the other is that, in most species, the male undertakes all parental duties. A rich food supply is important both for females to produce numerous eggs, and to enable the male to rear the young unaided. Because the male cares for the clutch the female is free to pursue other males and lay clutches for them too. This reversal of male and female roles has resulted in females of some polyandrous species being larger and brighter than the males.

It is thought that the evolutionary starting point for polyandry was the habit of 'double clutching'. This form of breeding occurs in the Red-legged Partridge (see Galliformes, p. 97) and certain waders. In Temminck's Stint for example, the female produces two clutches of eggs in rapid succession; the male looks after the first and the female the second. It is a relatively small step from this type of system to one in which the female mates with a second male and leaves him to look after the second clutch. Once this occurred the female would be free to lay a further clutch. This is basically what happens in the Spotted Sandpiper, in which the female may mate successively with up to four males laying a clutch for each of them, but usually helping only the final one to rear young.

Different jacana species may be monogamous or sequentially polyandrous, like the Indian jacana, or simultaneously polyandrous, like the American jacana. In the latter the males each defend a sub-territory within the female's territory, and she helps each of them defend their territory against her other males! The female mates with all males, sometimes in rapid succession, and lays eggs in each of their nests. Males can never, therefore, be sure whether they are the true father of the young they rear. This is a real problem for males of polyandrous birds, and in all species studied so far, females mate extraordinarily frequently. In the Galápagos Hawk for example, a female may be simultaneously bonded to four males. If one male mates with the female all the others do so in rapid succession, presumably in the hope of fathering at least some of the offspring.

People tend to think of particular species as being either monogamous or polygynous, but in fact there is quite a lot of flexibility in mating systems. Some species which are normally monogamous will breed polygynously if and when the opportunity arises. The ultimate example of this flexibility occurs in that most unobtrusive of European birds, the Dunnock. This species shows the full range of mating systems from polyandry (one female with three males), through monogamy, to polygyny (one male with two females).

*Other tactics*

Most birds are, as we have seen, monogamous, and this implies an exclusive mating relationship. However, close observation of individually marked birds in recent years has indicated that mating relationships among monogamous birds are less exclusive than was once thought, with some sneaky matings going on between neighbours. The advantage for males of such extra-pair copulations is that they can parasitize the parental care of other birds. The female equivalent is to dump eggs in the nests of conspecifics. Extra-pair copulations have been recorded in well over 100 species typically regarded as being monogamous, including raptors, ducks, seabirds and passerines. While it is clearly advantageous for a male to obtain an extra-pair copulation, it is clearly not to any males' advantage to be cuckolded. Males can minimize the risk of being cuckolded in two ways, either by mate-guarding, that is by closely following their partner everywhere during the days on which she can be fertilized, or by copulating very frequently with her. Many passerines, for example the Skylark, adopt the former strategy, while many colonial species and raptors adopt the latter. The goshawk for example, copulates 500–600 times per clutch. The reason for this is that, as in other raptors, the male goshawk invests a considerable amount of time and energy in reproduction, mainly by feeding the female at the nest. To be cuckolded would result in the male wasting his efforts, since he gains no genetic advantage from rearing offspring fathered by another male.

What is the likelihood of a single extra-pair copulation resulting in extra-pair paternity? This question has been examined in the Zebra Finch by a combination of field and laboratory study. Zebra Finches breed in the arid parts of Australia; they are basically monogamous but extra-pair copulations also occur. Zebra Finches breed readily in captivity, and in an aviary extra-pair copulations result in offspring. Controlled experiments with birds breeding in cages have been performed to show how the sperm from two different males compete to fertilize a single female's egg. If the extra-pair male is the last to copulate with the female before she lays, he will fertilize, on average, about 80% of the eggs. If however, the male partner copulates last, the best the extra-pair male can hope for is to fertilize about 20% of the eggs. In the few species studied so far in the wild, estimates of extra-pair paternity range from about 1% (White-throated Bee-eater) to over 20% (Indigo Bunting).

Until recently the techniques for examining paternity were rather crude (as in humans), but now the technique known as DNA-fingerprinting allows biologists to determine the parentage of all individuals in a population. The DNA of every individual is unique. After extraction from blood, semen or any tissue, the DNA is chemically cut into fragments which are then radiolabelled and separated on a gel, to produce a 'fingerprint' which is unique to each individual.

DNA-fingerprinting can also determine maternity, and can therefore be used to detect egg-dumping. The value of this type of technique is clear in studies of species like the Lesser Snow Goose, where both extra-pair copulation and egg-dumping occur.

Egg-dumping or intraspecific brood parasitism has recently been found to occur in a number of birds, including the European Starling, Moorhen, several ducks, Barn Swallow, and Cliff Swallow. There are several reasons why ornithologists have only recently become aware of, and interested in this behaviour. First, it is difficult to detect, much more so than interspecific brood parasitism, because eggs from two different females of the same species often look very similar. In addition, it is only in the past 10 or 15 years that the evolutionary consequences of such behaviour have become apparent.

The best-studied intraspecific brood parasite is the North American Cliff Swallow. This species makes gourd-shaped nests from mud in dense and sometimes large colonies (up to 3000 pairs). Charles Brown of Princeton University found that up to 25% of nests were parasitized by other swallows. Most eggs were dumped into nests which are within a few metres of the parasite's own nest. Parasitic females laid their eggs very rapidly, usually in less than a minute and sometimes as quickly as 15 seconds. This is similar to interspecific brood parasites like the European Cuckoo, but different from non-parasitic birds which generally take much longer to lay an egg.

It emerged that Cliff Swallows transferred eggs from their nest to another, carrying the egg in the bill. Parasitic female Cliff Swallows also increased their reproductive success by

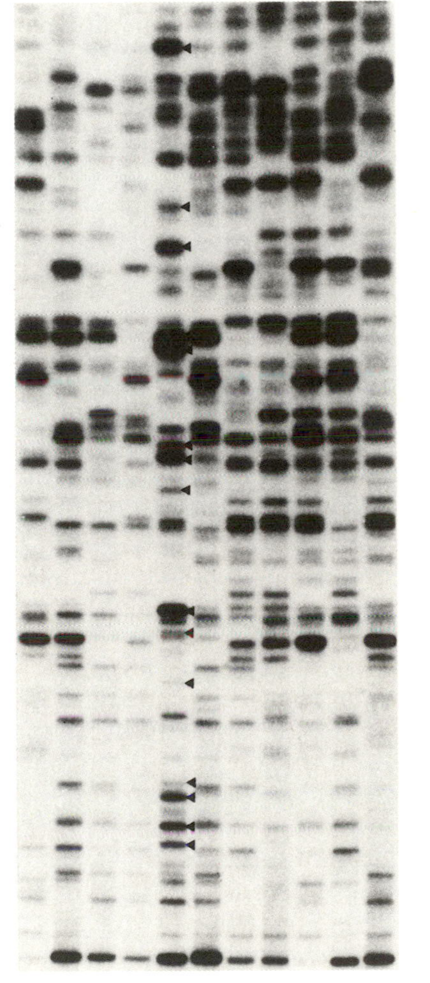

*DNA-fingerprints from a male and female Zebra Finch (Poephila guttata) and the nine offspring they reared from two successive broods. The nestling marked with an asterisk had many bands (indicated by arrows) absent from the male, indicating that it was unrelated and the result of an extra-pair copulation*

dumping extra eggs, since they reared their own brood, but also had one or two additional chicks reared by others. In contrast, the birds that were parasitized suffered, since they either laid fewer of their own eggs, or had one or more of their own eggs thrown out by the parasite when she laid, and because they incurred the energetic cost of rearing a chick which was not their own. The frequency of intraspecific brood parasitism in Cliff Swallows was relatively low in colonies of less than ten pairs, suggesting that parasitism is one of the costs of coloniality (for the parasitized pair).

TRB

## Communal breeding

In the previous section we considered monogamous birds in which the male and female worked together to rear their brood, and polygamous systems in which either the male or female alone reared the brood. In this section we discuss those cases where a breeding pair is assisted by one or more 'helpers' to produce young, a breeding system referred to as cooperative or communal breeding.

Communal breeding was discovered and studied in three tropical bird species, the Brown Jay, the Bushtit and the Banded-backed Wren in the 1930s. During the next 40 years communal breeding was simply considered to be one of several odd things that some birds did. The study of communal breeding was neglected largely because the geographic distribution of these birds and ornithologists did not coincide. Most communal breeders occur in Australia and the tropics, whereas most ornithologists occur in temperate regions. In the 1970s the central importance of evolutionary theory to animal behaviour was realized and studies of communal breeding in birds (and mammals and insects) suddenly became fashionable. Why was this? As discussed in Chapter 1, we expect individuals to behave selfishly in order to reproduce themselves; the occurrence of non-breeding 'helpers' appears at first sight to be an evolutionary paradox. Charles Darwin had been aware of this problem over 100 years ago with respect to the sterile castes within social insects. Since the 1970s there has been a rapid expansion of studies on communal breeders, reflecting an effort to try to understand how such apparently unselfish behaviour could evolve.

Communal breeding refers to a pair of birds (male and female) being assisted in their breeding by one or more helpers. Helpers are often offspring from previous breeding seasons and may be one, two or more years old. The numerous detailed studies of recent years have demonstrated the wide range of breeding and social systems encompassed by the term 'communal breeding'. About 220 species of birds (about 3% of all species) have been classed as communal breeders (although not all of these breed in this way all the time). The two main types of communal breeding are either group territorial or colonial. Group territorial species occupy large all-purpose territories, whereas colonial species breed in close proximity with very small territories. The group territorial species can be further subdivided into those in which only one pair breeds, or those where more than one pair breeds. These are referred to as singular or plural breeders respectively. A well-studied example of a singular breeder is the Australian Blue Wren. The helpers are

*The Blue Wren (Malurus cyaneus) is a cooperative breeder. The mated pair (female on right, male on left) are helped by other birds, often from their previous broods*

mainly male birds, offspring of previous breeding attempts, but the breeding female mates with only one male. In the Galápagos Hawk the helpers are again all male, but as mentioned above, they all mate with the female. In some species like the Stripe-backed Wren and several babblers, both male and females act as helpers. Among the plural breeders the females may either all lay in the same nest, as occurs in anis and the Acorn Woodpecker, or each breeding female has its own separate nest, as in the Mexican Jay.

Among the communal breeders that breed colonially the European Bee-eater has a rather different system. At the start of the breeding season all birds breed as pairs, but if a nesting attempt fails the birds either re-lay, or more often, go and help their relatives. By the time the chicks hatch a pair may therefore have several helpers.

Communal breeding occurs in such a wide range of bird species occupying such divergent habitats that we can be fairly certain that it has evolved several times. Communal breeders have relatively few features in common, other than (1) they are most abundant in warmer parts of the world – Australia has over 60 communally breeding species; (2) they tend to be non-migratory; and (3) they are normally resident on the same territory all year. An analysis of Australian communal breeders shows that most are ground-feeding insectivores breeding in dry woodland. This habitat typically shows a lack of seasonality and hence productivity varies little through the year. It is thought that since food may be difficult to find in such areas, communal breeding is the best way to produce any offspring.

The ecological conditions in which a species evolves can mould certain features of its population biology, such as how long it lives, and how old it is when it starts breeding (see Chapter 8). These features in turn can have important implications for that particular species' social organization.

One of the main features of communal breeders is that breeding is often delayed for one or more years. If this occurs and individuals tend to remain together on the same territory, then the opportunities for helping exist. Birds may delay breeding because opportunities for doing so are limited; territories might be in short supply (as in some Acorn Woodpecker populations), or females may be scarce (as occurs typically for Superb Blue Wrens). If breeding opportunities are restricted the best thing to do might be to remain at home in the natal territory with their parents. Once this happens the stage is set for helping to occur.

One of the central questions in the study of communal breeding is 'who actually benefits from the helping behaviour?' Are the helpers being selfish by remaining at home? Or do parents benefit from keeping their offspring in the territory? Or do both helpers and recipients benefit from the arrangement? It seems that in general helpers are acting in their own interests and at the same time they do really help their parents. Whether non-breeders help or not may be determined by food availability. This has been examined in the Moorhen: when extra

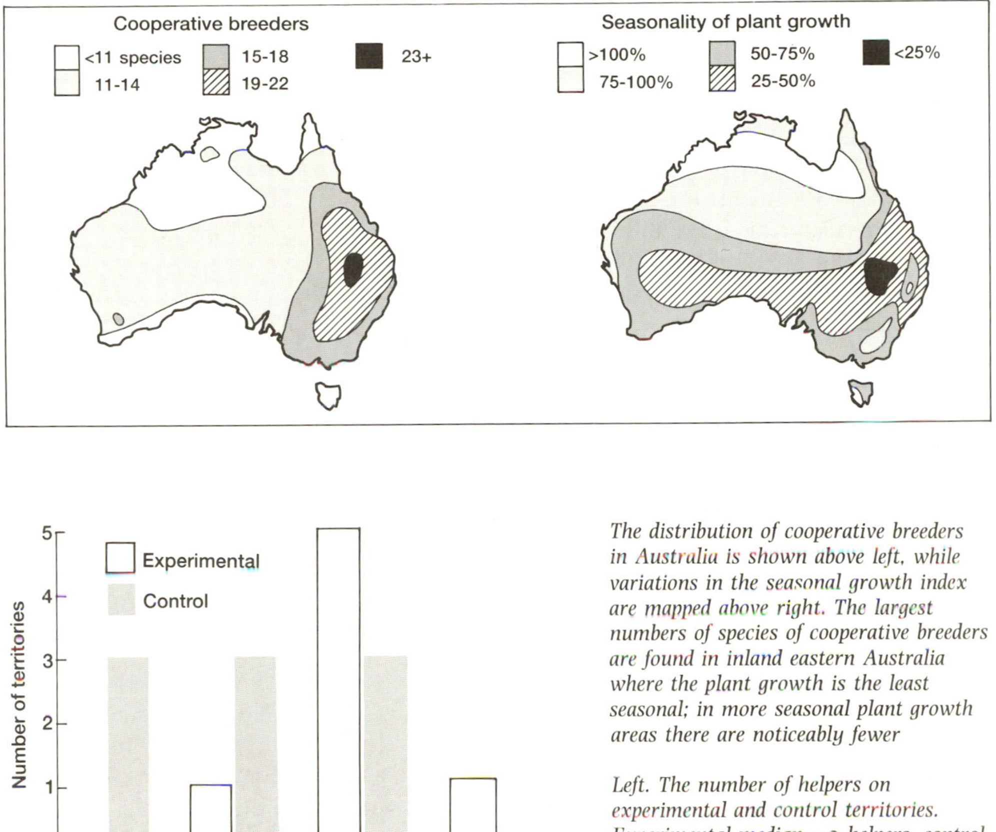


*The distribution of cooperative breeders in Australia is shown above left, while variations in the seasonal growth index are mapped above right. The largest numbers of species of cooperative breeders are found in inland eastern Australia where the plant growth is the least seasonal; in more seasonal plant growth areas there are noticeably fewer*

*Left. The number of helpers on experimental and control territories. Experimental median = 2 helpers, control median = 1 helper, Mann-Whitney* $U = 12$, $P < 0.05$ *(two-tailed)*

food was provided non-breeders were more likely to help feed their younger siblings.

Most studies have shown that helpers would do best by going off to breed themselves, but that the lack of opportunities (i.e. lack of territories or females) prevents them from doing so. Under these conditions the best option is to remain at home and help their parents rear offspring. In other words they help to rear their own brothers and sisters, their close genetic kin (see Chapter 1).

TRB

## Territory

It has been known for a long time that some birds defend territories. In fact most bird species are territorial at some stage in their lives. Both breeding and non-breeding territories occur, and within these broad categories several different types of territory exist. The term 'territory' has been defined in several different ways and the most straightforward and useful definition is 'any defended area'. Territorial defence usually results in individuals or pairs being more evenly spaced than one would expect by chance. A clear example of this is provided by the European Sparrowhawk.

Territories vary in size from just a few square centimetres in some seabirds and lekking species, to several hundred square kilometres in large raptors.

Territoriality may play a role in regulating populations (see Chapter 8). If all available breeding space is occupied by territories some individuals may be prevented from breeding. This has been demonstrated in several species. When Oystercatchers were removed from their territories on Skokholm Island, Wales, their places were rapidly taken by individuals from the non-breeding flock (referred to as 'floaters'). It is important to realize however, that territoriality did not evolve to regulate populations; any regulatory effect is a mere consequence (not a function) of territoriality.

The classical territory is the all-purpose one in which birds perform all their activities; feeding, sleeping and breeding. Such territories may be occupied by a monogamous pair of birds (e.g. Australian Magpie Lark, European

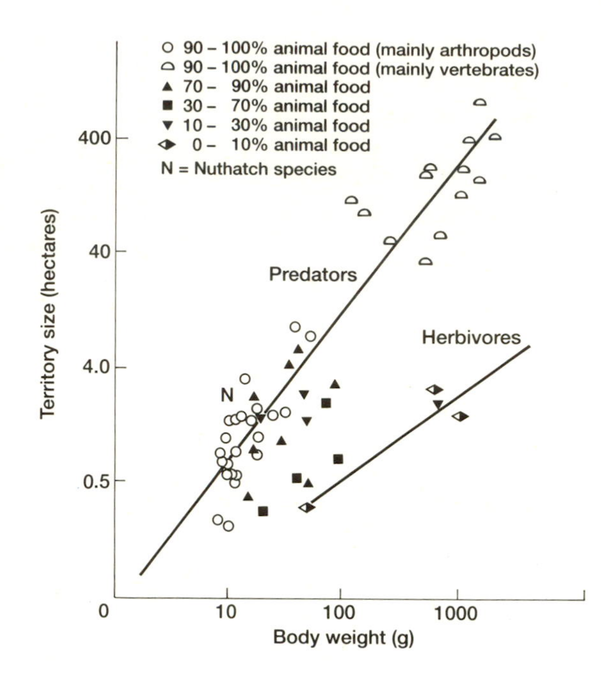


*Territory size of birds in relation to body size. Broadly speaking, larger birds occupy larger territories. Animal-eating birds of a given body size generally require a larger territory than herbivorous birds of a similar size*

Blackbird, American Robin) or a group of birds, such as the communally breeding Arabian Babbler. Territory size increases with the species' body size since larger birds require more food and hence a larger foraging area.

Multi-species territory ownership even occurs. In the Amazonian forests of South America groups of 8–12 pairs or families of birds (each of a different species) defend a common territory of about eight hectares, and feed together as a flock. The territory is defended against other similar flocks, and during territorial disputes each bird displays to its own species. All flock members nest within the joint territory.

The second territory type occurs in colonially breeding birds such as herons, ibises, swallows, bee-eaters and seabirds. It is usually a very small area around the nest or nest site. The Adélie Penguin, for example, breeds in colonies and defends only a tiny area around its nest.

Lekking species, already discussed, utilize the third type of territory. These are usually very small territories which contain no resources whatsoever; they are simply mating territories.

Non-breeding territories fall into two broad classes; roosting territories and feeding territories. Starlings defend small territories at their roost sites – most easily seen as they jostle

*Many birds defend their nests. Here a Sunbittern (Eurypyga helias) performs a spectacular nest-defence display, showing the two huge 'eye spots' on its wings*

and squabble as they prepare to sleep on city buildings. A relatively small number of species, such as the European Robin, Pied Wagtail, certain shrikes, hummingbirds and sunbirds defend non-breeding feeding territories. A few species, such as the Australian Magpie, occupy and defend their territories throughout the year, and some species, like the Tawny Owl, may stay within the same territory for their entire life.

Why do birds defend territories? The answer is that by doing so they get more of a particular commodity than they would if they were not territorial. Three main 'commodities' or benefits accrue from territory defence: food, mates and breeding sites. Clearly there must be a balance between the costs of territory defence (time, energy, risk of injury during defence) and these benefits. As yet relatively few studies have successfully quantified the costs and benefits of territory defence. However, in sunbirds and hummingbirds it has been possible to measure the energetic costs of defending territories, and the energetic reward, in the shape of nectar-yielding flowers within the territory, resulting from defence. These studies show that birds defend territories of a size that results in a net energetic benefit.

For many colonial birds the territory is the same as the breeding site. Common Guillemots (Murres) for example, breed in large colonies at high densities; up to 70 pairs per square metre have been recorded. Guillemots make no nest, and lay their single large egg directly onto bare rock. The territory they defend is just the few square centimetres around this breeding site. To the human eye one guillemot territory looks much like another, but to a guillemot the quality of its territory can mean the difference between rearing a chick or not.

The basic physical characteristics of the guillemot's territory are important in determining breeding success. Sites that are level rather than sloping and broad rather than narrow tend to be the most successful. In addition to these features, the greater the number of close neighbours a guillemot has, the better the territory and the greater the chances of rearing a chick successfully.

In the guillemot the close proximity of conspecifics increases success, but this is not true for all birds. Other species benefit from being spaced out. In both Great Tits and gulls well-spaced nests are less likely to be found and robbed by predators. The spacing is achieved through territorial behaviour.

*Common Guillemots (Uria aalge) breed colonially and at extraordinarily high densities to minimize the risks of predation on their eggs and young. Social interactions are frequent and much of the Guillemot's social behaviour is therefore designed to avoid aggression*

## Territory defence

Birds defend their territories in a variety of ways, the most familiar of which is through song. Most passerines (songbirds) sing for prolonged periods during the breeding season, and for many years it was taken for granted that song was a territorial 'keep out' signal. It is only comparatively recently that this idea has been tested experimentally. In fact, small birds use several different methods to defend their territories: singing is a long-range signal, visual displays are medium or short-range signals, and if these fail the ultimate 'signal' is actual fighting.

Birds other than passerines use postures, calls and aerial displays to signal territory ownership. European magpies for example, simply sit in the top-most branches of trees to do this.

In the majority of birds it is the male that performs most of the territory defence, although in some species males attack male intruders and females attack female intruders.

Some birds, as well as defending their territory from conspecifics, also exclude other species from their territory. What is the purpose of interspecific territoriality? There are two main explanations: (1) misidentification, and (2) resource overlap or a similarity of requirements, such as food. The first of these assumes that birds evict other species from their territories simply because they mistake them for conspecifics. If this were true interspecific territoriality should have evolved among similar looking species regardless of whether they compete for the same resource. Instead, the opposite appears to be true. Birds defend territories against species with similar ecological requirements, regardless of what they look like. There is therefore little evidence for the misidentification hypothesis, but some good evidence for the resource overlap idea. In several species the degree of interspecific territoriality is correlated with the degree of dietary overlap. In North America, mockingbirds defend winter territories around fruit bushes and are most aggressive towards other frugivores and least aggressive towards insectivores.

TRB

*Above. Visual threat displays are used to repel intruders as here with a Great Tit (Parus major)*

## Social groupings

Many birds are highly social and may feed, roost or breed together at some stage of their annual cycle. Birds like geese, finches and parrots regularly feed in flocks. There are two main benefits from flocking: finding food, and avoiding becoming food, i.e. avoiding predation. It has long been recognized that an association exists between the types or species of bird that forage in flocks and the types of food they consume. Birds that feed in flocks typically exploit foods that are patchily distributed in both space and time, such as seeds, fish, and aerial insects. Moreover, many of those bird species that feed socially also breed and roost socially, in colonies. In contrast, most solitary feeders consume well-dispersed animal prey, such as voles and foliage-eating caterpillars, and breed alone.

Foraging as part of a flock may be more efficient than feeding alone for a number of reasons. In a few species flock members may cooperate to capture difficult prey; a well

*Above right. Moving around in groups can help birds avoid predators. Red-and-Green and Scarlet Macaws (Ara chloroptera and Ara macao) here collect clay together at a clay-lick in the forests of Peru. The clay is eaten either as a mineral supplement or to help detoxify poisonous seeds*

*Right. These White Pelicans (Pelecanus onocrotalus) illustrate one advantage of social foraging. They surround a fish shoal and then simultaneously plunge their beaks into the shoal, causing chaos and confusion among the fish but enhancing their own chances of a catch*

known example is the synchronized fishing of pelicans. Cooperative hunting also occurs in Harris' Hawk in New Mexico, USA. This species lives and breeds socially, and groups of up to six individuals cooperate to hunt large prey such as Cottontails and Jack Rabbits. In other species flock members may help to flush prey, as probably occurs in Cattle Egrets foraging in long grass among cattle or other herbivores. A more complicated system operates in Barnacle Geese wintering in the Netherlands. The birds feed on salt marshes where their main food is sea plantain. Areas of sea plantain cannot be defended since they are submerged at high tide. Instead the geese visit feeding areas as a flock and at regular intervals. The geese return to patches of sea plaintain every four days, and this interval results in the most efficient exploitation of the plant. Experiments which involved cutting sea plaintain with scissors to simulate goose foraging showed that four days was the optimum return time. Earlier return by the geese would not have given the plant long enough to recover, while a longer interval

would have meant wasted feeding opportunities. This system works only if birds feed in a flock because this ensures no other geese interfere with the system.

The second main benefit of being part of a flock or other aggregation is protection from predators. A group of birds is likely to detect an approaching predator sooner, allowing birds to take cover. A flock of European Starlings attacked by a sparrowhawk will instantly become much more compact; this makes it more difficult for the predator to single out any particular individual. The predators may also be reluctant to attack a dense flock of birds in case they injure themselves, or possibly because they are confused by so many individuals. When a trained goshawk was flown at flocks of Woodpigeons, larger flocks responded sooner and the goshawk was much less likely to be successful in making a kill.

Obviously the feeding and anti-predator advantages of group living are not independent. Several studies have shown that individuals in flocks spend less time looking around and more time foraging than solitary individuals. Nevertheless, in most species individuals in flocks operate independently of each other; there is generally no coordinated system of predator surveillance. However, in a few highly organized communal breeders, like the Jungle Babbler, one individual undertakes sentinel duty while other group members forage.

## Coloniality

A large breeding colony of gannets, penguins, or ibises is an unforgettable sight. Colonially breeding birds are relatively easy to study compared with solitary breeders; since the birds are close together several pairs can be watched simultaneously. The birds also occur at a predictable location for much of the time, and studying cliff-nesting species like auks or fulmars is rather like going to a movie. You get into your hide, open the window and it all happens in front of you. Despite the relative ease with which colonial birds can be studied, and despite the large amount of information these studies have produced, we are still a long way from understanding why some birds breed colonially and others solitarily.

*The breeding colonies of some birds contain hundreds of thousands of pairs, as in this colony of King Penguins (Aptenodytes patagonica) on Marion Island*

There are several possible explanations of why some species should breed in colonies. For example, suitable sites may be in short supply, forcing birds to breed together. This may be one of the selection pressures favouring colonial breeding. In fact there is some evidence that breeding sites for certain species may be limited. When artificial nesting platforms were created for guano-producing Cape Cormorants in South Africa, numbers increased from no pairs breeding before platform construction to 500 000 pairs twenty years later! Exactly the same type of effect has been recorded in Peruvian seabird numbers, when the guano-managers there created new nesting sites.

The two other main explanations of why birds should breed in colonies are basically the same as those for why birds should feed in flocks: to gain protection from predators, and to increase their efficiency in finding food. Initially ornithologists considered only these possible advantages of colonial breeding. More recently however, it has been realized that greater insight might be gained by considering both the benefits and the costs of breeding colonially or alone. Obviously if coloniality is adaptive then we would expect the benefits to exceed the costs. Although this approach has provided

*If natural breeding sites are in short supply, some species, like these Black-legged Kittiwakes (Rissa tridactyla), will utilize man-made structures like wharfs*

clear new insights, it has not yet been possible to assess the costs and benefits in a really useful way. This is because there is, as yet, no easily accessible common currency that we can use to assess relative costs and benefits. The ideal currency is life-time reproductive success. If we could obtain this for individuals of the same species breeding colonially and solitarily while all other ecological conditions were kept constant, then we might be able to assess the selection pressures favouring different types of nesting dispersion. However, obtaining lifetime reproductive success is not particularly easy especially since many colonially breeding species are long-lived, and some, like seabirds and vultures, might require a 40-year long study to obtain the necessary information.

There are several automatic costs to coloniality: (1) competition for resources – a large number of birds breeding in a small area will almost inevitably result in competition for food, space (i.e. breeding sites), nest material and matings; (2) risk of disease and ectoparasite transmission; (3) intraspecific killing; and (4) misdirected parental care.

There is evidence that colony members deplete food supplies around their breeding colony, thus increasing competition for this vital resource. Several measures of reproductive output (clutch size, breeding success, chick growth) in seabirds, and some other colonial species, such as the Fieldfare, are negatively correlated with colony size. This suggests that the larger the colony the greater the degree of food depletion, and the lower the breeding success. Direct measures of fish abundance around Double-crested Cormorant colonies in Canada showed that fish were indeed less abundant close to the colony. When a Herring Gull population was decreased through culling by 75%, the body weight and egg size of the remaining birds increased, suggesting that previously their food supply had been limited by intraspecific competition.

Competition for space has been demonstrated among Shags on the Farne Islands, north-east England, through a natural experiment. Two 'red tides' (blooms of toxic dinoflagellates) in 1968 and 1975 killed off a large proportion of the Shag population each time, thus making a large number of breeding sites available. Remaining birds moved 'up-market' into good quality sites and in doing so increased their breeding success. There is also some indirect evidence that competition for space is important; the fact that in mixed colonies each species tends to have its preferred type of breeding site suggests that competition has been intense in the past. A good example is provided by the two kittiwake species; the familiar Black-legged Kittiwake and the much rarer, Red-legged Kittiwake. These two species breed almost side by side on the Pribilof Islands, Alaska, but the Red-legged Kittiwake has a distinct preference for breeding sites with an

*Competition for nesting sites has led to segregation of the sites used even by closely-related species. On St Paul Island off Alaska, Black-legged and Red-legged Kittiwakes (Rissa tridactyla and R. brevirostris) breed in the same colonies, but the Red-legged Kittiwake prefers sites with a slight overhang*

overhang. More direct evidence for interspecific competition for space comes from a study of puffins and Manx Shearwaters on Skomer Island, Wales. Both species suffered a decrease in breeding success through the interference caused by the other species as they fought over breeding burrows.

Competition for nest material is a further disadvantage of colonial breeding, and the theft of nest material has been recorded in many birds including penguins, gannets, herons, Rooks and Zebra Finches. Such theft will be disadvantageous for the victim, but beneficial for the thief. Stealing material from an adjacent nest may save birds long, energetically costly flights. However, the risk of theft has meant that in many species of colonial birds, one member of the pair must be at the nest at all times to defend it. This in turn creates other problems, one of which is that males cannot guard their females from extra-pair copulations by close following.

Competition for matings is expressed through extra-pair copulation attempts. These occur in solitary and colonial birds, but the risks (or potential benefits, depending upon your perspective) are greater among colonially breeding birds.

Disease and ectoparasites will be most readily transmitted between individuals in close proximity. In Sand Martins the average number of fleas per nesting burrow increases with the size of colony. Similarly, in colonial American blackbirds, such as Red-winged and Yellow-headed Blackbirds, body mites occur much more frequently than they do on solitary species. It is also known that outbreaks of ticks among colonial seabirds can cause individuals or even the entire colony to abandon breeding. However, in general the effect of disease or ectoparasites on birds is an understudied field.

Intraspecific egg and chick killing occur in gannets, terns, gulls and skuas. In some species, like gulls and skuas, chick killing is effectively predation since the victim is also eaten. Jasper Parsons, while at Durham University, studied this sort of cannibalism among Herring Gulls breeding on the Isle of May, Scotland. He found that the behaviour was restricted to just a few 'specialist' individuals.

Misdirected parental care can arise either through eggs or chicks getting into the wrong nest, or as a result of offspring produced by extra-pair copulations. The energetic cost of rearing offspring can be substantial and it is not usually in any individual's interest to rear chicks which are not its own. The chances of eggs or chicks becoming mixed up are far greater in colonial birds, particularly those which breed very close together. Common Guillemots usually breed in close proximity. They make no nest but lay their single egg on bare rock, so the chances of eggs or chicks getting mixed up are potentially high. However, guillemots have the most variably coloured and patterned eggs of any birds, and detailed studies by the Swiss ethologist Beat Tschanz, working in Norway, demonstrated that these markings enable guillemots to recognize their own egg. The risk of misdirected parental care does not end once the egg hatches – the same problem exists for chicks too. Guillemot chicks and their parents start to call to each other even before the chick has hatched from the egg! Parents and chicks recognize each other's calls, thereby minimizing the chances of a guillemot rearing the wrong chick. In fact, in virtually all colonial species studied, where there is a risk of eggs getting mixed up, either within the breeding colony or within creches, adults and their offspring are able to recognize each other in some way.

These are the costs of breeding colonies. Now let us consider the benefits. Colonial breeding can decrease the risk of predation in several ways. Simply by being part of a group each individual has a lower probability of being attacked by a predator. Providing the intensity of predation does not increase proportionately with the size of the colony, an individual in a colony will, on average, be less likely to have its eggs or chicks taken by a predator. In some species individuals may gain from being in a colony because the group can detect predators sooner, or because individuals are in the centre rather than at the edge of the group. Another way of reducing egg and chick predation is to breed synchronously, effectively 'swamping' the predators with food. This is known as the Fraser Darling effect, in honour of Frank Fraser

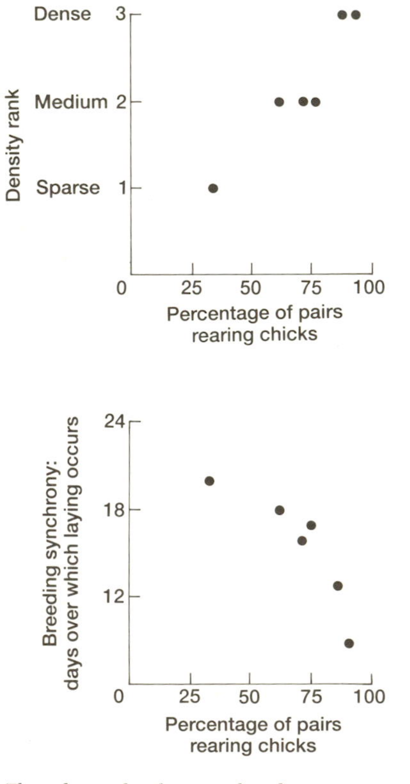


*The relationship between breeding success and population density (above) and breeding success and the time-span over which laying occurs (below)*

Darling, who first suggested it after studying gulls. Highly synchronous breeding, a common characteristic of colonial birds has indeed been found to reduce predation in gulls, terns and auks. Consequently, birds breeding early or late, out of phase with the rest, tend to be the most vulnerable to predation.

Many species also have a further line of defence: mobbing. This is particularly well-developed in gulls and terns. A dive-bombing gull that hits you on the back of the head with both its feet and simultaneously vomits and defecates on your head is a pretty effective deterrent.

The two main feeding advantages associated with breeding in colonies seem to be as follows: (1) the transfer of information regarding the location of food; and (2) a centrally placed breeding site. Birds breeding in colonies may gain a feeding advantage by obtaining information about the location of food.

It has long been thought that groups of birds, such as communal roosts or breeding colonies might serve as 'information centres'. Information on the location of food could be vital since many birds that feed, roost and breed socially utilize food that is difficult to locate because it is unpredictable in space and time. For example, many colonial seabirds feed on shoaling fish that are highly clumped, and rarely stay in the same place for long: obviously such prey is difficult to locate. Exactly the same is true for aerial insects.

Only recently has it been shown that birds can and do gather information regarding the whereabouts of food. Ospreys in eastern North America often breed in loose colonies. When feeding on shoaling fish they caught their next fish much more rapidly if they could see where other Ospreys had caught fish: that is if they had information on the location of the fish shoals. A similar exchange of information occurs in the colonial Cliff Swallow, which breeds in much of North America and feeds on swarms of aerial insects. Unsuccessful swallows return to the colony and wait to follow successful foragers back to good feeding areas.

All individuals are equally likely to be followers or to be followed so all colony members stand to benefit from this system. And as far as is known, information exchange is entirely passive. There is nothing equivalent to honey bee dances.

It is relatively simple to demonstrate numerically that if a bird species uses a food supply that is evenly distributed in space, the best nesting dispersion is also even. This would effectively distribute the breeding pairs over the food supply, and thereby minimize each individual's foraging range.

If a species' food supply is patchily distributed in space and time, as is the case with fish shoals, then it is most efficient to breed in the centre of the entire feeding area since this minimizes the foraging time. While this may seem intuitively obvious it has been a difficult idea to test, mainly because measuring the temporal and spatial aspects of a bird's food supply is difficult. The best demonstration of this effect to date has been in a study of Yellow-billed Magpies in California.

TRB

## Displays

Birds communicate with each other primarily by means of visual and vocal signals. Often these two occur simultaneously, but many visual signals are given in the absence of vocal signals. Vocal signals, songs and calls, are discussed later; the emphasis here is on visual signals. These are generally referred to as displays. They constitute specialized body postures or movements which are often enhanced by elaborate plumage, patches of coloured skin, or sometimes by specially built structures. Designed to be conspicuous, displays can be both beautiful and bizarre.

The main messages that birds need to communicate to each other are concerned with fighting and mating. In general, the more social contact a species has, the more extensive is its repertoire of displays. Gannets, for example, breed colonially and re-pair with the same partners for several years. They have a wide range of displays, many of which are highly stereotyped. In contrast, Golden Eagles, which are also long-lived but breed solitarily, have relatively few displays, none of them highly stereotyped.

## Threat displays

Virtually all species produce threat displays in disputes with conspecifics over valuable resources such as territory, food or mates. If a territory owner challenges an intruder by adopting a threat display, the intruder may adopt a submissive posture and then retreat. Often, however, the outcome of a quarrel is neither so rapid nor so clear cut, and both contestants may oscillate between being aggressive and fearful. The entire continuum of these behaviours is referred to as agonistic behaviour: at one extreme pure aggression occurs, at the other, pure submission. In-between, the individuals may show ambivalent behaviour reflecting these conflicting tendencies.

When this occurs the display may contain elements of aggression and submission simultaneously; for example, during fights several species turn their head away from their opponent. The turned head may reflect the tendency to run (or fly) away, while the stationary body reflects the tendency to stay. In some cases completely irrelevant behaviours may occur. A familiar human example is tie-straightening when men are in a slightly threatening situation such as an interview. Such behaviours are referred to as displacement activities, and there are many examples among birds. Probably the most widespread is preening during the course of a fight. One of the earliest ethologists, on seeing this behaviour among seabirds, suggested that they were infested with ectoparasites! Among domestic chickens and turkeys, contestants may suddenly turn and start feeding in the middle of a fight. Herring Gulls pull grass as a displacement activity during territorial interactions.

The threat displays of most bird species involve postures in which the head, beak and body are directed towards the opponent. In this position the bird can easily launch an all-out attack if the other individual does not back down.

Depending upon the intensity of the aggression, more or less extreme versions of the same posture may be used, or alternatively different displays may be utilized to signal different levels of aggression.

*Left. By displaying his monstrous, inflatable throat pouch, a male Great Frigatebird (Fregata minor) tries to attract a partner*

*Below. Two pairs of Blue-footed Boobies (Sula nebouxii) at their territorial boundary lunge and jab at each other to maintain their exclusive nesting area*

*Right above. By 'freezing' them, most photographs give a misleading idea of displays. These pictures of a Yellow-crowned Night Heron (Nycticorax violaceus) show the range of display postures that can be adopted within just a few seconds of each other*

*Right below. A pair of White Spoonbills (Platalea leucorodia) preen each other, concentrating on those areas which are difficult for each bird to preen itself. Such allopreening may be functional and help to remove dirt or parasites, or it may be ritualized; either way, it helps to develop and maintain social bonds*

Submissive postures can be just as variable, but in contrast to threat displays, their function is to reduce aggression. Accordingly, they often involve birds turning their beak (the main weapon) away from the opponent, or withdrawing the head and neck and sitting in a crouched or fluffed up position.

## Courtship displays

Sexual displays are, either immediately or eventually, concerned with mating and are usually known collectively as courtship displays. They have much in common with agonistic displays since they too involve internal conflicts. Here, the conflicting tendencies comprise a tendency to approach (in order to mate) and a tendency to retreat (because of the fear associated with close proximity). Sexual displays can also rapidly switch into agonistic displays. This can be confusing for the ethologist trying to describe and understand the sexual behaviour of a particular species.

The preening of one individual by another (allopreening), may help to maintain social bonds and is common among communal breeders and colonial species. Among the latter it most commonly occurs among those, like kittiwakes, noddies and herons, where pair members are forced into close proximity because they have very small nest sites. Such allopreening is also common among pigeons,

parrots, owls and passerines. In some species allopreening may do nothing more than provide a pleasant tactile sensation. But in others it may help to keep the plumage clean or free from ectoparasites. Most allopreening is directed at the head and neck, precisely those places where a bird cannot preen itself. In one study of Macaroni and Rockhopper Penguins, in which pair members allopreened each other, unpaired birds had significantly more ticks on their heads than paired birds.

## Ritualization

Many displays have evolved from behaviour patterns which have become ritualized. This process usually involves the formalizing and stereotyping of movements. The original function of allopreening, discussed above, was almost certainly to remove dirt and foreign bodies from the plumage. In many species it has now become ritualized and may no longer serve this purpose, but may instead help to alleviate aggression, and hence strengthen social bonds. In a conflict a bird may start to perform an irrelevant behaviour such as self-preening (above). In some species this sort of displacement activity has become ritualized and now forms an integral part of a display. Ritualized preening in the courtship displays of ducks has been studied in detail: in the shelduck preening of the wing feathers occurs during conflict situations; because the preening is genuine this is considered to be real displacement preening. In similar situations the Mallard raises one wing in order to preen its brightly coloured speculum. The preening here however is confined to the speculum. In the Garganey a further stage or ritualization is reached: one wing is raised, making the blue wing coverts conspicuous, but the 'preening' is completely ritualized as the feathers are not actually groomed by the bill. The most extreme ritualization occurs in the male Mandarin Duck: here the preening is reduced to touching an elaborately enlarged wing feather which is erected during the display.

Many ritualized displays are derived either from conflict behaviours or from what are called intention movements. Prior to taking off, a bird may crouch before springing into the air; the crouch is an intention movement. In several birds, including the Goldeneye duck, several herons, cormorants and boobies this type of movement has become a display in itself, or become part of a display.

*Self preening in four species of duck: the Mallard (Anas platyrhynchos), Garganey (Anas querquedula), Shelduck (Tadorna tadorna) and Mandarin (Aix galericulata)*

## Are displays honest?

It was originally thought that the function of ritualized displays was to transfer information as efficiently as possible and thus avoid ambiguity between the sender and recipient. In other words communication was seen as a cooperative affair between the interacting parties. Recently a more cynical 'selfish gene' view of communication has been proposed by Richard Dawkins and John Krebs, of Oxford University. Their basic idea is that we should not expect communication to be cooperative since it is not always in an animal's interest to indicate its intentions unambiguously. In fact, they suggest ritualized behaviours may have evolved precisely to avoid giving away certain types of information. In other words a ritualized display may constitute a 'poker face'. During an aggressive interaction birds may use a number of different displays. If each one unambiguously

## ADAPTIVE SIGNIFICANCE OF COLOURATION

Birds show a remarkable range of colour patterns and some are among the most brightly coloured of all organisms. Colours range from the cryptic browns and greys of species like the Marbled Murrelet, to the striking blue of the Splendid Fairy Wren and iridescent colours of hummingbirds and trogons. Why should such a range of colours and patterns exist? The two main explanations are that colours have evolved to make birds either more or less conspicuous.

Cryptic colouration hinders detection by prey, predators, and conspecifics. There are relatively few examples of cryptic colours helping birds to approach their prey. One possible example is the black dorsal and white ventral pattern of many diving birds, such as auks, which may render these birds less visible to the fish they prey upon. When a fish looks up towards the sky the pale undersurface of the bird is less visible than if it were dark in colour.

There are, however, numerous examples of birds showing a broad resemblance to their habitat; the Snowy Owl, Snow Bunting and winter plumaged Rock Ptarmigan are all predominantly white and match their background in the Arctic. Similarly, many desert species are buff, or sandy coloured, while green species are common in tropical rain forests. This matching may aid food capture in predatory species, but it is more likely to render the birds less conspicuous to predators.

Ground-nesting birds are especially vulnerable to predators, and some such as woodcock, quail, pheasants, grouse, certain ducks, waders (shorebirds), nightjars and thick-knees have evolved remarkable camouflage. In birds like the peacock, certain grouse, ducks and many passerines, where only the female incubates, she is cryptically coloured while her partner is often brightly marked.

*The superb camouflage of the White-necked Nightjar (Nyctidromus albicollis) protects it as it incubates on a forest floor in Central America*

Conspicuous colouration acts as an advertisement, either to predators or to conspecifics. But why should birds advertise themselves to predators? One suggestion is that bright colours serve as warning colouration associated with unpleasant taste. In a series of taste trials in which the palatability of different birds was assessed, cryptically coloured species such as quail were found to be highly palatable, while brightly, or strikingly coloured species such as the Pied Kingfisher were often unpleasant.

A further suggestion is that bright plumage signals a bird's unprofitability to a predator. In other words by being brightly coloured or conspicuously marked a bird advertises the fact that it is difficult to catch. An analysis of ringing (banding) recoveries provides some support for this idea. A significant negative correlation was found between predation rate and conspicuousness: in other words more conspicuous species were less vulnerable to predation. However, the unprofitable prey idea has not received much support elsewhere, and other workers have argued that sexual selection is the main factor responsible for bright colouration.

Colours and patterns may serve as signals between members of the same species. Birdwatchers use the colour patterns of birds to identify different species and there is every reason to suppose that birds can do the same. Colours, along with other features and behaviour, therefore serve as species recognition signals (see Chapter 10).

In many birds the colour and patterns of the male and female differ markedly, a phenomenon referred to as sexual dimorphism. In most cases males are more brightly coloured than females, but in polyandrous species the opposite is true. In most species sexual dimorphism is permanent, but in some, such as weaver birds like the Red Bishop, the difference between the sexes is apparent only during the breeding season. There are two main ideas for why sexual dismorphism should exist; predation and sexual selection. The concept of sexual selection, introduced by Darwin to explain why males are generally more brightly coloured than females, is discussed in the main text. TRB

indicated their intent, about the likelihood of launching an all-out attack for example, this would be equivalent to a poker player showing his opponents his hand at the start of a game. Dawkins and Krebs have gone one stage further and suggested that the idea that communication involves cooperation between participants is wrong, and that individuals might actually use displays to manipulate others.

One of the clearest examples of what might be manipulation involves alarm calls. Imagine a flock of sparrows or finches foraging in a stubble field. A hawk flips over a hedge into the field and is spotted by one of the flock. This bird gives a high pitched 'Wheee' call and it and the rest of the flock fly for cover. The traditional explanation for the behaviour of the caller is that it is telling the other flock members, unambiguously and efficiently, that a predator is approaching. This explanation suggests that the caller is behaving for the good of the flock. However, as we have seen, it is much more likely to be behaving selfishly and it may in fact be manipulating the rest of the flock to its own ends. The bird sees the predator, takes cover, and *then* calls, thereby exposing the others (not itself) to the hawk. Whether this explanation is correct is still debated, but it does illustrate how signals could be used to manipulate other individuals.

This raises a further question about the evolution of displays. When two birds are contesting a territory, a mate, or whatever, why do they bother displaying to each other at all? Why don't they simply fight over the resource in question? The traditional answer to this is that fighting is not in the species' interests. We now believe that this argument is wrong (see Chapter 1). The real reason is that it is not in the individual's interest: ritualized signals have evolved not for the good of the species, but to enable individuals to assess each other's fighting ability. Since fighting can be dangerous it is obviously in each individual's interest to assess the likelihood of winning before getting into a fight. The ritualized displays and contests we see birds performing allow them to do this.

## Displays and taxonomy

Many displays are at least partly under genetic control and the displays of closely related species are usually more similar than are those of more distantly related species. Displays used in courtship and threat may help to establish the phylogenetic, or evolutionary, relationships between species. This question was first investigated by the German zoo keeper Oskar Heinroth in the early 1900s, and later developed by Konrad Lorenz, both of them studying the displays of ducks. Similar comparative studies were undertaken by Niko Tinbergen on the displays of gulls, and by Bryan Nelson on gannets and boobies. The sexual and agonistic displays of the main orders of seabirds, the Sphenisciformes (penguins), Procellariiformes (the albatrosses and petrels), Pelecaniformes (cormorants and boobies) and Charadriiformes (gulls, terns, etc.), have little in common. This is not really surprising since these orders are not closely related. However, within each order, similarities start to appear. Within the Charadriiformes both gulls (Laridae) and skuas (Stercorariidae) use a similar display, the long-call, in territory defence. Similarly within the Pelecaniformes the cormorants (Phalacrocoracidae) and boobies (Sulidae) share a number of similar displays. The displays of the seven sulid species are much more similar than they are, say, to any gull, auk or penguin. Within families the adaptive significance of the differences between species in their displays can sometimes be surmized. The beautiful Abbot's Booby lacks many of the ritualized displays common to the other sulids and this may be associated with low density breeding and its habit of nesting precariously in trees.

## Current research directions

The study of bird displays has its roots in the field observations made between 1900 and 1930 by pioneers like Edmund Selous, Eliot Howard and Julian Huxley. But it was Konrad Lorenz and Niko Tinbergen who, between about 1945 and the 1960s, gave such study a robust scientific foundation. Their meticulous studies of birds (and many other animals), provided the basis for much modern ethology, and they were awarded the Nobel Prize in 1973 for their work. Since the birth of behavioural ecology in the mid 1970s there has been a change of emphasis. Purely descriptive studies of displays are no longer fashionable. Instead most field ethologists, presently called behavioural ecologists, are busy testing ideas about the adaptive significance of specific behaviours or social patterns.

TRB

# Sound production

The basic principle of sound production in birds is similar to that involved in human speech production. Airflow from the lungs sets a system of membranes into vibration and this causes sound. The bird's equivalent of the vocal cords, however, lies not in the larynx but in an organ not present in mammals, the syrinx, which is located at the base of the neck where the trachea divides into its two main bronchi. The syrinx is suspended in the clavicular air sac, one of a series of such sacs which ventilate the lung. The fact that it is surrounded by air is vital for its role as a sound-producing organ.

Sound among songbirds is produced by the extremely thin, paired internal tympaniform membranes which can be set into vibration by airstreams travelling from the lungs through the syrinx in an expiratory direction. The internal tympaniform membranes form the medial wall of the syrinx whilst the much thicker lateral wall contains a series of three cartilaginous bars, the bronchial semi-rings, to which are attached several small syringeal muscles. In addition, two much larger muscles, the tracheolateralis and the sternotrachealis, are attached to the trachea just towards the head from the syrinx.

There is still considerable dispute concerning the precise mechanism of sound production and even the classic theory that the membranes are the direct source of sound has not yet been verified by direct observation. The first event in the process leading up to sound production is a powerful expiratory effort by the bird which forces up pressure in the air sacs, including the clavicular one. This pressure has two effects: first, it forces the tympaniform membranes to bulge like sails into the lumen of the syrinx; second, it forces a jet of air past the membranes. The suction created by this jet assists the process of inward bulging of the membranes and a point is quickly reached where the membrane is so stretched that it 'pings' into vibration. There is also an active muscular component to this series of events. Contraction of the sternotrachealis muscle pulls the trachea towards the primary bronchi, in this way compressing the syrinx along its longitudinal axis and helping the concertina-like infolding of the tympaniform membranes. Finally, activity in the small syringeal muscles causes the third bronchial semi-ring to protrude into the lumen of the syrinx and this reinforces jet-stream formation.

The scheme outlined above applies mainly to passerine birds, which possess a full array of syringeal muscles, but many other birds lack such tiny muscles and appear to rely on the activity of the two larger muscles. Some birds lack an internal tympaniform membrane and instead possess an external membrane on the lateral side of the syrinx between two of the bronchial semi-rings.

Whilst the classic model appears to account for the production of the majority of bird sounds, there may also exist a second slightly different mechanism for pure, whistled sounds. The principle involved is similar to that we associate with human whistling through the lips except that, again, it involves the tympaniform membranes.

According to this model, as the jet-stream surges through the syrinx past the infolded tympaniform membranes it becomes broken up into a string of vortices which are themselves the source of sound. In this case the membranes appear to perform an enabling role, rather than making the sound directly through their own vibration.

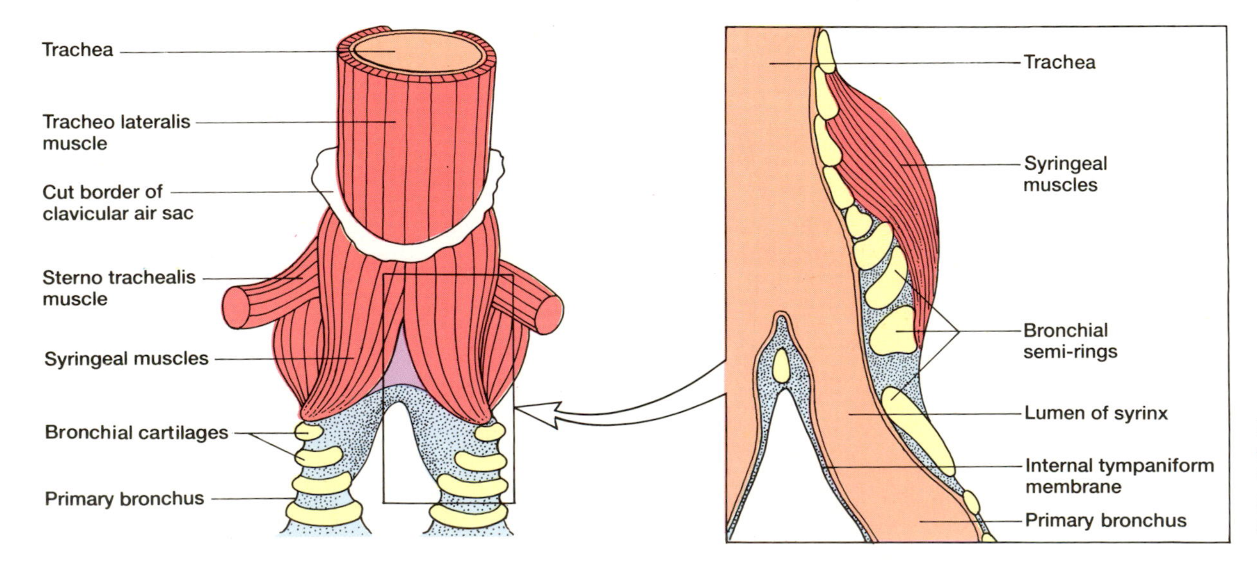


*How bird song is produced. This diagram of the passerine syrinx presents a surface view of the posterior aspect of the syrinx (left), and (right) a longitudinal cross-section*

Leaving aside the possibility that the trachea itself may contribute to sound quality, the models described contain all the ingredients for versatile pitch and loudness control. Loudness can be varied by altering the force of expiration as well as subtle alterations in syringeal configuration. Both aerodynamic and muscular forces can be used to control membrane tension and hence pitch. Rhythmical activity of the respiratory muscles, variously interspersed with that of the syringeal muscles, can result in a wide assortment of trills, warbles, churrs and the like. Extra resonance can be added in the form of a chamber or bulla attached to the syrinx, as occurs in ducks. The variety of sounds made possible by the avian voice-box is almost unlimited.

JHB

## Songs

### How calls differ from songs

Birds produce an immense variety of vocal sounds, and these are usually divided into calls and songs. Songs are the long, complicated sounds produced by some male birds during the breeding season, whereas calls are shorter, simpler sounds produced by both sexes throughout the year. Because of their relative simplicity calls may represent an earlier stage in the evolution of the more complex songs. Calls usually consist of a single sound or note, whereas songs are constructed from a series of simple sounds, organized into a more complicated structure.

Calls are often given in quite specific contexts, such as fighting or courtship, which makes interpreting their function relatively easy. The call itself may affect the behaviour of the receiving individual, again throwing some light upon its probable function. By studying different calls in relation to behaviour it is possible to build up a vocabulary of calls for a species. Passerines seem to have a vocabulary of up to 20 different calls, whereas gulls, for example, have about half that. There are calls for threat, courtship, flight, alarm, begging and so on, which cover most aspects of everyday social behaviour.

Although calls are relatively short and simple, there is now considerable evidence from a number of species that they contain enough information to transmit the identity of a particular individual. Many observers have noted the apparent ease with which parents returning to a densely packed seabird colony manage to locate their own young among thousands of apparently identical individuals. Although all the birds may look alike, each one produces his or her 'call-sign', a particular variation of the basic species call. The variation may be difficult for us to hear, but the subtle differences in duration, frequency or amplitude show up clearly with sonagraphic analysis. Playback experiments demonstrate that young birds respond only to their own parents' call, and that mates respond only to their partners'. Chicks hatched from eggs taken into the laboratory would only respond to calls which they had been exposed to. Individual recognition by voice has been found in a wide variety of seabirds such as gulls, gannets and terns, and is obviously an advantage to colonial species which breed in such crowded conditions.

### Why do birds give alarm calls?

A rather special call which has been the subject of considerable study is the alarm call given by many passerines when a predator such as a hawk flies over. It was Peter Marler who noticed that alarm calls given by many different species shared the same basic design. He reasoned that although it may be advantageous for a bird to warn its mate or offspring of impending danger, it was also dangerous for the calling bird to give away its position. The conflict has been resolved by the evolution of special flying predator calls which can be clearly heard, but which transmit a minimum of directional information and are difficult to locate.

Birds, like humans, are thought to locate sounds by comparisons of phase, intensity and time differences between the two ears. Phase differences are more effective at low frequencies, since the information becomes ambiguous when the wavelength is less than twice the distance between the two receiving ears.

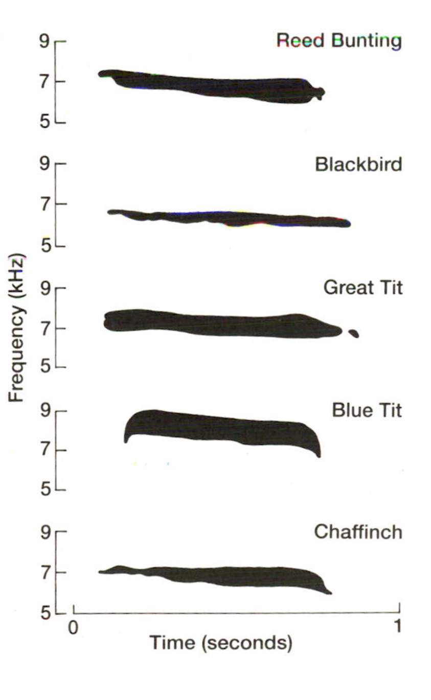


*Alarm calls given by different species as a hawk flies over*

Conversely, intensity differences are more effective at high frequencies, because the sound shadow formed by the head of the listening bird operates only when the head dimension exceeds the sound wavelength. Time differences are effective throughout the frequency range, but are enhanced by interruption, repetition and modulation. The further apart the ears of the listening bird, the more obvious time differences will become. Rough calculations suggest that for a medium-sized bird of prey, a call pitched at about seven kiloHertz would be too high for detectable phase differences, and too low for an appreciable sound shadow intensity effect. This is just the frequency range of the high, thin, whistle many passerines produce when a hawk flies over. Any possibility of time differences is minimized by the almost imperceptible gradual start and finish, modulations are minimal and the calls are not repeated close together. However, although these calls are theoretically extremely difficult to locate, recent research has shown that at least some birds of prey can orientate towards their source under experimental conditions.

But why should individuals who detect danger take even a minimal risk in warning others? Are they a rare case of true altruism, or are the calls primarily for the benefit of those who share a proportion of the caller's genes – their young or other kin? Most evolutionary biologists suspect that the caller somehow benefits by calling, and a very recent study provides some experimental evidence to support this view. A study of mixed-species feeding flocks in the Amazon forest found that two species of flycatching birds, the White-winged Shrike Tanager and Bluish-grey Antshrike, were engaged in a profitable line of deception using their alarm calls. The two species act as sentinels for the flock, and whenever a hawk approaches sound the alarm. The alarm was also occasionally sounded when another member of the flock had flushed out a large arthropod. The flycatcher then sounds a false alarm call, and rushes in to take the prey when the flock scatters for shelter. 'Crying wolf' seems an unlikely behaviour to evolve, but the penalty for ignoring an alarm could be death in the talons of a hawk.

## Song, an aid to species recognition

The additional length and complexity of songs provides even more potential for information transfer between individual birds. The most obvious function is that of species recognition. Ornithologists use songs to identify a species and it would be surprising if the birds themselves were incapable of similar powers of discrimination. Many experimental studies have confirmed that whilst songs do act as specific signals, the cues which listening birds use for recognition are not necessarily the same as we use. Species recognition is studied by playing back tapes of modified songs to territorial males, who react strongly to normal control songs by approaching the speaker. Normal songs of the European Robin are extremely complex and composed of alternating high and low pitched phrases. Artificial songs consisting of either all high or all low phrases obtain a much weaker response, but a completely synthetic song constructed of high and low phrases achieves a good response. Clearly the basic syntactical rule of high–low–high–low is more important than the actual structure of the notes or syllables themselves. But other species seem to have quite different cues to which they respond. In Bonelli's Warbler, the actual structure of the syllables themselves is more important, and any tampering with their fine structure renders an experimental song very ineffective. In the Indigo Bunting the very opposite applies, and individual syllables in songs can be completely reversed without any effect at all. In this species timing seems to be the most important factor, and if intervals between the syllables are lengthened or shortened, a very weak response is obtained during playback. It seems quite clear that there are no general rules of song structure for species recognition, but instead considerable variation exists from species to species.

Although the features of song which are important in species recognition must remain fairly constant, within this general framework there is still considerable scope for individual variation. In many studies it has been shown that each bird within a population may have its own peculiar version of the species song which is quite characteristic. When repeated, the characteristic song structure remains remarkably constant and usually stays so throughout the season, and even from year to year. Repetition of a constant signal gives every opportunity for other individuals to learn it and so possibly come to recognize a male as a particular individual within the community. There is some evidence from playback experiments that territorial males learn both the identity and positions of their neighbours from their individual songs and respond very quickly to a strange new bird, or even a neighbour shifting his territory boundary.

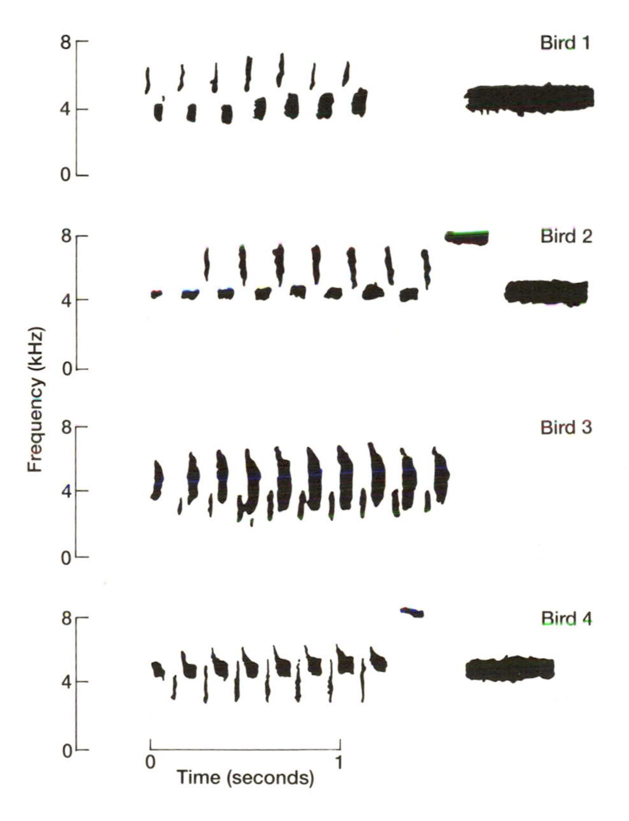


*Individual songs from four male Yellowhammers (Emberiza citrinella) in adjacent territories*

## Song and territory defence

Song can carry information concerning the specific and individual identity of the singing bird. Other types of information may also be contained within the signal, but to whom is this information directed and why is it transmitted, or put simply, what are the possible functions of song? It is common knowledge that territorial males sing against each other at the boundaries of their territories, and playback experiments show that song is produced when invasion from a rival male is simulated. Whether or not the song itself actually repels a rival male is rather a different question. John Krebs of Oxford University has attempted to answer this question by the use of ingenious experiments on the Great Tit.

Territorial males were caught and removed and their songs replaced with a sophisticated system of loudspeakers. The point of this experiment is that it isolates the effects of song itself. If any aversive effects are detected they can be due only to the songs played through the speakers, not the birds themselves. The experiments took place in a small copse, normally occupied by eight territorial males. Early one morning, the males were all captured, removed, and three of the territories were each 'occupied' by four speakers and a tape recorder. Each recorder had a continuous tape-loop containing eight minutes of normal Great Tit song, linked by a multiway switch to the four speakers placed in different parts of the territory. This pattern was designed to mimic the normal singing behaviour of a resident male, each speaker being active in turn for two minutes, followed by a pause before the sequence starts again. This is just as though a resident male is regularly patrolling his territory, singing from several positions and pausing occasionally to feed. The wood also had two control areas. One of these was left completely silent, as though empty. The other, to control for the possibility that any experimental effect is merely due to the very presence of speakers in the wood or any unusual noises they may make, was occupied by speakers playing a recording of a repetitive, two-note song from a tin whistle. The wood now contained three areas, the experimental, the control silent, and the control sound. A few hours after the system was switched on at dawn, only the control areas were occupied and the experimental area avoided. It could of course be that this is just the normal pattern of occupation in this wood. To allow for the possibility that the control end was in some way more attractive, the experiment was repeated with the position of the experimental area completely changed. This second experiment showed much the same result, the invading males avoided the experimental area, no matter which end of the wood it was located in.

These experiments suggest that song is an important first line of defence and does have some repelling effect upon rival males searching for a territory.

Great Tits like many species have a repertoire of different versions of the species song. Each song type is repeated several times before switching to another, and so on. Krebs has suggested that Great Tits may be practising a subtle form of deceit; by singing in this way the impression is created that there are many more males present in a given area. Listening prospecting males may well decide to try their luck in a less crowded area. Krebs has tested his Beau Geste theory by using a similar speaker replacement technique as before. This time he broadcast a repertoire of song types in one area, and found that it was re-occupied later than an area broadcasting just one song type. It does seem that a repertoire is a more effective territorial proclamation than just a single song.

The repertoire size of the Great Tit varies between two and seven whereas the Song Sparrow has between five and fifteen song types. The number of song types in the repertoire of the European Blackbird can exceed 100 while, at the other end of the scale, the White-crowned Sparrow has just one song type. There are many questions and few answers about the cause of this variation.

## Song and mate attraction

The idea that song has a mainly territorial as opposed to sexual function, has enjoyed great popularity amongst most ornithologists for two main reasons. First, singing males are extremely conspicuous due to their overt, aggressive territorial behaviour, and have been used a great deal in successful field experiments. Unfortunately, females are usually shy, retiring creatures who have proved to be difficult and elusive subjects for observations or experiments. Secondly, although Charles Darwin included bird song in his development of the theory of sexual selection (Chapter 1), the theory was slow to gain general acceptance, and indeed remains somewhat controversial to this day.

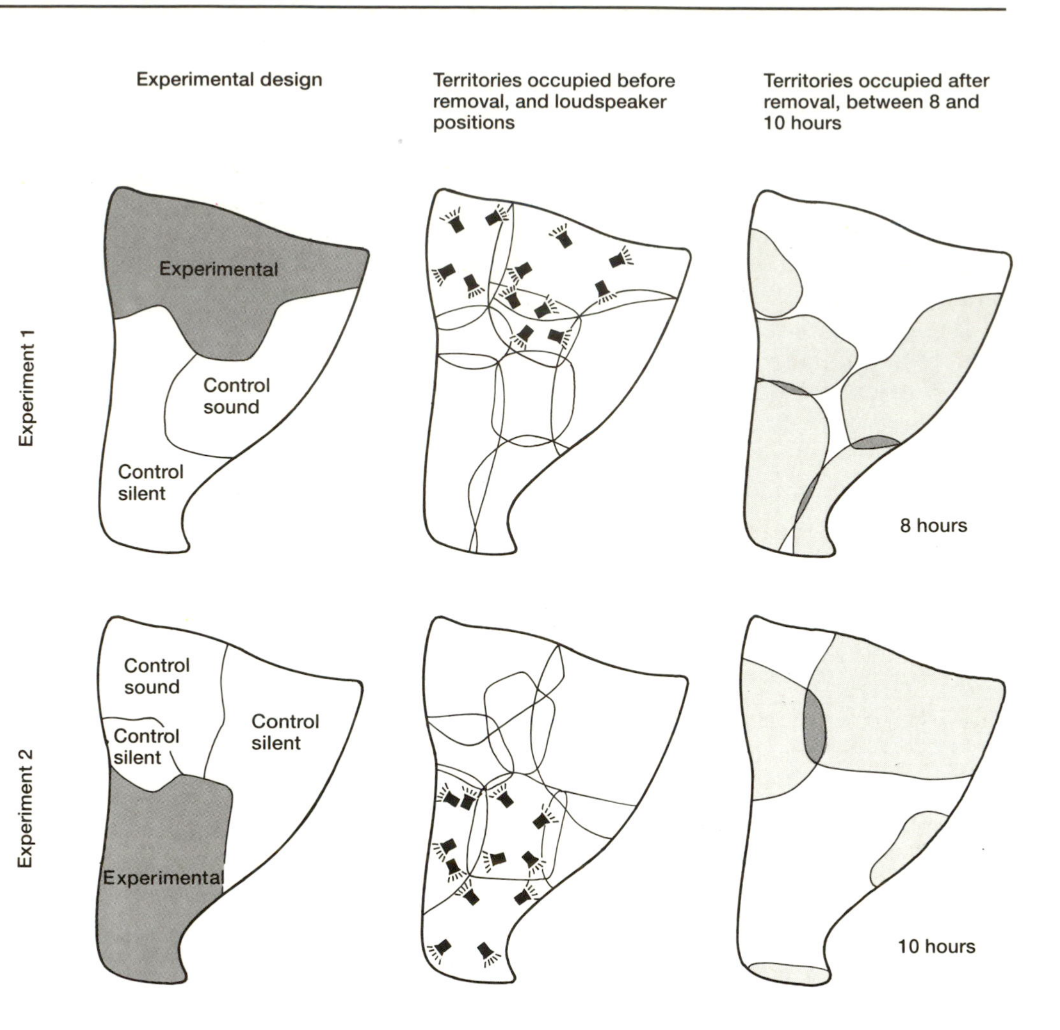


*John Krebs' experiments with Great Tit songs demonstrate the strong association between territoriality and song*

The Sedge Warbler is a small, brown, marshland bird, with one outstanding feature, an extremely long and elaborate song. It arrives back in Europe in spring, from its wintering area in Africa, and for most of the day and night pours out a continuous stream of song. Unlike most songbirds, the Sedge Warbler does not repeat a stereotyped song, but continues to compose an amazing variety of songs from a repertoire of basic syllables or building blocks. In many ways, the Sedge Warbler can be considered to have the acoustic equivalent of the peacock's tail. When the male has attracted a female and paired with her, the singing stops abruptly. It seems quite clear that the song functions in mate attraction, and so perhaps the elaboration is a result of continuing female choice and runaway sexual selection. To test this in the field, all the males in a population had their songs recorded and analysed by sonagram to establish the repertoire size of each male. The dates upon which they paired were also noted, and a strong inverse correlation between repertoire size and date of pairing was obtained. In other words, the males in the population with more complex songs attracted females before their rivals with more simple songs. This result is indeed powerful evidence for the effects of sexual selection upon song complexity, but there are a number of confounding variables which must also be considered. For example, it could be that males with more elaborate songs obtained better territories, and that females selected the males indirectly through territory rather than male quality. The only way to eliminate such variables completely, is by undertaking experiments under controlled laboratory conditions.

Until quite recently, captive females have been remarkably unresponsive to playback of tape-recorded songs. However, if primed with implants of the female sex hormone oestradiol, females perform a sexual display in response to male songs. This display can be timed and used as an accurate index of female response. Using tape-recorded songs from the original field study, it has now been shown that females respond significantly more to the higher repertoire males in the population. But there still remains the possibility that some other variation besides repertoire size might also be present in the test recordings. To control for this, artificial songs of varying repertoire size were constructed from a tape recording of one individual male. When the females were tested, a strong correlation between repertoire size and female response was obtained. The use of tape recordings effectively eliminates confounding variables from the male bird or his territory, and provides one of the very few experimental tests of Darwin's theory. Obtaining a female earlier may well be a considerable advantage to the male. Early breeders tend to produce more surviving young, and males who leave it too late may not obtain a female at all.

## How are songs learned?

It was W. H. Thorpe who pioneered the scientific study of song development in his classic works on the Chaffinch. Until then, it had often been assumed that birds inherited their complex song structures. Thorpe was able to show the importance of early learning by raising nestlings in sound-proof chambers where they were deprived of auditory feedback. The simple isolated song the birds eventually produced was nothing like the complex structure of normal adult song.

How Chaffinches normally develop their song structure was also studied with captive birds. After fledging, a rather loose, rambling kind of quiet song, known as early subsong, appears. It ceases in winter, but reappears in spring as a more complex version called late subsong.

When the young male takes up territory and is exposed to the songs of nearby males the song structure becomes more recognizable as a Chaffinch song – divided into three distinct phrases with a terminal flourish. At this stage it may still be changed and is called plastic song. When changes no longer occur and the structure becomes quite stereotyped it is called full song. In the Chaffinch and many other species the song will now remain unaltered for life.

By isolating his birds in this way, Thorpe had prevented them from learning any detailed structure from other birds throughout the important first year of life. Later work established that just hearing playback of recorded songs at this early sensitive stage is enough to enable later development of full song.

Another interesting point is that the bird is capable of storing some sort of model of full song structure as a neural template for several months before starting to sing in the following spring. The basic concept of an early sensitive period of song learning in the first year, and then no further significant changes is now well established, although there are some exceptions. There is also considerable variation in the length of the sensitive period for early learning. When and where a young bird actually learns its song may be controlled by a number of factors which confer considerable flexibility in

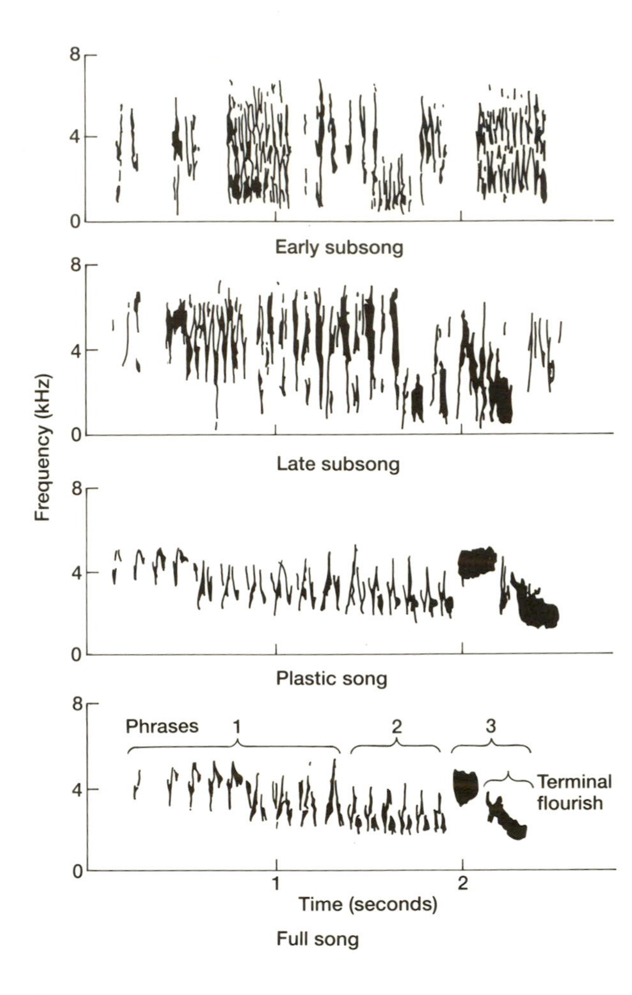


*Left. The development of song in the Chaffinch (Fringilla coelebs)*

the final composition of the song repertoire. For example, species or individuals which hatch early and are relatively sedentary will end up with songs learned from parents or neighbours in the natal area. Those which hatch later may be too late to hear local songs, and if they disperse to other areas will inevitably copy the songs of their new neighbours.

## Song dialects

The fact that birds learn their songs has a number of interesting consequences. One of these is that, just like humans, birds can develop pronounced geographic variations in the structure of their sounds. Local dialects occur when the songs within one population share certain structural characteristics which are different from other populations nearby. Although local variations have been reported for many species, the most studied is the White-crowned Sparrow. Birds recorded around the San Francisco Bay area show clear dialects based upon variations in the introductory whistles, the fine structure of the main trill and the position of the noisy buzz.

The population at Berkeley, for example, has three or more ascending introductory whistles, that at Marin has one or two with a terminal buzz, and that at Sunset Beach has two ascending or two followed by a buzz. There are other consistent differences in the fine structure of the trill which make this an excellent example of how individuals within each population conform to the dialect for their particular area.

To answer questions as to how this conformity comes about, nestlings and young birds were taken from the wild at up to 100 days old and raised within sound-proof chambers. The nestlings eventually produced abnormal songs which bore no relation to their home dialect, whereas the slightly older birds produced good copies of their dialect.

This suggests that the special features of dialects are not inherited but learned. This was confirmed by tutoring young birds during the sensitive period, i.e. the period when song learning could occur, and showing that they could learn their own or even a 'foreign' dialect at this time. With the sensitive period in the White-crowned Sparrow ending as soon as 50 days, it seems highly likely that the normal model will be either the song of the father, or a neighbouring male. Thus the local dialect is preserved and passed on by a form of cultural transmission.

The question which remains to be answered is: are dialects merely a functionless by-product of song learning, or do they have some additional role to play in communication? There are many theories concerning the significance of dialects, but perhaps the most interesting to emerge is that they may be involved in reducing gene flow between populations, as a result contributing to the eventual formation of new species. Dialects are sometimes associated with different habitats and it has been suggested that dialects might attract only those females best adapted to breed in local conditions.

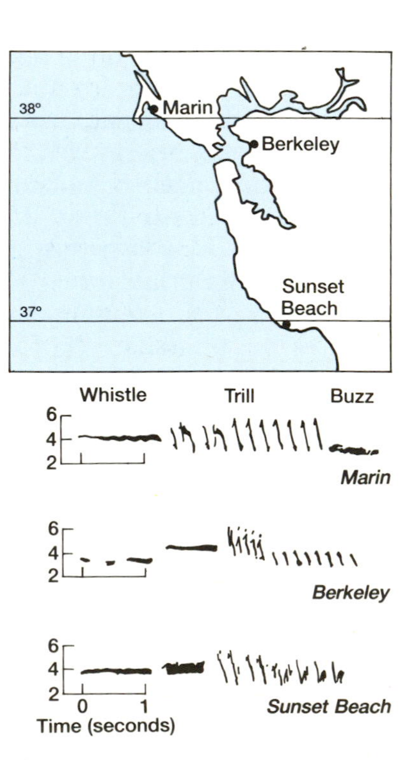


*Three song dialects in the White-crowned Sparrow (Zonotrichia leucophrys) around San Francisco Bay*

Several predictions follow from this theory: (1) local populations are genetically different; (2) there is reduced gene flow between populations; and (3) females select males which sing their own dialect.

In the White-crowned Sparrow there is some evidence from playback experiments that both males and females show a stronger response to their own dialect. Electrophoretic analysis of proteins has demonstrated that there are greater genetic differences between populations of White-crowned Sparrows which also have dialects. However, field studies have shown that some females disperse and select a mate from a different dialect area to their own. Conversely, experiments on captive females show that females will only display to songs from their own dialect area. Clearly more research is needed, perhaps on other species, before a clear view about the evolutionary significance of dialects will emerge.

## The origins and evolution of songs

Even though bird songs appear to be stereotyped and relatively unchanging, they have presumably evolved from older, ancestral forms. How this came about during evolution, perhaps by repeating simple calls, we can never be absolutely certain, but the answer lies to a certain extent in what we already know about song learning. Most species learn their songs from other individuals and songs are passed from generation to generation by a form of cultural transmission. During transmission there is always the possibility that errors will occur in copying the complex songs. This has been observed in an isolated island population of Saddlebacks in New Zealand. There were relatively few song types on the island and these were usually shared by neighbouring males grouped into dialect areas. When a young bird took up territory for the first time he learned his songs from neighbours, whether they were his parents or other individuals. New song types appeared to be the result of errors in the song-learning process, or cultural mutations. They arose quite abruptly in an individual, either by changes in frequency, addition or deletion of an element, or a new combination of existing elements. For example, one individual in the population suddenly produced a cultural mutation which added low-frequency components to the end element. This was then learned by other individuals and so started to spread through the population.

Long term study of the cultural transmission of songs has now been undertaken on several species. Some of the results are quite surprising and show that very few songs survive unchanged over the years. They either disappear or have become altered beyond recognition by cultural drift. In the Indigo Bunting, a few songs persist for over 15 years but most have a half-life of about six years. It is difficult to see why some songs survive for so long, and others disappear remarkably quickly. Whatever the reason, cultural transmission proceeds at an impressive rate, and what we have often regarded as stereotyped signals are clearly changing rapidly with time. The arrival of new song types brought in by males dispersing from other areas seems to be a particularly potent force in the Chaffinch. Such new song types can suddenly sweep through the new population.

It seems that through song learning, birds have acquired the ability to make continual, rapid changes to their song structures. Why they learn songs in the first place continues to be something of a puzzle. Is song structure so complicated that it cannot be transmitted genetically, or do the constant changes through copying errors confer some great advantage? This enormous potential for generating new variety is one of the reasons why bird songs are so remarkably complex and diverse in their structure. We have already seen how increasing complexity appears to be advantageous in both mate attraction and territorial defence. There is also an apparent conflict between stereotypy for species recognition, and variation for more subtle intraspecific communication. Some birds produce short simple songs, and others are truly the acoustic equivalent of the peacock or bird of paradise in their extravagance. Learning and cultural transmission certainly creates the potential for the immense variety, which is still both a puzzle and yet an enormous challenge, to all the evolutionary biologists who study bird song.

## Duetting and mimicry

Duetting and mimicry are both cases where evolution has favoured the development of complexity in song structure. In a duet, instead of the male singing alone, the female also contributes. The result is a specific song which contains within its structure elements from both individuals. The songs produced are usually quite stereotyped and the contributions of the two individuals so well coordinated and integrated that to the listener it seems to be produced by only one bird.

Duetting has been found to occur in many different groups of tropical birds such as African Bou-bou shrikes of the genus *Laniarius*. In some species, such as *L. erythrogaster*, the male produces one element and the female another in a simple alternating system called antiphonal singing. In other cases, such as *L. funebris*, elements are combined in rather more complex sequences. Each pair has a small repertoire of different elements and develop their own particular pattern of duets from them. Other pairs may share the elements, but the combination selected by each pair for their duets are usually distinct.

Whilst the actual control and mechanics of duetting is certainly fascinating, so too are its possible functions. Once again we are faced with the problem of why such a complicated system should have evolved at all. However, there may be clues to this in the distribution, ecology and behaviour of duetting birds. Although they are to be found in many different groups, they are almost exclusively tropical and often breed in fairly dense vegetation. They also tend to be resident rather than migratory, and so maintain a territory all the year round over several years. They are usually monogamous and maintain a pair-bond for several years and perhaps even for life. These correlations led Thorpe to suggest that duetting may play an important role in maintaining the pair-bond over long periods of time in dense habitats where visibility is difficult. But if we look carefully at the context in which most duets are performed, it is in territorial encounters with rival pairs. Furthermore, duetting pairs invariably sit close together and yet the songs are extremely loud. The conclusion appears inescapable that they are territorial proclamations directed at other pairs and not to each other.

An even more puzzling form of elaboration occurs in various forms of vocal mimicry. The Indian Hill Mynah is a well-known example, which in captivity will learn to imitate the spoken words of its human captors, although in the wild it never mimics the sounds of other species. Natural mimicry in the wild involves the regular or permanent incorporation of species-specific elements from the vocalization of one species into those of another. Why this has happened during evolution is difficult to explain, as in theory it would seem to promote confusion rather than efficiency in species recognition. However, in the case of brood parasites such as the widow-birds of Africa, the advantages of mimicry seem more clear. The widow-birds have evolved a close relationship with various species of estrildid grass finch whose nests they parasitize. Not only do they mimic the eggs, but also the young, which

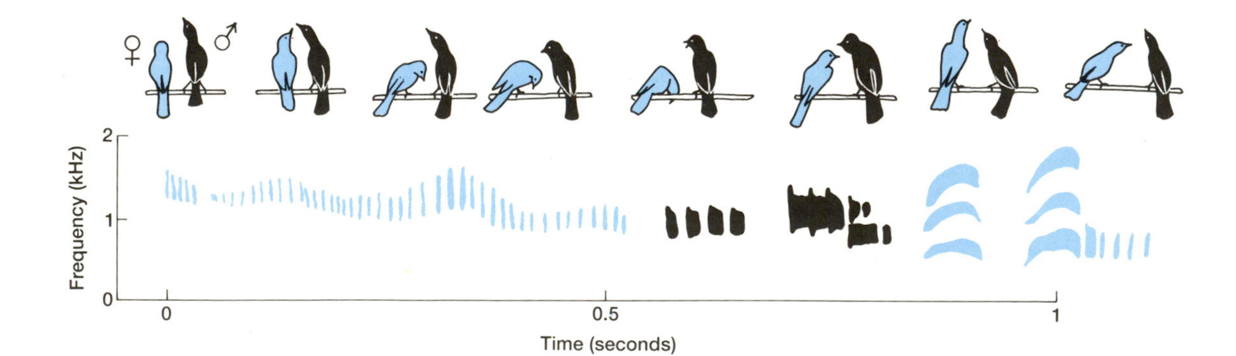


*A duet by a pair of African shrikes (Laniidae)*

appear identical in terms of appearance and behaviour to the host young with whom they share the nest. The young parasites also learn the songs of the host species and the eventual mimic is an extremely good copy of the host model. Young females also learn the host song and are attracted to it, whether model or mimic. This is not so inefficient as it seems, as a widow-bird female must be able to find nests of the host species and synchronize her laying pattern with them so that the eggs hatch together.

However, most cases of mimicry involve species which mimic a bewildering variety of other species with which there is no apparent relationship at all. This category includes such celebrated mimics as starlings, lyrebirds, mockingbirds and warblers.

The quite extraordinary song of the Marsh Warbler contains elements that are mimics of nearly 100 different European species. The elements are copied extensively from the local bird community, for example individuals from the coast have more maritime species in their repertoires than those breeding inland.

Nevertheless, Marsh Warblers sing songs that do not belong to any members of the local bird community. Where do these song types come from? Most of the warbler's 'non-European' elements are in fact 'borrowed' from over 100 different species of African avifauna. Because some of the species copied have an extremely local distribution, it may even be possible to locate more accurately the autumn and winter quarters of the Marsh Warbler. Only a small proportion of the total repertoire remains unaccounted for, and as Marsh Warblers appear to copy indiscriminately from local avifauna whenever they breed or travel it may be that we will eventually learn the intricate details of their migratory journeys from sonagraphic analysis of their repertoire.

It is difficult to see what the possible functions of mimicry in such species might be. Mimicry clearly increases the complexity of song, and there is some evidence in both mockingbirds and *Acrocephalus* warblers that males with the most complex songs attract females before their rivals. There are also observations that male mockingbirds produce the song of a particular species when it approaches their territory. Playback experiments have also shown that the other species cannot discriminate between their own and mimic songs. These may be slender clues that mimicry merely enhances the known dual functions of song in either mate attraction or territorial defence. However, much more work is needed before any firm conclusions concerning the functions of vocal mimicry can be reached.

CKC

## MECHANICAL SOUNDS

As well as uttering vocal sounds, some birds also produce mechanical sounds with their tails, wings or beaks. The wings of most bird species make little or no sound during normal flight. The silent flight of owls is well known. At the other extreme the flight of a few species is distinctly noisy. Grouse and pheasant wings 'whirr', Mute Swans in flight can be heard over a considerable distance, and each species of hummingbird hums at a slightly different pitch. All of these sounds are incidental and almost certainly play no part in communication.

In other bird species mechanical sounds, including those produced by the wings, are used as displays, or as part of a display. Male Woodpigeons and nightjars produce a loud 'clap' by beating their wings together during their aerial courtship displays. In the American Woodcock the first three primaries are specifically modified for sound production. During courtship flights, performed at dusk, the birds spiral upwards and then plummet towards the ground producing a musical whistling sound from their wings. During the breeding season pheasants 'crow' and immediately follow this with a noisy beating and rattling of their wings. In the Ruffed Grouse both sexes beat their wings across their breast producing a noise which has been likened to that of a distantly beating heart, followed by an accelerating roll of thunder! The complicated and bizarre courtship dances of several manakins are supplemented by buzzing and snapping sounds produced by specially modified wing feathers. Princess Stephanie's Bird of Paradise produces a loud rustling sound with its wings and tail during its displays. Like most other larks the African Flappet Lark defends its territory by singing high in the air. The song does not come from the syrinx but is instead produced by the wings. Flappet Larks ascend on rapidly beating wings and produce a loud, crisp burring sound unlike any other bird. The exact way in which the sound is created is not known. Brian Bertram studied Flappet Larks in the Serengeti National Park and found that each bird produced its own, individually distinct song.

Bill sounds are used in the pair maintenance displays of White Storks. On meeting, one or both members of a pair throw their heads over their backs and noisily clapper their mandibles together. Most owls click their mandibles together as a threat display. Many woodpeckers produce mechanical sounds by beating their beaks against trees. As a result their skulls are strongly reinforced to absorb the shocks of these blows. Great Spotted Woodpeckers beat their beaks very rapidly (up to 25 times per second) on the branches of dead trees to produce a loud, distinctive drumming sound. This drumming is used both in pair formation and maintenance and in territorial behaviour. Woodpeckers of various species sometimes drum on metal telegraph poles to produce notably loud signals.

Tails are used less often for producing mechanical sounds, and the best known examples occur in the snipe. In the European snipe a two-second burst of drumming or bleating is produced by the specially modified outermost (seventh) pair of rectrices. Drumming occurs as the bird descends during its aerial courtship displays early in the breeding season.

TRB

# 11 PEOPLE AND BIRDS

## The history of human interest in birds

'The eventful history of ornithology is much more intimately interwoven with the growth of philosophy than many people realise.' So wrote Stresemann in 1953 in his work on the development of ornithology. It is no surprise then that the first person to give bird study respectability as a science was the Greek philosopher Aristotle (384–322 BC). Although Aristotle recognized over a hundred species, he was just as interested in their biology and he compiled information from accounts given to him by fishermen and farmers, from his own observations and other written sources. In contrast to other philosophers of his time Aristotle developed his hypotheses in the light of known facts.

During Greek and Roman times much was written about birds. Yet little of it was of scientific value since it was based on fables and anecdotes passed from source to source and enriched by the imagination of the writers. Only early attempts at a system of classification, mostly based on physical or habitat characteristics, had any ornithological significance.

### Seventh to fifteenth centuries

Little emerges from the Dark Ages to suggest any major advance of knowledge, although this may be simply because few records survive. It is clear from the Anglo-Saxon poem 'The Seafarer', written about AD 630 that people were well aware of the habits of a number of species. However, it was not until the early period of European civilization that ornithology picked up from where Aristotle left off. The principal catalyst was the German Emperor Frederick II (1194–1250) who, being a disciple of Aristotle, extended knowledge by simple observation. Watching birds caged in a menagerie he conducted a number of simple experiments. For example, he alternately blindfolded and blocked the nostrils of a vulture in order to determine by what sense it found its food. He also spent time watching wild birds and realized that geese, and other species that migrate in a V formation, share the onerous duties of leading with each individual passing back along the V at every changeover.

Frederick was, however, an unpopular figure with the establishment of his day due to his conflicts with the Church, from which he was excommunicated. For this reason his work was not published until 1596 and was given no scientific attention until 1788, more than five hundred years after his death.

There followed another period of stagnation with a lack of any notable work until the sixteenth century when there was a considerable upsurge of interest, mainly as a result of the revolution in the production of books – printing.

### Sixteenth and seventeenth centuries

The first printed book devoted wholly to birds, *Avium praecipuarum*, was produced by the Englishman, William Turner (c. 1510–1568). His book, *A short and succinct account of the principal birds mentioned by Pliny and Aristotle*, was a summary of knowledge to that time and was based largely on his own observations. Its accuracy assured him of a place among the principal ornithologists. Turner, a Protestant convert, was compelled to live abroad and while in Switzerland he made the acquaintance of Conrad Gesner.

Gesner (1516–1565), a resident of Zurich, was physically infirm and thus condemned to academic study. His treatise *Historia animalium* was an attempt to draw together the knowledge on all animals. It was published in five volumes, the third being devoted to birds. The work was compiled as a result of considerable correspondence with practising European naturalists such as Turner and John Kaye or Caius (1511–1573), the founder of Gonville and Caius College, Cambridge. Kaye was himself the author of a book, *A Natural History of rare animals and plants* which contained much of ornithological interest based on such observations as keeping a puffin in captivity for several months.

The sixteenth century saw the beginning of comprehensive and coordinated bird study. The acceptance, at last, of the Aristotelian view of basing written work on fact meant that ornithologists had established a *modus operandi* and a network of communication which centred at that time on Gesner.

With the age of exploration in the sixteenth and seventeenth centuries many new species were introduced to the centres of civilization and much of the ornithological interest in the Renaissance and Tudor periods concentrated on collections of birds. As these collections expanded it became essential to establish some order for their presentation and thus began the interest in classification. Differentiation was established by separating birds according to whether they were found on land or water until Francis Willughby (1635–72), with help from his companion John Ray, wrote his study of birds *Ornithologia libri tres* (1678), which suggested physical characteristics as a basis for classification.

This was followed by the system developed by the Swede, Carolus Linnaeus (1707–78), who was considered to be an organizational genius. Linnaeus divided the natural world up into sections and sub-sections: classes, orders, genera and species. While much opposition greeted the publication of his *Systema Naturae* (1735), he was assured of acceptance of his subsequent suggestion of binomial nomenclature. Thus thrushes were given the generic name *Turdus* and the Song Thrush, for example, was allocated the specific name *philomelos* to become *Turdus philomelos*.

## Eighteenth and nineteenth centuries

Throughout the eighteenth and nineteenth centuries considerable activity centred around the ornithological collections that titled and wealthy patrons amassed in the European countries. Specimens were shot, skinned and stored for transportation, observations were made in the field, new species discovered and much energy was expended in cataloguing and publishing all the gathered information, often lavishly illustrated with paintings. Peter Pallas (1741–1811) in northern Asia and in Russia, Johann Förster (1728–98) in Russia, Australasia and America, François Levaillant (1753–1824) in South Africa, and Alexander Wilson (1766–1813) and Charles Bonaparte (1803–57) in America all pursued such work to the highest standards.

In addition to the collectors there were those whose sole interest was the study of birds in the wild. In Germany, Johann von Pernau (1660–1731) was perhaps the first. Studying birds around his home he began to outline the ecology of various species and made many fundamental observations such as the importance of learning in the development of song. In England, Gilbert White (1720–93) occupied his time studying the avifauna of Selborne in great depth, revealing his discoveries in letters to two other naturalists. White was, for instance, the person who realized that the Chiffchaff and the Willow Warbler were two distinct species and his careful description of their characteristics could not be improved upon today.

The nineteenth century saw tremendous expansion in the study of birds. While the systematics were still contentious, particularly after the 1859 publication of Darwin's *On the Origin of Species by means of natural selection*, the interest widened to encompass behaviour, migration, and populations. Learned societies for the publication and discussion of research were formed: the British Ornithologists' Union in 1858 and, across the Atlantic, the American Ornithologists Union in 1883.

As the readership became more widespread there was a new generation of publications with Naumann in Germany, Audubon in America, and Gould and Bewick in England producing avifaunal folios.

## Nineteenth and twentieth centuries

At the turn of the century protection of birds became a political topic and in England a society (now the Royal Society for the Protection of Birds) was formed in 1889 to promote

*Carolus Linnaeus (born Carl von Linné) (1707–78), the Swedish founder of the binomial system of scientific naming*

*John James Audubon (1785–1851)*

*David Lack (1910–73), perhaps the most influential ornithologist who has ever lived*

legal protection for persecuted species. In the United States the establishment in 1886 of the US Biological Survey (now the US Fish and Wildlife Service) served a similar function.

Ornithology branched out. In established universities and centres of learning many different aspects of birds now came under investigation. In the museums, where most collections were housed, taxonomists continued their work separating and characterizing species. Both Richard Sharpe (1847–1909) and Ernst Hartert (1859–1933), working on the British Museum collection and the Rothschild collection at Tring respectively were active in this area; similarly in America Robert Ridgeway (1850–1929) concerned himself with separating species from varieties.

In the field of behaviour, Francis Herrick (1858–1940) in America, Selous (1858–1934) and Howard (1873–1940) in England, together with Oskar Heinroth (1871–1945) in Germany, set new standards for field ornithology with very careful observation translated into skilled interpretations of the activities of birds.

Migration was studied by the use of ringing, first introduced by a Dane, Hans Mortensen (1856–1921), and by the turn of the century several centres in both Europe and America were busily ringing birds.

The returns were almost immediate with short-range movements reported from larger birds such as crows and in 1909 a White Stork ringed in Hungary was shot in South Africa, marking the beginning of the long-range recoveries.

## Up to the present

As the twentieth century has progressed expansion has continued through the entrepreneurial enthusiasm of the likes of Harry Witherby (1873–1943) in England and Frank Chapman (1864–1945) in the United States. Institutions specially for the study of birds have been established: The Edward Grey Institute for Field Ornithology in Oxford, The British Trust for Ornithology at Tring, and The American Museum of Natural History in New York are just some of the many focal points.

There has also been an exceptionally strong amateur input into many areas of bird biology: population monitoring, measuring distribution through atlas projects and migration studies involving bird ringing and observatory work.

PMB

# Birds in art, literature and music

## Art

From the earliest times birds have featured in the culture of man. Some 12 000–15 000 years ago Stone Age men painted birds on the walls of caves in France and Spain. About 2500 BC the ancient Egyptians used bird shapes among the many gods that they worshipped. Thus Thoth, god of wisdom, had the head of an ibis and Horus, the sky-god, was appropriately given the head of a falcon. In addition, paintings from Theban tombs of the 18th dynasty (1550–1340 BC) show ducks and doves.

*Birds in art: an Egyptian wall painting of a heron in the tomb of Sennedjem at Thebes (c. 1280 BC)*

Early illustrated manuscripts often depicted birds, but perhaps the earliest work of which we know in which birds formed an integral part was the triptych called 'The Garden of Earthly Delights' by Hieronymous Bosch (1450–1516) which shows several species such as goldfinch, kingfisher, and woodpecker in great detail.

*Detail from the central panel of Hieronymus Bosch's triptych, The Garden of Earthly Delights, showing the meticulous detail with which he painted the Goldfinch (Carduelis carduelis), European Robin (Erithacus rubecula), Green Woodpecker (Picus viridis), Mallard (Anas platyrhynchos), Common Kingfisher (Alcedo atthis) and Hoopoe (Upupa epops)*

Much of the early scientific study was carried out by people whose first interest was artistic. The books of birds needed to have pictures and most of the great bird artists were illustrators. George Edwards (1694–1773) was perhaps the first, followed by many others, including Naumann, Bewick, Audubon, Gould and Wolf. All these great bird painters attempted to bring to their work a combination of accuracy and authenticity.

Some, such as Audubon, sought their subjects in their natural environment, but others used specimens. John Gould (1804–81) was perhaps different in that he employed a team of painters including his wife Elizabeth, W. Hart, J. Richter and Edward Lear, the nonsense poet.

In the early years of this century Archibald Thorburn (1860–1935) illustrated Coward's *Birds of the British Isles* (1920–26) and his work became greatly prized.

In modern times with the interest in birds so widespread artists have found their work to be in great private demand and many ornithologists prize the realistic pictures by Ennion, Gillmor, or Shackleton, together with the more impressionistic works of Michael Warren.

Perhaps the greatest modern contribution artistically has come from the American Roger Tory Peterson (b. 1908) who used a whole new technique when illustrating his field guides in both Europe and in the United States where his first *Field Guide to the Birds* appeared in 1934. Highlighting, by means of a pointer line, the diagnostic characteristics of each species, he produced pictures that have become part of every bird watcher's armoury.

*Left. A page from Roger Tory Peterson's Field Guide to the Birds East of the Rockies. Peterson's stylized but attractive illustrations with key identification features indicated by arrows were a major breakthrough in bird identification. Virtually all subsequent field guides have followed a comparable format*

*Below left. Audubon's painting of Eider ducks (Somateria mollissima). Audubon was one of the first bird artists to place birds in realistic, life-like postures*

## Literature

In religious works like the Bible and the Koran there is mention of birds, such as the dove that brought the olive branch to tell Noah that land had emerged from the flood. However, it is in more recent times that birds' characteristics have been drawn into the literature, music and art of the world.

In literature most mentions appear in poetry; for example in the early work 'The Seafarer' (AD 630) many marine species such as gannet, kittiwake, and White-tailed Sea Eagle are referred to. Yet it was in the Middle Ages, towards the end of which printing developed, that poets began to extol birds with regularity. William Langland (*Piers Plowman*, 1362) and Geoffrey Chaucer (*Parlement of Foules*, 1382) provide two contrasting works. The former mentions crows, sparrows, doves, and falcons, and the latter bullfinch, siskin and 'feldefare' (Fieldfare). In the *Parlement of Foules* there is a clear

*Medieval artists were often excellent observers. Here a Peacock and several other species are illustrated in the Imago Mundi Bestiarium of c. 1230*

indication that Chaucer was aware of the parasitic habits of the cuckoo; the host he mentions was in fact the Dunnock which is an observation of some significance. Cuckoo eggs generally mimic those of the selected host in colouration except in the case of the Dunnock. One explanation for this might be that the Dunnock is a relatively recent host and that the cuckoo has not yet evolved the ability to mimic the colour of its eggs. However, since the Dunnock is mentioned as a host by Chaucer over six hundred years ago this explanation could seem rather tenuous. Thus we have an example of the way in which literature can play a part in scientific investigation.

William Shakespeare, perhaps the most quoted of all writers, was also able to include many ornithological references such as 'Hark, hark the lark at heavens gate sings'. His contemporary John Donne wrote of 'the lyric lark and the grave mournful dove'. Dove, skylark, cuckoo and nightingale have continually captured the imaginations of both writers and musicians. Shelley's 'To a Skylark', and Keats' ode 'To a Nightingale' are perhaps two of the best known poems devoted entirely to a single species. Both contain the magic that is possible only through familiarity with the subject. William Blake however, was well before his time with his work for wildlife protection that begins 'A robin redbreast in a cage puts all heaven in a rage'. John Clare ('The Autumn Robin'), Alfred Lord Tennyson ('The Eagle') and William Wordsworth ('To the Cuckoo') wrote less well-known examples of bird poems and made reference to several species in their work.

In the Victorian era two writers were able to capture the atmosphere of the countryside in prose and their work is still considered among the finest of its kind. W. H. Hudson and Richard Jefferies wrote about birds as an integral part of the environment and recounted their tales under such titles as *The Life of the Fields* and *Field and Hedgerow* (Jefferies) and *Far Away and Long Ago* (Hudson).

In the present century Edward Thomas, W. H. Davies, W. B. Yeats and Walter de la Mare have all written poems with a high ornithological flavour. Henry Williamson, a devotee of Richard Jefferies, produced two novels based on the life of individual birds: *The Phasian Bird* about a pheasant, and *Scandaloon* about a racing pigeon. Both showed his deep love for, and knowledge of, his subjects.

### Music

In the world of music it is, as in literature, bird song that has provided special inspiration. Thus the lark, cuckoo, and nightingale emerge pre-eminent once again. Was there ever a more descriptive work than Vaughan Williams' 'The lark ascending', the music spiralling up and down with the bird's song? And perhaps Delius had the last word on cuckoos; 'On hearing the first cuckoo in spring' says all that a bird-watcher feels at that moment. Beethoven in his Pastoral Symphony also uses cuckoo calls, while Couperin's harpsichord sounded of warblers, linnets and of course nightingales.

Nevertheless, some music has been wholly devoted to other species and to very great effect. The serious end-piece to Saint-Saens' 'Carnival of the Animals', 'The Swan', is an excellent example. And Sibelius took the same subject for his beautiful 'Swan of Tuonela'.

It cannot be said that modern music ignores birds either. In Duke Ellington's Far East Suite, under the title 'Bluebird of Delhi', the clarinet plays a perfect imitation of the Indian mynah. John Lennon and Paul McCartney went further and in their song 'Blackbird' they used a recording of the bird itself as the background to their statement, a device which perhaps those who have gone before might have wished to have at their disposal.

PMB

## Domestic birds

### Domestic and domesticated birds

The process of domestication started some 5000 to 8000 years ago and still continues today. It represents one of the largest biological experiments ever performed by man.

Domestication is the process of isolating individuals of free living populations, selectively breeding them for many generations with a more or less defined goal, and producing, as a result, a breed which is *genetically* different from the original populations. As a rule, domesticated birds can be recognized by the changes in their morphology, physiology or behaviour when compared with their ancestors. The White Leghorn Chicken for example, is several times heavier, differs in shape, has no feather pigmentation, is sexually hyperactive throughout the year and utters a prolonged low pitched crow compared with its wild jungle-fowl ancestor.

Not all domestic birds are domesticated. Birds taken from the wild and kept in captivity or in the house (Latin: *domus*), or bred for just a few generations in captivity are domestic, but not domesticated. The transition between the two states is a continuous one. The Australian Zebra Finch for example, took less than 50 years to become domesticated in several characters. The Peach-faced Lovebird has been domesticated even more rapidly, with yellow faced forms appearing after just 15 or 20 years. The process of domestication is still going on; in some cases simply for the pleasure, as in bird fancy; in others for profit, since domesticated birds are commercially important.

## History of domestication

It is not known what first motivated people to keep and breed birds in captivity. It may have been an attempt to keep edible species alive until they were ready for eating, or to save them in times of plenty against seasons of shortage. It could have been an affectionate bond to some dainty and droll chick that induced parental feelings. Another possible cause was the need for an available, handsome victim to offer as a sacrifice to a god.

Despite our uncertainty regarding motives, the dates and places of domestication are known for several species. From old writings or pictures, or from the discovery of their bones at archaeological sites, we can identify three epochs of bird domestication: an ancient, intermediate and a modern one.

The ancient epoch started in the Oriental region, centred on India, long before 3000 BC. The first species to be domesticated here were the Greylag Goose, Rock Dove and Red Junglefowl. Some centuries later the Mallard was domesticated in the Near East, probably in Mesopotamia. At about the same time the Swan Goose in eastern Asia, and the Muscovy Duck and Turkey in Central America, were also domesticated.

The domestication of birds did not therefore start at the same time everywhere. For domestication to occur two things had to coincide: (1) the presence of a highly civilized human culture which had already possessed domesticated mammals for several thousand years previously, and (2) an abundance of wild birds characterized by a wide geographic distribution with at least one subspecies adapted to unpredictable local climate. Such bird populations would contain abundant genetic variation, rendering them more suitable to domestication. Other predisposing factors were certain behavioural features, including precocial young, unspecialized ground feeding, and a tendency to imprint upon their parents, mate or surroundings.

The intermediate epoch lasted from the classical Greek civilization until one hundred years ago. It was characterized by an increase in domestication processes and by the spreading of different breeds. Thus peafowl and pheasants were bred in China, Temminck's Cormorants were trained to catch fish in Asia, and in Japan quail were bred for their 'song'. In several places the same species was domesticated independently, using either the same or different races. For example, the Helmeted Guineafowl, originally from North Africa, was bred in Greece more than 2500 years ago. However, the West African subspecies was reimported by the Portuguese and these replaced the extinct classical stocks.

Pets, for example the Canary, were imported by European sailors in the sixteenth century. One thousand years before that, the Bengalese Finch and the Java Sparrow had been selected for white plumage and good singing ability in eastern Asia. Domesticated birds were also exchanged between newly discovered continents, such as North America, and Europe.

In some parts of the world no domestication took place. Instead wild birds were hunted and harvested. During the past 2000 years the Polynesians have captured nesting parrots in

*Artificial selection and careful breeding have transformed the Rock Dove (Columba livia) into a remarkable variety of forms. The four types of domestic pigeon shown here are from The Illustrated Book of Pigeons with Standards for Judging, published in 1876. The types are the Pouter, Fantail, Trumpeter and Jacobin*

order to acquire their red feathers for ritual decorations. Also, the Indians of South American rain forests still collect nestling parrots to keep as pets.

In the modern epoch of domestication more goal-orientated breeding has resulted in an enormous increase in the number of bird species being domesticated. The outcome of domestication has also increased, partly due to increased understanding of nutrition, diseases and especially genetics.

The many birds with breeds modified by artificial selection include gamebirds (e.g. California Quail, Golden Pheasant), wildfowl (e.g. Canada Goose, Wood Duck), parrots (e.g. Budgerigar, Cockatiel) and finches (e.g. Gouldian Finch, Red Siskin). While some domestic stocks are still quite rare, others easily outnumber their wild ancestors (e.g. Turkey and Red Junglefowl). In some species a huge number of released domesticated individuals endanger the genetic composition of wild populations, for example the Mallard in North America where domesticated ducks are released for game shooting. In others the wild population is already extinct (e.g. the Imperial Pheasant). Domesticated individuals which escape in a foreign country may form a new 'species' or at least acquire a new scientific name, as is the case with the Turtle Dove. In California and Texas this species is erroneously referred to as *Streptopelia risoria*, but is in fact derived from the African Collared Dove *S. roseogrisea*. In most cases however, domesticated stocks do not survive for long in the wild, due to the changes they have undergone during artificial selection.

## Characteristics of domestic and domesticated birds

The differences between domestic and wild birds may be due to either environmental or genetic causes. These causes are difficult to distinguish because they often produce similar effects. Thus a given character might be affected by both. The most frequent *environmental* effects include an increase in size caused by better nourishment, especially during the period of growth. This has been demonstrated in the Zebra Finch; chicks fed on a high protein diet were structurally larger as adults, compared to those reared on a poor quality diet. Domestic birds may also be larger than wild ones simply because they get fat through lack of exercise. Changes in colouration can result from dietary changes, as in flamingos and spoonbills where carotenoids, which control the birds' pink colour, may be lacking in an artificial diet. The lack of ultra-violet light results in species like the Zebra Finch having a brown iris when kept indoors, rather than the red one of wild birds. Among parrots a humid environment sometimes results in birds developing melanistic plumage.

The fecundity of domestic birds may be increased by food quality, the lack of stress and by artificial light regimes. In addition, learning can play an important role: domestic birds learn to accept food their wild ancestors would never touch. Some species, typically reared in incubators, never learn to build a nest. Others, if reared by a foster parent (e.g. Gouldian Finches which are frequently reared by Bengalese Finches), imprint on their foster parent and subsequently try to mate with them in preference to their own species. Domesticated birds also tend to become tame, partly due to a multitude of learning processes, and partly due to a somewhat degenerated nervous system caused by an impoverished environment during their development.

Similar effects may be caused by the genetic changes resulting from artificial selection and inter-breeding. Only from experiments with birds reared under identical conditions can the effects of nature and nurture be distinguished. In many cases though, the changes are so dramatic that they must be the result of persistent artificial selection. For example many domestic ducks and fowl show virtually no fear of strange objects which would certainly have frightened their ancestors. Another example is the cockerel's 'crow': the domesticated cockerel gives a very different, and much longer call than its wild ancestor.

There are often differences in the size of wild and domesticated birds. In the chicken, domesticated breeds vary in size and weight by a factor of ten. The English exhibition Budgerigar is three times the weight of wild birds; in some

cases the males are even too heavy to copulate successfully with the females. There are also often genetic changes in many morphological characters, often involving changes in body proportions. These include short legs in Kruper Chickens, small wings causing flightlessness in ducks and chickens, a short skull and bill producing a pug-like head in the Owl Dove, and a crippled spine in hunchback Belgian Canaries.

Crests have been selected for in almost all domesticated birds, from geese to canaries. Interestingly, in almost all species the homozygous form is lethal. In other words, inbreeding of crested individuals will result in a low number of offspring. In the chicken the comb can be duplicated, in doves the bill cere protruded and in poultry the shape can be altered by dewlaps.

Plumage changes are frequently manifest, and the production of different colour forms invariably involve the loss of pigments. It is ironic for example that one of the world's most colourful birds, the Gouldian Finch, now occurs in a variety of dilute forms including a completely colourless form. Changes in the distribution and character of the feathers in other species have caused bizarre results: feathers all over the legs, lack of feathers on the neck, swirls, crowns, silky feathers, increased numbers of feathers or even extreme tail length, up to 7 metres long in the Yokohama Cock, caused by permanent feather growth.

In addition to these morphological effects there have also been physiological changes. Some of these are relatively simple and consist of a change in enzyme or hormone level. Others are more complex, such as a loss of sexual dimorphism, more efficient food utilization or reduced resistance to certain diseases. Fecundity has been subject to intense selection and in many breeds has been increased dramatically by precocity and an extended breeding season.

The intensity and success of this selection is obvious; in 1940 chickens were laying an average of 120 eggs per year, but just forty years later the average was over 260. Some domesticated chickens now lay more than 365 eggs a year, compared to a dozen in the Red Jungle-fowl.

Many behavioural peculiarities result from domestication. In several breeds the reaction to various stimuli is reduced, resulting in extreme tameness. In others the birds maintain extreme postures, as in Indian Runner Ducks and the Fantail Pigeon. The Pouter Pigeon maintains a permanently inflated crop, and the Rhine Wingbeater persistently beats its wings, wearing out the feathers until it becomes flightless. The Birmingham Roller Pigeon cannot stop somersaulting! All these movements occur naturally during the courtship of wild Rock Doves and their excessive frequency has been artificially selected by breeders.

Selection has also altered the frequency of courtship and sexual behaviour in some species. Roller Canaries have been selected for their song, some chickens for the duration of their crowing: in some races it lasts 20 seconds instead of the 1.2 seconds of the Red Jungle-fowl. Broodiness also differs, from a complete lack of incubation behaviour in Leghorn Chickens and Japanese Quail, to the over zealous brooding of other species by bantams.

A genetically determined increase in aggressiveness is most evident in fighting cocks. These breeds have been selected for their fighting ability since the time of Egyptian pharaohs. Cock-fighting was popular in Great Britain between pre-Roman times and the nineteenth century, and still is in places like South America.

## Aviculture and bird fancy

Selective breeding still occurs, in poultry farming and among cage bird enthusiasts. The number of people that keep cage birds is unknown; it varies among countries and centuries. In the past aviculture was restricted to noblemen in civilized cultures. Today, however, it is widespread among all social classes. Cage bird breeding is often directed towards goals which appear incomprehensible to the uninitiated: canaries are bred for song or posture, pigeons for colour, shape, speed and homing ability. Breeders are often organized into societies, and being quite numerous in Europe they are commercially important, requiring food and specialized equipment. Often the experience of bird breeders has played an

*Cage birds are extremely popular, and breeders compete selectively to breed the best birds according to set standards. Here an exhibition of canaries is being judged*

*Chicken numbers across the world (millions)*

| Continent or country | 1975 | 1981 | 1985 |
|---|---|---|---|
| Africa | 480 | 602 | 746 |
| China | 698 | 941 | 1361 |
| Germany | 92 | 85 | 79 |
| France | 164 | 186 | 188 |
| Great Britain | 133 | 122 | 112 |
| Japan | 244 | 286 | 337 |
| Soviet Union | 727 | 967 | 1090 |
| World | 5841 | 6621 | 8592 |

*The commercial production of poultry usually results in the birds being kept in appallingly crowded conditions, as in the battery chickens being bred for eggs shown here*

important role in stimulating scientific research. There are also some negative aspects of aviculture. These include the constant removal of some species from the wild, the development of crippled canaries, and birds unable to reproduce because they are imprinted upon foster parents.

In commercial aviculture and farming much has changed in recent years. Whereas farmers used to have their chickens running free in meadows they are now confined to battery farms containing tens of thousands of birds. Two or three birds are kept in a cage, each one with as much floor space as a single page of this book. As long as the majority of people in the west are more interested in a supply of cheap meat and eggs than in the welfare of millions of birds, this situation will persist.

The numbers of domestic and domesticated birds all round the world is not known, but they greatly outnumber human beings. In Britain in 1985, as the table above indicates, there were an estimated 1 million ducks, 11 million turkeys and 112 million chickens. The number of eggs laid by the world's chickens each week is about 11 600 million – two for every person in the world. To visualize this number imagine these eggs lying end to end. They would reach from earth to the moon and back again!

RS

## Falconry

One special form of domestication is falconry or hawking, the sport of taking wild game (or 'quarry') in its natural state and habitat by means of trained hawks.

The earliest records appear to be in China about 2000 BC. The first known record in Britain is in the eighth century AD. It was immensely popular in Europe, the Middle East, India, China and Japan until the eighteenth and nineteenth centuries when sporting guns provided an easier method of killing game. It has never been extinct in Britain, and at present is enjoying a substantial revival world-wide. There are falconry clubs in most European and American countries, which regulate the sport and help beginners.

The birds flown by falconers fall into three basic classes – 'long-wings' (falcons) such as the Peregrine, 'short-wings' (Accipiters) such as the goshawk, and eagles and buzzards which are sometimes trained and known as 'round-wings'.

A trained hawk is nowadays flown at quarry it would normally kill in the wild (provided it is not a protected species), though in Europe and India, over a century ago, Peregrines and Gyrfalcons were specially trained to fly at kites and herons. Arab falconers still fly these hawks at bustards. There seems to be no historical evidence to support the legend that in

*Far left. A trained Northern Goshawk (Accipiter gentilis) with his falconer and assistant, in Pakistan*

*Left. A falconer's Peregrine Falcon (Falco peregrinus) flies in to the lure. Jesses can be seen trailing from the bird's legs*

mediaeval times, certain hawks were allocated to certain social ranks; except perhaps in old Japan, where only the nobility were allowed to have hawks.

Up to ten years ago, falconers' hawks were taken from the wild – either from the nest ('eyas'), or on first year migration ('passagers') or, rarely, as fully adult birds ('haggards'). But recently a great change has taken place and most trained hawks are now bred in captivity. In most Western countries a government licence is needed to take a hawk from the wild and this is very rarely granted.

Trained hawks permanently carry light leather straps ('jesses') round their legs. When a hawk is not flying, the jesses are attached to a swivel and leash about three feet long, by which it is held on the falconer's gloved fist or tethered to a perch. A hawk also carries light bells which can be heard if it flies into a wood or undergrowth. A modern development is to equip the hawk with a tiny radio transmitter, by which it can be traced for several miles. A light leather cap (the 'hood') can be put over the hawk's head which acts as a blindfold when the falconer does not want it to see wild quarry or when it is near crowds of people, flashing cameras, etc. which would otherwise alarm it. There are several kinds of hood. The Dutch pattern is probably the most widely known and is often used as a symbol of falconry. When not being flown, hawks are tethered to perches in the open air. At night or in bad weather they are kept in a warm building. They moult into fresh plumage in the summer, when quarry is out of season.

The falconer starts training a hawk by teaching it to perch on his or her gloved fist and feed from meat held in the fingers. A short-winged hawk is then persuaded to jump a few inches from a perch to food held in the falconer's fist. This distance is increased daily until it will fly 50 yards to the fist. A long-wing is fed on a 'lure', which is a dummy of the quarry the hawk is intended to fly at – rooks', pheasants' or grouse's wings sewn onto a pad with meat tied to it. The lure is swung on the end of a line and dropped on the ground. The falcon is taught to come increasing distances to it, day by day. At the same time it is tamed to people, dogs, machinery and so on.

When the trainer is sure that the hawk will come promptly to the lure or fist, he starts to show it quarry in the wild, which it will fly after with increasing enthusiasm until it makes its first kill.

Long-wings are 'slipped from the fist' at rooks and similar quarry. The falconer approaches the quarry on the ground with the falcon hooded on his fist. As the quarry takes to the air, the falconer unhoods the hawk which flies after the rook. The rook will make various evasions in the air, often 'ringing up' in spirals with the hawk ringing after it. If the rook outflies the hawk, it will drift downwind. The falconer swings the lure and the bird returns to it. If the hawk catches the rook it will grasp it in the air ('bind to it') and bring it to the ground where the hawk kills it with a bite at the nape of the neck. The falconer picks up the hawk and gives it a mouthful of the quarry's flesh.

When the quarry is grouse or other game, a pointer dog ranges the moor until it stands motionless, showing that it has scented grouse ahead of it. The hawk leaves the falconer's fist, circling up high above him ('waiting on'). The grouse are flushed and the hawk hurtles downwards ('stoops') in a dive of several hundreds of feet, during which she reaches a very high speed. The hawk hits the grouse in mid-air with its talons, killing it.

The flights described above need open country. Long-wings cannot manoeuvre in woodland and the quarry will dive in a thicket and be lost.

Short-wings will follow the quarry into cover and can be flown in wooded country. They are usually held unhooded on the fist and slipped when a rabbit or pheasant is flushed from a hedgerow. They bind to the quarry and do not stoop. If they do not kill they will perch in a nearby tree or on a fence and return to the falconer's fist. Eagles will sometimes wait on like a long-wing, but they are usually flown from the fist like a short-wing.

A hawk in proper flying condition must be healthy, hard muscled and moderately hungry. A starved hawk is too weak to fly properly. A fat and flabby hawk does not want to fly. A hawk that is too much of a pet will simply wait for the falconer to offer it food. One that is too wild will not let a falconer near it. A hawk is completely unamenable to any sort of discipline or punishment. The falconer's chief control lies in the food. Fresh bird's flesh is the best, butcher's meat being a poor substitute. Each hawk has its own individual weight (to within an ounce) at which it flies best, and it must be fed the correct amount to bring it to, or keep it at, that weight – and each kind of food has a different effect. As well as having a thorough understanding of all that, the falconer needs a knowledge of bird flight, the effect of the wind and a deep grasp of the character and ability of both hawk and quarry. It takes a great deal of application and study to fly a hawk properly. Falconry is thus a sport for the dedicated enthusiast.

TAMJ

## The impact of birds on people

### Impact on crops

Birds and humans share many common preoccupations. They sometimes choose the same food and the same place to live. Where this leads to a lowering of man's well-being, the birds involved are generally regarded as pests. In places where man has opted for permanent settlements based on agriculture, the general tendency has been the removal of forest and its replacement with open country, in which many of the crops grown resemble grassland. This has taken place over huge areas of the earth's surface, such as the savannah regions of Africa and the temperate zones of Eurasia and North America. While populations of many forest birds must have declined, some birds that like open habitats have benefited enormously. Notable examples are the Red-billed Quelea in Africa, the Red-winged Blackbird in North America and the European Starling in Europe and central Asia. For much of the year these birds are able to subsist on a diet largely of seeds and this predisposes them to attack crops grown in artificially open areas such as fields.

Cereal crops are thus especially vulnerable and barley, wheat, sorghum, millet, rice and maize can suffer extensive losses. Other non-cereal crops, such as brassicas, are also eaten by birds. For example, Woodpigeon depredations on oilseed rape crops in Britain in winter are so widespread that most fields suffer damage. However, severe bird damage to crops is generally localized so that, on a national basis, damage of this kind rarely has a significant impact on a country's crop production. On the other hand, for an individual farmer or even an African village, severe local damage can deplete a year's food production, thereby lowering profit or even leading to a requirement for food importation or aid.

This localized nature of bird damage to crops makes estimation of the losses very difficult, for in any given area losses may range from zero to total yield depletion. Further difficulties can arise because some plants are, to greater or lesser extents, able to compensate for damage inflicted by birds. In general, the earlier during plant growth that birds attack, the greater are the possibilities of compensatory growth. Thus bird attack on sown or germinating cereal often has little influence on the yield of the crop, although it may delay ripening, while bird consumption of ripening grain or fruit can lead to losses in proportion to the extent of bird activity. For example, the removal of up to 37% of winter wheat plants shortly after germination by starlings in England had no effect on subsequent yield, but the consumption of ripening seeds of sorghum, millet and rice by quelea and other African seed-eating birds can cause total crop failure on some farms.

Bird damage to crops is subject to many environmental influences and its extent varies in time and space. Some forms of damage, such as that by Bullfinches to fruit-tree buds in winter, has been known for centuries. On the other hand, many forms of damage change in importance as crops or farming practices change. Rooks used to cause considerable losses to oats as they dried in stooks in the fields and were subsequently stored in farmyard stacks in winter. Stooking and stacking were consequences of harvesting by binder, and the replacement of this with the combine harvester led to the disappearance of this food source and hence a cessation of this kind of damage. Other damage problems have arisen through recent agricultural developments. Perhaps the most costly is that following the introduction of intensive rearing of cattle, where the stock are retained in buildings, yards or corrals and their food, usually containing cereals, is brought to them. Such systems provide huge quantities of readily accessible food for starlings, which cause considerable losses in North America, France and Britain where intensive husbandry is practised. 'New' crops can also lead to new forms of damage. The advent in Britain of winter sown varieties of barley has resulted in these crops ripening very early in the summer, providing a grain source for seed-eating birds that was not previously available at this time of year. The enormous rate of spread of oilseed rape in Britain since the early 1970s has provided Woodpigeons with a new source of winter food; this has led to an increase in their population size.

Other recent changes in the status of birds as pests depend upon less direct influences of farming practice. For example, the use of herbicides has reduced the number of weed seeds in arable fields. In sugar beet fields, where such seeds used to form a valuable food source for Skylarks, the birds have had to resort to grazing the leaves of the sugar beet itself, thereby depleting yields. In other instances, the presence of alternative foods within a crop can attract birds to that crop. North American maize fields that support dense populations of insects are more prone to damage by Red-winged Blackbirds than fields that lack insects; here, the insects form the initial attraction to the birds, which later remain in the same field but switch to eating the corn.

Birds may also become pests through changes in their population size. In the northern hemisphere, many species of Arctic-breeding goose are currently undergoing population expansion. This is due, at least in part, to a reduction in human persecution and disturbance, although changes in farming procedures on the birds' migration routes and on their wintering areas have also contributed. Snow Geese, Pink-footed Geese, Greylag Geese and Barnacle Geese have all increased in numbers and their larger populations have caused conflict with man.

The Dark-bellied Brent Goose is a notable example since, in addition to undergoing spectacular population growth, this bird has also apparently changed its feeding habits. From a world population of around 16 500 in the late 1950s, numbers increased to around 200 000 in 1985. Initially, the small population seemed to be dependent upon eel grass, a plant of mud flats, as its winter diet. As the population expanded, however, flocks moved inland to graze winter wheat and, to a lesser extent, grass and oilseed rape.

After harvest, crops may still be vulnerable to bird damage although the range of species involved is small. Nevertheless, large numbers of feral pigeons, House Sparrows, Spanish Sparrows and others can inhabit or live near to warehouses and food stores. Here, the problems involve both food consumption and fouling by the birds. In the developed world, even the

*Birds in large numbers can become an economic problem or health risk. European Starlings here have homed in on an English pig farm; in 1979 the estimated 1.8 million birds caused £7000 of damage*

merest trace of bird droppings can render a consignment of food unsaleable. While the appearance of the food is important from the public's point of view, the main concern is the risk of disease.

## Diseases carried by birds

Wild birds are known to carry a wide variety of bacteria, viruses, fungi and parasites but only a small proportion of these infect man or his domestic stock and instances of birds' involvement in disease transmission are rare. Nevertheless, the possibility that House Sparrows that feed and breed inside a food warehouse might be carrying one of the species of *Salmonella* bacteria that can cause food poisoning in man is sufficient to necessitate the removal of the birds. This possibility is made plausible following demonstrations that birds have transmitted diseases. For example, in Liverpool, England, grain destined for poultry food was stored in large warehouses that were infested with feral pigeons. The pigeons, which were fouling and dying in the grain, were infected with a paramyxovirus that causes Newcastle Disease in poultry, and subsequently over 20 poultry flocks, given food containing grain from the warehouses, contracted the disease. In North America, the huge quantities of droppings that accumulate under the night-time roosts of Red-winged Blackbirds and their allies provide a growth medium for a fungus whose spores cause histoplasmosis, a potentially serious lung disease, in humans. Children attending schools near such roosts are known to have become infected from this source. It is well known that people who maintain close contact with birds of the parrot family are prone to psittacosis, a particular form of the disease ornithosis that is carried by many birds. Generally, however, the evidence that birds are implicated in disease transmission is circumstantial and while it is believed that European Starlings are implicated in the spread of transmissible gastroenteritis among pig herds in Britain and America, it now seems unlikely that starlings were significant disseminators of foot-and-mouth disease in Britain, as was believed in the 1960s. On the other hand, gulls, many of whose populations have increased this century in response to man's activities, now roost in large numbers at night on reservoirs and the water in such reservoirs has been found to have elevated bacterial concentrations, including a number of salmonellae.

Some birds live in close proximity to people for a large proportion of their time by inhabiting towns and villages. Their close association with humans sometimes does give rise to concern over diseases, but more often the birds' fouling of buildings and other artefacts is regarded as the main aspect of damage. In central London the amount of feral pigeon and starling droppings removed from buildings is measured in tonnes and cleaning operations cost thousands of pounds each year. Droppings, nest material, feathers and dead bodies block drainage gutters, leading to damp and causing building deterioration, while droppings on pavements, roads and industrial walkways can be slippery and pose a direct danger to the public. The tendency for starlings to roost in city centres has increased this century and at present we are experiencing a rise in the number of gulls that choose to nest on buildings in coastal towns. Here, in addition to the building maintenance problems mentioned above, the raucous calls of the birds in the early morning can cause considerable annoyance to residents.

## Bird strikes

Gulls are also the group of birds most frequently involved in collisions with aircraft. Aerodromes are often situated on the outskirts of cities and at some distance from housing, in areas that are thus also suitable for water storage in reservoirs and for refuse disposal. These features prove attractive to gulls and large numbers are therefore attracted to the vicinity of many airports. Collisions between birds and aircraft, called bird strikes, vary in their severity but larger birds, such as gulls, crows and some waders, are capable of inflicting greater damage to airframes than smaller birds. Damage ranges from minor dents, through more severe damage that necessitates expensive repair, to total engine failure that can lead to

*Birdstrike: this mid-air collision between bird and Hunter jet resulted in the plane having to land with nose wheel retracted*

loss of the aeroplane and the death of its occupants. Most bird strikes occur around airfields, because most bird activity occurs within a few hundred metres of the ground, but there are records of high altitude strikes, the highest being that involving a Rüppell's Vulture at 12 000 metres over West Africa.

## Damage prevention

Some of the earliest records of bird damage are of birds eating germinating cereals and around 200 BC Babrius documented the use of catapults to kill and scare the offending animals. These two approaches, killing and scaring, are still those most commonly used in bird damage control. The aim of scaring is to reduce the attractiveness of particular feeding areas so that birds feed elsewhere, whereas the killing of birds has a dual role, involving scaring and population reduction.

Population reduction is sometimes attempted on individual farms by shooting and trapping. In general, only small numbers are killed and these are quickly replaced by immigrants from elsewhere. This control is thus ineffective in reducing bird pest numbers but shooting, which involves the presence of a human being and a loud bang which birds learn to associate with danger, can act as an efficient deterrent to return to that site. Birds have also been killed on a much larger scale, using a variety of techniques. The number of species that can be subjected to this level of mortality is limited by their social behaviour, for potential targets must feed in large flocks, breed in large, dense colonies or roost communally in vast assemblages at night. For these reasons, mass-killing has been aimed principally at Red-winged Blackbirds in North America, quelea in Africa and starlings in Europe and North Africa. The armoury that has been used against these birds includes petrol bombs, flame throwers, dynamite, poisons as sprays or bait and detergents which are sprayed on to roosting birds' feathers; this destroys the insulation properties of the plumage and leads to death by hypothermia. By these means, large numbers of birds have been killed. For example, it has been estimated that up to 180 million quelea have been killed annually in Africa, while in North America, several million Red-winged Blackbirds and associated species have been killed in a single night's operation. Despite these massive kills, however, bird numbers, and often bird damage, have not fallen in the long term, indicating that these techniques have failed to exert the intended limit on populations. It is now realized that in most cases the levels of mortality achieved are insufficient to overcome the natural recruitment rates of the birds; reductions in number are restricted to the immediate locality of the operations and even these reductions are only temporary. Nevertheless, temporary reductions in numbers can protect vulnerable farms in Africa from quelea during the critical period of crop ripening, while in North America birds are still killed in their night roosts near towns in order to prevent dispersal to new areas that might become foci of histoplasmosis infection for the local human community.

Killing on such a large scale can be impracticable where birds fail to form sufficiently large assemblages, as occurs with those weaver birds and munias that eat crops, especially rice, in South-east Asia. In addition, the techniques required for such mass-killing are illegal in many European countries, and in these instances other forms of damage prevention are employed.

A wide variety of bird scarers has been developed, largely based on what man finds objectionable or on what man thinks birds ought to find objectionable. Effigies of a man, noise-making devices and brightly coloured material that flaps in the wind are those most commonly used but birds rapidly become accustomed to them. Attempts have been made to combine the signals conveyed to pest birds, for example by incorporating into a moving scarecrow a device that produces a loud bang, thereby more closely mimicking a man with a gun. More recently, the signals used by the birds in their own communication have been employed in the manufacture of scarers. For example, tape-recordings of the distress calls of a species can be broadcast at flocks of that species in order to frighten them away; this practice is commonly used against roosting

*The damage some birds cause to crops has resulted in the development of a variety of bird scarers, from the traditional scarecrow to sophisticated machines that transmit loud bangs or the pest species' alarm calls*

starlings, especially in woodland but also occasionally in towns, and against gulls on airfields. Some pest species have particular postures or markings that are used to signal anxiety or imminent flight. These have been incorporated into models of birds that have proved to be effective in manipulating the behaviour of their natural counterparts. These scarers, using signals that the birds themselves have evolved, retain their aversive properties longer than scarers based on people's perception of the unpleasant, but even so birds eventually become accustomed to these 'biological' scarers as well.

The idea of using naturally occurring signals to frighten troublesome birds away is taken a step further by employing falconry. This kind of scaring is sometimes used on airfields, where it can be effective in driving birds away. However, the number of days on which the raptors can be flown is limited by weather and it is felt that the principal deterrent is often the person accompanying the bird of prey.

The search for chemicals that will render foods unpalatable to birds has also been based on materials that humans find distasteful or on pesticides that are thought to possess aversive properties.

One of these pesticides, methiocarb, is now widely used in some developed countries, in spite of being highly toxic and giving rise to food contamination concerns, and it sometimes gives valuable protection to vulnerable crops. In order to avoid the use of poisons and to find more effective repellents, naturally-occurring plant defence chemicals are, however, now being sought to offer a humane and environmentally safe alternative.

Physical barriers also play an important part in bird damage prevention. On buildings, nets, wires, spikes and gels, smeared on to ledges to produce an unstable surface, deter feral pigeons and starlings, and wires have been strung across reservoirs to prevent gulls from roosting there. This is an extension of the ancient practice of stringing cotton strands over plants in small plots, such as gardens, and the method has been applied with some success on a larger scale on farms. Similarly, the garden fruit cage principle has been adopted on a farm scale in Britain and New Zealand, where cherry orchards of three or four hectares have been completely enclosed with protective netting. In autumn, European vineyards can alter the colour of the countryside when huge areas are draped with yellow netting to exclude migrating starlings and thrushes.

Bird damage is essentially an ecological problem, that results from the environmental changes wrought by humans and which prove favourable to particular birds. A reversal of these changes would solve many bird problems but this is clearly impracticable in most cases. Nevertheless, the replacement of vulnerable crops with those that birds do not eat can prevent damage; some British farmers have stopped growing oilseed rape on account of extensive Woodpigeon damage, switching to alternative crops that these birds do not eat. In Africa, some farmers change from the small seed cereals like sorghum and millet to maize, which is less susceptible to quelea damage, even though maize sometimes fails under drought conditions. Cultural and economic considerations clearly influence the extent to which alternative crops can be contemplated.

Within a particular crop, varieties often have differing susceptibilities to bird damage. Protection of non-preferred varieties may be conferred by physical aspects of the plant, as in sunflowers where varieties with seed heads that hang down are eaten less frequently than those whose heads remain upright. Chemical constituents of plants also influence bird preferences: some varieties of sorghum contain high concentrations of tannins and are resistant to bird attack, but unfortunately some of these varieties are also less palatable to man.

In many parts of the world, both crop growth and bird food selection are seasonal events and bird damage depends upon a coincidence of the bird's need for food of a particular type and the availability of a crop that provides that need. Woodpigeons eat oilseed rape in winter and thus the replacement of autumn-sown with spring-sown rape could considerably reduce this kind of damage. Similarly, in parts of Africa quelea migrations are more or less predictable and it might be possible to time the crop's ripening with the absence of the birds. However, weather vagaries render crop sowing and ripening and bird movements somewhat variable and it is therefore difficult to achieve the necessary precision to eliminate damage by birds.

There are thus many possible approaches to the alleviation of bird damage but none of them is perfect. Increasingly, it is being realized by scientists and farmers that the best solution is to be found through integrating a wide variety of protective measures to suit individual situations. This requires a flexible approach, coupled with a thorough understanding of the problem in terms of crop growth and its responses to bird attack, and of bird behaviour and food requirements. This is the goal to which we must aspire.

CJF

## Human exploitation of wild birds

People use birds for a wide variety of reasons, satisfying their culinary, agricultural, recreational and aesthetic needs.

### Eggs and nests

Many species of wild bird, including their nests and eggs, are eaten throughout the world although in many parts the practice is declining as alternative foods, such as farm produce and imported meats, become more readily available. The only birds' nests to be eaten are those of some swiftlets that nest in caves of South-East Asia. The most favoured are those of the Edible Nest Swiftlet, whose nest is built almost entirely of salivary secretion. The nests can be made into a gelatinous soup which is essentially a protein–sugar solution. To the European palate the soup is almost tasteless but it is regarded as a great delicacy by the Chinese. Other species incorporate feathers into their nests and these are regarded as inferior as they need cleaning before they can be made into soup.

Birds that suffer the heaviest predation by man are those that are most accessible in large numbers. Thus ground-nesting colonial seabirds have borne the brunt of human attention. This is particularly true with regard to birds' eggs, where ground-nesting gulls and terns have been, and in some places still are, heavily exploited. Throughout the tropics, Sooty Terns yield well over a million eggs per year. In some areas, like the Seychelles, this has developed into a small industry with associated management and government control, so that the eggs constitute a renewable resource. Elsewhere, uncontrolled human predation has decimated populations, as in Indonesia and the West Indies, although habitat changes have also contributed to colony extinctions. Egg collecting presumably had a significant effect on gull populations in Europe, since the huge increase in their numbers this century is at least partly attributed to relaxation of human persecution. Similarly, populations of penguins that were once heavily exploited are recovering. On Arabian offshore islands human predation on seabird eggs, especially those of the Sooty Gull, Crested Tern, Roseate Tern, White-cheeked Tern and Bridled Tern, is so intense that it is difficult to see how the populations survive, and yet breeding colonies are still found. Lapwing eggs also used to be harvested in large quantities but the cessation of this practice in Britain, as a result of increased bird protection, has not led to an increase in their population. This is because their numbers have been limited by various aspects of agricultural intensification. In Sulawesi, the highly nutritious eggs of the Mallee Fowl have been over-exploited to such an extent that this species is seriously threatened unless protective measures can be quickly put into practice.

### Wild birds as food

The chicks of many species are prized as food and this was the prime factor behind the domestication of the Rock Dove. In some parts of the world, and especially in the Mediterranean, colonies are still maintained for their chick production. In many cities, however, feral pigeons are regarded as both ornamental and as a pest but their chicks are no longer harvested.

Before fledging, the chicks of many of the shearwater family become heavier than their parents. This extra weight is fat, which acts as an energy store to tide them over the last few days in the nest when they are deserted by their parents. These large young are harvested, often to be salted for later consumption. In the Seychelles, Wedge-tailed Shearwater chicks formed the basis of a cottage industry in which tens of thousands were collected annually. This had no noticeable impact on the population but the practice is now in decline, largely as a result of the increased availability of imported meats. Around Tasmania, over half a million chicks of the Short-tailed Shearwater, which is locally called the 'mutton bird', are taken annually by a well organized industry. Studies have shown that even this predation does not affect population size, but the harvest of chicks by unorganized groups of people, largely as a sport and accompanied by much revelry, is leading to habitat destruction and consequent reduction in the size of some colonies.

The young of many other species are eaten by man on an incidental basis. Young boobies are harvested along with adults from many tropical islands and it is the predation of adults, especially of such long-lived birds with a low reproductive rate, that poses the greatest threat to these seabird populations.

Boobies living on tropical islands are relatively fearless of humans and they also settle readily on ships, where they can be easily captured. It is from this apparent stupidity that their name derives (the French treat them with even less respect, giving the family the name 'fou') and their fearlessness has rendered them extremely vulnerable to human persecution. As a result, many of their colonies have disappeared and most surviving ones are in decline. The temperate counterparts of the boobies, the gannets, similarly suffered persecution during the last century. In addition to being taken for food, many were used as bait for fishing. Relaxation of human predation, coupled more recently with active protection of colonies, has led to a recovery of numbers. Unfortunately, the tropical boobies are unlikely to share this good fortune, since human interference does not seem to be waning and any proposed protection measures are difficult or impossible to enforce on remote oceanic islands.

The huge auk colonies of the North Atlantic suffered heavily from predation of their eggs, chicks and adults for food, feathers, fertilizer and bait. The harvest often involved daring exploits, including leaving a man suspended down vertical cliffs overnight so that he could catch the earliest arriving guillemots the following dawn. A wide variety of nets, snares and other implements was invented for bird catching, upon which many islanders depended for their survival. This dependence in some areas imposed a need for a sustainable yield and on St Kilda, the Faeroes and possibly elsewhere, the number of birds taken was regulated through locally evolved customs. Where no such regulation was applied, exploitation was intense and colony survival depended upon the inaccessibility of a proportion of the birds. It seems that the largest of the auks, the Great Auk, was able to withstand human predation on migration and on its wintering areas, but when European boats and navigation improved sufficiently for people to reach the breeding colonies, extinction of this flightless bird rapidly followed. Technological improvements in boats and firearms pose the greatest current threat to auks in the western North Atlantic. When bird harvests served primarily for subsistence, limits on numbers were imposed by mobility and catching methods. Now, the sporting element of bird killing has increased and the availability of small, fast powered craft and high powered rifles is leading to wanton destruction of auks, especially Brünnich's Guillemots in Greenland.

*In the nineteenth century and before, the people of St Kilda in the Outer Hebrides harvested seabirds, like Fulmars (Fulmarus glacialis), for their meat and feathers*

It is often difficult to divorce the collection of birds for food from the sporting implications of bird-catching. While many of the birds that are killed on migration through the Mediterranean are eaten by the catchers or sold for food, much of the present day killing is for 'fun'. Many techniques are used. Shooting, trapping, liming (smearing very sticky substances on to branches so that small birds adhere to them and can be collected) and netting are widely practised and the development of mist nets (nets made from very fine black thread which is almost invisible against a background of vegetation) in the 1950s must have greatly increased the toll. This kind of killing is indiscriminate and it is estimated that hundreds of millions of birds, representing about 15% of those migrating through the area, are killed annually. Turtle Doves and thrushes are the birds most prized for the table, while birds of prey and decorative birds like Hoopoes are stuffed for subsequent admiration. In West Africa, the snaring of birds is a recreational activity mainly of children and while most bird populations are probably unaffected, the coast is the main wintering area of European Roseate Terns, whose small population cannot withstand the mortality inflicted. Similarly, in the Seychelles the endemic Paradise Flycatcher proved an easy target for children with catapults. In both of these cases, education of the children is likely to provide the best long-term solution to these activities.

The use of birds for sport can, however, have much wider implications since the management of habitats for quarry populations often provides benefits for non-target species. In North America, wetlands are managed to maximize habitats available to wildfowl that are hunted extensively: the habitats thus safeguarded are used also by other birds such as waders and cranes. Birds of the pheasant and partridge family are valued for both their palatability and sport, and to increase their availability to hunters they have been introduced to many parts of the world, for example in Europe and North America. Within their new range they are reared artificially for release and suitable habitat is maintained for them. In the past, many raptors were killed by gamekeepers in attempts to maximize production for the gun and the absence of buzzards in much of England is a legacy of this activity. However, there seems little doubt that were it not for the value of game shooting, much of the woodland left for these birds, and widely used by other species and contributing to landscape diversity, would have been destroyed in the name of agricultural improvement.

In Britain, the use of Woodpigeons for food and as a sporting target is combined with pest control. Hunters are encouraged to concentrate their activities on farms where vulnerable crops are grown. Pest control activities are also used to provide food elsewhere where large numbers of birds are killed. For example in Africa, after vast numbers of quelea have been killed in roosts or breeding colonies, local villagers gather dead birds for cooking and eating: this practice should be discouraged, however, where the birds have been killed with poisons.

## Guano harvesting

The major bird by-product that man has exploited is guano, notably from the breeding colonies of oceanic birds, although off Arabian shores roosts of non-breeding Socotra Cormorants have also provided large accumulations. In South-East Asia, cave-nesting swiftlets provide fertilizer for local consumption. The main guano-producing region, however, has been the seabird colonies of the Humboldt Current in coastal Peru. Here, the guano is deposited in such quantities that it constitutes a renewable resource. Colonies are visited in rotation every two to three years and yield up to 300 000 tonnes annually. The birds involved are predominantly Guanay Cormorants, Peruvian Boobies and Brown Pelicans. Their populations fluctuate according to the abundance of their main prey, anchovies, and in El Niño years, when the coastal waters warm and the anchovies fail to appear near the colonies, the seabird populations crash. Recently, man's overfishing has depleted anchovy numbers and consequently seabird numbers and guano production are much reduced. In many other parts of the world, annual guano deposition is inadequate to produce a sustained yield and

*Left. The sun-dried droppings, or guano, of some tropical seabirds such as Cape Gannets (Morus capensis) are harvested and used as fertilizer*

guano that has accumulated over thousands of years has been extracted in a destructive way. This has led to total destruction, in ecological terms, of some islands such as Assumption and St Pierre in the western Indian Ocean. A similar fate might have befallen Christmas Island, in the eastern Indian Ocean, but for the efforts of conservationists. Here, tall forest was being destroyed in order to extract the guano below. These trees are the only breeding place of Abbott's Booby, and continued guano exploitation was threatening this species with extinction. However, forest destruction has now been arrested in order to preserve the island's remaining unique ecosystems.

The exploitation of guano has produced both riches and hardship. On Nauru, in the western Pacific, the export of about two million tonnes of phosphate annually, providing employment for over 1000 of the 8000 inhabitants, has enabled the Nauruans to make considerable investments abroad in order to provide for the time when the guano runs out. On Ocean Island, on the other hand, the inhabitants, the Banabans, were forced to leave the island to allow further guano extraction in 1942. They bought an island in Fiji and settled there but in 1976 brought an action against the British Government and the Phosphate Commission for the rehabilitation of Ocean Island and for higher royalties. This action, costing £750 000 and lasting 225 days, was at that time the longest and most expensive lawsuit in British judicial history.

Birds' bodies have also been used as fertilizer and once again colonial seabirds have been the primary source. Such birds have additionally been used to produce oil for lighting and fuel, as well as food, and on southern ocean islands the remains of vats (called trypots) in which penguins were rendered down can still be found.

### *Birds that assist humans*

In some instances, birds are used as aids to human exploitation of other resources. Perhaps the most widespread is the fisherman's use of seabird flocks as pointers to the location of shoals of fish. In South-East Asia this is taken a stage further through the use of captive and tethered cormorants, some of which are bred in captivity. The birds' necks are fitted with collars to prevent swallowing. Set to work at dusk, when a boat-borne light will attract fish, the cormorant brings a fish to the surface. It is then hauled in and the fish taken for human consumption, although the birds are rewarded with a proportion of the catch. This practice was adopted as a sport by the British aristocracy in the Middle Ages. In Africa, honeyguides have a symbiotic relationship with certain mammals, notably the honey badger or ratel. The birds, through their behaviour, guide the mammals to bees' nests so that the mammals can break open the nests and eat most of the contents. The honeyguides then eat the remaining larvae and wax, which would have been inaccessible without the mammals' assistance. Humans also play the mammal role, following the birds and using the honey, but

*Trained Cormorants are used to catch fish in China. The birds are prevented from swallowing their catch by a ring placed around their necks. Cormorants have been used in this way since the Song dynasty (960–1298 AD)*

leaving some of the nest contents for them. A most unusual use of birds in this regard has been to locate leaks in oil pipelines in the Americas. When a leak is suspected, foul smelling ethyl mercaptan can be pumped into the pipeline. Some of this will seep through any leaks and Turkey Vultures are attracted to the odour, thereby indicating the location of the seepage. Increasingly, the importance of bird populations as indicators of environmental welfare is being appreciated and bird population monitoring is one of the array of techniques used to detect people's deleterious effects on their planet.

In many cultures, birds find a place in people's emotional well-being. Many species of bird are admired for their colour or behaviour, and adornment with particular feathers can convey important meaning to different societies. Bird of paradise plumage has been used for tribal decoration in Papua New Guinea, and North American Indians prized eagle feathers, the wearing of which was an indicator of social status. Egret plumes, among other feathers, were heavily exploited for use in the Victorian millinery trade but this fashion has, thankfully, largely disappeared. Bird plumage has been employed for other forms of human comfort, notably in the harvest of down to fill bed clothes for warmth: feathers from eider ducks, used to make eiderdowns, have now been largely superseded by artificial fibres, although feathers from domestic waterfowl are still used for filling pillows.

An unusual instance of the use of captive birds relates to a Dutch cattle farm in which Purple Glossy Starlings are kept in the fully enclosed cattle sheds in order to control flies, apparently with some success. Attempts to introduce insectivorous birds as an aid to insect control have usually misfired, with the birds themselves becoming pests, as with the Indian mynah on many tropical islands. The naturalization of European birds, for aesthetic reasons, by European settlers in colonial times also produced some ecological disasters, the most notable of which is the European Starling in North America: in less than eighty years it became one of America's most numerous birds and a significant agricultural pest.

Waterfowl perhaps constitute, along with peafowl, the birds most introduced to gardens and parks for ornamental reasons. Most of these birds have remained captive but some, like the Mute Swan and Mandarin Duck, have established feral populations. In Britain, the introduced Canada Goose has become well established, even developing an extensive moult migration within the country, and in some areas has become a pest of growing cereals and a nuisance, owing to its fouling of pathways and grassland in public parks.

In the developed world the major present-day use of wild birds, and one which forms the basis of a growth industry that is spreading to developing countries, is for recreation. Bird watching has arguably been the driving force behind popular environmentalism and tourism based on bird watching provides important funding for conservation in some countries and, perhaps more important, encouragement to people of such countries to take a greater interest in their local natural history.

CJF

*Bird feathers have frequently been used for decoration and adornment. This New Guinea tribesman wears a headdress made from the feathers of a number of species, including birds of paradise*

## CATCHING METHODS

The desire to catch living birds for scientific study received a boost in the early part of this century with the development of ways of permanently marking birds individually, with leg rings. Many catching methods used were developed from those that had been earlier employed to obtain birds for food. The simplest form involves a small cage, supported off the ground by a stick which can be pulled away by a length of string, by an observer when he sees a bird enter to eat bait placed within. Other traps, also baited and sometimes containing decoys, rely on birds finding their way in through funnels and being unable to find their way out again. These range from small traps that can be placed in gardens or buildings for small birds, through 'crow' traps for catching larger numbers of medium sized to larger birds, to huge structures that are used in North America for catching New World blackbirds. These are developed from traps used for bird pest control. In 'Chardonneret' traps an entering bird triggers a trap-door release and this technique is often used in traps placed on nest-boxes for catching hole-nesting birds.

Some traps are very large and require birds to be driven in by human or animal disturbance. Duck decoys are classical examples of these and Heligoland traps were designed to catch large numbers of a wide variety of species, especially during migration.

More recently, nets have been developed for bird catching. 'Clap nets' are small nets that can be propelled, usually by elastic, over bird flocks attracted to bait. Larger nets, used mainly over feeding or roosting flocks of geese, waders and gulls, are propelled by rockets or projectiles fired from mortar-like devices.

The advent of 'mist' nets, fine nets that birds find difficult to see against a background of vegetation, revolutionized bird-catching in the 1950s. These nets are now the mainstay of ringing (banding) and other scientific study and have opened up new areas of research, for example the study of forest canopy birds in the tropics. Unfortunately, they have also proved invaluable to the bird-hunting fraternity in the Mediterranean and elsewhere.

Those birds that moult all of their flight feathers simultaneously, such as wildfowl, can be caught by shepherding them into corrals during this flightless period. While breeding, some seabirds, such as Sooty Terns, can be simply picked up off their eggs. Others can be caught using running nylon or wire nooses on the end of poles. Nooses are also employed in the 'Bal-chatri' trap, which consists of an array of nooses surrounding a small cage containing a decoy item of prey: the feet of raptors and owls become caught in these nooses. Some predominantly diurnal birds can be temporarily immobilized at night by dazzling them with bright lights. It is also possible to immobilize birds by putting narcotics in their food, releasing them after a brief period of recovery.

It is important to note that the species that can be caught and the methods that can be employed are legally restricted in many countries and people embarking on research should ensure that they do not contravene local laws or that they obtain necessary licences. That apart, the catching of birds for scientific purposes necessitates the ingenuity and fieldcraft that our forbears must have employed for their sustenance, and generates similar satisfaction.

CJF

*A large bird trap, known as a Heligoland trap, is used for catching migrant birds for ringing purposes. This one is located on Fair Isle, Shetland, and is run by Britain's most important bird observatory*

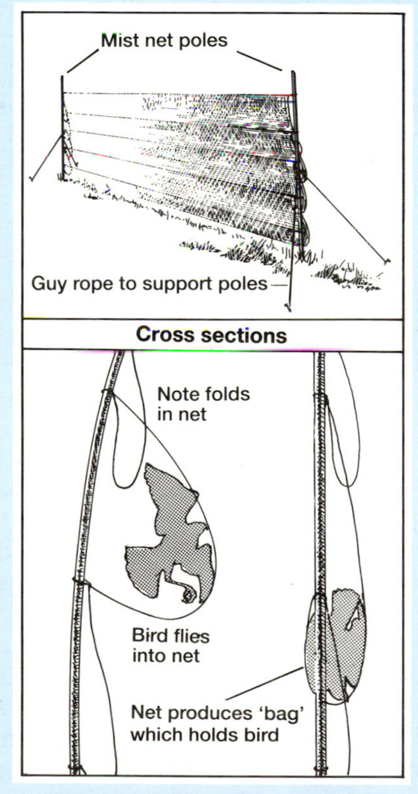


*A mist net in position. Mist-netting is most successful when there are bushes nearby to camouflage the net and make it less obvious to birds*

## Conservation

Although humankind has long been interested in birds, it is only relatively recently that an interest in their conservation has arisen: conservation is often defined as 'wise use of natural resources'. The word 'wise' in this definition begs all the important questions, but is generally held to embody the concept of 'sustainability', or use which allows the population concerned to perpetuate itself indefinitely. This concept has replaced an earlier ethic of 'protection' which dominated attitudes towards restoring exploited bird populations until quite recently.

### Past extinctions

Since the year 1600, over 150 kinds of bird (including both subspecies and full species) are known to have become extinct. Many more must have vanished without trace; the ones we know about were the large, colourful or edible species – like the Dodo and Great Auk – which ordinary people noticed, and for every one of those there must have been several small brown birds whose disappearance went quite unrecorded.

In historic times, we know that birds have become extinct at the rate of at least one species every two years; the true rate, including those species we do not know about, would certainly be higher.

Until recently, it was widely believed that Europeans expanding across the globe in the sixteenth, seventeenth and eighteenth centuries caused many more extinctions than other races; people native to a region were assumed to live in harmony and ecological balance with birds and other wildlife. There was no factual basis for this assumption, which can be traced more clearly to the Victorian concept of the 'noble savage' than to any biological evidence. Modern studies of recent sub-fossil remains associated with human occupation on oceanic islands – especially in the Caribbean, Hawaii and the South Pacific – have demonstrated that the peoples who occupied those islands before the Europeans arrived were responsible for at least as many extinctions as their successors.

However, the timing and the geographical patterns of extinction since 1600 do suggest a marked effect by Europeans. For example, the Indian Ocean was explored relatively early and most extinctions took place before 1900. In contrast, as the table below shows, European exploration and extinction occurred in the Pacific Ocean rather later, with most losses occurring since 1850.

About 90% of the species and subspecies of birds which we know to have become extinct, were confined to oceanic islands; some archipelagos, such as Hawaii and the Mascarenes, have lost large proportions of their original bird faunas. In Hawaii for example, about one third of native species has been lost in the last 200

*Once among the world's most abundant birds, the last Passenger Pigeon died in 1914 in Cincinnati Zoo. Although heavily persecuted, the precise cause of the Passenger Pigeon's extinction remains a mystery*

*Chronology of extinction of birds on islands*

| | Date of extinction | | | | | | | | |
|---|---|---|---|---|---|---|---|---|---|
| Ocean | 1601–1650 | 1651–1700 | 1701–1750 | 1751–1800 | 1801–1850 | 1851–1900 | 1901–1950 | 1951–present | Totals |
| Indian | 2 | 2 | 4 | 4 | 4 | 5 | 0 | 0 | 21 |
| Atlantic | 0 | 1 | 1 | 4 | 0 | 12 | 9 | 0 | 27 |
| Pacific | 2 | 1 | 0 | 8 | 12 | 36 | 45 | 10 | 113 |
| Total | 4 | 4 | 5 | 16 | 16 | 53 | 54 | 10 | |

years. Island species often have small populations, because many islands themselves are small, and small populations can lose fewer individuals before becoming extinct. Island populations also are more vulnerable than mainland ones to catastrophes such as invasions of predators, disease outbreaks, or violent storms, because there is nowhere for the birds to flee to. These difficulties affect all island populations whether the catastrophes are natural or man-made.

One powerful and overlooked reason why so many more species have become extinct on islands than on continents, is that a population which becomes extinct on an island is much more likely to contain all the members of a species.

Islands are crucibles of evolution, containing far more unique species in proportion to their area than an equivalent area of continent. Although islands make up only about 5% of the earth's land area, about 20% of all bird species live only on islands, and nearly half the species currently regarded as threatened are confined to islands; thus an island species is four times more likely than a mainland species to be threatened with extinction.

The future hardly looks any better; Hawaii currently has 29 endangered species, the Mascarenes 11, New Zealand 16, and on the Antilles no less than 39 species and subspecies are endangered.

## Causes of past extinctions

### *Killing for food*

Many islands lost species through human interference even before people settled on them. Throughout the period of Europeans' exploration of the world by sea, when meat was pickled or salted, not frozen, islands which were on routes regularly travelled by sailing ships became important sources for re-supply of fresh produce.

Native wildlife was exploited for food – Great Auks were eliminated from Funk Island in this way – and when these indigenous food supplies ran out, sailors would stock islands with goats to assure their future supplies of fresh meat. Goats have proved notoriously destructive of native vegetation, on continents as well as on islands, and have seriously damaged the habitats of many islands. Thus the need for fresh meat led to bird extinctions not only directly – through killing them to eat – but also indirectly, through destruction of habitat by animals introduced to provide meat.

### *Introduced predators*

It was a habit among seamen to moor close enough to land to run a gangplank ashore. In this way, as well as through shipwrecks, rats colonized many oceanic islands in advance of permanent human settlement, and have proved devastatingly efficient as predators of island birds.

Many of the predators and competitors to which mainland birds are adapted, are mammals; but because birds can fly, they are much better than mammals at colonizing islands. Because of this difference in dispersal ability, bird communities on islands usually evolved without mammals, or at least among many fewer species of mammal, than continental bird communities. Consequently, island birds are unused to mammalian predators and competitors. So, when people help rats ashore, or settle on islands bringing with them domesticated mammals such as cats, dogs, and pigs, the indigenous birds have no defences against them. (Pigs are not usually considered predators, but have played this role with the eggs and young of ground-nesting birds on many islands.)

The most widespread predators are rats, which often become pests to people as well as to native birds and so precipitated the introduction of cats. Cats were not always successful in controlling rats, but were usually successful in destroying any native ground-nesting birds which the rats had left. Mongooses were also often introduced to control rats – and, on some West Indian islands, snakes – and have accounted for their share of extinctions of island birds. Compared to ground-nesting birds, tree-nesting birds have suffered less from introduced predators. Nevertheless, considerable losses have been inflicted by mammals that can climb – including ship rats and monkeys.

### *Other introductions*

Not only have mammals been introduced to new countries to control pests; so too have birds. Cattle Egrets and mynahs were introduced to Pacific and Indian Ocean islands to control insect pests; House Sparrows to control moths in Argentina, mosquitoes in Brazil, and dropworms in North America; and Barn Owls to combat rats in the Seychelle Islands; many of these have themselves become pests in their new countries, just as mammals have done.

One of the more insidious effects of introducing exotic birds has been hybridization. Some close relatives of native species, introduced by accident or design, have interbred with the native forms and hybridized them out of existence. Madagascar Turtle Doves accompanied early settlers from Mauritius to the Seychelles, where they crossed with the endemic subspecies of the same dove, which is now effectively extinct as a result. A similar fate befell the Boobook Owl on Lord Howe Island when a close relative was introduced from Australia.

People settling on remote islands have often taken with them familiar songbirds to remind them of home. As a result, many continental species – especially of European birds – have become established on distant oceanic islands; of nearly 1200 known introductions of birds, over 70% have been to islands. Two island groups – Hawaii and New Zealand – have been particularly seriously affected in this way, with 162 and 133 introductions respectively. These species have hundreds of years of adaptation to human habitats behind them, and successfully establish themselves in the settled areas of their new homes, thus denying native birds the opportunity to adapt to them. In New Zealand, the native forests have been altered by introduced herbivorous mammals and are now inhabited mainly by European species such as Dunnocks, Song Thrushes and Blackbirds, rather than by forest birds native to New Zealand.

The birds which were introduced to Hawaii were accompanied by mosquitoes which carried several diseases, including avian malaria. The incoming birds were immune to malaria, but the native birds were not; the mosquitoes spread through the lowland forest, to an altitude of about 600 m, and many of the native birds of the lowlands became extinct. Disease is extremely difficult to establish as a cause of a species' extinction, so it may have contributed to the extinction of many more species than those of which we are aware.

### *Habitat loss*

Most oceanic islands have lost large numbers of species within a very short time of being settled by people. The earliest losses were to introduced predators (including people); later, as human populations expanded and the species which were vulnerable to predation were wiped out, others fell victim to less obvious causes. Chief among these was loss of habitat. Birds are adapted to a particular habitat or range of habitats, and most species cannot maintain a viable population in other environments. This applies even to species which migrate over large distances; the habitats which they use along the way, and to breed and winter at either end of the migration route, are distinct and necessary to that species. When an area of habitat is lost, or irreversibly altered, the birds which it supported are doomed. They cannot simply 'move somewhere else' because other areas of suitable habitat will already be occupied by other individuals of the appropriate species; and in the case of islands, there is nowhere else to go anyway. As islands are settled, the coasts and lowlands are the first to be turned into agricultural and urban habitats; this is why these parts of islands are usually populated by species which are widely distributed throughout a geographic region. Native habitat survives longest in the interior and at high altitudes, which is where remnant populations of endemic and other native species survive longest. Habitats are changed directly – by conversion to agricultural use and settlements – and indirectly, by introducing herbivorous mammals which invade the forest. From a plant's point of view, large herbivores (which are usually mammals) are predators; island plant communities evolved without these predators just as island bird communities did, and proved just as defenceless. The forests of New Zealand and Hawaii, for example, have probably suffered as much damage from being eaten and eroded by introduced mammals as they have from clearing by people, and their capacity to support their original bird communities has declined accordingly.

Not all the species which have been extinguished lived on islands, and the causes of extinction on islands differ from those on continents in ways that reflect the differences in biology and history between islands and continents. There are, for example, no known cases of extinction of continental birds by introduced predators; however it is also true that introductions of predators on continents are unlikely to be documented as well as those on islands. North America was, as far as European civilization was concerned, essentially an enormous island when first 'discovered'; the record of bird extinctions there is in many ways reminiscent of the record on oceanic islands. Birds as a source of food played a significant part in the successful establishment of Europeans on the east coast, and in their relentless expansion westwards; there are graphic eye-witness descriptions of the carnage that was inflicted on spectacularly dense and extensive flocks of Passenger Pigeons and Eskimo Curlews barely a century ago. The last Passenger Pigeon died in captivity in 1914, and Eskimo Curlews have been balanced precariously on the knife-edge of oblivion for many decades.

However, while the toll inflicted by hunters on both these species was unquestionably enormous, we cannot be sure that hunting alone was responsible for their extinction, because at the same time that hunters were filling wagons with the bodies of pigeons and curlews, settlers were felling the vast stands of forest where the pigeons and parakeets nested. Passenger Pigeons bred in huge colonies, and may have needed a large nucleus of birds to maintain the viability of a colony; fragmentation of the forest might have played as central a part in their demise as hunting. Such habitat destruction by the remorseless expansion of human populations, has proved too much for a number of other species of continental birds, including the spectacularly beautiful Pink-headed Duck of Bangladesh and the Mountain Quail of the Himalayas.

## Current threats to birds

The various causes of extinction can all be attributed to the expansion of human populations into places where they had either not been before, or had lived at much lower densities. The relative importance of these causes has changed, and is continuing to change, but all are still operating, and some others which have not yet caused any extinctions are threatening to do so now. For convenience, threats are treated separately, but they rarely act in isolation and the most seriously threatened species are usually under attack from several causes at once.

### *Habitat loss*

Of the species currently at risk of extinction, about 60% are threatened by loss of their habitat. Rising human populations have ensured that increasing areas of the earth's surface have suffered habitat modifications of varying degrees of severity; in parts of the world which have been densely populated for many human generations – such as much of western Europe, many parts of North America, and most inhabited oceanic islands – little or no habitat remains unaffected by human activities. Probably the only natural habitats which are not declining in extent are deserts; all the others are shrinking or being seriously modified by man.

The most important habitat – in terms of the number of species which depend on it, the speed at which it is being destroyed, and the global impact of that destruction – is tropical broadleaved forest.

The area of tropical forest cut down each year would cover England once according to the most conservative estimates, and twice according to others. Since the tropical forest harbours more species than any other habitat on earth, the threats to this vegetation type are the most serious faced by birds as a whole. About 2500 species live in tropical forest, and 1500–2000 of these could probably live nowhere else; a substantial fraction of the world's avifauna is therefore at risk from the threats to this habitat. Between 40% and 50% of the world's 1000 or so threatened species are found in the tropical forest environment.

Left. The loss of habitat is one of the main reasons for extinctions. Deforestation is occurring throughout the world but is nowhere more damaging than in the tropics. Here tree-clearing takes place on the periphery of the Kerinci-Seblat National Park Rainforest in Sumatra

Wetlands are more limited in area than tropical forests, but are equally threatened. They are widely regarded as wastelands and sources of disease, and are being drained for agriculture, building land, and water control at an accelerating pace.

Much natural habitat is lost not to clearing, but through the impact of other forms of land exploitation. Mining operations account for much lost habitat, especially forest, and are a factor in the precarious situation of the endemic Kagu in New Caledonia and several species confined to Mount Nimba in West Africa. The construction of dams for generating hydro-electric power, controlling floods and providing irrigation, inundates vast areas of land up-stream and drastically changes riverine habitats downstream. Grasslands, heathlands and forests in many parts of the world, especially at higher altitudes in the tropics, are being converted into monocultures of imported fast-growing trees (especially conifers and eucalyptus); often these exotic plantations are more susceptible than the native vegetation to natural disasters such as hurricanes and disease, and a haven for indigenous birds is destroyed in return for little discernible benefit. The Blue Swallow (*Hirundo atrocaerulea*) of central and southern Africa has lost much of its habitat of montane grassland to such afforestation projects in the recent past and its range has shrunk dangerously as a result.

*Agriculture*

People have been clearing forest and grassland to grow food ever since agriculture was invented. In many countries of the developed world, the area of farmland has not increased greatly in recent years; but the intensity of land use has increased, and fewer birds can use these impoverished agricultural habitats. The former pattern of mixed farmland, with woodland copses, hedgerows and thickly vegetated roadside verges separating fields planted to a variety of crops, has been replaced by vast fields planted to one or two highly profitable species. There is no habitat for tree-nesting species, and even those that nest on the ground cannot breed successfully because of disturbance and destruction by heavy harvesting machinery.

Intensive application of pesticides on farmland not only kills birds, but removes a wide variety of insects, severely depleting the food supply for insectivorous birds. The problem was particularly severe between the 1950s and 1960s, the heyday of DDT and dieldrin. These are particularly persistent chemicals and their effects on birds are cumulative. Although farmers tended to apply them at fairly low concentrations, they accumulated through the food chain so that top predators like birds of prey were particularly affected. High doses of DDT will kill birds outright, but lower doses cause a number of effects such as sterility, embryo death and eggshell thinning. Thin-shelled eggs are accidentally broken by the adults during

Shell thickness index of British Sparrowhawks (Accipiter nisus), 1870–1980. In 1947 shells abruptly became thinner, coinciding with the widespread introduction of DDT as an agricultural pesticide. Each dot represents the mean shell index of a clutch; nearly 2000 clutches are represented from all regions of Britain. The shell index is calculated as shell weight (mg)/ length × breadth (mm)

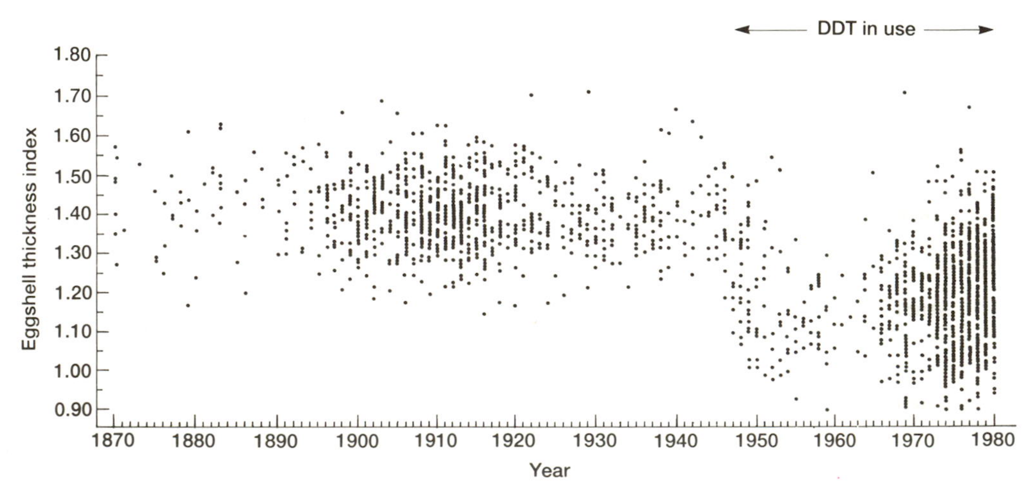

incubation and continued breeding failure eventually results in a population decline. The amount of DDT in a bird is closely correlated with the extent of eggshell thinning, and therefore provides a good indicator of the level of environmental contamination. DDT caused widespread reductions in several raptor populations, but particularly among species that fed on birds or fish, like the Peregrine Falcon, Sparrowhawk and Osprey. These species lived at the top of a longer food chain than mammal eaters and therefore accumulated more pesticide.

Today these chemicals are either banned or applied much more sparingly in the developed world, but they are still widely used in many parts of the tropics. They have partly been replaced by chemicals which are less persistent, and often less directly toxic to birds, but which can kill enough insects to deplete the food chain seriously. Such chemicals may not kill birds outright, but they surely cause them to die.

Modern agricultural practice involves heavy applications of chemical fertilizers which may inflict long-term damage on natural soil fertility before being washed into streams and lakes. Here they often stimulate the growth of algae, which reduces the oxygen content of the water and can kill much of the original aquatic life. This process of 'eutrophication' causes very widespread damage to wetland habitats in agricultural areas.

Habitats can cease to support birds even if they are neither cleared nor chemically contaminated; human activity alone can make some habitats unsuitable for breeding. Beaches, for example, are the breeding habitat of Least Terns and Piping Plovers in North America, but are so heavily disturbed by tourists and holiday-makers in the summer that the number of breeding sites suitable for these species is shrinking year by year.

The deliberate or accidental dumping of fuel or crude oil at sea has killed many thousands of marine birds, particularly auks and penguins. If ingested this oil is toxic, and if it contaminates the feathers the birds lose their waterproofing and die of hypothermia. Efforts have been made to clean badly oiled birds but with limited success.

*Each year thousands of marine birds like this Western Grebe (Aechmophorus occidentalis) die lingering deaths because of oil pollution. Both single major incidents, such as the Exxon Valdez in Alaska, and smaller, less obvious occurences which result in chronic pollution, can have devastating effects on bird populations*

### *Hunting*

People kill birds for food, for sport, and to protect crops. Some species – such as the Great Auk – were, quite clearly, made extinct by being killed for food, and others – including a number of species of parrot on islands from the Caribbean to the Indian Ocean – through being destroyed as crop pests. No species – with the possible exception of the Pink-headed Duck – is known to have been wiped out by sport hunting, and hunters are often a major force for habitat conservation. Hunting for food is not controlled or regulated nearly as strictly as hunting for sport, and poses a more serious threat. In several Mediterranean countries, songbirds are shot and trapped indiscriminately for food, and the practice may threaten a number of species, especially those under pressure from other factors as well.

A side-effect of sport hunting may be more dangerous than the hunting itself. Bird shot is made of lead pellets, which are highly toxic, and accumulate in the mud at the bottom of marshes where hunting is carried out. Lead shot from anglers' weights has killed a great many Mute Swans in southern Britain, and it has been estimated that as many waterfowl in the United States of America die from lead poisoning as are shot. The USA has moved to replace lead shot with steel, which is less toxic, but this applies only to waterfowl; and the lead already in the environment will continue to take its toll of waterfowl for generations to come.

A further side-effect of hunting, this time for fish, has been the accidental capture and death of thousands of seabirds. The use of monofilament drift nets to catch Atlantic salmon and cod during the 1960s off west Greenland and eastern Canada resulted in the drowning of many guillemots. At one stage this fishery was catching more birds than fish; an estimated 500 000 per year for several years. Similar numbers of birds, this time mostly shearwaters, have been killed in the North Pacific.

### *The cage-bird trade*

People love to keep birds in cages. This need is a symptom of deep aesthetic and spiritual bonds between birds and mankind, which are also manifested in a burgeoning world-wide interest in birds and their conservation, and probably led to your reading this book. A vividly-coloured songbird in a cage brightens many an urban living-room, and supports a thriving trade in live birds. While the trade is thriving, the birds are not; for every healthy bird in a living-room cage, tens die in capture, transport and quarantine. Large but uncounted numbers die in the home country in the process of capture, and many more in transport; about four per cent of those arriving at Heathrow Airport in London are dead on arrival, and similar or larger numbers die in quarantine.

Europeans and Americans keep birds in cages for their song, colour and companionship; the high prices paid for some species, such as tens of thousands of dollars for the rarer macaws, suggests less worthy motives such as the need to signal wealth and status. In other cultures, birds are often caged for different or additional reasons; in some oriental countries religion plays a part in bird-catching, and in South America, parts of the Caribbean and Far East small birds are kept chiefly for their songs and there are keen competitions among owners for the best songster.

The numbers handled in this trade are staggering. Over 7 million birds were transported between countries in 1975, 80% of them songbirds; the USA, Japan and western Europe imported more than half of these. International trade accounts for only a fraction of the total trade, since much takes place within national boundaries; the total trade world-wide involves probably at least 20 million birds a year. It seems certain that far more die as a result of this trade than are killed by hunters. More than one species in five of all the world's birds are involved in the cage-bird trade, which is the major threat faced by about 6% of all threatened species.

The capture and supply of birds for the captive trade is a large industry that has led to the over-exploitation of certain species. The resultant rarity of some of them, such as the Bali Starling and some parrots, can enhance their market value, exacerbating the problem and raising their possession to that of a status symbol. In a few cases, attempts are made to increase production of potential captives by 'farming' them. Hill Mynahs, renowned for their ability to mimic human speech, are encouraged to breed in baskets erected as nest boxes so that the chicks can be easily harvested at the appropriate time.

It must be remembered, however, that captive populations can have value. Aviculturists and scientists are able to describe aspects of behaviour that are difficult to study in the wild, and the re-establishment of a free-living population of the formerly endangered Néné or Hawaiian Goose illustrates that at least some captive populations may have potential in terms of species conservation.

*The illegal traffic in cage-birds is a major threat to certain populations. The conditions in which birds are transported are usually atrocious, as is the case with these Senegal Parrots (Poicephalus senegalus) photographed at Dakar airport en route to West Germany*

*The ultimate threat to birds – human populations*

Most birds of whose extinction we have any record were exterminated as a result of a human population expanding into places where no people – or very few – had lived before. The invaders destroyed habitats, killed birds for food or to protect crops, and introduced predators, competitors and diseases to which the native birds were not adapted.

Human populations continue to spread into new places, and to increase exponentially in places which are already settled. The processes of extinction accelerate in direct relation to the expansion of human populations. These are swelling fastest in the tropics, where people are invading tropical forests and other habitats at speeds which put at risk the largest number of bird species ever threatened in the course of human history. Human populations are much more stable, and have already caused the kind of devastation to other natural habitats which is about to be visited on the tropics; but industrial societies are indirectly responsible for much of the damage to tropical environments through international trade. Tropical forests are logged for timber, which is exported to Europe, North America, and Japan. They are also cleared for agriculture. This may be to grow products exported to industrial countries, especially food crops – including beef, sugar and bananas, and drugs, particularly tea and coffee.

The destination of much of the beef raised in Central and South America is North America. Alternatively, subsequent agriculture may produce subsistence crops for peasant farmers who cannot grow food for their families on the best agricultural land because that is already allocated to growing cash crops for export. In both cases, the loss of forest is attributable to the demands of distant industrial societies.

Developed economies are also responsible for a growing danger to birds and all life, whose extent is only beginning to be appreciated. This is the prospect of significant changes to global climates, caused by rising world temperatures as the 'greenhouse effect' takes hold of the atmosphere. These changes arise from carbon dioxide, methane and other gases produced by burning fossil fuels in vehicles, power stations and factories – and by burning forests. The extent of likely climate changes is not clear, but it is doubtful if any habitat will escape its effects, and there could well be significant changes in the distribution of many of the world's major habitats. The ecology and distribution of breeding grounds, migration routes, and wintering grounds, will all change, probably faster than at any time in the earth's history, and these new relationships will test the adaptability of birds throughout the world.

## Conservation measures

*Protective legislation and international action*

The first approach to bird conservation was to enact laws which protected individual species and groups of species from hunting and other forms of exploitation. In Britain, birds and other wildlife have traditionally been treated in law as the property of the landowner; accordingly, early bird-protection laws were designed to confirm the landowner's exclusive right to hunt game on his land. This concept of birds as private property remained the basis of conservation legislation in Britain until 1954, when the Wild Birds Protection Act protected all birds and their eggs, with exceptions for hunting, agriculture and research.

The concept of birds as a common heritage had been introduced 40 years before in the Netherlands, as well as in the first international conservation legislation – the Migratory Birds Convention of 1916 between the USA and Canada. This convention was enlightened for its time, especially in recognizing the need for international action to protect species which migrate across national boundaries, but still reflects two concepts which now seem archaic; the divisions between game and non-game birds, and between 'harmful' and 'useful' birds. The gamebird concept has dominated the implementation of the convention in the signatory nations, which have traditionally concentrated their activities on waterfowl, and especially the regulation of hunting. The harmful/useful concept is reflected in the exclusion of raptors and most fish-eating birds from

*Below and overleaf. The logos of various ornithological societies across the world: the American Ornithologists' Union (AOU), the International Council for Bird Preservation (ICBP), the Royal Society for the Protection of Birds (RSPB), the British Ornithologists' Union (BOU), the Royal Australasian Ornithologists Union (RAOU), the National Audubon Society (NAS), and the British Trust for Ornithology (BTO)*

the convention, with the result that these groups, together with non-migratory gamebirds (mainly species of grouse), are the responsibility of state (US) or provincial (Canada) governments, while those covered by the convention are the responsibility of the federal (national) governments.

The most important recent international legislation is the Convention on Trade in Endangered Species of Fauna and Flora (CITES), signed by 10 nations in 1975 and over 90 by 1986. Unfortunately not all signatories exercise adequate policing of the convention. Other important international agreements to protect birds include the Council of Europe Convention on the Conservation of European Wildlife and Natural Habitats (the 'Berne Convention'), and the Convention on the Conservation of Migratory Species of Wild Animals (the 'Bonn Convention').

*Education and public action*

Legislation is ineffective unless it is enforced, and wildlife legislation is generally not a high priority for police forces. The enforcement of bird conservation legislation is weak everywhere, making effective education doubly important. Education has been stimulated by voluntary societies such as the Royal Society for the Protection of Birds (in Britain), and the National Audubon Society and its state affiliates in the USA. These bodies have undertaken a great deal of conservation work, including public education, in their own countries, and have also contributed to international efforts through their support of the International Council for Bird Preservation (ICBP) which was formed in 1922. Much of ICBP's early interest was in waterfowl and led to the formation of the International Waterfowl Research Bureau (IWRB); ICBP and IWRB now act as specialist ornithological advisers to the International Union for the Conservation of Nature and Natural Resources (IUCN) and the World-Wide Fund for Nature (WWF), which were formed in 1948 and 1961 respectively. One of IUCN's most important activities is the production of Red Data Books, listing endangered species, of which the bird volumes are produced by ICBP. IUCN, WWF and the United Nations Environment Programme cooperated to produce a World Conservation Strategy in 1980; this document has had a profound effect on conservation in all those countries which have followed it in preparing their own conservation strategies.

*Reserves and habitat protection*

Many countries have established nature reserves and national parks to protect wildlife habitat. Reserves of international importance have been set up under the Convention on Wetlands of International Importance, signed in Ramsar, Iran, in 1971, and under the Berne Convention and the European Community Bird Directive. One reserve – Cousin Island, in the Republic of Seychelles, Indian Ocean – is owned internationally, by the International Council for Bird Preservation.

In North America, much breeding habitat for waterfowl is scheduled to be protected or restored under the North American Waterfowl Management Plan signed by Canada and the USA in 1987; and a private organization funded by hunters, Ducks Unlimited, has restored, preserved or created impressive areas of wetland.

Mere protection of habitats is rarely sufficient to maintain their value for birds; active management is often necessary, and indeed can often create good habitat from unpromising land; the RSPB's Minsmere Reserve in eastern England is an outstanding example. Management can include maintenance of vegetation at the required stage of succession by cutting, burning, or grazing, controlling water levels, and providing artificial nest sites. This latter has been particularly successful for waterfowl and can be used effectively even in habitats which are not protected.

Management of protected areas may also have to include control of predators, particularly on islands where the predators have been introduced by man.

*Captive breeding*

Some species are so extremely rare or gravely threatened in the wild that their chances of survival depend on the maintenance of a breeding stock in captivity. Not only can the

captive individuals be effectively protected, but if they can be bred successfully, the population may be increased sufficiently to restock the wild range. This strategy depends for its success both on the captive breeding, and on the effective removal of the threats operating in the wild; clearly the habitat must be restored to a condition in which it can support the birds before they are released back into it. Captive breeding can buy valuable time for habitat to be restored.

Captive breeding has played – or promises to play – a critical part in the survival of a variety of species, including the Hawaiian Goose or Néné, the Crested Ibis of China and Japan, and the Pink Pigeon of Mauritius. The technique has become the last resort for the recovery of the Californian Condor of the United States, of which the remaining 17 birds are all in captivity awaiting the adequate protection and management of sufficient areas of their rangeland habitat.

### *Fostering*

A variant of captive breeding involves removing eggs from wild birds and raising them under foster parents of a closely related wild species. This technique has been used with varying success for Whooping Cranes in North America (raised by Sandhill Cranes), New Zealand Black Stilts (raised by Pied Stilts), and Peregrine Falcons (raised by Prairie Falcons) in North America. This method is most appropriate in species which will either re-lay to replace lost eggs or, like Whooping Cranes, lay more eggs than they raise young; in these species, the young which are raised by foster parents represented an increased production over and above that produced by the parents on their own. There is apparently no problem in these species of the young imprinting on their foster parents, as is known to occur in some others, such as Zebra Finches reared by Bengalese Finches.

### *Translocation*

Species successfully bred in captivity can be reintroduced to their original habitats. In other cases where the problem lies in the original habitat, the birds must be moved elsewhere for their own protection. Vegetation changes and introduced predators on small islands off New Zealand have reduced populations of some small endemic songbirds to critical levels; three species, Black Robins, South Island Robins, and Saddlebacks, have been successfully moved to islands with more suitable vegetation and no predators.

These intensive techniques for restoring and preserving critically threatened species, are last-ditch measures. They are equivalent to fighting fires; in conservation, as with fire fighting, preventive measures are preferable where possible.

Bird conservation crises are often regarded as biological problems, with biological solutions; but in most cases, the bird's status is a biological symptom of a problem arising from human misuse of the natural environment. The use of canaries to warn coal-miners of poison gas is thus a graphic metaphor for the value of threatened birds as indicators of mankind's misuse of the planet. To that extent, the threatened status of 11% of the world's bird fauna is a chilling measure of the environmental status of the planet.

AWD

*Conservation at its most expensive. The attempt to save the California Condor (Gymnogyps californianus) from extinction has been among the most costly for any species. The attempt has included the hand-rearing of chicks, using a condor glove-puppet to ensure that the young bird imprints on a condor and not its human foster parent*

## RADIO TRACKING – METHODOLOGY AND RESEARCH APPLICATIONS

Radio tracking is a research technique for finding a tagged bird at will, without having to see, hear or trap it, and with minimal disturbance to the bird and its habitat. Regular tracking can reveal home ranges and habitat preferences, or let the researcher make systematic observations of a bird's foraging and social behaviour or diet. For shy or nocturnal species, especially in forests and marshes, short-term radio tracking is often the most practical way to find nests, roosts and feeding sites, whereas long-term studies provide data on causes of mortality and dispersal of young.

Radio tags which contain circuitry to vary pulse rate are used to indicate activity. For example, by listening to the signal from an activity-sensing tag on the tail of a goshawk, a researcher can distinguish between feeding, flying, perching, incubation and death. In Swedish studies of goshawk predation, the activities of tagged hawks were monitored by radio signal alone. When the signal indicated that a bird was feeding, the researchers radio-located it and identified the prey. Home ranges were mapped and habitat preferences determined. Knowledge of which species were prey, and where and how often they were caught, indicated why particular habitats were favoured, and what impact goshawk predation might have on prey populations.

A radio tag must not affect the behaviour or survival of its bearer. The main consideration is the tag weight, expressed as a proportion of the bird's body weight. As a general rule, large birds should not be loaded with more than about 3% of body weight, although the smallest species can safely carry up to 10%. The radio tag may be attached with a harness or neckband, glued to back feathers, fixed to the bases of tail feathers or secured by a strap to one leg. Radios have been mounted as wing tags on condors.

A radio tag contains a transmitter, usually crystal-controlled, which typically emits a 30 millisecond pulse every second. The transmitter is powered

*Radio tracking has greatly enhanced our ability to follow individual birds and determine their activity budgets. Here an Antarctic Blue-eyed Cormorant (Phalacrocorax atriceps) carries a tracking device on its back*

by a small battery or sometimes by solar cells, which generally contribute most to the weight of the tag. A short, flexible 'whip' antenna broadcasts the signal. A tagged bird is recognized by the radio frequency of its transmitter, usually on an authorized band of up to 2 MHz within the range 142–174 MHz. The smallest tags weigh slightly less than one gram and last for five to seven days with a range of several hundred metres. In contrast, a radio tag for a large raptor could weigh 60–90 g and last two to three years, giving a range of 5–10 km on the ground and more than 50 km in the air.

For tracking, a hand held antenna (usually a directional '3-element Yagi' or 'H' antenna) is used to collect the tag's signal, which is amplified and converted into audible 'bleeps' by a special receiver. The loudest signal is heard when the antenna is pointing towards the radio tag. At the start of a tracking session, the receiver is tuned to the tag frequency and the researcher listens for the signal while sweeping the antenna slowly through a full circle. If the tag is within range, the direction of the loudest signal is recorded as a compass bearing, and the procedure is then repeated from one or more other sites. When the bearings are plotted as lines on a map, their intersection shows the tag's position.

Signal range is extended by increasing receiver antenna height, so if a tagged bird cannot be found by a ground-based search, an aircraft or antenna mast (static or vehicle-mounted) may be needed. For birds which travel great distances, for instance during migration, the 'Doppler' principle can be used to estimate tag position from a satellite. A relatively powerful tag with sophisticated frequency stabilization is required. At present the smallest such tags weigh 150 grams, so the technique is confined to studying the wanderings of albatrosses, eagles and other large birds.

BC

# *Ornithological organizations*

Ornithology has at its roots amateur naturalists – the collectors who in the eighteenth and nineteenth century enthusiastically provided material for bird taxonomists, and the observers who studied bird behaviour and patterns of distribution. The pioneers of the science were mostly field ornithologists. They operated very much as individuals, perhaps in correspondence with their contemporaries but usually working alone. Amateur interest expanded in the mid-nineteenth century at which time sufficient impetus was achieved, especially in Europe and North America, for individuals to group together into specialist societies.

The purpose and structure of these societies varied widely. The Society for the Protection of Birds (founded 1889; now the Royal Society for the Protection of Birds – RSPB) in Britain and the Massachusetts Audubon Society (founded 1896) in the United States were both formed as pressure groups to halt the international trade and use of bird plumes in the millinery industry. Contemporary advances in the academic study of birds and development of ornithology as a science were promoted through early scientific societies such as the Deutsche Ornithologen-Gesellschaft (founded 1853) and the British Ornithologists' Union (founded 1859). During the middle years of the present century a greater specialization evolved, and societies focusing on a particular family or order were established, such as the Severn Wildfowl Trust (founded 1948; now known as the Wildfowl and Wetlands Trust) in Britain and, a little later, the Raptor Research Foundation (founded 1966) in the United States of America.

The post World War II period witnessed an expansion in number of bird conservation societies, such as the Ornithological Society of Malta (founded 1962) and Hellenic Ornithological Society (founded 1982) in Greece, and in the membership of established societies (RSPB membership increased from 5,800 in 1945 to 390,000 in 1985). A more recent trend, notably in the late 1980s, has been for membership-based societies – both those which started as more or less pure bird protection societies and those of a more scientific nature – to move towards bird conservation or towards promoting a broader protection of the natural environment. The Spanish Ornithological Society – Sociedad Española de Ornitologia – has expanded from a scientific organization with a largely academic membership to add a new dimension to its activities and membership through high-profile lobbying for bird conservation. Similarly the German Federation for Bird Protection – Deutscher Bund für Vogelschutz – in 1990 became the German Nature Protection Federation – Naturschutzbund Deutschland – reflecting the perceived need to cater for a better-informed public which increasingly recognizes that bird protection is only a part of the conservation of wildlife and its habitat.

The earliest organizations established to promote the study and understanding of birds tended to publish journals and other works to disseminate the expanding reservoir of ornithological knowledge; many continue today to publish papers at the frontiers of ornithology – the products of investigations by a growing number of professional scientists in this field. Equally, many of the long-established bird protection societies are now leaders in the modern environmental conservation movement; they devote considerable resources to enable the general public to encounter and enjoy birds, and they engage in practical conservation work, frequently aimed at the protection of threatened species and their habitats.

There are so many international, regional and national associations and societies covering all aspects of ornithology that the task of teasing out a representative selection is daunting. In compiling the following list an attempt has been made to identify scientific associations which publish key ornithological journals, particularly those which advance the science through harnessing the interest and motivation of the enthusiastic amateur. To complement these, conservation organizations which encourage public participation in less scientific study and appreciation of birds are also included.

What follows therefore is not a directory; it is a selective list of independent, and mostly non-profit or charitable, associations which seek to advance knowledge in ornithology or the conservation of birds. More information on ornithological activities – from specialist research to birdwatching locations – in Europe, North America, South Africa, Australia and New Zealand can be obtained from the associations and publications listed. The focus on these areas reflects the basic distribution of well-established ornithological associations (though there are, of course, exceptions like the Bombay Natural History Society, established in 1886) and the present concentration of active professional and amateur ornithologists.

The list is organized by country, with key international organizations identified separately at the end. At the risk of an over-simple categorization of the various societies, those of a more scientific nature include a contact address and any key serial publication they produce; those societies active in bird conservation have telephone and facsimile numbers. Inevitably, some organizations fall into both categories, a fact reflected in the presentation of the information.

IH

**AUSTRALIA**

Royal Australasian Ornithologists Union (RAOU)
21 Gladstone Street
Moonee
Victoria 3039

Tel (03) 370 1422
Fax (03) 370 9194

Publishes: *Emu* (1901–)

**AUSTRIA**

Österreichische Gesellschaft für Vogelkunde
Naturhistorisches Museum
Burgring 7, Postfach 417
A-1014 Wien

Publishes: *Egretta* (1958–)

**BELGIUM**

Institut Royal des Sciences Naturelles de Belgique (IRSNB)
Rue Vautier, 31
1040 Brussels

Publishes: *Le Gerfaut/De Giervalk* (1919–)

Ligue Royale Belge pour la Protection des Oiseaux (LRBPO)
Rue de Veeweyde, 43
1070 Brussels

Tel (02) 521 28 50
Fax (02) 527 09 89

Reserves Naturelles et Ornithologique de Belgique (RNOB)
Rue Royal Ste-Marie, 105
1030 Brussels

Tel (02) 245 55 00
Fax (02) 245 39 33

Société d'Etudes Ornithologiques
Secretary:
Rue de la Cambre 16, Bte 2
1200 Brussels

Publishes: *Aves* (1964–)

**CANADA**

Canadian Nature Federation
453 Sussex Drive
Ottawa K1N 6Z4

Tel (613) 238-6154
Fax (613) 230-2054

For a comprehensive list of North American national and provincial associations, see the National Wildlife Federation's annual Conservation Directory, available from NWF, 8925 Leesburg Place, Vienna, Virginia 22184, USA

Tel (703) 790-4000
Fax (703) 442-7332

**CYPRUS**

Cyprus Ornithological Society
Secretary:
Kanaris Street 4
Strovolos 105

**DENMARK**

Dansk Ornitologisk Forening
Vesterbrogade 140
DK-1620 København V

Tel (31) 31 44 04
Fax (31) 24 75 99

Publishes: *Dansk Ornitologisk Forenings Tidsskrift* (1906–)

**FINLAND**

Ornitologiska Föreningen
P. Rautatiekatu 13
SF-00100 Helsinki

Tel (90) 191 7388

Publishes: *Ornis Fennica* (1924–)

**FRANCE**

Ligue Française pour la Protection des Oiseaux (LPO)
La Corderie Royale
BP263
F-17315 Rochefort

Tel (46) 99 59 97
Fax (46) 83 95 86

Société d'Etudes Ornithologiques
MNHN Laboratoire d'Ecologie
4 avenue du Petit Château
F-91800 Brunoy

Publishes: *Alauda* (1929–)

Société Ornithologique de France
Muséum National d'Histoire Naturelle
55 rue de Buffon
F-75005 Paris

Publishes: *L'Oiseau et la Revue Française d'Ornithologie* (1929–; formerly *Oiseaux, Paris*, 1920–28)

**GERMANY**

Deutsche Ornithologen-Gesellschaft
Hardenbergplatz 8
1000 Berlin 30

Publishes: *Journal für Ornithologie* (1853–)

Naturschutzbund Deutschland
Am Michaelshof 8–10
5300 Bonn 2

Tel 0228-358031
Fax 0228-358036

**GIBRALTAR**

Gibraltar Ornithological and Natural History Society
The Gibraltar Museum
18–20 Bomb House Lane
Gibraltar

Tel 74289

**GREECE**

Hellenic Ornithological Society (HOS)
PO Box 64052
GR-15701 Zographos

Tel (1) 522 55 06
Fax (1) 362 93 38

**HUNGARY**

Hungarian Ornithological and Nature Conservation Society
Költő u. 21
1121 Budapest

Tel 156-2133, 156-2927

Instituti Ornithologici Hungarici
Költő u. 21
1121 Budapest

Publishes: *Aquila* (1894–)

**ICELAND**

Museum of Natural History
PO Box 5320
125 Reykjavik

Publishes: *Bliki* (1982–)

**IRELAND**

Irish Wildbird Conservancy (IWC)
Ruttledge House
8 Longford Place
Monkstown
Co. Dublin

Tel (01) 280 4322
Fax (01) 284 4407

Publishes: *Irish Birds* (1977–)

**ITALY**

Centro Italiano Studi Ornitologici
Secretary:
Museo di Storia Naturale della Lunigiana
Fortezza della Brunella
Aulla

Publishes: *Avocetta* (1978–)

Lega Italiana Protezione Uccelli (LIPU)
Vicolo San Tiburzio 5ª
43100 Parma

Tel (0521) 233413
Fax (0521) 287116

Stazione Ornitologica Lombarda
Dipartimento di Biologia
Università degli Studi di Milano
20129 Milano

Publishes: *Sitta* (1987–)

**LUXEMBURG**

Lëtzebuerger Natur- a Vulleschutzliga (LNVL)
BP 709
2017 Luxemburg

Tel 47 23 69
Fax 47 47 27

**MALTA**

Ornithological Society of Malta (MOS)
PO Box 498
Valletta
Malta

Tel 230684

**NETHERLANDS**

Nederlandse Ornithologische Unie (NOU)
Rikjsinstituut voor Natuurbeheer
Postbus 9201
6800-HB Arnhem

Publishes: *Ardea* (1912–) and *Limosa* (1937–; formerly *Jaarbericht der Club van Nederlandsche Vogelkundigen*, 1917–28 and *Orgaan der Club van Nederlandsche Vogelkundigen*, 1928–36)

Nederlandse Vereniging tot Bescherming van Vogels (NVBV)
Driebergseweg 16c
3708-JB Zeist

Tel (03404) 25406
Fax (03404) 18844

## NEW ZEALAND

Ornithological Society of New Zealand
Secretary:
PO Box 316
Drury
South Auckland

Publishes: *Notornis* (1950–; formerly *New Zealand Bird Notes*, 1941–42 and *Bulletin of the Ornithological Society of New Zealand*, 1943–50)

Royal Forest and Bird Protection Society of New Zealand
PO Box 631
Wellington

Tel (04) 728 154

## NORWAY

Norsk Ornitologisk Forening (NOF)
PO Box 2207
7001 Trondheim

Tel (07) 52 51 42
Fax (07) 59 22 77

Publishes: *Fauna Norvegica, Ser. C, Cinclus* (1979–; formerly *Cinclus*, 1978)

## PORTUGAL

Liga Para a Proteccao da Natureza (LPN)
Estrada do Calhariz de Benfica 187
1500 Lisboa

Tel (01) 78 00 97

## SOUTH AFRICA

Southern African Ornithological Society (SAOS)
PO Box 87234
Houghton
Johannesburg 2041

Tel (011) 782 1547

Publishes: *Ostrich* (1930–)

## SPAIN

Sociedad Española de Ornitologia (SEO)
Facultad de Biologia
Pl. 9
Ciudad Universitaria
28040 Madrid

Tel (01) 549 3554

Publishes: *Ardeola* (1954–)

## SWEDEN

Sveriges Ornitologiska Förening (SOF)
Box 14219
104 40 Stockholm

Tel (08) 662 64 34

Publishes: *Vår Fågelvärld* (1942–)

## SWITZERLAND

Schweizer Vogelschutz (SVS/ASPO)
PO Box 8521
8036 Zürich

Tel (01) 463 72 71

Schweizerische Gesellschaft für Vogelkunde und Vogelschutz
Secretary:
Krähenbergstraße, 53
2543 Lengnau

Publishes: *Der Ornithologische Beobachter* (1902–)

Société Romande pour l'Étude et la Protection des Oiseaux
PO Box 54
1197 Prangins

Publishes: *Nos Oiseaux* (1947–)

## TURKEY

The Society for the Protection of Nature
PK 18
80810 Bebek-Istanbul

Tel (1) 1790139

## UK

British Ornithologists' Union (BOU)
Secretary:
Natural History Museum
Sub-department of Ornithology
Tring
Hertfordshire
HP23 6AP

Publishes: *Ibis* (1859–)

British Trust for Ornithology (BTO)
The Nunnery
Nunnery Place
Thetford
Norfolk
IP24 2PU

Tel (0842) 750050
Fax (0842) 750030

Publishes: *Bird Study* (1954–; formerly *Bulletin of the British Trust for Ornithology*, 1934–54) and *Ringing and Migration* (1975–; formerly annual *Report on Bird-ringing*, supplement to the *Bulletin* and *Bird Study*, 1937–74)

Royal Society for the Protection of Birds (RSPB)
The Lodge
Sandy
Bedfordshire
SG19 2DL

Tel (0767) 680551
Fax (0767) 692365

Publishes: *RSPB Conservation Review* (1987–)

Wildfowl and Wetlands Trust
Slimbridge
Gloucestershire
GL2 7BT

Tel (0453) 890333
Fax (0453) 890827

Publishes: *Wildfowl* (1968–; formerly *Report of the Severn Wildfowl Trust*, 1948–52, and *Report of the Wildfowl Trust*, 1953–66)

## USA

National Audubon Society (NAS)
950 Third Avenue
New York
NY 10022

Tel (212) 832-3200
Fax (212) 593-6254

Ornithological Societies of North America
PO Box 1897
Lawrence
Kansas 66044-8897

This is a general mailing address for four key American ornithological organizations; they are listed here with their associated publications:

American Ornithologists' Union

Publishes: *Auk* (1884–; formerly *Bulletin of the Nuttall Ornithological Club*, 1876–83) and *Ornithological Monographs* (1964–)

Association of Field Ornithologists

Publishes: *Journal of Field Ornithology* (1979–; formerly *Bird Banding*, 1930–78)

Cooper Ornithological Society

Publishes: *Condor* (1900–; formerly *Bulletin of the Cooper Ornithological Club*, 1899) and *Studies in Avian Biology* (1978–; formerly *Pacific Coast Avifauna*, 1900–74)

Wilson Ornithological Society

Publishes: *Wilson Bulletin* (1902–)

For a comprehensive list of North American national and state associations, see the National Wildlife Federation's annual Conservation Directory, available from NWF, 8925 Leesburg Place, Vienna, Virginia 22184

Tel (703) 790-4000
Fax (703) 442-7332

## INTERNATIONAL ORGANIZATIONS

International Council for Bird Preservation (ICBP)
Secretariat:
32 Cambridge Road
Girton
Cambridge CB3 0PJ
UK

Tel (0223) 277318
Fax (0223) 277200

International Waterfowl and Wetlands Research Bureau (IWRB)
Secretariat:
Slimbridge
Gloucestershire GL2 7BX
UK

Tel (0453) 890697
Fax (0453) 890827

Scandinavian Ornithologists' Union
Secretary:
Department of Animal Ecology
Ecology Building
S-223 62 Lund
Sweden

Publishes: *Ornis Scandinavica* (1970–)

**aberrant plumage** Abnormality in plumage colour (the result of changes in the amount and distribution of pigments) or changes in feather structure which may be caused by disease or poor diet, or which may have a genetic basis.

**agonistic behaviour** A combination of threat, attack and withdrawal postures and actions, used as an alternative to overt aggression in an attempt to repel or intimidate an opponent of the same species.

**albinism** Aberrant plumage caused by an absence of pigment, resulting in, for example, a white blackbird.

**allopatric** Species or populations occupying separate geographical areas.

**allopatric speciation** Differentiation of populations of a single species that occupy separate geographical areas. The process of speciation (the creation of new species) is complete when the separated groups can no longer interbreed.

**allopreening** The preening of one individual by another.

**altricial** Hatching of chicks as very dependent young – usually blind and naked – so they need extensive parental care. See also semi-altricial.

**anting** Behaviour of passerines in which the bird squats on the ground and ants are allowed and encouraged to move freely through the plumage and over the skin. The reason for this behaviour is not known, but it may help maintain healthy feathers and skin.

**antiphonal singing** A simple alternating system in which the male produces one element of a song and the female another, as in the African bou-bou shrike *Laniarius erythrogaster*.

**Australian realm** The most arid of the six biogeographic realms of the Earth, containing nearly 1600 species, over 1000 of which are endemic to the Australian plate and neighbouring islands. The realm includes Australia, New Zealand, Papua New Guinea and adjacent islands in the East Indies and Polynesia. See also Ethiopian realm.

**baroreception** Ability to detect changes in air pressure, proved by experiment to be present in pigeons.

**Beau Geste hypothesis** The idea that male songbirds may deceive others of their species by singing in such a way as to simulate the presence of many birds in their territory, so discouraging entry by other territory-seeking males.

**brood parasitism** The habit of laying eggs in the nests of other birds, often of a different species, effectively forcing the host parents to raise the chick from the introduced egg. Usually the term refers to the regular use of another host species to raise young, as with the European Cuckoo (*Cuculus canorus*) laying eggs in the nests of Meadow Pipits (*Anthus pratensis*), Reed Warblers (*Acrocephalus scirpaceus*), or Dunnocks (*Prunella modularis*). Such behaviour is distinct from the occasional egg-dumping or transfer that may occur among, for example, Cliff Swallows (*Petrochelidon pyrrhonota*). An *obligate brood parasite* is a bird like the European Cuckoo wholly dependent on birds of other species to raise its young.

**brood patch** An area of ventral skin that loses most or all of its feathers during the breeding season and becomes engorged with blood vessels for the purpose of incubating eggs.

**brood reduction** The loss of late chicks from a brood as a result of inter-chick competition, especially when food is scarce. The pattern is characteristic of altricial species in which there is a gap of several days between the hatching of the first and last eggs, an interval induced by starting incubation with the laying of the first egg.

**character displacement** The observation that competition may cause two related species to be more different at sites where they live side by side than at sites where one species lives in the absence of the other.

**cline** The graded sequence of differences, created by gene flow, in a single species spread over a large geographical area: for example, the increasing darkness of plumage observed in many European bird species as one moves from east to west.

**cooperative breeding** Breeding behaviour in which a pair of birds is assisted in rearing their young by one or more helpers, often related to them.

**conspecific** Considered to belong to the same species; can include taxa having enough differences to justify division into subspecies.

**convergent evolution** The independent evolution of similar characteristics or adaptation in unrelated species, for instance flightlessness in penguins and the now extinct Great Auk (*Pinguinis impennis*).

**courtship feeding** Feeding of one member of a breeding pair by its mate, as with a male bringing food to the female in the days before egg laying.

**density-dependent factor** Any regulating factor, such as predation, disease or food shortage, which acts more severely to reduce numbers as population rises.

**density-independent factor** Any factor, such as severe weather or other natural disaster, which can kill large numbers of birds regardless of population density; the effect is to de-stabilize numbers, causing large and random fluctuations.

**dimorphism**, *see* **polymorphism**

**displacement activity** Behaviour during a bird's threat display that is completely irrelevant to the display itself. Examples are preening during the course of a fight, or the pulling of grass by Herring Gulls during territorial interaction.

**eclipse plumage** Dull, inconspicuous plumage that precedes and follows the brighter breeding plumage.

**edge effect** The observation that birds are more numerous at the edge of a habitat patch than in the middle.

**egg dumping** Intraspecific brood parasitism in which birds lay or 'dump' eggs in the nests of fellow members of their own species.

**endemic** A taxon restricted to a defined geographical area. Henderson Island in the South Pacific, for example, has four endemic bird species – the Henderson Fruit Dove, Henderson Rail, Henderson Lorikeet and Henderson Warbler – and these species occur nowhere else.

**erythrism** An abnormality of pigmentation in which chestnut-red replaces other dark pigments.

**Ethiopian realm** Biogeographic region consisting of Africa south of the Sahara, Madagascar and South Arabia. Excluding Madagascar, it has eight endemic families of few species; Madagascar itself has five endemic families. The region occupies similar latitudes to the Neotropical realm, but contains fewer than half as many species, most of them passerines. See also Palearctic realm.

**eutrophic** Nutrient-rich, normally with reference to fresh water. An excessive loading of nutrients, for example from agricultural fertilizers washed into a lake, may result in abnormally low levels of oxygen.

**extra-pair copulation** Mating between the male or female of a pair and a bird other than its mate: such behaviour may increase the number of offspring an individual male can produce.

**flavism (xanthochroism)** An abnormality of pigmentation in which there is an excess of yellow.

**fledging period** The interval between hatching and being able to fly; in flightless species the equivalent interval is between hatching and independence of the parents.

**functional response** A predator response in which individuals include more of a particular prey in their diet as its population numbers rise.

**gene flow** The regular exchange of genetic material between separate populations.

**hatching asynchrony** Eggs in a clutch hatching with a significant interval between first and last. If food is in short supply, the youngest – and therefore smallest – chicks will die, but some of the brood will survive. In this way the size of the brood is decreased without starving and weakening all the chicks.

**hatching synchrony** The hatching of different eggs in a clutch at the same time. This can be achieved by deliberately leaving the early eggs in the clutch without incubating them until the last or last but one egg has been laid. The eggs then commence their embryonic development at the same time. In many precocial species the eggs hatch together, within as little as an hour of each other, as a result of communication amongst the developing eggs; those laid later must accelerate their development and early ones must slow down to achieve synchronous hatching.

**heterochroism** An increase or decrease in the normal amount of a pigment, particularly in plumage.

**Holarctic region** The Palearctic and Nearctic realms combined. Though each of these realms has its own distinct species, a large number of species are common to both.

**holotype** The type specimen of a species, usually chosen by the scientific describer from those specimens collected at the locality where the species was first identified, at the time of the original discovery. All other specimens collected at that time are designated paratypes.

**imprinting** Attachment to an individual – usually of the same species and generally the young bird's parent – as a result of exposure to that species during the first few hours or days of life. The young bird appears to imprint on the species in general rather than on the individual representing the species.

**inverse density dependence** Situation in which a mortality factor acts more strongly at low population densities than at high ones: for example, the predation of constant numbers of a prey species regardless of population size. The destabilizing effect that results causes a population to fall at low densities and rise at high ones.

**irruption** A form of migration common to several bird species, in which, forced by changing conditions to leave their usual territory, they invade another in search of food.

**kleptoparasite** A bird which obtains much of its food by stealing from others, for example the Great Frigatebird (*Fregata minor*) stealing from boobies.

**leap-frog migration** Migration pattern in which, in the northern hemisphere, the most northerly breeding population of a species winters furthest south, and the most southerly breeding population migrates little, if at all. Leap-frog migration also occurs between ecologically related pairs of species: for example, the Whimbrel (*Numenius phaeopus*) breeds chiefly to the north of the larger Curlew (*Numenius arquata*) in Europe, but winters further south, in Africa.

**lek** Communal display ground where, in the breeding season, males compete with each other for status, and mate with as many females as possible.

**leucism (dilution)** A reduction in the intensity of all pigments, particularly in plumage.

**magnetoreception** The sense that gives birds the ability to read compass directions from the Earth's magnetic field.

**magnetoreceptor** Specialized sense organ present in birds – probably in the head – that allows them to read compass directions from the Earth's magnetic field.

**melanism** An excess of black or brown pigment, particularly in the plumage.

**migratory restlessness** Unsettled behaviour of migratory species as the time for migration approaches. It is manifested in caged birds by wing whirring, attempts to fly in the direction of the usual migration, and generally restless behaviour.

**mobbing** A collective attack on a predator – often a bird of another species – by a group of birds, sometimes of more than one species. Usually there is no physical contact, just harassment.

**monogamy** Term used to cover a wide range of mating relationships, generally referring to a breeding pattern in which a male and female remain together for at least one breeding cycle, have a more or less exclusive mating relationship, and work together to rear their young. The Barn Swallow (*Hirundo rustica*) is a good example.

**Nearctic realm** Composed of Greenland and North America north of the tropics, this biogeographic region has no endemic families. Because several families occur only in the Palearctic and Nearctic realms, the two areas are sometimes treated by ornithologists as a single region, the Holarctic. Many species in the Holarctic are migrants. See also Oriental realm.

**Neotropical realm** Biogeographic region including South and Central America, southern Mexico and the West Indies which shares many species with the Nearctic. It is by far the richest of the Earth's six realms, containing over 3000 species, and some 31 endemic families. See also Ethiopian realm.

**nidicolous** Needing prolonged parental care before being capable of leaving the nest.

**nidifugous** Leaving the nest immediately or soon after hatching.

**numerical response** Predator response in which individuals concentrate temporarily in areas where prey are plentiful.

**oligotrophic** Nutrient-poor, usually referring to fresh water.

**optimal foraging** Choosing the best feeding strategy.

**Oriental realm** Biogeographic area, separated from the Palearctic realm by the Himalayas, comprising South and South-East Asia and adjacent islands. Nearly 1000 species breed here, though there is only one endemic family, the Irenidae. The region has strong links with Palearctic and Australian faunas as well as with tropical Africa. See also Neotropical realm.

**Palearctic realm** Biogeographic region made up of Europe, Africa north of the Sahara and all but the south-eastern parts of Asia. Only one family, the hedge-sparrows (Accentors), is endemic to the Palearctic. See also Nearctic realm, Holarctic region.

**partial migrants** Migratory species in which part of the population leaves the breeding ground and flies to winter quarters elsewhere, and part stays within the home range and winters there. The Blackcap (*Sylvia atricapilla*) is an example.

**passerine** Term commonly used to refer to birds in the order Passeriformes, the perching birds. Birds of all other orders are referred to as 'non-passerines'.

**philopatry** Fidelity to a particular area; often applied to species whose fledglings return to breed in the area where they were raised.

**phylogenetic relationship** Evolutionary relationship.

**plumage dimorphism** Differences in plumage, either between a male and female of the same species (for example, between a male and female Mallard, *Anas platyrhynchos*) or as a result of genetically controlled colour phases (morphs) in individuals of a species (for example, among Snow Geese, *Anser caerulescens*, which are either blue or white). These differences are unrelated to age or season of the year, the other principal factors causing plumage variation.

**polyandry** Mating system, much rarer than polygyny, in which the female forms a bond and mates with several males. A female mating with several males during a single breeding attempt is known as *simultaneous polyandry* (as in several jacana species); *sequential polyandry* is the female mating with one male, laying eggs for him to tend, and then courting another male (as in the Spotted Sandpiper, *Actitis macularia*).

**polygamy** A general term for breeding patterns other than monogamy. More specifically, polygamy can be divided into *polygyny*, *polyandry* and *polygynandry*. Promiscuity is a mixture of polygyny and polyandry and usually means there is no lasting pair bond, as in some hummingbirds.

**polygynandry** Mating system in which several males mate with several females, as in the Dunnock (*Prunellus modularis*).

**polygyny** Mating system in which a male mates with several females, either by defending a large territory and attempting to attract as many females as possible (as in the Red-winged Blackbird, *Agelaius phoeniceus*); or competing with other males for status at a lek and mating there with as many visiting females as possible (as in the Black Grouse, *Tetrao tetrix*). *Male deceptive polygyny* is a breeding strategy in which the male pairs with one female and leaves her to lay the eggs while he flies to a different area and mates with a second female; the second female is then abandoned and the male returns to help his original partner near the chicks. This pattern is characteristic, for example, of the Pied Flycatcher (*Ficedula hypoleuca*).

**polymorphism** The occurrence of two distinct forms (dimorphism), or more (polymorphism), within a single species. An example would be distinct variants in plumage colour, as with the dark and light plumage of the adult Arctic Skua (*Stercorarius parasiticus*). See also plumage dimorphism.

**population study** The study of a number of birds in a defined area at a specified time like the breeding season.

**precocial** Independently active immediately or soon after hatching. See also semi-precocial.

**pterylae (singular: pteryla)** Feather tracts from which the contour feathers grow.

**pterylosis** The arrangement of contour feathers, sometimes interspersed with other feathers, in pterylae or feather tracts on the skin. Only penguins (Spheniscidae), screamers (Anhimidae) and adult ratites have uniform contour feather distribution – most birds have large areas (about half the skin area in land species) bare of feather growth, but covered by overlapping feathers from the pterylae.

**pullus (plural: pulli)** Term for a young bird from hatching to full growth and flight; the word is usually used in a ringing (banding) context.

**radiation** Dispersal of a species or group of related species from the area of evolutionary origin to a wider geographical region.

**ratites** Large flightless birds like kiwis (Apterygidae) or the Ostrich (*Struthio camelus*) belonging to six separate orders, two of which are extinct.

***retia mirabilia* (singular: *rete mirabile*)** Latin, 'wonderful nets'. A network of blood vessels – both arterial and venous – in the head and legs of many birds, allowing heat to be exchanged along the length of the artery and vein travelling to and from the legs or brain and the body. This allows cooler blood to reach the brain and feet, so conserving heat within the body.

**rhamphotheca** Horny plates covering the beak; usually the shape of the rhamphotheca follows that of the underlying bone, but it may be modified by local thickening, as in the Atlantic Puffin (*Fratercula arctica*) or toucans (Ramphastidae).

**ring species** A series of interbreeding races (subspecies) spread across a wide geographical area so that neighbouring races are reproductively compatible but the forms at either end are sufficiently different from each other to prevent interbreeding, even though they may occupy the same region. Examples are the Herring Gull (*Larus argentatus*) and the Lesser Black-backed Gull (*Larus fuscus*).

**schizochroism** The absence of one or more normally occurring pigments; rarely, one colour may be replaced by another.

**semi-altricial** Partial development of the ability to regulate body temperature before walking and flight are possible, as in falcons (Falconidae) and owls (Strigidae). See also altricial.

**semi-precocial** Newly-hatched young that are down-covered and have eyes open, but are not advanced enough to leave the nest immediately after hatching and are fed by their parents. In some species such as gulls (Laridae), ability to walk develops early, thermoregulation developing in intermediate fashion and flight coming late. See also precocial.

**semipalmate** Half-webbed feet.

**sexual dimorphism** Readily distinguishable differences – usually in plumage and/or size – between the male and female of a species, as in the European Sparrowhawk (*Accipiter nisus*).

**sexual selection** A type of natural selection in which one sex – usually the male – competes with others of the species in order to obtain more mating opportunities. This process is responsible for the evolution of brightly coloured and elaborate plumages. Sexual selection has two main elements: *intrasexual selection* is competition among males to secure a mate; *intersexual selection* is the display of characters that females find attractive.

**sigmoidal growth** Growth pattern in which young birds initially increase in weight rather slowly, then accelerate rapidly before levelling off as they near adult size.

**sympatric** Populations or species occurring in the same geographical area; in *biotic sympatry* they occupy the same habitat, while *neighbouring sympatry* implies merely living in the same geographical area.

**totipalmate** Having all four toes connected by webs, as in pelicans and their allies.

**tube noses** Birds of the order Procellariiforms (sometimes referred to as petrels), which have paired tubular nostrils surmounting the bill. The exact function of these tubes is unknown, but may be related to the application of stomach oil on the feathers or the marked development of the olfactory section of the brain.

**windhovering** Flying tactic in which a bird flies into the wind in order to remain stationary relative to the ground while scanning for prey, as in the Common Kestrel (*Falco tinnunculus*).

**xanthochroism (flavism)** An abnormality of pigmentation in which there is an excess of yellow.

# Further reading


## Chapter 1

Darwin, C. (1859). *On the Origin of Species.* London: Murray.

Grant, P. R. (1986). *Ecology and Evolution of Darwin's Finches.* Princeton: Princeton University Press.

Hamilton, W. D. & Zuk, M. (1982). Heritable true fitness and bright birds: a role for parasites? *Science*, **218**, 1, 384–7.

Parker, G. A., Baker, R. R. & Smith, V. G. F. (1972). The origin and evolution of gamete dimorphism and the male–female phenomenon. *J. Theor. Biol.*, **36**, 529–53.

Tinbergen, N. (1953). *The Herring Gull's World.* London: Collins.

## Chapter 2

Bock, W. J. (1974). The avian skeletomuscular system. In *Avian Biology*, vol. IV, ed. D. S. Farner & J. R. King, pp. 120–257. New York: Academic Press.

Burton, R. (1987). *Egg: Nature's Miracle of Packaging.* London: Collins.

George, J. C. & Berger, A. J. (1966). *Avian Myology.* New York: Academic Press.

Ginn, H. B. & Melville, D. S. (1983). *Moult in Birds.* BTO Guide 19. Tring.

Hainsworth, F. R., Collins, B. G. & Wolf, L. L. (1977). The function of torpor in hummingbirds. *Physiological Zoology*, **50**, 215–22.

Harrison, C. J. O. (1985). Plumage, abnormal. In *A Dictionary of Birds*, ed. B. Campbell & E. Lack, pp. 472–4. Calton, England: Poyser.

Jones, D. R. & Johansen, K. (1972). The blood vascular system of birds. In *Avian Biology*, vol. II, ed. D. S. Farner & J. R. King, pp. 157–285. New York: Academic Press.

King, A. S. & McLelland, J. (1979–1985). *Form and Function in Birds*, vols. I, II and III. London: Academic Press.

Rahn, H., Ar, A. & Paganelli, C. V. (1979). How bird eggs breathe. *Scientific American*, **240**, 38–47.

Rawls, M. E. (1960). The integumentary system. In *Biology and Comparative Physiology of Birds*, vol. I, ed. R. J. Marshall, pp. 189–240. New York: Academic Press.

Romanoff, A. L. & Romanoff, A. J. (1949). *The Avian Egg.* New York: Wiley.

Schmidt-Nielsen, K. (1983). *Animal Physiology: Adaptation and Environment*, 3rd edn. Cambridge: Cambridge University Press.

Seller, T. J. (ed.) (1986). *Bird Respiration*, vols. I & II. Boca Raton: CRC Press.

Stettenheim, P. (1972). The integument of birds. In *Avian Biology*, vol. II, ed. D. S. Farner & J. R. King, pp. 1–63. New York: Academic Press.

Storer, R. W. (1971). Adaptive radiation of birds. In *Avian Biology*, vol. I, ed. D. S. Farner & J. R. King, pp. 150–88. New York: Academic Press.

## Chapter 3

Alexander, R. McN. (1982). *Locomotion of Animals.* Glasgow: Blackie.

French, M. J. (1988). *Invention and Evolution: Design in Nature and Engineering.* Cambridge: Cambridge University Press.

Greenewalt, C. H. (1962). Dimensional relationships for flying animals. *Smithson. Misc. Collns*, **144**, part 2.

Norberg, U. M. (1985). Flying, gliding, soaring. In *Functional vertebrate morphology*, ed. M. Hildebrand, D. M. Bramble, K. F. Liem & D. B. Wake, pp. 129–58. Cambridge, Mass: Harvard University Press.

Norberg, U. M. (1989). *Vertebrate Flight.* Zoophysiology Series, vol. 27. Berlin, Heidelberg and New York: Springer Verlag.

Pennycuick, C. J. (1972). *Animal Flight.* London: Edward Arnold.

Pennycuick, C. J. (1972). Soaring behaviour of East African birds, observed from a motor-glider. *Ibis*, **114**, 178–218.

Pennycuick, C. J. (1975). Mechanics of flight. In *Avian Biology*, vol. V, ed. D. S. Farner & J. R. King, pp. 1–75. New York: Academic Press.

Phillips, J. G., Butler, P. J. & Sharp, P. J. (1985). *Physiological Strategies in Avian Biology.* Glasgow: Blackie.

Rayner, J. M. V. (1985). Bounding and undulating flight in birds. *J. Theor. Biol.*, **117**, 47–77.

Rayner, J. M. V. (1988). Form and function in avian flight. *Current Ornithology*, **5**, 1–77.

Rüppell, G. (1977). *Bird Flight.* New York: van Nostrand Rheinhold.

Storer, R. W. (1960). Evolution in the diving birds. *Proc. 12th Int. Orn. Congr.*, pp. 694–707.

Ward-Smith, A. J. (1984). *Biophysical Aerodynamics and the Natural Environment.* London: John Wiley.

## Chapter 4

Campbell, B. & Lack, E. (eds.) (1985). *A Dictionary of Birds.* Calton, England: Poyser.

Cracraft, J. (1983). Species concepts and speciation analysis. In *Current Ornithology*, vol. I. ed. R. F. Johnston, pp. 159–87. New York: Plenum.

James, F. C. (1983). Environmental component of morphological differentiation in birds. *Science*, **221**, 184–6.

Mayr, E. (1970). *Populations, Species, and Evolution.* Cambridge, Mass: Belknap Press.

Perrins, C.M. (ed.) (1990). *The Illustrated Encyclopedia of Birds.* London: Headline.

Perrins, C. M. & Middleton, A. L. A. (1985). *The Encyclopedia of Birds.* London: Allen & Unwin.

Sibley, C. G. & Monroe, B. L. (1990). *Distribution and Taxonomy of Birds of the World.* New Haven: Yale University Press.

van Tyne, J. & Berger, A. J. (1971). *Fundamentals of Ornithology.* New York: Dover.

## Chapter 5

Amlaner, C. J. & Ball, N. J. (1983). A synthesis of sleep in wild birds. *Behaviour*, **87**, 85–119.

Bruggers, R. (1990). *The Quelea – Africa's Bird Pest*. Oxford: Oxford University Press.

Bryant, D. M. & Tatner, P. (1988). Energetics of the annual cycle of Dippers, *Cinclus cinclus*. *Ibis*, **130**, 17–38.

Davis, J. W., Anderson, R. C., Karstad, L. & Tainer, D. O. (eds.) (1971). *Infectious and Parasitic Diseases of Wild Birds*. Ames, Iowa: Iowa State University Press.

Higuchi, H. (1986). Bait-fishing by the Green-backed Heron *Ardeola striata* in Japan. *Ibis*, **128**, 285–90.

Kamil, A. C., Krebs, J. R. & Pulliam, H. R. (eds.) (1987). *Foraging Behavior*. New York: Plenum.

Kear, J. (1985). Food Selection. In *A Dictionary of Birds*, ed. B. Campbell & E. Lack. pp. 234–5. Calton, England: Poyser.

Krebs, J. R. (1978). Optimal foraging: decision rules for predators. In *Behavioural Ecology: an Evolutionary Approach*, ed. J. R. Krebs & N. B. Davies. pp. 23–63. Oxford: Blackwell.

Lucas, A. M. & Stettenheim, P. K. (1972). *Avian Anatomy. Integument. Parts I & II*. Agricultural Handbook 362, Washington, DC: US Dept. of Agriculture.

Lyon, B. E. & Montgomerie, R. D. (1986). Delayed plumage maturation in passerine birds: reliable signalling by subordinate males? *Evolution*, **40**, 605–15.

Marshall, A. G. (1981). *The Ecology of Ectoparasitic Insects*. London: Academic Press.

Masman, D., Daan, S. & Beldhuis, H. J. A. (1988). Ecological energetics of the kestrel: daily energy expenditure throughout the year based on time-energy budget, food intake and doubly labelled water methods. *Ardea*, **76**, 64–81.

Payne, R. B. (1972). Mechanisms and Control of Molt. In *Avian Biology*, vol. II, ed. D. S. Farner & J. R. King, pp. 104–55. New York: Academic Press.

Potter, E. F. & Hauser, D. C. (1974). Relationship of anting and sunbathing to molting in wild birds. *Auk*, **91**, 537–63.

Rothschild, M. & Clay, T. (1952). *Fleas, Flukes and Cuckoos*. London: Collins.

Snow, B. & Snow, D. (1988). *Birds and Berries*. Calton, England: Poyser.

Terres, J. K. (1980). *The Audubon Society Encyclopedia of North American Birds*. New York: Knopf.

Walsberg, G. E. (1983). Avian ecological energetics. In *Avian Biology*, vol. VII, ed. D. S. Farner, J. R. King & K. C. Parkes, pp. 161–220. New York: Academic Press.

Welty, J. C. (1982). *The Life of Birds*, 3rd edn. Philadelphia: College Publications.

## Chapter 6

Cracraft, J. (1986). Origin and evolution of continental biotas: speciation and historical congruence within the Australian avifauna. *Evolution*, **40**, 977–96.

Haffer, J. (1974). *Avian Speciation in Tropical South America*. Nuttall Ornithological Club Publ. No. 14.

Harrison, C. (1982). *An Atlas of the Birds of the Western Palaearctic*. Princeton: Princeton University Press.

Lack, D. (1976). *Island Biology Illustrated by the Land Birds of Jamaica*. Oxford: Blackwell Scientific Publications.

MacArthur, R. H. (1972). *Geographical Ecology*. New York: Harper & Row.

Root, T. (1988). Environmental factors associated with avian distributional boundaries. *Journal of Biogeography*, **15**, 489–505.

Sibley, C. G. & Ahlquist, J. E. (1985). The phylogeny and classification of the Australo-Papuan passerine birds. *The Emu*, **85**, 1–14.

Sibley, C. G. & Ahlquist, J. E. (1990). *Phylogeny and Classification of Birds: A Study in Molecular Evolution*. New Haven: Yale University Press.

Wiens, J. A. (1989). *The Ecology of Bird Communities*, vols. I and II. Cambridge: Cambridge University Press.

## Chapter 7

Baker, R. R. (1984). *Bird Navigation: The Solution of a Mystery?* London: Hodder & Stoughton.

Berthold, P. (1975). Migration: Control and metabolic physiology. In *Avian Biology*, vol. V, ed. D. S. Farner, J. R. King & K. C. Parkes, pp. 77–128. New York: Academic Press.

Berthold, P. (1984). The endogenous control of bird migration: a survey of experimental evidence. *Bird Study*, **31**, 19–27.

Berthold, P. (1985). Main aspects of bird migration research. *Universitas*, **27**, 117–24.

Eastwood, E. (1967). *Radar Ornithology*. London: Methuen.

Emlen, S. T. (1970). Celestial rotation: its importance in the development of migratory orientation. *Science*, **170**, 1198–201.

Farner, D. S. (1955). The annual stimulus for migration: experimental and physiological aspects. In *Recent Studies of Avian Biology*, ed. A. Wolfson, pp. 198–237. Urbana: University of Illinois Press.

Gauthreaux, S. A. Jr. (ed.) (1980). *Animal Migration, Navigation and Orientation*. London and New York: Academic Press.

Gwinner, E. (1986). *Circannual Rhythms*. Heidelberg: Springer.

Keeton, W. T. (1971). Magnets interfere with pigeon homing. *Proc. Nat. Acad. Sci.*, **68**, 102–6.

Mead, C. J. (1983). *Bird Migration*. Feltham, Middlesex: Country Life Books.

Papi, F. (1982). Olfaction and homing in pigeons: ten years of experiments. In *Avian Navigation*, ed. F. Papi & H. G. Wallraff, pp. 149–59. Heidelberg: Springer.

Schmidt-Koenig, K. (1979). *Avian Orientation and Navigation*. London: Academic Press.

Wiltschko, W. & Wiltschko, R. (1972). Magnetic compass of European robins. *Science*, **176**, 62–4.

## Chapter 8

Haartman, L. von. (1971). Population dynamics. In *Avian Biology*, vol. I, ed. D. S. Farner & J. R. King, pp. 392–459. New York: Academic Press.

Lack, D. (1954). *The Natural Regulation of Animal Numbers.* Oxford: Oxford University Press.
Lack, D. (1966). *Population Studies of Birds.* Oxford: Oxford University Press.
Newton, I. (1979). *Population Ecology of Raptors.* Berkhamsted: Poyser.
Perrins, C. M. (1979). *British Tits.* London: Collins.
Sharrock, T. (1976). *The Atlas of Breeding Birds in Britain and Ireland.* Berkhamsted: Poyser.

## Chapter 9

Lack, D. (1968). *Ecological Adaptations for Breeding in Birds.* London: Chapman and Hall.
O'Connor, R. J. (1984). *The Growth and Development of Birds.* Chichester: John Wiley.
Wyllie, I. (1981). *The Cuckoo.* London: Batsford.

## Chapter 10

Birkhead, T. R. (1988). Behavioural aspects of sperm competition. *Adv. Stud. Behav.*, 18, 35–71.
Birkhead, T. R., Atkin, L., & Moller, A.P. (1987). Copulation behaviour of birds. *Behaviour*, 101, 101–38.
Birkhead, T.R., Pellatt, J. E. & Hunter, F. M. Extra pair copulation and sperm competition in the zebra finch. *Nature*, 334, 60–2.
Borgia, G. (1986). Sexual selection in bower birds. *Scientific American*, 254, 70–9.
Brown, C. R. (1986). Cliff swallow colonies as information centres. *Science*, 234, 83–5.
Brown, C. R. & Brown, M. B. (1988). A new form of reproductive parasitism in cliff swallows. *Nature*, 331, 66–8.
Brown, J. L. (1987). *Helping and Communal Breeding in Birds.* Princeton: Princeton University Press.
Catchpole, C. K. (1980). Sexual selection and the evolution of complex songs among European warblers of the genus *Acrocephalus. Behaviour*, 74, 149–66.
Davies, N.B. & Houston, A. I. (1984). Territory Economics. In: *Behavioural Ecology: an Evolutionary Approach.* ed J. R. Krebs and N. B. Davies. pp. 148–69. Oxford: Blackwell.
Koenig, W. D. & Mumme, R. L. (1987). *Population ecology of the cooperatively breeding Acorn Woodpecker.* Princeton: Princeton University Press.
Krebs, J. R. & Davies, N.B. (eds.) (1987). *An Introduction to Behavioural Ecology.* Oxford: Blackwell Scientific Publications.
Krebs, J. R. & Kroodsma, D. E. (1980). Repertoires and geographical variation in bird song. *Adv. Stud. Behav.*, 11, 143–77.
Kroodsma, D. E. & Miller, E. H. (1983). *Acoustic Communication in Birds.* New York: Academic Press.
Moller, A. P. (1988). Female choice selects for male sexual tail ornaments in the monogamous swallow. *Nature*, 332, 640–42.
Newton, I. (ed.) (1989). *Lifetime Reproduction in Birds.* London: Academic Press.
Oring, L. W. (1982). Avian Mating Systems. In *Avian Biology*, vol. VI, ed. D. S. Farner, J. R. King & K. C. Parkes, pp. 1–91. New York: Academic Press.
Perrins, C. M. & Birkhead, T. R. (1983). *Avian Ecology.* Glasgow: Blackie.
Vehrencamp, S. L. (1977). Relative fecundity and parental effort in communally nesting anis *Crotophaga sulcirostris. Science* 197, pp. 403–5.
Westneat, D. F., Sherman, P. W. & Morton, M. L. (1990). The ecology and evolution of extra-pair copulation in birds. *Current Ornithology*, 7.
Wittenberger, J. F. & Hunt, G. L. (1985). The adaptive significance of coloniality in birds. In *Avian Biology*, vol. VIII, ed. D. S. Farner, J. R. King, and K. C. Parkes, pp. 1–78. New York: Academic Press.

## Chapter 11

Atkinson, I. A. E. (1985). The spread of commensal species of *Rattus* to oceanic islands and their effects on island avifaunas. In *Conservation of Island Birds*, ed. P. J. Moors, pp. 35–81. Cambridge: ICBP Technical Publication No. 3.
Collar, N. J. & Andrew, P. (1988). *Birds to Watch.* Cambridge: ICBP Technical Publication No. 8.
Croxall, J., Schreiber, R. W. & Evans, P. G. H. (1983). *Status and Conservation of the World's Seabirds.* Cambridge: ICBP Technical Publication No. 2.
Diamond, A. W. & Filion, F. (1987). *The Value of Birds.* Cambridge: ICBP Technical Publication No. 6.
Diamond, A. W., Schreiber, R. L., Attenborough, D. & Prestt, I. (1987). *Save the Birds.* Cambridge: Cambridge University Press.
Ehrlich, P. & Ehrlich, A. (1981). *Extinction.* New York: Ballantine.
Feare, C. J. (1976). The exploitation of Sooty Tern eggs in the Seychelles. *Biological Conservation*, 101, 169–82.
Feare, C. J. (1984). *The Starling.* Oxford: Oxford University Press.
Halliday, T. (1980). *Vanishing Birds.* London: Pelican.
Kenward, R. E. (1987). *Wildlife Radio Tagging.* London: Academic Press.
King, W. B. (1981). *Endangered Birds of the World: The ICBP Bird Red Data Book.* Washington, DC: Smithsonian Institution.
Stesemann, E. (1975). *Ornithology from Aristotle to the present.* Cambridge, Mass: Harvard University Press.
Woldhek, S. (1980). *Bird-killing in the Mediterranean.* European Committee for the Prevention of Mass Destruction of Migratory Birds.
Wright, E. N., Inglis, I. R. & Feare, C. J. (1980). *Bird Problems in Agriculture.* Croydon: British Crop Protection Council Publications.

# Acknowledgements

Cambridge University Press gratefully acknowledges the help of many individuals and organizations in collecting the illustrations for this volume. In particular they would like to thank Dr M. L. Birch, Alexander Library, EGI; the British Trust for Ornithology; Tim Harris, NHPA; Alison Stattersfield, ICBP; Ann Stonehouse; Mark Tasker and Dr. R. Wiltschko. Every effort has been made to obtain permission to use copyright materials and the publishers would appreciate any errors or omissions being brought to their attention.

**1** Francois Gohier/Ardea, London; **3** L. Campbell/NHPA; **4** after Flux, J. E. C. & Flux, M. M. (1982). *Naturwissenschaften* 69, 96–97, Heidelberg: Springer-Verlag; **5** photo: M. P. Harris; **8** after Pettingill, O. S., Jr. (1970). *Ornithology in laboratory and field*, 4th edn. Minneapolis: Burgess Publishing Co.; **9** after Bock, 1974; **11** after Jollie, M. T. (1957). The head skeleton of the chicken and remarks on the anatomy of this region in other birds. *J. Morphology*, 100, 389–436; **12** *far left*: after Campbell & Lack, 1985; *bottom right*: J-M. Labat/Ardea, London; *top right*: after Bühler, P. (1981). Functional anatomy of the avian jaw apparatus. *In*: King & McLelland, vol. II, pp. 439–68; **13** *top*: after Campbell & Lack, 1985; *bottom*: after Bellairs, A. d'A. & Jenkin, C. R. (1960). The skeleton of birds. *In*: *Biology and comparative physiology of birds*, vol. I, ed. A. J. Marshall, pp. 241–300. London: Academic Press; **15** *(a)* after Van Tyne & Berger, 1976; *(b)* after Bellairs & Jenkin, 1960; **16** *left*: *(a)* after Van Tyne & Berger, 1976; *(b)* after Bellairs & Jenkin, 1960; *right*: after Bühler, 1981; **17**: Julian Smith; **18** *left*: Paul Richardson, after Pettingill, 1970; *cross-section*: Paul Richardson, after Welty, 1982; *right*: Paul Richardson, after Berger, A. J. (1953). On the locomotor anatomy of the Blue Coua, *Coua caerulea*. *Auk* 70, 49–83; **19** *left*: Paul Richardson, after Marshall, 1960; *right*: Paul Richardson, after Welty, 1982; **20** *left*: Julian Smith; *right*: after Lucas & Stettenheim, 1972; **21** after Ginn & Melville, 1983; **22** Paul Richardson, after Thomsan, A. L. (ed.) (1964). *A new dictionary of birds*. London: Nelson; **23** W. Peckover/VIREO; **24** after Clench, M. H. (1985). Pterylosis. *In*: Campbell & Lack, 1985, pp. 487–8; **25** Julian Smith, after Van Tyne & Berger, 1976 (*top*); and after Welty, 1982 (*bottom*); **26** *top*: W. S. Paton/Aquila; *bottom*: © ANT/NHPA; **27** Paul Richardson, after Evans, H. E. (1969). Anatomy of the budgerigar. *In*: *Diseases of Cage and Aviary Birds*, ed. M. Petrak. Philadelphia: Lea & Febiger; **28** after McLelland, J. (1979). Digestive system. *In*: King & McLelland, vol. I; **29** *top*: after McLelland, 1979; *bottom*: Melvin Grey/NHPA; **33** *(a)* Redrawn with modifications from Akester, A. R. (1971). The blood vascular system. *In*: *Physiology and biochemistry of the domestic fowl*, vol. II, ed. D. J. Bell & B. M. Freeman, pp. 783–839. London: Academic Press; *(b)* Compiled and redrawn from Portman, A. (1950). Les organes de la circulation sanguine. *In*: *Traité de zoologie*, vol. XV, ed. P. P. Grasse. Paris: Masson; and Akester, 1971; **34** *far left*: photo: © Orion Press/NHPA; *left*: Wendy Shattil & Bob Rozinski/OSF; *right*: after West, N. H., Langille, L. and Jones, D. R. (1981). Cardiovascular system. *In*: King & McLelland, vol. II, pp. 253–339; **35** Jos Korenromp/OSF; **37** *top*: after Sperber, I. (1960). Excretion. *In*: *Biology and Comparative Physiology of Birds*, vol. I, ed. A. J. Marshall. London: Academic Press; *bottom*: Michael Brooke; **38** *(c)* after Getty, R. (1975). *In*: Sisson & Grossman: *The anatomy of the domestic animals*, vol. II, 5th edn., ed. R. Getty. Philadelphia: Saunders; **42** after O'Connor, R. J. (1984). *The growth and development of birds*. Chichester: J. Wiley & Sons; **43** after Rahn, H., Ar, A. & Paganelli, C. V., 1979; **44** *top*: Days 4–16: after LaFarge, M. *In*: Terres, J. K. (1980). *The Audubon Society Encyclopedia of North American Birds*. New York: Alfred A. Knopf; Day 19: after Rahn, H., Ar, A. & Paganelli, C. V., 1979; *bottom*: Jane Burton/BC; **45** *(a)* after Baumel, J. J. (1975). *In*: Sisson & Grossman's: *The anatomy of the domestic animals*, vol. II, 5th edn., ed. R. Getty. Philadelphia: Saunders; *(b)* after King, A. S. & McLelland, J. (1984). *Birds: their structure and function*. London: Ballière Tindall; **46** *top*: *(c)* after Kappers, C. U. A., Huber, C. C. & Crosby, E. C. (1936). *The comparative anatomy of the nervous system in vertebrates*. New York: Macmillan; *bottom*: after Bang, B. G. (1971). Functional anatomy of the olfactory system in 23 orders of birds. *Acta Anat* 79, Suppl. 1–76; **47** *far left*: after Evans, H. E. (1979). Organa sensoria. *In*: *Nomina Anatomica Avium*, ed. J. J. Baumel. London: Academic Press; *top*: flamingo: after Bang, B. G., 1971; *bottom*: after Ward, C. A. (1917). *The fundus oculi of birds especially as viewed by the opthalmoscope*. Chicago: Lakeside Press; **49** *top*: R. Van Nostrand/FLPA; *centre*: J. B. Blossom/Aquila; *bottom*: after Norberg, R. A. (1978). Skull asymmetry, ear structure and function, and auditory localization in Tengmalm's Owl, *Aegolius funereus*. *Phil. Trans. Royal Soc. 282B*, 325–410; **50** *(a)* after King & McLelland, 1984; *(b)* after Bang, B. G. & Wenzel, B. M. (1985). Nasal cavity and olfactory system. *In*: King & McLelland, 1985; **51** after Bang and Wenzel, 1985; **62** Julian Smith; photo: Stephen Dalton/NHPA; **63** after Mead, C. (1983). *Bird migration*. Feltham, Middlesex: Country Life Books; **64** *top*: Mike Danzenbaker; *bottom*: Manfred Danegger/NHPA; **66** Anthony Bannister/NHPA; **67** ©ACT/NHPA; **68** after Butler, P. J. & Stephenson, R. (1986). Diving behaviour and heart rate in Tufted Ducks (*Aythya fuligula*). *J. Exp. Biol.* 126, 341–59. Company of Biologists, Cambridge; **69** Courtesy of Museum für Naturkunde an der Humboldt-Universität zu Berlin, Paläontologisches Museum; **71** Julian Smith; **74** Julian Smith; **76** Julian Smith; **77** *top*: L. C. Marigo/BC; *bottom*: after van Tyne & Berger, 1971; **78** after van Tyne & Berger, 1971; **80** *top*: Julian Smith; *bottom*: Mike Danzenbaker; **81** after Mayr, E. (1942). *Systematics and the origins of species from the viewpoint of a zoologist*. New York: Columbia University Press; **82** after Mayr, E. (1963). Animal species and evolution. Cambridge, Mass.: Belknap Press; **83** *top to bottom*: Charlie Ott/BC; Erwin & Peggy Bauer/BC; W. S. Clark/FLPA; C. H. Greenewalt/VIREO; **84** *top*: S. J. Lang/VIREO; *bottom*: T. Parker/VIREO; **85** Simon Trevor/BC; **86–120** Julian Smith; **121** © Thomas D. Mangelsen/Images of Nature®; **122** *top*: Leonard Lee Rue III/OSF; *bottom left*: Paul Sterry/Nature Photographers; *bottom right*: Robert A. Tyrrell/OSF; **124** *top*: Robert Kenward; *bottom*: Julian Smith; **126** G. Moon/FLPA; **127** *top*: R. Thompson/FLPA; *bottom*: © Thomas D. Mangelsen/Images of Nature®; **129** after Zach, R. (1979). *Behaviour* 68, 106–17; **130** after Zach, 1979; **131** modified from Krebs, J. R. & Davies, N. B. (1987). *An introduction to behavioural ecology*. Oxford: Blackwell; **133** Nigel Adams; **134** *left*: after Kendeigh, S. C., Dolnik, V. R. & Gavrilov, V. M. (1977). Avian energetics. *In*: *Granivorous birds in ecosystems* ed. J. Pinowsky & S. C. Kendeigh, pp. 127–204. Cambridge: Cambridge University Press; *right*: after Adams, N. J. & Brown, C. R. (1984). Metabolic rates of sub-Antarctic Procellariiformes: a comparative study. *Comp. Biochem. Physiol.* 77A, 169–73; **136** Kestrel data from Masman, D., Gordijn, M., Dean, S. & Dijkstra, C. (1986). Ecological energetics of the Kestrel *Falco tinnunculus*: field estimates of energy intake throughout the year. *Ardea* 74, 24–39; Dipper data from Bryant, D.M. & Tatner, P. (1987). Energetics of the annual cycle in dippers *Cinclus cinclus*. *Ibis* 130, 17–38; **137** *top*: Jeff Foott/Survival Anglia; *bottom*: after Lucas & Stettenheim, 1972; **138** S. Nielsen/BC; **140** A. T. Moffett/Aquila; **143** Dieter & Mary Plage/BC; **144** Wardene Weisser/Ardea, London; **145** *top*: Kenneth Day; *centre*: M. P. Harris; *bottom*: after Amlaner, C. J. & McFarland, D. J. (1981). Sleep in the Herring Gull (*Larus argentatus*). *Animal Behaviour* 29, 551–6; **146** after Goodman I. J. (1974). The study of sleep in birds. *In*: *Birds: brain and behaviour*, ed. I. J. Goodman & M. W. Schein, pp. 133–52. New York: Academic Press; **147** Michael and Patricia Fogden; **150** *left*: Eric & David Hosking; *right*: W. S. Paton/RSPB; **151** courtesy of C. M. Perrins; **152** courtesy of Peter Hudson; **153–154** after Matheson, R. (1944). *Entomology for introductory courses*. London: Constable & Co; **157** after Welty, 1982; **161** *top*: after Wiens, 1989; *bottom*: *world map*: after Goodwin, D. (1976). *Crows of the World*. Ithaca: Cornell University Press; *US map*: after Robbins, C. S., Bystrak, D. & Geissler, P. H. (1986). The breeding bird survey: its first fifteen years. 1965–1979, p. 49. U.S. Dept Interior, Fish and Wildlife Service. *Resource Pub.* 157. Washington D.C.; **162** *top*: after Root, J. B. (1900). *J. Biogeography* 15, 489–505; *bottom*: after Haffer, J. (1985). *Ornithological Monographs* 36, pp. 113–46; **163** *top*: after Harrison, C. (1982). *An atlas of the birds of the western Palearctic*. Princeton University Press; *bottom*: after Robbins, C. S. & Van Velzen, W. T. (1969). The breeding bird survey 1967 and 1968. U.S. Dept of the Interior, Bureau of Sport Fisheries & Wildlife. *Special Scientific Report: Wildlife No.* 124, pp. 62–3; **165** after Wiens, 1989; **168** *top*: (*left*): after Cook, R. E. (1969). *Systematic Zoology* 18, 63–84; (*centre*): after Schall, J. J. & Pianka, E. R. (1978). *Science* 210, 679–86; (*right*): after Rabinovich, J. E. & Rappoport, E. H. (1975). *J. Biogeography* 2, 141–57; *bottom left*: after Scot, J. M. et al. (1987). *BioScience* (December issue); *bottom right*: after Wiens, 1989; **173** Julian Smith; after Robert Gillmor *in* Lack, D. (1971). *Ecological isolation in birds*. Oxford: Blackwell Scientific Publications; **177** Stephen Dalton/NHPA; **181** *top*: after Berthold, P., Gwinner, E. & Klein, H. (1971). Circannuale Periodik bei Grasmücken (*Sylvia*). *Experientia* 27, 399; *bottom*: after Berthold, P. & Querner, U. (1981). Genetic basis of migratory behaviour in European warblers. *Science* 212, 77–9; **182** *left*: after Gwinner, E. & Wiltschko, W. (1978). Endogenously controlled changes in migratory direction of the Garden Warbler, *Sylvia borin*. *J. Comp Physiol.* 125, 267–73; **183** Michael & Patricia Fogden; **189** Michael

Brooke; **190** *left*: Hugh Miles; *right*: courtesy of Point Reyes Bird Observatory; **191** David M. Cottridge/OSF; **192** Courtesy of the Swiss Ringing Centre; **193** *left*: Imperial War Museum; *right*: Martyn F. Chillmaid/OSF; **194** Courtesy of Gerlinde Orth-Tocha; **196** J. B. Blossom/NHPA; **199** photo: Cambridge University Collection of Air Photos, 144/226367; **202** Stephen Dalton/NHPA; **203** Jeff Foott/Survival Anglia; **204** Jen & Des Bartlett/Survival Anglia; **205** Courtesy of Anne Hudson; **206** *right*: after Lack, 1954; **207** *left*: (*top*) after Newton, 1979; (*centre*) after Newton, I. (1989); (*bottom*): after Newton, I. (1972). *Finches*. London: Collins; (*bottom right*) after Mayr, E. (1926). Die Ausbreitung des Girlitz (*Serinus canaria serinus* L.). *J. Orn.* 74, 571–671. **211** *top*: after Shelford, V.E. (1945). The relation of Snowy Owl migration to the abundance of the Collared Lemming. *Auk* 62, 592–6; *bottom*: Ted Levin/OSF; **216** Hans Reinhard/BC; **220** Manfred Danegger/NHPA; **222** after Newton, 1979; **224** Babs & Bert Wells/OSF; **225** *left*: © A.N.T./NHPA; *top*: G. Ziesler/BC; *bottom right*: Jany Sauvanet/NHPA; **226** Michael Brooke; **227** *top*: Leonard Lee Rue/BC; *bottom*: D. Freeman/RSPB; **230** *top*: Peter Steyn/Ardea, London; *bottom*: Doug Allan/OSF; **231** *top to bottom*: Jean-Paul Ferrero/Ardea, London; Geoff Dore/BC; C.M. Perrins/OSF; Hans Dieter Brandl/FLPA; **233** By kind permission of Otorohanga Zoological Society Inc.; **240** Alan Root/Survival Anglia; **242** after O'Connor, 1984; **244** *left & bottom*: after O'Connor, 1984; *centre & right*: Peter Castell/Aquila; **245** O.S. Pettingill, Jr./VIREO; **246** Michael & Patricia Fogden; **248** after O'Connor, 1984; **249** Hellio & Van Inger/NHPA; **250** *top*: Wilf Schurig; *bottom*: Stan Osolinski/OSF; **251** Mike Wilkes; **253** Cyril Laubscher/C & L Nature World; **256** Ian Wyllie/Survival Anglia; **257** Godfrey Merlan/OSF; **258** Norbert Rosing; **261** *(a)* after Lill, 1974; *(b)* after Wiley, 1973; *(c)* after Kruijt & Hogan, 1967; **262** *top*: John Hawkins/Eric & David Hosking; *bottom*: Leonard Lee Rue III/BC; **263** *top left*: Fritz Pölking/FLPA; *top right*: after Anderson, M. (1982). Female choice selects for extra tail length in a widow-bird. *Nature* 299, 818–20; *bottom*: T. P. Gardner/FLPA; **265** Courtesy of Terry Burke; **266** W. R. Moore/National Photo Index, Australian Museum; **268** after Schoener, T. W. (1968). Sizes of feeding territories among birds. *Ecology* 49, 127; **269** *top*: Michael & Patricia Fogden; *bottom*: David & Katie Urry/Ardea, London; **270** D. J. Saunders/OSF; **271** *top*: Michael & Patricia Fogden; *bottom*: Peter Davey/BC; **272** *top*: Michael Brooke; *bottom*: Brian J. Brown; **273** Mike Danzenbaker; **276** *top*: J. & D. Bartlett/Survival Anglia; *bottom*: Jeff Foott/BC; **277** *top*: Erwin & Peggy Bauer/BC; *bottom*: Udo Kirsch/BC; **278** Julian Smith; **279** Michael Fogden/OSF; **283** after Marler, P. (1957). *Behaviour*, 11, 13–39; **285** after Catchpole, C. K. (1979). *Vocal communication in birds*. London: Edward Arnold; **286** after Krebs, J. R. (1976). *New Scientist* 70, 534–6; **288** *left*: after Thorpe, W.H. (1958). *Ibis* 100, 535–70; *right*: after Marler, P. & Temura, M. (1964). *Science* 146, 1483–6; **290** after Seibt, V. & Wickler, W. (1977). *Tierpsychologie* 43, 180–7; **294** *top*: Artis-Bibliotheek, Amsterdam; *bottom*: By courtesy of E. Lack; **295** *left*: Museo Nacional del Prado; *right*: Ardea, London; **296** *left*: By courtesy of Sotheby's; *right*: from *Imago Mundi Bestiarium*, by permission of the Syndics of the Cambridge University Library; **298** from Fulton, R. (1876). *The illustrated book of pigeons with standards for judging*; **300** J.B. Blossom/NHPA; **301** Jean-Paul Ferrero/Ardea, London; **302** Eric & David Hosking; **304** Mike Marshall-Hollingsworth; **305** Courtesy of RNAS Yeovilton; **306** Tick Ahearn; **309** Aberdeen University Library, George Washington Wilson Collection; **311** *left*: G.J. Broekhuysen/Ardea, London; *right*: John Hatt/Hutchison Library; **312** G. I. Bernard/NHPA; **313** photo: Ronald Toms/OSF; **314** P. Morris; **318** *top*: Gerald Cubitt/BC; *bottom*: after Newton, 1979; **319** Norman Myers/BC; **320** Dave Currey/Environmental Investigation Agency; **323** © Zoological Society/San Diego; **324** M. P. Harris. All other illustrations courtesy individual authors.

Abbreviations used: NHPA = Natural History Photographic Agency; OSF = Oxford Scientific Films; FLPA = Frank Lane Picture Agency; VIREO = Visual Resources in Ornithology; BC = Bruce Coleman Limited. Short titles refer to titles in the list of further reading.

# Indexes

**Numerals in italics refer to illustrations or tables**

## SCIENTIFIC NAMES

### A



### B



### C

## D



## E



## F



## G



## H

## Q



## R



## S



## T

## COMMON NAMES

## F



## G



## H



## I



## J



## K

## R



## S



## T



## V



## W

## Y

## SUBJECTS

## A



## B

## C

## N



## O

# T

## U



## V



## W

# X



# Y



# Z